Springer Water

The book series Springer Water comprises a broad portfolio of multi- and interdisciplinary scientific books, aiming at researchers, students, and everyone interested in water-related science. The series includes peer-reviewed monographs, edited volumes, textbooks, and conference proceedings. Its volumes combine all kinds of water-related research areas, such as: the movement, distribution and quality of freshwater; water resources; the quality and pollution of water and its influence on health; the water industry including drinking water, wastewater, and desalination services and technologies; water history; as well as water management and the governmental, political, developmental, and ethical aspects of water.

Matthew Chidozie Ogwu · Sylvester Chibueze Izah
Editors

Water Quality and Safety in the Global South

Challenges, Solutions and Future Directions

Springer

Editors
Matthew Chidozie Ogwu
Appalachian State University
Boone, NC, USA

Sylvester Chibueze Izah
Department of Community Medicine
Bayelsa Medical University
Yenagoa, Nigeria

ISSN 2364-6934 ISSN 2364-8198 (electronic)
Springer Water
ISBN 978-3-032-10601-8 ISBN 978-3-032-10602-5 (eBook)
https://doi.org/10.1007/978-3-032-10602-5

This Springer imprint is published by the registered company Springer Nature Switzerland AG
The registered company address is: Gewerbestrasse 11, 6330 Cham, Switzerland

Contents

Chapter 1
Water Scarcity and Insecurity: Causes and Consequences

Sylvester Chibueze Izah and Matthew Chidozie Ogwu

Abstract Water scarcity and insecurity are escalating global challenges with significant implications for human health, food production, environmental sustainability, and socio-economic development. As both natural and anthropogenic factors (such as climate change, population growth, industrial pollution, and poor water governance) exacerbate these issues, the need for integrated and interdisciplinary solutions becomes more urgent. This chapter explores causes and consequences of water scarcity. The major drivers of water insecurity, including emerging pollutants, mismanagement of water resources, and the degradation of aquatic ecosystems, are critically examined. The chapter highlights the role of integrated resource management, surveillance and early warning systems, and community engagement in mitigating the risks associated with water scarcity. Special attention is given to the long-term implications of water insecurity, such as antimicrobial resistance, declining food security, and the impact on vulnerable populations. Policy recommendations are discussed, advocating for sustainable, inclusive, and climate-resilient water governance frameworks that align with the United Nations Sustainable Development Goals, particularly SDG 6. The One Health approach serves as a comprehensive strategy for addressing water scarcity, promoting cross-sector collaboration, and fostering innovation in water use and conservation. Thus, this chapter stresses the

The original version of the chapter has been revised. A correction to this chapter can be found at https://doi.org/10.1007/978-3-032-10602-5_15

S. C. Izah (✉)
Department of Community Medicine, Faculty of Clinical Sciences, Bayelsa Medical University, Yenagoa, Bayelsa, Nigeria
e-mail: Sylvester.izah@bmu.edu.ng

Department of Microbiology, Faculty of Science, Bayelsa Medical University, Yenagoa, Bayelsa, Nigeria

M. C. Ogwu
Goodnight Family Department of Sustainable Development, Appalachian State University, Living Learning Center, Boone, NC, USA
e-mail: ogwumc@appstate.edu

M. C. Ogwu and S. Chibueze Izah (eds.), *Water Quality and Safety in the Global South*, Springer Water, https://doi.org/10.1007/978-3-032-10602-5_1

urgency of adopting holistic and equitable water management solutions that safeguard public and environmental health, ensure future water availability, and build resilient communities.

Keywords Water scarcity · One health · Water insecurity · Sustainable water management · Emerging pollutants · Antimicrobial resistance · Climate change · Public health · Environmental governance

1.1 Introduction

Water scarcity and insecurity have become critical issues globally, impacting millions of people each year. Water scarcity is defined as the lack of sufficient water resources to meet the demands of water usage within a region, while water insecurity refers to a lack of reliable access to sufficient water for health, livelihoods, and well-being [52]. The terms are increasingly relevant, considering that approximately 2.1 billion people lack access to safe drinking water, as reported by the United Nations [88]. This disparity translates into severe consequences for human health, agricultural productivity, and economic development, underscoring the essentiality of water as a human right and a cornerstone for sustainable development [8, 80].

The scale of water stress affects various regions uniquely, but its global implications are alarming. Reports from the Global Risks report highlighted that water supply crises represent significant challenges to both the economy and the alleviation of poverty [113]. As climate change exacerbates these crises, particularly in agricultural areas where water is critical, the discrepancies between supply and demand are becoming more pronounced [107]. The paradigm of water management has also transformed, demonstrating the urgent need for integrated water resource management (IWRM) approaches that consider both supply-side and demand-side challenges [97].

The recognition of water as a human right has gained momentum since the United Nations General Assembly explicitly declared it in 2010, framing access to safe and clean drinking water and sanitation as essential to the enjoyment of all human rights [80]. This right implies not only availability but also accessibility, acceptability, and quality of water [8]. A holistic understanding of water rights acknowledges its historical and cultural significance, aligning with perspectives that advocate for water as a common resource regulated by local communities while ensuring universal access. The implications of non-recognition of water as a human right are severe, as this leads to inequities in access and exacerbates social and economic vulnerabilities [96].

Currently, the mismatch between water demand and supply is increasingly concerning. Rapid urbanization, population growth, and changing consumption patterns [47–49] have further exacerbated this mismatch, with estimates suggesting that by 2025, two-thirds of the world's population could be living in water-stressed conditions [103]. As sectors such as agriculture, which consume about 70% of freshwater resources, continue to demand more water, it is crucial to implement sustainable practices and innovations, such as rainwater harvesting and wastewater management [108, 114]. Effective governance and sustainable policies in the management of water resources remain imperative for meeting the rising demand and addressing the urgent global water crisis [46].

This chapter focuses on understanding the root causes and far-reaching consequences of water scarcity and insecurity across environmental, social, and economic dimensions. It explores both natural and human-induced drivers, including climate change, population growth, and poor governance, while examining the severe impacts on health, livelihoods, ecosystems, and global stability. Ultimately, the chapter aims to highlight the urgent need for sustainable and inclusive strategies to mitigate water scarcity and ensure equitable access to this vital resource.

1.2 Types and Dimensions of Water Scarcity and Insecurity

Water scarcity and insecurity poses a critical challenge globally, characterized by varying dimensions and implications that affect populations differently based on socioeconomic, geographic, and structural factors [84, 85]. This lead to water poverty. Water poverty is a complex, multidimensional issue shaped not only by natural scarcity but also by disparities in access, management capacity, and usage patterns. In semiarid regions, where low rainfall and drought are common, socioeconomic factors (such as income, education, and health infrastructure) often play a more decisive role in deepening water insecurity than environmental conditions alone (Fig. 1.1, [84]). This cycle of water poverty reinforces itself, as limited access to safe water exacerbates poor living conditions, which in turn hinders communities' ability to improve water access, creating a persistent and self-perpetuating loop. This section of the chapter outlines the complex nature of water scarcity by categorizing its types into physical, economic, temporal, and spatial variations, while also examining rural–urban dynamics and relevant metrics for assessment (Table 1.1).

Physical Water Scarcity and Insecurity is predominantly recognized in arid and semi-arid regions, where insufficient natural water resources fail to meet the demands of the population due to factors such as low rainfall and high rates of evaporation. Environmental aspects such as over-extraction of groundwater, desertification, and declining river flows exacerbate these conditions [84, 85]. In regions like Mexico, the overexploitation of aquifers due to agricultural demands illustrates the severe competition for water across sectors, leading to critical shortages that impact both rural and urban communities [27, 82]. As outlined by Klassert et al. [53], the intensification of human-nature systems variability contributes to the increasing frequency and severity of extreme droughts, thus aggravating physical water scarcity.

Economic Water Scarcity arises when water resources are available but remain inaccessible due to inadequate infrastructure and poor governance. This is especially prevalent in low-income countries where communities often rely on unsafe

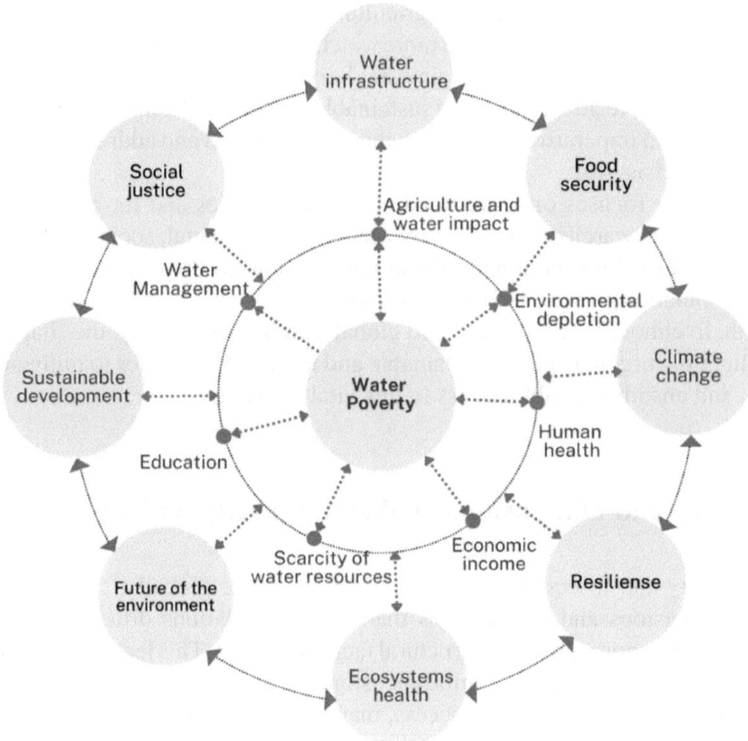

Fig. 1.1 Cycle of water poverty. *Source* Pacheco-Treviño and Manzano-Camarillo [84]

water sources due to the lack of investment in water supply systems [14, 38]. Poor management practices further complicate the situation, as institutional failures lead to a lack of equitable water distribution, disproportionately affecting marginalized populations and perpetuating cycles of poverty [61]. For instance, in rural Nigeria, inadequate physical infrastructure leads to significant barriers in water accessibility, influencing household economic stability and promoting a heightened vulnerability to water-related health issues.

The temporal and spatial variations in water scarcity significantly influence water availability. Temporal fluctuations, such as seasonal droughts and climate-induced shifts, often lead to unpredictable water supply, affecting agricultural planning and productivity [53]. Spatial disparities highlight the unequal distribution of resources, as certain regions experience abundance while others are left to navigate acute shortages. Effective management strategies must, therefore, incorporate adaptive planning that accounts for these variabilities, ensuring that interventions are both context-sensitive and forward-looking.

Table 1.1 Types and dimensions of water scarcity and insecurity

Type/dimension	Narrative	Major features
Physical water scarcity and insecurity	Occurs when natural water resources are insufficient to meet demand	Common in arid regions, linked to overuse of groundwater and diminished river/lake levels
Economic water scarcity and insecurity	Water exists, but cannot be accessed due to a lack of infrastructure or investment	Poor water governance, infrastructure deficits, and effects primarily low-income communities
Temporal and spatial variations	Water availability fluctuates seasonally and geographically	Seasonal droughts or floods, regional inequality, and climate-driven variability
Rural versus Urban water scarcity and insecurity	Differences in how water scarcity impacts urban and rural areas	Rural areas lack infrastructure, reliance on natural sources, whereas in urban areas, overpopulation and infrastructure strain result
Principal metrics and indices	Tools to assess and quantify water scarcity and insecurity	Falkenmark Index, Water Poverty Index, Water Exploitation Index (WEI) and IWRM Index

The rural versus urban dimensions of water challenges further complicate the landscape of water insecurity. Rural areas typically suffer from infrastructural deficiencies, relying heavily on rain-fed agriculture, which makes them particularly vulnerable to droughts [98]. Conversely, urban centers face new challenges marked by rapid population growth and poor sanitation, which increase overall water demand and stress existing resources [61]. Informal settlements often lack efficient water supply systems, forcing residents to depend on costly and unsafe alternatives. This tension between urban demands and rural supply needs can exacerbate inequalities, as urbanization often leads to the diversion of essential resources away from surrounding rural communities [84–86].

To evaluate and address these diverse challenges, key metrics such as the Falkenmark Water Stress Index and the Water Poverty Index have been developed. These tools assess per capita water availability, access, governance capacity, and resource demand pressures. Integrating such metrics allows policymakers to identify areas at risk of water scarcity, facilitating targeted interventions designed to promote sustainability and equity in water management. The effectiveness of these indices is critical in guiding the development of policies that can mitigate the multifaceted impacts of water scarcity across different contexts.

1.3 Environmental and Human-Induced Causes of Water Scarcity and Insecurity

Water scarcity is increasingly recognized as a critical global challenge, with both environmental and human-induced factors intensifying its prevalence. Understanding the multifaceted nature of water scarcity requires examining climatic and ecological drivers (Table 1.1), alongside human and institutional influences that exacerbate or mitigate the issue (Table 1.2).

1.3.1 Climatic and Ecological Drivers

One prominent driver of water scarcity is climate change [39], which disrupts traditional rainfall patterns and contributes to a decline in snowpack and accelerated glacial melt. Alterations in precipitation not only decrease freshwater availability but also heighten the risks associated with droughts and extreme weather events. Schewe et al. [89] present evidence that climate change results in increased exposure to water scarcity for many populations, with certain regions experiencing prolonged periods of drought that diminish surface and groundwater resources. Shemer et al. [93] reinforce this perspective, noting that climate change will more significantly impact the number of individuals facing escalating water shortages than those experiencing relief from prior scarcity. Furthermore, as temperatures rise, evaporation rates increase, compounding the challenges by exacerbating existing drought conditions. Table 1.2 present climatic and ecological drivers of water scarcity and insecurity.

In addition to changing climate patterns, desertification and land degradation play a crucial role in diminishing water availability. The transformation of arable land into non-arable desert areas not only reduces natural water retention capacities but also leads to vegetation loss, which is vital for aquifer replenishment [102]. Furthermore,

Table 1.2 Climatic and ecological drivers of water scarcity and insecurity

Driver	Narrative
Climate change	Alters rainfall patterns, reduces snowpack, and accelerates glacial melt; increases temperatures and evaporation rates, intensifying droughts
Droughts and extreme weather events	More frequent and intense droughts reduce surface and groundwater; floods contaminate freshwater sources; unpredictability hampers planning
Desertification and land degradation	Loss of arable land and vegetation reduces natural water retention and aquifer recharge capacity
Changing precipitation patterns	Uneven rainfall distribution creates regional imbalances; shorter or delayed wet seasons disrupt agriculture and water planning

shifting precipitation patterns complicate water management, leading to scenarios where some regions experience water abundance while others suffer acute scarcity [89]. For example, the uneven distribution of rainfall can disrupt agricultural cycles, particularly through shorter rainy seasons or delays in monsoon arrivals, which are critical for both domestic and commercial water supply [60].

1.3.2 Human and Institutional Drivers

Human activities significantly compound the risks associated with water scarcity (Table 1.3). Over-extraction of groundwater for irrigation or industrial use strains aquifers and diminishes their ability to recharge, leading to long-term depletion [65]. Agriculture accounts for more than 70% of global freshwater use, and inefficient irrigation practices aggravate this situation [35]. Furthermore, the acceleration of population growth and urban expansion has created unprecedented demand for domestic and industrial water supplies, which often outpaces the development of appropriate infrastructure [1]. This urban growth typically results in increased impervious surfaces, which limit natural water recharge, further intensifying water scarcity issues [1].

Moreover, inadequate policy frameworks and weak governance structures hinder effective water management. A lack of integrated water resource management (IWRM) often leads to fragmented decision-making processes, poor coordination, and insufficient investment in necessary infrastructure [37]. Liu et al. [59] highlight that assessing water scarcity based solely on water quantity without accounting for water quality and environmental flow requirements is inadequate for developing responsive management strategies. The assessment frameworks that consider both qualitative and quantitative aspects of water resources are essential to inform policymakers and manage this critical resource sustainably [111].

Table 1.3 Human and institutional drivers of water scarcity and insecurity

Driver	Narrative
Over-extraction and unsustainable water use	Excessive groundwater withdrawal depletes aquifers; lack of water-saving technologies increases waste
Rapid population growth and urban expansion	Rising domestic demand outpaces infrastructure; urbanization reduces natural recharge zones
Industrial and agricultural water demands	Agriculture consumes over 70% of freshwater, often inefficiently; industries pollute and consume large volumes
Policy gaps and weak governance structures	Inadequate regulations and enforcement lead to misuse and pollution; poor coordination and investment hinder water management

1.4 Health, Environmental, and Socioeconomic Consequences of Water Scarcity and Insecurity

Water scarcity and insecurity have profound health, socioeconomic, environmental, and geopolitical consequences that compound existing vulnerabilities globally (Table 1.4).

1.4.1 Health Consequences of Water Scarcity and Insecurity

Contaminated water sources directly result in the proliferation of waterborne diseases such as cholera and dysentery [82]. This is particularly alarming in regions where access to safe drinking water is severely limited, exacerbating public health crises. The lack of proper sanitation and hygiene facilities (often a direct consequence of

Table 1.4 Consequences of water scarcity and insecurity

Category	Impact area	Associated issues
Health	Spread of waterborne diseases	Use of contaminated water leads to outbreaks of cholera, typhoid, dysentery, hepatitis A, etc
	Poor hygiene and sanitation	Limited water access hinders hygiene, causing skin infections, eye diseases, and poor sanitation
Socioeconomic	Food insecurity and malnutrition	Crop failure and reduced livestock output lower food availability, worsening malnutrition
	Livelihood disruption and poverty	Farmers and water-dependent industries suffer income loss, deepening poverty in communities
	Rural–urban migration and displacement	Drought and water stress cause population displacement and strain on urban infrastructure
Environmental	Depletion of natural water sources	Rivers, lakes, and aquifers dry up from overuse and mismanagement
	Loss of ecosystems and biodiversity	Aquatic habitats degrade, species decline, and ecosystem services are lost
Geopolitical/security	Rising tensions over shared water	Competing demands for transboundary waters increase interstate and intercommunity tensions
	Water as a driver of conflict	Water scarcity fuels local and cross-border conflicts, exploited by non-state actors

water scarcity) leads to increased instances of infections and hygiene-related issues [11, 12]. For instance, a study indicates that poor sanitation directly correlates with higher disease prevalence due to inadequate water access to maintain hygiene [101]. The overlap between water scarcity and inadequate healthcare resources ensures that these health risks persist, thereby trapping vulnerable populations in a continuous cycle of poverty and ill health [34, 106].

1.4.2 Socioeconomic Consequences of Water Scarcity and Insecurity

Water scarcity significantly impacts food security and malnutrition, ultimately reducing both crop yields and livestock productivity, which are vital for subsistence in many communities [58, 89]. This decrease in agricultural output leads to higher food prices and limited availability, exacerbating malnutrition among affected populations. Furthermore, water scarcity disrupts livelihoods not only for farmers but also for industries that depend on reliable water sources, thereby increasing poverty levels [81]. The forced migration from rural to urban areas due to declining agricultural prospects further strains urban resources, as cities grapple with influxes of displaced populations seeking better living conditions under already stressed capacities [106].

1.4.3 Environmental Consequences of Water Scarcity and Insecurity

The environmental ramifications of water scarcity are equally critical. The depletion of natural water sources (such as rivers, lakes, and aquifers) results from extensive overuse and mismanagement, leading to ecological degradation [50, 112]. Consequently, the decline in freshwater habitats adversely affects biodiversity, as numerous species rely on these ecosystems for their survival. Pollution exacerbates this situation, as reduced water flow can concentrate toxins and pollutants detrimental to aquatic life [58, 81]. Thus, the interdependence between water availability and ecosystem health cannot be overstated, as environmental degradation only serves to further undermine the already precarious water supplies available to humans and wildlife alike [112].

1.4.4 Security and Geopolitical Consequences of Water Scarcity and Insecurity

The geopolitical landscape is increasingly fraught with tensions arising from competition over shared water resources. Conflicting claims over transboundary waters can escalate into diplomatic strife, especially in regions where water is scarce and critical for sustaining livelihoods [4, 34]. Water insecurity has become a catalyst for local and cross-border disputes, which extremist groups may exploit to further their agendas in fragile states [34]. As countries grapple with the implications of climate change and population growth on water resources, international cooperation and effective governance mechanisms are essential for managing these conflicts constructively [4, 50].

1.5 Studies on Water Scarcity and Insecurity

Case studies on water scarcity and insecurity present a stark picture of challenges faced by various regions globally, particularly Sub-Saharan Africa, the Middle East and North Africa (MENA), South Asia, Latin America, and urban centers. In the context of Sub-Saharan Africa, countries such as Nigeria, Ethiopia, and Kenya illustrate the intersection of limited access to safe drinking water and community adaptation struggles. Rural areas epitomize this crisis, grappling with inadequate infrastructure that hampers access to clean water. The work of Baba-Adamu and Jajere [14] highlights the environmental and socioeconomic challenges exacerbating this situation, especially in rural populations vulnerable to water scarcity. Recent studies confirm that such scarcity catalyzes migration, compelling communities to employ strategies like rainwater harvesting and borehole drilling as means of adaptation amid growing tensions over water access, particularly between farmers and herders [32, 51].

Government and NGO interventions are also crucial in this dynamic, with initiatives implemented to alleviate water scarcity. Efforts, such as the installation of solar-powered water pumps and education programs, play significant roles in enhancing water security in rural communities. Wakhungu [104] research on water reuse policy in Kenya underscores the potential of reused water to address water scarcity and improve productivity in agriculture and industry. Moreover, community engagement processes highlight a pathway for sustained management of water resources, fostering collaborative actions among stakeholders [100]. Effective management, particularly in arid and semi-arid regions, is vital for mitigating the effects of climate variability, especially as dependency on seasonal rainfall increases local populations' exposure to drought conditions [51].

Examining the MENA region reveals intricate water politics and scarcity. Countries here often endure extreme water scarcity due to arid climates and high evaporation rates. Additionally, reliance on transboundary rivers, such as the Nile and Tigris-Euphrates, fuels geopolitical tensions among nations, which Tilmant and Kinzelbach [99] explore regarding non-cooperation's costs in international river basins. Desalination presents a practical, albeit costly, solution for water-scarce nations like Saudi Arabia and the UAE, where significant infrastructure investments aim to bolster water supplies, despite associated environmental and economic drawbacks [25]. The severe impact of mismanagement and drought in nations like Syria and Yemen underscores the cascading effects of water insecurity on regional stability, reinforcing the need for policies focusing on cooperation and water pricing [33].

In South Asia, countries such as India, Pakistan, and Bangladesh wrestle with groundwater depletion and urban water crises exacerbated by poor planning and infrastructure. Over-extraction for agricultural purposes has resulted in alarming declines in water tables, notably in Punjab, India [10]. Urban centers like Delhi face acute shortages due to rapid population growth and aging infrastructure, as emphasized by the significant challenges faced in Chennai, including immediate crises necessitating emergency measures like water rationing [33]. Water source pollution, driven by industrial discharge and untreated sewage, compounds these challenges and necessitates comprehensive governmental responses that include regulating groundwater use and promoting sustainable agricultural practices.

Latin America reflects a diverse array of water-related issues, especially affecting indigenous communities facing environmental degradation from deforestation and mining activities. The Amazon basin exemplifies how industrial activities compromise water quality and biodiversity, inciting resistance from indigenous groups advocating for water rights and ecosystem preservation [23]. Legal victories in recognizing water as a human right clash with ongoing industrial pollution, emphasizing the need for social and environmental justice in water development [23]. Community efforts aimed at restoring river health are essential for movements advocating better water quality management [32].

Urban water crises are starkly evident in cities like Cape Town and Chennai, both experiencing acute scarcity due to climatic changes and resource mismanagement. Cape Town's "Day Zero" incident in 2018 epitomized the consequences of prolonged drought and ineffective demand management strategies [16]. In Chennai, 2019 shortages emerged from failed monsoon rains and excessive groundwater extraction, highlighting vulnerabilities within urban water systems [16]. A comprehensive approach must incorporate lessons learned from these crises,thus, integrated water resource management strategies emphasizing sustainability and community involvement gain prominence as frameworks for addressing urban water insecurity [16, 100].

1.6 Global and Regional Responses to Water Scarcity and Insecurity

The issue of water scarcity and insecurity presents significant challenges worldwide, making it an essential area of focus under the United Nations Sustainable Development Goals (SDG 6). Goal 6 specifically aims to ensure access to clean water and sanitation for all by 2030, promoting universal and equitable access to safe and affordable drinking water [21]. Achieving this goal necessitates a multidimensional approach that encompasses improving sanitation, hygiene, and significantly reducing pollution as well as the discharge of untreated wastewater [21]. Furthermore, the role of community participation in managing water resources has garnered attention, emphasizing the need for inclusive policy designs that engage stakeholders at all levels [21]. Monitoring progress towards these goals relies on the assessment of global indicators related to service coverage, water quality, and the sustainable management of water resources, highlighting the interlinked nature of water accessibility and sanitation [21]. Table 1.5 presents possible responses to water scarcity and insecurity.

Table 1.5 Possible responses to water scarcity and insecurity

Response strategy	Application
SDG 6: Ensuring access to clean water and sanitation	Aims for universal access to safe, affordable water and sanitation by 2030, promotion of hygiene, reduces pollution, and increases community participation and tracking of global progress using standardized indicators
Water diplomacy and international treaties	Facilitates cooperation on shared water bodies (e.g., Nile Basin, Indus Waters), enables peaceful negotiation and equitable water-sharing and prevents conflict and supports joint resource management
Role of international development agencies and NGOs	Agencies like World Bank, UNICEF, and WaterAid support water projects, fund infrastructure, emergency response, and capacity building, and advocate for policy reforms and community involvement
Innovative technologies: desalination, water recycling, smart irrigation	Desalination to convert seawater to drinking water (e.g., Middle East), recycling for treating wastewater for reuse (e.g., Singapore's NEWater) and smart irrigation for optimizing water use in farming via sensors and automation
Integrated water resources management	Coordinates water, land, and resource use across sectors, promotes sustainable and equitable water allocation and encourages stakeholder engagement and ecosystem-based planning

Water scarcity has also galvanized the need for comprehensive water diplomacy and international treaties, illustrating how nations can better manage shared water resources. Frameworks such as the Nile Basin Initiative and the Indus Waters Treaty serve as examples of cooperative strategies for transboundary water resource management [24, 56]. These treaties establish mechanisms not only to prevent conflict but also to ensure fair and equitable distribution of water resources across nations [3, 56]. Effective water diplomacy fosters dialogue, especially in politically sensitive regions where competition for water often exacerbates tensions [109]. Moreover, strategies promoting integrated basin-level planning can enhance collective infrastructure development and resource sharing, contributing to long-term stability amidst evolving climate challenges [13].

International development agencies and non-governmental organizations (NGOs) play crucial roles in addressing water insecurity. Organizations such as the World Bank, UNICEF, and WaterAid implement projects aimed at improving water access, particularly in drought-prone and low-income areas [21, 56]. These entities not only fund infrastructure development but also emphasize capacity-building for sustainable water governance and resilience in communities facing water crises [13, 21]. Furthermore, during emergencies caused by droughts or conflicts, these organizations facilitate swift responses, demonstrating the necessity of their involvement in global water management [21]. They promote participatory approaches to water resource management, advocating for policy reforms that resonate with local needs and context [13].

Innovative technologies have emerged as critical tools in enhancing water access and sustainability. Desalination technologies, which convert seawater into potable water, present viable solutions for arid regions grappling with chronic water scarcity [9]. Countries like Singapore exemplify successful water recycling initiatives, such as the NEWater system, which treats and reuses wastewater for diverse applications [9]. Similarly, smart irrigation technologies have the potential to significantly reduce water consumption in agriculture by utilizing sensors and data analytics [83]. The adoption of greywater systems and strategies for rainwater harvesting also illustrate practical responses in urban contexts aiming for sustainable water use [13]. These technological advancements underscore the importance of integrating various approaches in combating water-related challenges while maximizing resource efficiency. IWRM provides a holistic framework that coordinates the multifaceted use of water, land, and related resources to enhance social and economic welfare sustainably [13, 21]. Stakeholder engagement remains a focal point of IWRM, advocating for the active involvement of local, national, and international actors in water governance processes [9]. This approach promotes policy integration across various sectors, such as agriculture, energy, and urban planning, ultimately supporting the equitable distribution of water resources [13, 21]. Adopting ecosystem-based practices and watershed management strategies further contributes to sustainable water allocation, ensuring that water management systems can adapt to changing climatic and socio-economic conditions [13].

1.7 One Health Perspective on Water Scarcity and Insecurity

1.7.1 Interconnected Causes of Water Scarcity

Climate Change Impacts: Climate change significantly affects water availability through altered rainfall patterns, prolonged droughts, and glacial melt. These phenomena reduce freshwater resources, making water scarcity a pressing issue globally [92, 115]. Moreover, the increased frequency and intensity of extreme weather events, such as floods and cyclones, can damage existing water infrastructure, exacerbating both water supply and quality management challenges [92]. This detrimental cycle illustrates the urgent need for robust water management practices that can adapt to changing climatic conditions.

Population Growth and Urbanization: The impact of population growth and urbanization is profound, as higher demand for water in domestic, agricultural, and industrial sectors strains available resources [59, 84]. Furthermore, urban sprawl tends to encroach upon natural recharge zones, leading to their degradation and, in many cases, pollution of water sources [84]. This trend underscores the necessity for integrated urban planning and water management policies that can address both the growing population demands and the protection of water resources.

Pollution and Degradation of Water Sources: Pollution arises from multiple sources, including agricultural runoff, industrial waste, and untreated sewage, significantly impacting both water quality and availability [17, 18, 69–71, 90, 91]. The contamination of water bodies with chemical pollutants such as heavy metals [5–6, 43, 45], pharmaceuticals, and microplastics presents serious risks to human health and aquatic ecosystems [2, 84]. The emergence of such contaminants emphasizes the need for integrated monitoring and management of water quality through a One Health approach that connects environmental health with human and animal health [68, 84].

Unsustainable Water Management: Unsustainable practices, such as the over-extraction of groundwater and inefficient irrigation methods, contribute to significant depletion of aquifers and water resources [84, 115]. The lack of integrated planning across agricultural, environmental, and public health sectors heightens vulnerability to water scarcity [2]. To address these challenges, comprehensive strategies that incorporate sustainable management, such as promoting smart water metering and responsible irrigation practices, must be pursued diligently Liu et al. [59, 92].

1.7.2 Health Consequences Across Humans, Animals, and Ecosystems

Human Health: The ramifications of water scarcity extend directly to human health, particularly through an increased risk of waterborne diseases such as cholera, typhoid, and hepatitis A [2, 22]. Additionally, inadequate water supply compromises hygiene and sanitation, which contributes to malnutrition and increased mortality rates, especially among vulnerable populations, such as children [2]. Understanding these interconnections is critical for developing health policies that prioritize both water access and public health.

Animal Health: The availability of water significantly influences animal health and productivity. Livestock suffering from water shortages exhibit decreased productivity and higher susceptibility to disease, while wildlife may be forced into closer proximity at limited water sources, increasing the transmission of zoonotic diseases [2, 22]. This reflects the interconnectedness of human, animal, and environmental health within the One Health framework, emphasizing the need for cohesive policies that protect all elements of health.

Environmental Health: Water scarcity and pollution severely threaten aquatic biodiversity, leading to the degradation of ecosystems that provide essential services such as water purification and soil fertility [36, 68]. The disruption of food chains and loss of species further highlights the importance of a holistic approach to water resource management, where environmental health is prioritized alongside human and animal health [68, 110]. This ecosystem perspective is essential for fostering resilience and sustainability within communities facing water challenges.

1.7.3 One Health Approach to Mitigation

Integrated Resource Management: A One Health approach emphasizes the the relationship between human, environmental and biodiversity health [41, 45, 66]. With regard to water security its importance in cross-sector collaboration among public health, veterinary, and environmental agencies. Effective resource management strategies must be developed to promote sustainable water use practices [67] in agriculture and industry, which can mitigate health risks and enhance the resilience of communities facing water scarcity [57, 59, 115]. This integrated approach is fundamental for advancing our understanding of the interdependencies among human, animal, and environmental health.

Surveillance and Early Warning Systems: Implementing comprehensive monitoring systems to track water quality and availability is vital for identifying emerging threats to health. Predictive models can assist in guiding timely interventions during droughts or contamination events, improving public health preparedness and response capabilities [22, 75, 77, 92]. Such early-warning systems need robust data and regional context to ensure their efficacy.

Community Engagement and Education: Engaging communities in conservation efforts, sanitation, and hygiene practices (WASH initiatives) is crucial for raising awareness of the health impacts stemming from water scarcity and pollution [44, 42–43, 46]. Educational activities can empower communities to participate in resource management, promoting stewardship and fostering a culture of water conservation that aligns with One Health principles [2, 76, 84].

Policy and Governance: Strengthening policies related to water safety management, particularly in the context of reuse and sustainability, is critical for securing public health [2, 110]. Supporting international cooperation on transboundary water issues and fostering climate-resilient water planning can enhance governance frameworks that prioritize the One Health approach, ensuring equitable access to water resources across populations [84].

1.7.4 Long-Term Implications and Global Priorities

Food and Water Security: The connection of water scarcity with agricultural output poses a significant threat to food security, exacerbating issues such as malnutrition and social instability [2, 22]. Proactive measures to combat water scarcity must be integrated into food security strategies to ensure that agricultural practices remain sustainable and resilient.

Antimicrobial Resistance (AMR): Contaminated water sources can facilitate the dissemination of antimicrobial-resistant pathogens, complicating public health challenges. Addressing this issue necessitates an integrated view of water quality management as part of a broader health strategy that involves safe agricultural practices and public health safeguards.

Climate Adaptation and Resilience: Adopting climate-resilient infrastructure and investing in green technologies can enhance the sustainability of water resources while supporting broader adaptation strategies. This approach aligns with One Health principles and acknowledges the necessity of a cohesive response to climate change, which directly influences water availability and security.

1.8 Future Perspectives on Water Scarcity and Insecurity

Water scarcity and insecurity are critical global challenges, heavily influenced by projections of climate change. The increasing frequency and intensity of extreme weather events, such as droughts, floods, and heatwaves, significantly disrupt the availability of freshwater. According to Woodhouse et al. [105], warming temperatures in regions like Southern California threaten water supply by shifting reservoir inflow timing and quantity, creating greater mismatches between water demand and supply timing. This phenomenon is supported by various climate models that project prolonged droughts will become increasingly common and will affect groundwater recharge and surface water supplies [19].

As precipitation patterns change, the impacts on groundwater levels and surface water availability are expected to worsen. Increased evaporation due to higher temperatures, combined with erratic rainfall patterns, can hinder the natural replenishment of aquifers [95]. Consequently, regions characterized by arid and semi-arid climates, including Sub-Saharan Africa and South Asia, are particularly vulnerable to heightened water stress [63]. Anticipating these long-term risks necessitates advanced climate model integration into water resource planning, allowing for strategic responses to mitigate projected water shortages.

Emerging technologies are reshaping water management and conservation [73]. The development of smart water meters and advanced leak detection systems aids households and industries in optimizing water use, demonstrating a proactive approach to management [29]. These technologies serve as crucial interventions during scarcity by enhancing awareness, which is crucial for community adaptation and sustainability. Furthermore, the expansion of satellite-based remote sensing technology facilitates real-time monitoring of water availability and distribution [85]. Additionally, the rise of artificial intelligence and machine learning tools assists in predicting consumption patterns, optimizing supply chains and resource allocation for enhanced efficiency [87].

Innovations in desalination processes provide promising solutions for addressing water scarcity. Mauter and Fiske [64] discuss how modern desalination technologies can extract potable water from various saline sources, effectively increasing freshwater availability and contributing to a circular economy. This approach is reinforced by research emphasizing the need for enhanced water recycling and reuse initiatives, particularly in urban settings where traditional freshwater sources are strained [20]. Digital platforms that enable transparency and data sharing among stakeholders are vital for fostering collaborative governance structures, a necessary step towards equitable water management solutions.

As global water demand continues to rise (primarily driven by urbanization, population growth, and agriculture) competition for finite resources intensifies [19]. Integrated water resource management (IWRM) strategies are emerging as vital frameworks to confront these challenges, indicating a paradigm shift toward sustainable practices and circular economic models [62]. This shift includes policies promoting water reuse and establishing water markets, as observed in regions like Southern Africa, where improved governance structures aim to enhance water-use efficiency [63, 75]. To effectively address these complexities, robust policy directions advocating for proactive water governance are essential. Strengthening frameworks for transboundary water cooperation is crucial to mitigate conflicts arising from shared resources. Adaptive governance models that respond to environmental changes while ensuring equitable access and affordability of water services are also critical [31].

Encouraging private sector investments in infrastructure can drive innovations necessary for enhanced water security metrics. Moreover, integrating water security within broader national security and climate adaptation policies can ensure that water management is approached holistically, balancing both supply and the escalating demand for freshwater [94].

1.9 Conclusion

Addressing the challenges of water scarcity and insecurity demands urgent, multifaceted action rooted in a comprehensive understanding of its diverse causes and far-reaching impacts. From climate change and population growth to pollution and inefficient governance, water scarcity is driven by both natural and human-induced factors that undermine health, food security, environmental sustainability, and economic stability. The One Health framework offers a valuable lens for approaching these complexities, emphasizing the interconnection between human, animal, and environmental well-being. Central to this strategy is the recognition of water as a fundamental human right, one that should be safeguarded through inclusive and participatory approaches that involve governments, local communities, civil society, and the private sector. By fostering cross-sectoral collaboration and integrating evidence-based policies with traditional knowledge, stakeholders can co-develop equitable solutions that address regional disparities and ensure the resilience of vulnerable populations.

There is a need to build a resilient and just response system to water scarcity that prioritizes sustainability, innovation, and equity. Technological advances in water conservation, surveillance systems, and climate-resilient infrastructure should be paired with robust policy reforms and transparent governance frameworks. Equally important is community engagement (particularly in water-stressed regions) where local ownership of water management efforts can enhance effectiveness and cultural relevance. Achieving Sustainable Development Goal 6 (Clean Water and Sanitation for All) will require coordinated global efforts that bridge environmental science, public health, and social justice. Thus, addressing water scarcity is not merely a technical or environmental challenge but a moral imperative to uphold human dignity, preserve ecosystems, and ensure a secure and healthy future for all.

References

1. Aawada H, Chamseddine S, Kayel S, El-Jarf A (2024) Water stress: between farmers' needs and public policies the case of smallholder farmers in the villages of Nabatieh Governorate. J Sci Res Rep 30(7):33–47
2. Abidola TB, Folorunso AE, Akib MO, Olaoye OJ, Adewale AO, Etoniru S, Oladipupo AA (2025) Sustainable water management and one health. J Multidiscip Sci Mikailalsys 3(2):899–912. https://doi.org/10.58578/mikailalsys.v3i2.5863

3. Abigail S (2024) Sharing water across borders: assessing the hydro-diplomacy process in the resolution of indus river disputes. Jurnal Hubungan Internasional 17(1):233–247. https://doi.org/10.20473/jhi.v17i1.53167
4. Acosta-González HN, Chafla P (2024) Measuring water poverty based on SDG 6. World Water Policy 10(2):424–441. https://doi.org/10.1002/wwp2.12191
5. Aghoghovwia OA, Izah SC, Miri FA (2018) Environmental risk assessment of heavy metals in sediment of Nun River around Gbarantoru and Tombia Towns, Bayelsa State, Nigeria. Biol Evid 8(3):21–31
6. Aghoghovwia OA, Miri FA, Izah SC (2018) Impacts of anthropogenic activities on heavy metal levels in surface water of Nun River AROUND Gbarantoru and Tombia Towns, Bayelsa State, Nigeria. Annal Ecol Environ Sci 2(2):1–8
7. Aghoghovwia OA, Umoru OD, Izah SC (2018) Physicochemical characteristics of nun river at Gbarantoru and Tombia Axis in Bayelsa State, Nigeria. Biosci Methods 9(1):1–11
8. Ahmad N, Lilienthal G (2021) Right to water as a human right: a critical overview of international instruments. Environ Policy Law 50(4–5):299–308. https://doi.org/10.3233/epl-200232
9. Aktar F (2021) Water diplomacy and water sharing problem between Bangladesh and India: a quest for solution. Int J Res Sci Innov 08(01):223–229. https://doi.org/10.51244/ijrsi.2021.8111
10. Alam K (2015) Farmers' adaptation to water scarcity in drought-prone environments: a case study of rajshahi district, bangladesh. Agric Water Manag 148:196–206. https://doi.org/10.1016/j.agwat.2014.10.011
11. Apelabi GO, Mbarep DPP, Nurjayati R, Sasto IHS, dan Bolly YY (2023) People conditions from socio-economic and health aspects on access to improved drinking water in East Nusa Tenggara Province. Jurnal Ilmu Lingkungan 21(4):1002–1008
12. Armah FA, Ekumah B, Yawson DO, Odoi JO, Afitiri AR, Nyieku FE (2018) Access to improved water and sanitation in sub-Saharan Africa in a quarter century. Heliyon 4(11)
13. Aslam B (2022) Hydro-diplomacy and the prospects of environmental peacebuilding between Pakistan and India. Biol Clin Sci Res J 3(1):204–216. https://ideapublishers.org/index.php/jhsms/article/view/680/286
14. Baba-Adamu M, Jajere IA (2020) Water scarcity measurement in the Yobe region of Nigeria. KIU Interdiscip J Hum Soc Sci 1(2):265–280
15. Baba-Adamu M, Jajere I (2021) Environmental determinants of rural households' vulnerability to water scarcity in semi-arid Nigeria. Malays J Comput 6(2):778. https://doi.org/10.24191/mjoc.v6i2.10423
16. Bamgboye O, Chrysoulas C, Liu X, Watt T, Sodiya A, Oyeleye M, Kalutharage S (2024) Towards building a smart water management system (SWAMS) in Nigeria. In: 2024 IEEE 22nd mediterranean electrotechnical conference (MELECON). IEEE, pp 1133–1138
17. Ben-Eledo VN, Kigigha LT, Izah SC, Eledo BO (2017) Bacteriological quality assessment of Epie Creek, Niger Delta Region of Nigeria. Int J Ecotoxicol Ecobiol 2(3):102–108
18. Ben-Eledo VN, Kigigha LT, Izah SC, Eledo BO (2017) Water quality assessment of Epie Creek in Yenagoa Metropolis, Bayelsa State. Nigeria. Arch Curr Res Int 8(2):1–24
19. Brown TC, Mahat V, Ramirez JA (2019) Adaptation to future water shortages in the United States caused by population growth and climate change. Earth's Futur 7(3):219–234
20. Burgess J, Meeker M, Minton J, O'Donohue M (2015) International research agency perspectives on potable water reuse. Environ Sci Water Res Technol 1(5):563–580
21. Chan NW, Roy R, Chaffin BC (2016) Water governance in Bangladesh: an evaluation of institutional and political context. Water 8(9):403
22. Dembe J (2024) One health approaches to disease management. Anim Health J 5(1):13–25. https://doi.org/10.47941/ahj.1777
23. Deresse T (2023) The vulnerability of communities to climate change-induced water scarcity in Ethiopia and Kenya: a systematic review. https://doi.org/10.21203/rs.3.rs-3812532/v1
24. Desai BH (2021) Sixty years of the indus waters treaty in the era of climate change: a look ahead in hydro-diplomacy and treaty law. Environ Policy Law 51(3):175–184. https://doi.org/10.3233/epl-210013

25. Dhakal N, Salinas-Rodriguez SG, Hamdani J, Abushaban A, Sawalha H, Schippers JC, Kennedy MD (2022) Is desalination a solution to freshwater scarcity in developing countries? Membranes 12(4):381
26. Fantini E (2020) An introduction to the human right to water: Law, politics, and beyond. Wiley Interdiscip Rev Water 7(2):e1405
27. Flores-Vázquez C, Losa E, Espinoza A, Morales-Téllez A, Tentori M (2024) Design of a serious game to teach water scarcity concepts in rural communities. Avances en Interacción Humano-Computadora 9(1):117–121. https://doi.org/10.47756/aihc.y9i1.156
28. García-López M, Cuadrado-Quesada G, Montano B (2024) Untangling the vicious cycle around water and poverty. Sustain Dev 32(3):1845–1860
29. Gonzales P, Ajami N (2017) Social and structural patterns of drought-related water conservation and rebound. Water Resour Res 53(12):10619–10634. https://doi.org/10.1002/2017wr021852
30. Gosling SN, Arnell NW (2016) A global assessment of the impact of climate change on water scarcity. Clim Change 134(3):371–385. https://doi.org/10.1007/s10584-013-0853-x
31. Grafton RQ, Garrick D, Manero A, Do TN (2019) The water governance reform framework: overview and applications to Australia, Mexico, Tanzania, USA and Vietnam. Water 11(1):137
32. Gulumbe BH, Yusuf ZM, Faggo AA, Yahaya TO, Manga SS (2023) The interplay among conflict, water scarcity, and cholera in Northern Nigeria. Public Health Chall 2(3):e118
33. Haas S, Gianoli A, Van Eerd M (2021) The roles of community resilience and risk appraisal in climate change adaptation: the experience of the Kannagi Nagar resettlement in Chennai. Environ Urban 33(2):560–578
34. He C, Liu Z, Wu J, Pan X, Fang Z, Li J, Bryan BA (2021) Future global urban water scarcity and potential solutions. Nat Commun 12(1):4667
35. Hoekstra AY, Mekonnen MM, Chapagain AK, Mathews RE, Richter BD (2012) Global monthly water scarcity: blue water footprints versus blue water availability. PLoS ONE 7(2):e32688
36. Horwitz P, Finlayson CM (2011) Wetlands as settings for human health: incorporating ecosystem services and health impact assessment into water resource management. Bioscience 61(9):678–688
37. Ioris AA (2012) The geography of multiple scarcities: urban development and water problems in lima, peru. Geoforum 43(3):612–622. https://doi.org/10.1016/j.geoforum.2011.12.005
38. Israilova E, Voronina A, Shatila K (2023) Impact of water scarcity on socio-economic development. In E3S web of conferences, vol 458. EDP Sciences, p 08027
39. Iyiola AO, Afolabi OA, Alimi SK, Akingba OO, Izah SC, Ogwu MC (2024) Climate change and water crisis in the Global South. In: Izah SC, Ogwu MC, Loukas A, Hamidifar H (eds) Water crises and sustainable management in the Global South. Springer, Singapore, pp 111–140. https://doi.org/10.1007/978-981-97-4966-9_4
40. Izah SC, Ogwu MC (2025) Modeling solutions for microbial water contamination in the global south for public health protection. Front Microbiol 16:1504829. https://doi.org/10.3389/fmicb.2025.1504829
41. Izah SC, Richard G, Stanley HO, Sawyer WE, Ogwu MC, Uwaeme OR (2023) Integrating the one health approach and statistical analysis for sustainable aquatic ecosystem management and trace metal contamination mitigation. ES Food Agrofor 14:1012. https://doi.org/10.30919/esfaf1012
42. Izah SC, Richard G, Stanley HO, Sawyer WE, Ogwu MC, Uwaeme OR (2024) Prospects and application of multivariate and reliability analyses to one health risk assessments of toxic elements. Toxicol Environ Heal Sci 16(2):127–134
43. Izah SC, Stanley HO, Richard G, Sawyer WE, Uwaeme OR (2024) Environmental health risks of trace elements in sediment using multivariate approaches and contamination indices. Int J Environ Sci Technol. https://doi.org/10.1007/s13762-024-05974-1
44. Izah SC, Etim NG, Alaka OO, Nwachukwu BC, Joshua G (2024) Public health interventions for cholera control and prevention: water, sanitation, and hygiene (WASH) perspective. Greener J Epidemiol Public Health 12(1):1–15

45. Izah SC, Stanley HO, Richard G, Sawyer WE, Uwaeme OR (2024) Surface water quality: a statistical perspective on the efficacy of environmental and human health assessment tools. Water Air Soil Pollut 235(3). https://doi.org/10.1007/s11270-024-06965-1
46. Izah SC, Jacob DE, Nelson IU, Avez S (2024) Urban water crisis in the Global South. In: Izah SC, Ogwu MC, Loukas A, Hamidifar H (eds) Water crises and sustainable management in the Global South. Springer, Singapore, pp 45–83. https://doi.org/10.1007/978-981-97-4966-9_2
47. Jacob DE, Nelson IU, Izah SC (2024) Indigenous water management strategies in the Global South. In: Izah SC, Ogwu MC, Loukas A, Hamidifar H (eds) Water crises and sustainable management in the Global South. Springer, pp 487–525. https://link.springer.com/chapter/https://doi.org/10.1007/978-981-97-4966-9_16
48. Jacob DE, Nelson IU, Eniang EA, Izah SC (2024) Advancing water security and resilience in the Global South through recreational development. In: Izah SC, Ogwu MC, Loukas A, Hamidifar, H (eds) Water crises and sustainable management in the Global South. Springer, pp 527–554. https://link.springer.com/chapter/https://doi.org/10.1007/978-981-97-4966-9_17
49. Jacob DE, Nelson IU, Izah SC, Daniel KS (2024) Rural water crises in the Global South: understanding the scope and impact. In: Izah SC, Ogwu MC, Loukas A, Hamidifar H (eds) Water crises and sustainable management in the Global South. Springer, Singapore, pp 3–44. https://doi.org/10.1007/978-981-97-4966-9_1
50. Jafari Shalamzari M, Zhang W (2018) Assessing water scarcity using the Water Poverty Index (WPI) in Golestan province of Iran. Water 10(8):1079
51. Kamta FN, Schilling J, Scheffran J (2021) Water resources, forced migration, and tensions with host communities in the Nigerian part of the Lake Chad basin. Resources 10(4):27
52. Khalid RM, Kadhim SA, Dahalan WSAW (2020) Revisiting the human right to water in contemporary international law. UUM J Legal Stud 11(1):37–49
53. Klassert C, Heilemann J, Werner S, Nagpal M, Digman E, Klauer B, Gawel E (2025) Rural-urban water scarcity risks in historically water-abundant regions: the role of intensifying human-natural systems variability (No. EGU25–18310). Copernicus Meetings
54. Kosoe EA, Osumanu IK, Ogwu MC (2025) Reimagining rural and small-town water supply in Ghana through inclusive governance and sustainable models. J Sustain 1(2):1–21. https://doi.org/10.55845/jos-2025-1234
55. Kosoe EA, Ogwu MC, Duniah J, Ndigun FB (2025) Sustaining indigenous water governance systems in Ayimaye and Lungni, Northern Ghana, through Hybrid Models. J Sustain 1(2). https://doi.org/10.55845/jos-2025-1235
56. Krzymowski A (2021) Water diplomacy and its strategic significance for sustainable development goals and global security architecture. Sustainability 13(24):13898. https://doi.org/10.3390/su132413898
57. Liu A, Giurco D, Mukheibir P (2017) Advancing household water-use feedback to inform customer behaviour for sustainable urban water. Water Sci Technol Water Supply 17(1):198–205
58. Liu Z, Liu W (2021) Spatial-temporal relationship between water resources and economic development in rural China from a poverty perspective. Int J Environ Res Public Health 18(4):1540. https://doi.org/10.3390/ijerph18041540
59. Liu J, Liu Q, Yang H (2016) Assessing water scarcity by simultaneously considering environmental flow requirements, water quantity, and water quality. Ecol Ind 60:434–441
60. Liu J, Yang H, Gosling SN, Kummu M, Flörke M, Pfister S, Hanasaki N, Wada Y, Zhang X, Zheng C et al (2017) Water scarcity assessments in the past, present, and future. Earth's Futur 5(6):545–559
61. Luque-Villa M, Granda-Rodríguez H, Garza-Tatis C, González-Méndez M (2024) Applying bourdieu's theory to public perceptions of water scarcity during el niño: a case study of santa marta, colombia. https://doi.org/10.20944/preprints202409.0015.v1
62. Marston L, Cai X (2016) An overview of water reallocation and the barriers to its implementation. Wiley Interdiscip Rev Water 3(5):658–677. https://doi.org/10.1002/wat2.1159

63. Matchaya G, Nhamo L, Nhlengethwa S, Nhemachena C (2019) An overview of water markets in southern africa: an option for water management in times of scarcity. Water 11(5):1006. https://doi.org/10.3390/w11051006
64. Mauter M, Fiske P (2020) Desalination for a circular water economy. Energy Environ Sci 13(10):3180–3184. https://doi.org/10.1039/d0ee01653e
65. Mekonnen M, Hoekstra A (2016) Four billion people facing severe water scarcity. Sci Adv 2(2). https://doi.org/10.1126/sciadv.1500323
66. Nabebe G, Ogamba EN, Izah SC (2024) Aspects of polycyclic aromatic hydrocarbons in aquatic ecosystems: a one health perspective. Greener J Environ Manag Pub Safety 12(1):22–43
67. Numbere AO, Kika EP, Izah SC (2024) Challenges and mitigation of water resource management in rural areas in Nigeria. In: Izah SC, Ogwu MC, Loukas A, Hamidifar H (eds) Water crises and sustainable management in the Global South. Springer, Singapore, pp 373–390. https://doi.org/10.1007/978-981-97-4966-9_12
68. Odume O (2017). Ecosystem approach to managing water quality. https://doi.org/10.5772/65707
69. Ogamba EN, Charles EE, Izah SC (2021) Distributions, pollution evaluation and health risk of selected heavy metal in surface water of Taylor creek, Bayelsa State, Nigeria. Toxicol Environ Health Sci 13(2):109–121. https://doi.org/10.1007/s13530-020-00076-0
70. Ogamba EN, Izah SC, Oribu T (2015) Water quality and proximate analysis of *Eichhornia crassipes* from River Nun, Amassoma Axis, Nigeria. Res J Phytomedicine 1(1):43–48
71. Ogamba EN, Izah SC, Toikumo BP (2015) Water quality and levels of lead and mercury in *Eichhornia crassipes* from a tidal creek receiving abattoir effluent, in the Niger Delta, Nigeria. Cont J Environ Sci 9(1):13–25
72. Ogamba EN, Seiyaboh EI, Izah SC, Ogbugo I, Demedongha FK (2015) Water quality, phytochemistry and proximate constituents of *Eichhornia crassipes* from Kolo creek, Niger Delta, Nigeria. Int J Appl Res Technol 4(9):77–84
73. Ogidi OI, Izah SC (2024) Water contamination by municipal solid wastes and sustainable management strategies. In: Izah SC, Ogwu MC, Loukas A, Hamidifar H (eds) Water crises and sustainable management in the Global South. Springer, pp 313–339. https://link.springer.com/chapter/https://doi.org/10.1007/978-981-97-4966-9_10
74. Ogwu MC, Kosoe EA (2025) Integrating green infrastructure into sustainable agriculture to enhance soil health, biodiversity, and microclimate resilience. Sustainability 17(9):3838. https://doi.org/10.3390/su17093838
75. Ogwu MC (2025) Sustainable development in Africa: challenges, innovations, and policy pathways for the future. Afr Res Rep 1(2):55–69. https://doi.org/10.5897/ARR2025.0005
76. Ogwu MC, Kosoe EA (2024) Place of cultural diversity in sustainable water resource management in Ghana. In: Izah SC, Ogwu MC, Loukas A, Hamidifar H (eds) Water crises and sustainable management in the Global South. Springer, Singapore. https://doi.org/10.1007/978-981-97-4966-9_14
77. Ogwu MC, Izah SC, Sawyer WE, Amabie T (2025) Environmental risk assessment of trace metal pollution: a statistical perspective. Environ Geochem Health 47:94. https://doi.org/10.1007/s10653-025-02405-z
78. Pacheco-Treviño S, Manzano-Camarillo MG (2024) The socioeconomic dimensions of water scarcity in urban and rural Mexico: a comprehensive assessment of sustainable development. Sustainability 16(3):1011
79. Pacheco-Treviño S, Manzano-Camarillo MGF (2024) Review of water scarcity assessments: highlights of Mexico's water situation. Wiley Interdiscip Rev Water 11(4):e1721
80. Palmer RC, Short D, Auch WET (2018) The human right to water and unconventional energy. Int J Environ Res Public Health 15(9):1858
81. Pandey S, Mohapatra G, Arora R (2024) Water Poverty Index and its changing trend in India. J Econ Adm Sci 40(5):1142–1168
82. Pasanen T, Lakkala H, Yliluoma R, Tuominen V, Jusi S, Luukkanen J, Kaivo-oja J (2017) Poverty–environment nexus in the Lao PDR: analysis of household survey data. Dev Policy Rev 35(3):349–371

83. Paudel J (2022) The role of military diplomacy in Nepal's foreign policy. Unity J 3(01):183–194. https://doi.org/10.3126/unityj.v3i01.43324
84. Prata JC (2022) A one health perspective on water contaminants. Water Emerg Contam Nanoplastics 1(3):15. https://doi.org/10.20517/wecn.2022.14
85. Puangkaew N, Ongsomwang S (2021) Remote sensing and geospatial models to simulate land use and land cover and estimate water supply and demand for water balancing in phuket island, thailand. Appl Sci 11(22):10553. https://doi.org/10.3390/app112210553
86. Sabbil AAS, Abdulrahman BMA (2022) Effects of water scarcity on rural household economy. J Econ Adm Sci 28(131):109–118
87. Saleem A, Mahmood I, Sarjoughian H, Nasir HA, Malik AW (2021) A Water evaluation and planning-based framework for the long-term prediction of urban water demand and supply. Simulation 97(5):323–345. https://doi.org/10.1177/0037549720984250
88. Sancin V, Juhart M (2023) The right to safe drinking water in international law and in Slovenia's legal framework and implementation. J Agric Env't L 18(34):106–124. https://doi.org/10.21029/jael.2023.34.106
89. Schewe J, Heinke J, Gerten D, Haddeland I, Arnell NW, Clark DB, Dankers R, Eisner S, Fekete B, Colón-González FJ, Gosling SN, Kim H, Liu X, Masaki Y, Portmann FT, Satoh Y, Stacke T, Tang Q, Wada Y, Wisser D, Albrecht T, Frieler K, Piontek F, Warszawski L, Kabat P (2014) Multimodel assessment of water scarcity under climate change. Proc Natl Acad Sci 111(9):3245–3250. https://doi.org/10.1073/pnas.1222460110
90. Seiyaboh, EI, Izah SC, Bokolo JE (2017) Bacteriological quality of water from river nun at Amassoma Axises, Niger Delta, Nigeria. ASIO J Microbiol Food Sci Biotechnol Innov 3(1):22–26
91. Seiyaboh EI, Izah SC (2017) Review of impact of anthropogenic activities in surface water resources in the Niger Delta region of Nigeria: a case of Bayelsa state. Int J Ecotoxicol Ecobiol 2(2):61–73
92. Shahsavari-Pour N, Bahador S, Heydari A, Fekih A (2023) Water shortage simulation using a system dynamics approach: a case study of the rafsanjan city. Sustainability 15(7):6225. https://doi.org/10.3390/su15076225
93. Shemer H, Wald S, Semiat R (2023) Challenges and solutions for global water scarcity. Membranes 13(6):612
94. Simonovic SP, Breach PA (2020) The role of water supply development in the earth system. Water 12(12):3349
95. Smith R, Li J, Grote K, Butler J (2023) Estimating aquifer system storage loss with water levels, pumping and InSAR data in the Parowan Valley, Utah. Water Resour Res 59(4):e2022WR034095
96. Starzyk KB, Neufeld KH, Gaucher D, Vorauer JD, Fontaine AS, Quesnel MS, Yakubovich AR (2021) Is water a human right?: priming water as a human right increases support for government action. Int Indig Policy J 12(3):1–28
97. Sullivan CA (2010) Quantifying water vulnerability: a multi-dimensional approach. Stoch Env Res Risk Assess 25(4):627–640. https://doi.org/10.1007/s00477-010-0426-8
98. Šulyová D, Kubina M (2022) Integrated water resources management in an urban concept: results from water smart cities and water management institutions in slovakia. J Entrep Sustain Issues 10(2):294–319. https://doi.org/10.9770/jesi.2022.10.2(18)
99. Tilmant A, Kinzelbach W (2012) The cost of noncooperation in international river basins. Water Resour Res 48(1). https://doi.org/10.1029/2011wr011034
100. Ugbah R, Meldrum A, Ehiwario K (2017) Water access and community engagement: creating the right environment for maximizing the benefits of community engagement processes and increasing water participation in Nigeria. WIT Trans Ecol Environ 220:129–139
101. Valero D, Cook J, Lee A, Browne A, Ellis R, Pancholi VS, Hoolohan C (2023) Addressing water poverty under climate crisis: implications for social policy. Soc Policy Soc 22(4):747–762. https://doi.org/10.1017/s1474746423000258
102. Veldkamp TI, Eisner S, Wada Y, Aerts JC, Ward PJ (2015) Sensitivity of water scarcity events to ENSO-driven climate variability at the global scale. Hydrol Earth Syst Sci 19(10):4081–4098

103. Wahi N (2022) The evolution of the right to water in India. Water 14(3):398. https://doi.org/10.3390/w14030398
104. Wakhungu MJ (2019) An ethnography of policy: water reuse policy in kenya. Water Policy 21(2):436–448. https://doi.org/10.2166/wp.2019.160
105. Woodhouse CA, Meko DM, Bigio ER (2020) A long view of southern California water supply: perfect droughts revisited. JAWRA J Am Water Resour Assoc 56(2):212–229
106. Yahaya S, Scholz M (2021) Water-energy-food accessibility and tracking progress towards achieving sustainable development goals in the savannah region of katsina state, nigeria. Water 13(24):3595. https://doi.org/10.3390/w13243595
107. Yan Y, Li J, Li J, Jiang T (2023) Spatiotemporal changes in the supply and demand of ecosystem services in the kaidu-kongque river basin, china. Sustainability 15(11):8949. https://doi.org/10.3390/su15118949
108. Yang H (2016) Assessment and prediction of water resources based on Grey Theory. In: 2016 2nd Workshop on advanced research and technology in industry applications (WARTIA-16). Atlantis Press, pp 778–783
109. Yeganeh Y, Bakhshandeh E (2022) Iran's model of water diplomacy to promote cooperation and prevent conflict over transboundary rivers in southwest Asia. World Affairs 185(2):331–358. https://doi.org/10.1177/00438200221081210
110. Yehia AG, Mehany MA, Fareed AM, El-sayed WH, Taman MS (2024) The role of water safety plan (WSP) to enhance the compatibility in water sector, Egypy. World Water Policy 10(2):524–552
111. Zeng Z, Liu J, Savenije HH (2013) A simple approach to assess water scarcity integrating water quantity and quality. Ecol Ind 34:441–449
112. Zhang Q, Liu B, Zhang W, Jin G, Li Z (2015) Assessing the regional spatio-temporal pattern of water stress: a case study in Zhangye City of China. Phys Chem Earth Parts A/B/C 79:20–28
113. Zhu R, Fang Y (2022) Application of a water supply-demand balance model to set priorities for improvements in water supply systems: a case study from the koshi river basin, nepal. Int J Environ Res Public Health 19(3):1606. https://doi.org/10.3390/ijerph19031606
114. Zimmer A, Winkler IT, De Albuquerque C (2014) Governing wastewater, curbing pollution, and improving water quality for the realization of human rights. Waterlines 337–356
115. Zyadin A (2013) Water shortage in mena region: an interdisciplinary overview and a suite of practical solutions. J Water Resour Prot 05(04):49–58. https://doi.org/10.4236/jwarp.2013.54a008

Chapter 2
Climate Change and Water Resources: Challenges and Adaptation Strategies

Chidinma Daokoru-Olukole, Daniel Etim Jacob, Sylvester Chibueze Izah, Ahmed El-Harairy, Moses Fayiah, Matthew Chidozie Ogwu, and Amged El-Harairy

Abstract Climate change poses a significant threat to the world's water resources since it alters the quantity, quality, and distribution of freshwater in various locations. This chapter examines the complex interactions between climate change and water systems. It also discusses the major issues and how to adapt for long-term

C. Daokoru-Olukole
Department of Microbiology, Faculty of Science, Niger Delta University, Wilberforce Island, Amassoma, Bayelsa, Nigeria

D. E. Jacob
Department of Forestry and Wildlife, University of Uyo, Uyo, Nigeria
e-mail: danieljacob@uniuyo.edu.ng

S. C. Izah
Department of Community Medicine, Faculty of Clinical Sciences, Bayelsa Medical University, Yenagoa, Bayelsa, Nigeria
e-mail: sylvester.izah@bmu.edu.ng

Department of Microbiology, Faculty of Science, Bayelsa Medical University, Yenagoa, Bayelsa, Nigeria

A. El-Harairy
Department of Chemical and Biomolecular Engineering, College of Engineering, University of Nebraska-Lincoln, Lincoln, NE, USA
e-mail: ael-harairy2@huskers.unl.edu

Department of Soils and Water, Faculty of Agriculture, Damietta University, Damietta, Egypt

M. Fayiah
Department of Forestry, School of Natural Resources Management, Njala University, Freetown, West Africa, Sierra Leone
e-mail: mfayiah@njala.edu.sl

M. C. Ogwu
Goodnight Family Department of Sustainable Development, Appalachian State University, Boone, NC, USA
e-mail: ogwumc@appstate.edu

A. El-Harairy (✉)
Department of Crop and Animal Sciences, Albrecht Daniel Thaer-Institute of Agricultural and Horticultural Sciences, Faculty of Life Sciences, Humboldt-Universität zu Berlin, Berlin,

M. C. Ogwu and S. Chibueze Izah (eds.), *Water Quality and Safety in the Global South*, Springer Water, https://doi.org/10.1007/978-3-032-10602-5_2

water resource management. The rapid rate of global warming has altered how water moves through the environment, increased the quantity of rain that falls, exacerbated droughts and floods, and caused glaciers to melt and sea levels to rise. These consequences disproportionately affect disadvantaged groups and key sectors including agriculture, energy, and public health, particularly in low- and middle-income nations. Some of the most important topics of concern are how climate change affects surface and groundwater systems, how vulnerable water-dependent industries are, and how diminishing water security affects the economy and society. The report also discusses technological and policy solutions, such as integrated water resource management (IWRM), climate-ready infrastructure, water-saving technologies, and community-based adaptations. The chapter provides carefully chosen case studies to show how to employ effective implementation models and stresses how important it is to adopt inclusive, multi-stakeholder adaptation strategies. It also points up problems that are still there, like not having enough money, poor institutional frameworks, and not having enough climate-water data. Last but not least, it's important to invest in adaptation measures, capacity building, and international cooperation ahead of time to conserve water resources in a changing climate. Making water systems more resilient to climate change is important for the environment and for long-term social and economic security.

Keywords Climate change · Water resources · Adaptation strategies · Hydrological cycle · Water scarcity · Resilience · Integrated water management · Water security

2.1 Introduction

Climate change has emerged as a defining environmental challenge of the twenty-first century, with profound implications for global water resources. In the Global South (where climatic extremes intersect with socio-economic vulnerability) the impacts are particularly acute, leading to increased water scarcity, deteriorating quality, and heightened competition over freshwater resources [55, 64]. A growing body of research has also underscored the compounding role of water contamination, including microbial pathogens, heavy metals, and chemical pollutants. Studies in the Niger Delta, for instance, have revealed alarming levels of microbial pollution in groundwater and surface waters, posing significant health risks to local populations [29, 59, 57, 124, 130, 132]. Furthermore, toxic metals such as lead, cadmium, and arsenic have been detected in drinking water sources, exacerbated by poor sanitation, industrial discharge, and waste mismanagement [5, 60, 99]. These contaminants,

Germany
e-mail: elharaam@hu-berlin.de

Unit of Entomology, Plant Protection Department, Desert Research Center, El-Matariya, Cairo, Egypt often worsened by climate-induced flooding and inadequate infrastructure, threaten both ecosystem health and water security [31, 62, 63, 90].

The interplay between climate change and water resources has emerged as a central concern globally, reflecting a nuanced relationship fraught with various socio-economic, ecological, and policy implications. Climate change, characterised by global warming and shifting precipitation patterns, poses significant threats to water availability and quality, which are increasingly recognised as critical components of sustainable development and public health [102]. The Intergovernmental Panel on Climate Change (IPCC) has underscored that changing climatic patterns disrupt hydrological cycles, leading to increased occurrences of droughts and floods, which exacerbate existing challenges in water management systems [20, 21]. Furthermore, there is growing evidence indicating that climate variability adversely affects water supply and demand dynamics, particularly in vulnerable regions where socio-economic stability is already tenuous [139].

Moreover, as populations grow and industrial activities expand, the utility of existing water resources is further stressed, leading to what has been termed a global water crisis [58, 61]. This crisis is characterized not only by the physical scarcity of water but also by its mismanagement and pollution, which are, in part, exacerbated by climate change [92]. Traditional models of water management, often predicated on the assumption of stable climatic conditions, are increasingly shown to be inadequate when faced with the variability introduced by climate change [49, 50]. Therefore, a reassessment of prevailing water governance strategies is necessary, incorporating adaptive management frameworks that recognize both climatic uncertainties and socio-ecological dynamics [145].

The interconnections between climate variability and water systems are intricate and multifaceted. Climatic shifts lead to alterations in precipitation and evaporation rates, subsequently affecting river flows, groundwater levels, and overall water availability [76]. Specific regions, such as sub-Saharan Africa and Southeast Asia, are particularly susceptible to these changes, where agricultural dependence on predictable rainfall patterns is paramount. As temperatures rise, the hydrological implications include diminished snowpack, accelerated glacier melt, and more frequent droughts, which are set to dramatically alter water supply patterns [83].

Research has illustrated that climate variability impacts the entire water cycle, from precipitation to runoff and groundwater recharge, necessitating a holistic understanding of these interactions [146]. Seasonal shifts in hydrometeorological variables significantly influence water management, affecting both the quantity and quality of water resources [76]. Consequently, water resource managers are increasingly being called upon to develop adaptive management strategies that incorporate probabilistic scenarios of climate-related risks and uncertainties, emphasizing the importance of sustainable resource allocation to mitigate adverse impacts [91]. Integrating climate change adaptation into water resources management is essential not only for maintaining water availability but also for ensuring the resilience of ecosystems and communities reliant on these critical resources [148].

The ongoing impacts of climate change on water resources management necessitate a reevaluation of existing frameworks and practices. Traditional regulatory structures have often failed to account for the complexities of climate variability, leading to inefficient water use and heightened vulnerabilities during periods of extreme weather events [49, 50]. The necessity for adaptive strategies that can accommodate the uncertainty associated with climate change is underscored by recent studies demonstrating that regions employing proactive management frameworks exhibit greater resilience to climatic shocks than those adhering to static models [69]. Critical to this adaptation process is the recognition that water management cannot occur in isolation, it intersects with numerous sectors, including agriculture, industry, and urban planning, each affected by shifts in water and climate dynamics [16].

Moreover, the integration of technological innovations offers promising pathways toward improved water resource management. Tools such as remote sensing and predictive modeling, including the SIE-Climate tool for local climate forecasting, provide vital information that can aid in decision-making processes related to water management [135]. These advancements underscore the potential for data-driven approaches to mitigate the risks associated with climate variability; however, they must be coupled with sound governance structures to ensure accountability and responsiveness to local needs [54]. Consequently, a multifaceted approach that embraces technological innovations while fostering participatory governance will be key to sustaining water resources under climate change [92].

This chapter explores the intricate relationship between climate change and water resources, highlighting how changing climatic conditions disrupt the availability, quality, and distribution of water globally. It examines the challenges posed by extreme weather events, shifting precipitation patterns, and rising temperatures on water systems. The chapter further outlines adaptation strategies, technologies, and policy frameworks aimed at building resilience in water resource management. Readers will gain a comprehensive understanding of how climate change impacts water resources across various sectors and scales. The insights provided will support informed decision-making and the development of sustainable, climate-resilient water management practices.

2.2 Brief Overview of Climate Change

Climate change is characterized by long-term alterations in temperature, precipitation, wind patterns, and other elements of Earth's climate system. The Intergovernmental Panel on Climate Change (IPCC) classifies changes into anthropogenic (human-induced) and natural phenomena, with anthropogenic climate change primarily driven by greenhouse gas emissions from human activities, such as the burning of fossil fuels and deforestation [82]. While climate variability refers to the natural fluctuations experienced in the climate system, climate change denotes a systemic shift towards distinct patterns in climate over longer periods [109]. In this context, global warming serves as a critical indicator of climate change, showcasing an average increase in global temperatures, predominantly attributed to human actions, which have led to unprecedented levels of carbon dioxide (CO_2) concentration in Earth's atmosphere [36].

A variety of climate metrics are essential for understanding the extent of climate change. Average global temperature, carbon dioxide concentration, and sea-level rise serve as vital indicators that illustrate the accelerating impact of climate change [82]. For instance, the rise in global average temperatures exemplifies the changes that have transpired post-industrialization and highlights the urgent need for adaptation and mitigation strategies [120]. The IPCC plays a pivotal role in synthesizing scientific research to reach a consensus on climate science, thereby guiding policy-making efforts at the global level [36].

The Earth's energy balance, influenced by the greenhouse effect, represents another critical aspect of climate dynamics. Greenhouse gases (GHGs) such as carbon dioxide (CO_2), methane (CH_4), nitrous oxide (N_2O), and fluorinated gases trap heat in the atmosphere, thus contributing to climate change [78, 82]. Radiative forcing serves as a significant measure of the influence various factors have on altering the energy balance within Earth's atmosphere. Notably, scientific projections via climate modeling are essential in anticipating future climate scenarios, which can help shape effective response strategies for mitigating the adverse effects of climate change [109].

2.2.1 Historical Trends and Projections of Global Warming

Analyzing paleoclimate data from various sources, such as ice cores and tree rings, indicates that many natural climate fluctuations have occurred over geological timescales [82]. However, observed changes since the pre-industrial era, particularly following the industrial revolution, have shown a dramatic increase in global temperatures. The past century has witnessed a rise of approximately 1.1 °C since the late nineteenth century, primarily due to anthropogenic influences [109]. Additionally, regional disparities exist in warming rates, with phenomena such as Arctic amplification demonstrating particularly severe impacts in polar regions compared to temperate zones [36].

Sea-level rise trends coupled with glacial retreat globally present further evidence of climate change. An increase in the frequency and intensity of extreme weather events (ranging from heatwaves to droughts) stresses the vulnerabilities exacerbated by climate change [82, 109]. Projections based on different greenhouse gas emission scenarios illustrate varying potential outcomes under the Representative Concentration Pathways (RCPs) and Shared Socioeconomic Pathways (SSPs), emphasizing the urgency to adhere to global warming thresholds set at 1.5 and 2 °C to prevent catastrophic impacts on ecosystems and human systems alike [36].

Exceeding these temperature thresholds could lead to irreversible impacts on global biodiversity and ecosystem services, significantly affecting freshwater resources. Additionally, future climate risks depend largely on current emission trajectories, underscoring the necessity for coordinated international climate action [106].

2.2.2 Climate Change Drivers and Feedback Mechanisms

Numerous drivers contribute to climate change, ranging from anthropogenic activities such as fossil fuel combustion and land-use change to natural drivers including volcanic eruptions and variations in solar radiation and oceanic circulation patterns [109]. Notably, the impact of climate change on water resources can be significant. For example, the fluctuations in hydrological regimes attributed to climate change have been shown to directly influence water availability, with complex interdependencies between human activities and natural systems complicating these dynamics [36].

Feedback mechanisms also play a crucial role in amplifying or mitigating the impacts of climate change. Positive feedback mechanisms (such as the ice-albedo feedback, which leads to increased warming due to reduced reflectivity when ice melts) are noteworthy. Water vapor feedback behaves similarly, whereby warmer air retains moisture, enhancing pre-existing greenhouse effects [44]. In contrast, negative feedback mechanisms, such as increased cloud cover, can contribute to reflecting sunlight, thereby having a cooling effect. Understanding the interactions among land, ocean, and atmospheric systems is essential for capturing the full scope of climate change dynamics [36, 82].

The carbon cycle intricately connects to these discussions, as both sinks and sources of carbon dioxide (CO_2) significantly impact freshwater availability. Human activities have disrupted natural carbon cycling, leading to enhanced greenhouse gas effects and consequent alterations in hydrology [106]. Furthermore, the concept of tipping points signifies potential irreversible shifts, such as the collapse of significant ice sheets or abrupt changes in rainforest ecosystems, that could pose profound risks to freshwater resources and overall environmental balance [78, 109].

2.2.3 Implications of Water Resource Availability

The implications of climate change on water resources are profound, as increasing temperatures accelerate the global hydrological cycle, thereby affecting the distribution and timing of freshwater availability across regions [36]. Research indicates that regions already vulnerable to water stress will experience further declines in water availability due to altered precipitation patterns and increased evaporation rates. For example, studies have shown that low-lying Pacific islands face notable sensitivity to climatic changes, thereby stressing community reliance on freshwater resources [35].

As freshwater availability diminishes, the impacts on agriculture, urban planning, and ecosystem integrity become increasingly pronounced. In many areas, agricultural practices exacerbate the challenges posed by climate change, with water-intensive crop cultivation leading to additional stresses on local water supplies [120]. Similarly, rapid urbanization and population growth can compound the pressures on freshwater systems, as newly developed areas often fail to consider sustainable water management practices, thereby accelerating the depletion of these vital resources [105].

Decentralized water management strategies, including harvesting rainwater and implementing integrated water resources management, have been proposed as solutions to mitigate these pressures while also enhancing flood resilience [128]. Adapting such measures in urban planning can create a win–win scenario whereby climate impacts are managed proactively while simultaneously ensuring sustainable water resource availability [36].

2.3 Impact of Climate Change on Water Resources

The impacts of climate change on water resources constitute a multifaceted issue requiring comprehensive analysis. Under the category of water resource management in this section, it focuses on changes in rainfall patterns and distribution, shifts in surface and groundwater availability, melting glaciers and sea-level rise, water temperature and quality degradation, as well as the intensification of water extremes such as droughts and floods. Each of these factors interconnects to shape the resilience and sustainability of global freshwater resources [8].

2.3.1 Changes in Rainfall Patterns and Distribution

Global climate change has manifested in altered rainfall regimes, which are critical determinants of surface and groundwater replenishment. According to Al-Qawasmi et al. [8], the variability in precipitation due to climatic changes poses severe risks to freshwater systems, as changes can lead to unexpected droughts or floods, fundamentally disrupting the hydrologic cycle. Rainfall patterns have become increasingly erratic, rendering agricultural planning and water resource management more challenging. Studies, including those by Ahmed et al. [6], highlight the profound implications of these changes, including a direct correlation to growing instances of water scarcity and intense flooding events in various regions due to shifting climate patterns.

Analysis of hydrological processes illustrates that increased atmospheric temperatures elevate evaporation rates, consequently reducing available water for ecosystems and agriculture. This conclusion is corroborated by research demonstrating that anthropogenic climate change has expanded the range and intensity of hydrometeorological extremes, leading many regions to experience unprecedented flooding juxtaposed with severe droughts [123]. For instance, regions such as Iran have reported significant alterations in both streamflow and rainfall intensities as a direct consequence of global climate variability [1]. The growing consensus within hydrological studies points out the need to implement adaptive management strategies that account for these shifting hydrological extremes.

2.3.2 Shifts in Surface and Groundwater Availability

The interplay between climate change and water availability is underscored by the collective changes in both surface and groundwater resources. As per studies conducted by Gulley et al. [42], intensified rainfall variability directly impacts groundwater recharge, altering the flow dynamics within aquifer systems. Increased evaporation combined with reduced rainfall diminishes groundwater levels, posing challenges for both natural ecosystems and human uses. Particularly in coastal regions, where aquifers are also at risk of salinization due to sea-level rise, the challenge of maintaining freshwater availability becomes even more complex. For example, jurisdictions along the southeastern coast of the United States are observing increased salinity levels in groundwater aquifers, which significantly diminishes the usability of these water sources for public supply and agriculture [74].

Furthermore, explicit modeling studies reveal that groundwater systems framed by historical hydrological contexts may now find themselves rendered obsolete under projected future climatic conditions [43, 56, 71]. As groundwater is the primary resource for drinking water and irrigation, changes in its availability can precipitate adverse socio-economic consequences, particularly in developing regions where such resources are vital for subsistence [95]. The socio-economic implications extend beyond simple water access. They encompass food security, health, and broader economic stability, necessitating a robust socio-ecological framework to address these impacts.

2.3.3 Melting Glaciers, Snowpack Decline, and Sea-Level Rise

Melting glaciers and declines in snowpack represent critical indicators of climate change emissions. These phenomena contribute directly to changes in both water quantity and quality. Research by Thomas et al. [138] illustrates how accelerated glacier melt alters seasonal runoff patterns, posing risks to water supplies dependent on seasonal snowpack. The decline of snowpacks particularly affects regions relying on meltwater for agricultural cycles, thus threatening food production stability. In many areas, melting glaciers are expected to provide supplemental water in the short term. However, as these glaciers continue to recede, they will ultimately lead to significant long-term deficits in water supply, complicating efforts to maintain agricultural viability [95].

Moreover, sea-level rise, which is projected to displace vast freshwater resources through increased salinity intrusion into coastal aquifers, adds another layer of complexity [136]. The implications on water quality can be drastic, as seawater intrusion contaminates the freshwater supplies upon which millions depend [77]. In vulnerable regions such as island nations, the dual impact of rising seas and retreating freshwater supplies presents existential threats, potentially rendering large areas uninhabitable [86]. Understanding these interdependencies is crucial for developing effective adaptation and mitigation strategies.

2.3.4 Water Temperature and Quality Degradation

An often-overlooked aspect of climate change is its influence on water temperature and quality, which directly affects both aquatic ecosystems and human health. Warming water temperatures increase the risk of harmful algal blooms and negatively impact the biodiversity of freshwater species [78]. According to Mastrocicco et al. [87], freshwater quality deterioration arises as higher temperatures can diminish dissolved oxygen levels, forming hypoxic conditions detrimental to aquatic life. Prolonged periods of elevated temperatures can also exacerbate the effects of pollutants entering water systems, leading to broader ecosystem degradation and posing health risks for populations relying on directly sourced water [114].

This degradation, compounded by sediment runoff and increased nutrient loading from both agricultural and urban areas, further complicates the problem. As ecosystems struggle to maintain homeostasis, the cascading effects on food webs become evident, leading to a potential collapse of local fisheries and decline of biodiversity [123]. Collaborative watershed management strategies focused on maintaining water quality, alongside aggressive climate action, are imperative to mitigate these ongoing threats.

2.3.5 Intensification of Droughts Floods, and Water Extremes

The enhanced frequency and intensity of droughts and floods epitomize the extremes brought about by climate change, which pose severe risks to water resources. While models project increased average temperatures and altered precipitation patterns, the resultant hydrological extremes occur with troubling frequency, leading to significant socio-economic disruptions [95]. The rise in drought incidence has upgraded the urgency for water conservation practices and has prompted widespread reevaluation of water resource management techniques [138]. Conversely, increased rainfall intensity contributes to flooding, overwhelming existing drainage systems, and leading to infrastructure damage while also threatening potable water supplies through contamination [6, 101, 103]. Integrated watershed management approaches emphasizing flood risk assessment and drought preparedness have proven crucial in addressing such water challenges, as evidenced by initiatives launched in regions recovering from severe climate-induced water events [9]. This necessitates a flexible approach to policy-making, grounded in scientific evidence, to effectively respond to the multifaceted nature of water resource challenges presented by climate change.

Hydrological climate change impact assessments typically follow a three-step framework: climate modelling, tailoring and hydrological modelling, and impact assessment (Fig. 2.1, [107]). The process begins with global and regional climate projections using AOGCMs and RCMs. These outputs are then tailored—often through bias correction or downscaling—before being input into hydrological models to simulate responses [107]. Finally, impact assessment involves post-processing the results for decision-making, process attribution, and adaptation planning [107].

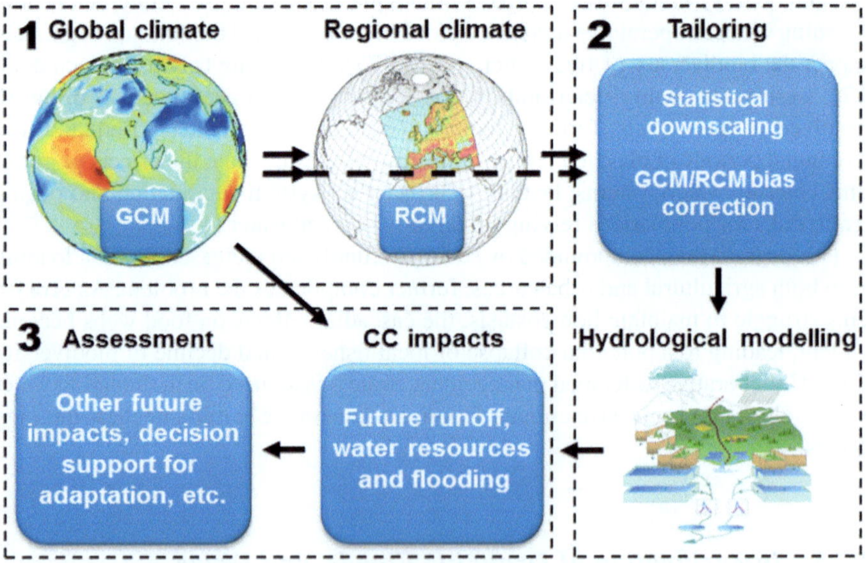

Fig. 2.1 Steps in hydrological climate change impact assessment. *Source* Olsson et al. [107]

2.4 Sectoral Vulnerability to Water Resource Challenges

Climate change-induced water resource challenges significantly impact key sectors, including agriculture, urban infrastructure, energy, and ecosystems. Reduced water availability and quality threaten food security through disrupted irrigation, compromise urban water supply and sanitation, impair hydropower generation, and endanger aquatic ecosystems and biodiversity. Table 2.1 presents sectoral vulnerability to water resource challenges.

2.4.1 Agricultural Irrigation and Food Security

Climate change poses significant challenges to agricultural irrigation and food security, primarily through increased variability in precipitation and intensified occurrences of droughts. These climatic shifts lead to reduced soil moisture and diminished

Table 2.1 Sectoral vulnerability to water resource challenges under climate change

Sector	Vulnerabilities and impacts
Agricultural irrigation and food security	– Reduced soil moisture and irrigation water due to droughts – Unpredictable crop yields threaten the food supply – High vulnerability of rain-fed agriculture – Disrupted planting/harvesting cycles – Competition for water among sectors
Urban water supply and public health	– Strained urban water infrastructure – Water scarcity intensifies inequality in access – Increased contamination from floods – Hygiene-related disease outbreaks – Public health burden from waterborne illnesses
Energy generation and hydropower reliability	– Declining hydropower output from reduced river flow – Thermal plants constrained by lack of cooling water – Reservoir evaporation reduces capacity – Power supply disruptions – Conflicts over water use between sectors
Aquatic ecosystems and biodiversity loss	– Disturbed habitats and migration routes – Higher water temperatures and pollution stress species – Shrinking wetlands and freshwater ecosystems – Invasive species proliferation – Loss of fisheries and natural purification services

availability of irrigation water, thereby threatening crop yields. Rosa et al. [121, 122] identify the crucial link between climate change and agricultural outputs, noting that the projected increase in temperature and alteration of precipitation patterns will exacerbate stress on water resources crucial for irrigation and agriculture. The alarming frequency of droughts, which is increasing globally, severely compromises soil moisture levels critical for crop production, particularly in regions heavily reliant on irrigation [110].

Unpredictable crop yields become particularly pertinent in regions that are already experiencing water stress due to climate change. Trnka et al. [140] argue that vulnerabilities in agricultural productivity will likely escalate without significant adaptation measures, with their findings indicating that wheat-producing areas may see increased water scarcity that will compromise yields. Moreover, dependence on rain-fed agriculture heightens the vulnerability of developing countries to climate impacts, where the capacity to adapt to such changes is limited due to economic constraints and infrastructural inadequacies [47, 100]. This dependency signifies a precarious condition for food security, as changes in precipitation can severely disrupt planting and harvesting cycles, directly affecting agricultural output [110].

Changing precipitation patterns further complicate the agricultural landscape by misaligning traditional planting and harvesting calendars. The work of Kourgialas [72] emphasizes that farmers must be increasingly adaptive to these shifts, as altered precipitation dynamics can impact crop cycles and reduce overall yield predictability, particularly in already vulnerable regions. The competition for water resources escalates as agricultural demands clash with those from urbanization and industrial needs. The increasing strain on limited water supplies fosters conflicts both within the agricultural sector and between agriculture and other sectors, leading to further instability in food production systems [121, 141].

In the face of these challenges, there is a pressing need for a shift towards more sustainable irrigation practices. Rosa argues for the potential of sustainable irrigation expansion coupled with innovative agricultural methodologies to enhance resilience against climatic fluctuations [121]. This adaptation strategy could mitigate the anticipated reversion of irrigated land back to rain-fed production as freshwater limitations become more evident. Integrated approaches that promote efficient water use while maintaining ecological integrity are essential for securing long-term food security in an era of climate unpredictability [72, 122].

2.4.2 Urban Water Supply and Public Health

Rapid urbanization, an inevitable consequence of demographic shifts and economic development, complicates water resource management by imposing excessive pressure on existing urban infrastructures. The increasing water demands coincide with climate-induced scarcity, which exacerbates inequities in access to clean water. This disparity is critical in urban settings where effective water distribution systems are already strained. As observed by Khamidov et al., urban areas face enhanced vulnerabilities, as extreme weather events can severely disrupt urban water supplies, leading to potential public health crises [68].

The contamination of urban water sources is another grave concern amplified by climate extremes such as flooding and storm surges. Such events often overwhelm urban drainage systems, leading to the contamination of drinking water supplies with pollutants and pathogens [2]. This compromised water quality substantially affects public health, as access to clean water is pivotal in preventing waterborne diseases. The implications are dire. Studies highlight that regions experiencing extreme water scarcity and contamination witness a significant rise in diseases such as cholera and typhoid, further straining public health infrastructures designed to manage these outbreaks [68, 100].

Clean water access is increasingly becoming a social justice issue. Marginalized communities disproportionately bear the brunt of water supply breakdowns. Disparities in water availability can exacerbate preexisting health inequities, leading to adverse outcomes, particularly among vulnerable populations [68]. Ogundeji et al. [100] elucidate the necessity for robust management practices that can bolster community resilience against climate-related disruptions, emphasizing that effective monitoring and infrastructure investment are crucial for urban water supply sustainability.

As urban public health systems struggle to cope with the dual challenges of climate change and growing populations, it becomes evident that adopting climate-smart water management strategies is imperative for safeguarding public wellness. There is an urgent need for comprehensive planning that incorporates climate projections into urban water supply frameworks, ensuring that cities are equipped to deal with both existing and future challenges [2, 140]. Continuous engagement with local communities in the development of these strategies is essential to enhance public health outcomes and ensure equitable access to safe water resources [47, 100].

2.4.3 Energy Generation and Hydropower Reliability

The intersection of climate change and energy generation presents significant challenges, particularly for hydropower, a critical renewable energy source in many regions. Hydropower reliability is increasingly jeopardized by reduced river flows during dry seasons, directly affecting hydropower efficiency and output [67]. As climate change alters hydrological cycles, energy generation systems that rely on consistent water flow must adapt to diminishing freshwater reserves that threaten their operational capacity [100].

Thermal power plants also struggle under these changing hydrological conditions. Limited cooling water availability can constrain their operational effectiveness, resulting in decreased energy output and compromised energy security [2]. The projected increase in evaporation from reservoirs further reduces potential energy generation capacity, exacerbating concerns regarding energy supply stability [67].

The result is an environment of intermittent power supply that threatens industries and households alike, particularly in regions already dependent on limited water resources for energy generation [140].

Furthermore, conflicts may arise as competition between energy production and agricultural demands intensifies. As urban areas expand and energy needs grow, the prioritization of water usage for energy generation could lead to further destabilization in agricultural water supply, where irrigation needs become increasingly desperate due to climate-induced limitations [100]. This dynamic presents a complex challenge requiring integrative policies that balance energy needs with agricultural water demands to prevent adverse repercussions across sectors [121].

The development and implementation of adaptive strategies for energy generation are imperative for mitigating climate impacts. A shift towards diversified energy sources alongside improved water management practices could enhance resilience against climatic variability and safeguard energy security [72, 85]. Additionally, advancing technologies for water recycling and conservation can address the challenges posed by water scarcity in energy generation, ensuring a sustainable framework for future energy demands [140].

2.4.4 Aquatic Ecosystems and Biodiversity Loss

Climate change has profound impacts on aquatic ecosystems, leading to disruptions in habitat conditions and species migration patterns. Altered flow regimes resulting from changing weather patterns disrupt the ecological balance within freshwater systems, adversely affecting biodiversity [34]. Increased water temperatures, a direct consequence of climate fluctuations, place additional stress on aquatic species, escalating the risk of mortality across various fish and invertebrate populations [121].

Moreover, as wetland and freshwater habitats shrink due to anthropogenic pressures and climate effects, endemic and migratory species face heightened threats. The loss of these critical habitats compromises the ecological integrity of these ecosystems, leading to reduced biodiversity and a decline in the availability of ecosystem services that communities rely on for livelihoods and resilience [72]. Invasive species, thriving in disturbed ecosystems, may outcompete native species for resources, further accelerating biodiversity loss [34, 129].

The degradation of aquatic ecosystems affects not only biodiversity but also the essential services they provide, such as water filtration and fisheries. Rivera et al. emphasize that local communities depending on these ecosystems for their livelihoods will face increased challenges as the integrity of these resources declines [14]. Enhanced pressures on fisheries can result in economic instability for communities that rely on these resources for their food security and economic well-being [121].

Conservation efforts must be prioritized to maintain aquatic ecosystems in the face of climate change. Sustainable management practices aimed at restoring and protecting habitats can help bolster resilience against climatic impacts, ensuring that biodiversity is safeguarded. Monitoring changes and restoring ecological functions within aquatic systems are urgently needed to maintain ecosystem services crucial for human survival [34].

The integration of ecological considerations into water management policies is essential for fostering both biodiversity and human resilience in water-scarce regions. Promoting sustainable practices that consider the interconnections between human activities and aquatic ecosystems can catalyze efforts to combat the adverse effects of climate change on biodiversity [34, 121].

2.5 Adaptation Strategies for Climate Change and Water Resources

Climate change represents an unprecedented challenge for water resources management, necessitating comprehensive adaptation strategies that take into account both new scientific insights and historical wisdom gained through local practices. Integrated Water Resources Management (IWRM) emerges as a fundamental approach designed to harmoniously manage water, land, and related resources, aiming to maximize economic and social welfare without compromising environmental sustainability. The IWRM framework is positioned to address the complexities prompted by climate change, considering stakeholder engagement and ecological balance [24, 111]. By incorporating climate change considerations into water management strategies, regions can enhance their resilience to adverse climate effects, such as prolonged droughts and severe flooding events [40, 148].

As water infrastructure plays a critical role in the overall responsiveness of water resource systems, the development of climate-smart water infrastructure becomes vital. This infrastructure must be resilient to the changing climate, including the implementation of technologies that can withstand extreme weather events [126]. Climate-smart infrastructure incorporates principles of sustainability, efficiency, and flexibility, enabling adaptive responses to unexpected changes in water availability and quality. For instance, the construction of adaptable dam operations can mitigate the impacts of altered hydrological patterns due to climate change, allowing for dynamic adjustments in water storage and distribution [51]. Furthermore, innovative materials and construction practices that account for projected climate scenarios can reduce long-term vulnerability [49, 50].

Improving water use efficiency and adopting conservation practices are crucial in enhancing adaptation strategies. As climate change leads to altered precipitation patterns, it exacerbates challenges related to water scarcity, making efficient usage imperative [22, 118]. Techniques such as precision agriculture, advanced irrigation systems, and water recycling can dramatically reduce water consumption in agricultural settings [48, 108]. Participatory approaches that involve stakeholders, especially local farmers, in the decision-making process can further bolster effective water conservation practices, promoting equitable resource distribution and fostering community stewardship [25]. Empirical cases demonstrate that where such participatory mechanisms are implemented, enhanced water management outcomes are often observed, which reaffirm the importance of community engagement [108].

Strengthening institutional and policy frameworks for adaptation is foundational to addressing the multifaceted issues arising from climate change impacts on water resources. Effective governance structures that incorporate scientific insights, community needs, and environmental considerations can guide comprehensive adaptation efforts [20]. Policies that prioritize investments in research and development, education, and technology transfer will empower water management authorities to implement innovative solutions tailored to local conditions [20, 137]. Institutional arrangements must also enable dynamic policy adaptation, recognizing that the uncertainties surrounding climate forecasts require agile governance capable of navigating emerging challenges [148]. The development of resilient legal and regulatory frameworks is paramount for enforcing norms and practices that mitigate climate impacts on water resources [24].

Local knowledge and participatory approaches can serve as a powerful resource in enhancing the effectiveness of adaptation strategies to climate change. Evidence from diverse cultural contexts highlights the importance of indigenous practices and traditional ecological knowledge in water management [25, 108]. Local communities often possess invaluable insights regarding resource utilization and climate patterns that have been honed over generations [113]. By integrating this local expertise into formal water management practices, regions can bolster their adaptive capacity while ensuring that the solutions developed are culturally relevant and broadly supported [51, 52, 118]. Successful examples from various locales demonstrate that stakeholder engagement not only improves outcomes but also legitimizes decision-making processes, thereby fostering long-term commitment and compliance with adaptive strategies [25].

The need for a holistic understanding of climate-induced changes in water resources dynamics is increasingly evident. Water resource management must pivot towards an interdisciplinary paradigm that captures the interplay between climate variables, ecological integrity, and socio-economic factors [83, 147]. This necessitates the development of advanced modeling tools and assessment methods that can simulate various climate scenarios and their implications for both surface and groundwater resources [49, 50]. These tools can aid in evaluating the efficacy of adaptation measures and allow for real-time adjustments based on ongoing observations and data collection [13]. As climate models improve, they provide critical insights that inform future policy decisions and operational practices, enabling proactive rather than reactive management strategies [37].

Broadly, climate change is expected to intensify existing hydrological challenges across the globe, exacerbating water scarcity and influencing water quality significantly [51, 52, 118]. As temperatures rise, higher evaporation rates will further strain water supplies, while altered rainfall patterns can lead to both flooding and drought conditions [24, 53]. It necessitates that adaptation strategies also focus on protecting water quality from emerging threats, including pollutants resulting from overflow events or loss of filtration capacity in natural ecosystems [22, 51, 52]. Addressing these challenges through integrated approaches will help ensure that water resources remain viable for future generations.

Furthermore, cities, where water resources face particularly acute pressures from climate change, must adopt urban water management adaptations. Urban environments often experience unique challenges such as increased surface runoff and heat island effects, which necessitate targeted strategies for sustainable water management [51, 53]. Integration of green infrastructure, stormwater management systems, and public education campaigns can reduce the urban water footprint while enhancing the capacity of cities to cope with climate variability [48, 108]. Such innovations not only improve urban resilience but also support broader ecological goals, fostering ecosystems that can thrive despite climatic shifts [126].

2.6 Innovative Technologies for Climate-Resilient Water Management

The advent of climate change has necessitated a paradigm shift in the management of water resources, particularly as the frequency of droughts and floods increases. The future of effective water management hinges upon innovative technologies that can enhance resilience to climate impacts. This is crucial in regions where water scarcity threatens agricultural productivity and human well-being. Technologies that have emerged in this context can be classified into several key areas, including climate forecasting and monitoring, advanced irrigation systems, desalination and wastewater reuse, and digital decision support tools.

2.6.1 Climate Forecasting, Monitoring, and Early Warning Systems

Climate forecasting and monitoring systems are essential in water resource management, as they provide critical data that helps in anticipating water supply fluctuations due to climate change. These systems utilize advanced meteorological models and satellite data to predict weather patterns, enabling water managers to make informed decisions. For example, the integration of various models can increase the accuracy of precipitation forecasts, which is vital for irrigation scheduling and reservoir management [75]. Furthermore, the development of early warning systems has empowered stakeholders to take preemptive actions to mitigate the adverse impacts of extreme weather events, thus enhancing the resilience of water systems against climate variability [33].

Recent advancements have shown that combining data from multiple sources can enhance the reliability of climate projections. These systems not only focus on hydrological data but also consider socio-political factors that could affect water management, fostering a holistic approach to decision making [89]. The establishment of comprehensive databases and shared platforms facilitates collaboration between different sectors, which is crucial for effective adaptation strategies in the face of climate uncertainty [81].

2.6.2 Advanced Irrigation, Rainwater Harvesting, and Recharge Systems

The efficient use of water in agriculture is pivotal for food security, particularly in arid and semi-arid regions affected by climate change. Advanced irrigation techniques, such as smart irrigation systems, play a significant role in conserving water while maximizing agricultural output. These systems utilize real-time data and automation technologies to optimize water use based on soil moisture levels and crop water requirements. Research indicates that implementing smart irrigation can lead to a significant reduction in water waste while improving crop yields [19, 25].

In addition to smart irrigation, rainwater harvesting technologies are gaining traction as viable solutions for variable water supply. These systems collect and store rainwater for agricultural and domestic use, thereby reducing dependence on traditional water sources. The integration of rainwater harvesting with existing irrigation practices can lead to a more sustainable and resilient agricultural framework [144]. Groundwater recharge systems are essential in areas facing declining aquifer levels due to overextraction. Techniques such as managed aquifer recharge can replenish groundwater reserves, ensuring long-term water availability [23].

2.6.3 Desalination, Wastewater Reuse, and Smart Water Grids

Desalination has emerged as a critical technology in addressing water scarcity, especially in coastal regions where freshwater resources are limited. Advances in desalination technology have improved efficiency and reduced operational costs, making it a more viable option for many countries [17]. However, while desalination presents a promising solution for augmenting water supply, its environmental impacts, particularly energy consumption and seawater intake, must be carefully managed to minimize ecological damage [125].

Wastewater treatment and reuse are paramount in developing sustainable water management strategies. The application of advanced techniques, such as membrane bioreactors and advanced oxidation processes, has enhanced the quality of treated wastewater, enabling its safe reuse for irrigation, industrial processes, and even potable purposes [4]. A multi-criteria decision support tool has been proposed to assist in evaluating the feasibility of wastewater reuse projects, accounting for economic, environmental, and social factors to ensure holistic sustainability [84]. This integration conserves freshwater resources and promotes circular water economies.

Smart water grids represent a significant innovation in water management, enabling the integration of information and communication technologies in water distribution systems. By utilizing sensors and data analytics, smart grids enhance the efficiency of water delivery, reduce losses through leak detection, and improve overall management practices [3, 94]. These systems foster real-time monitoring and enable adaptive responses to operational challenges, thereby strengthening the resilience of urban water supplies amid climate-driven uncertainties.

2.6.4 Digital Tools for Decision Support and Real-Time Water Management

The use of digital tools for decision support has revolutionized water management by providing stakeholders with sophisticated techniques to analyze complex data and optimize decision-making processes. Decision Support Systems (DSS), which incorporate models and analytical tools, have been instrumental in enabling water managers to evaluate numerous scenarios and make informed choices based on varying climatic and operational conditions [143]. These tools have become increasingly relevant in the context of unpredictable climate patterns, allowing for dynamic management strategies that can adapt to ongoing changes [39].

Moreover, the combination of artificial intelligence and machine learning algorithms within these DSS can enhance predictive capabilities, enabling water managers to anticipate demand fluctuations and resource availability more accurately. This capability is essential for efficient water allocation and planning, especially during drought conditions [45]. The role of stakeholder engagement in the development and application of these digital tools is critical, as active participation and input can significantly improve tool effectiveness and user acceptance [81]. Digital tools also facilitate improved communication and collaboration across different sectors involved in water resource management. By creating shared platforms that aggregate and disseminate critical data, stakeholders can align their actions and strategies more effectively [88]. This interconnection between various actors enhances operational efficiencies and fosters resilience by creating a diversified approach to managing water resources in an era of climate change.

2.7 Case Studies on Climate Change and Water Resource Adaptation

2.7.1 Community-Based Water Adaptation in Vulnerable Regions

Community-based adaptation strategies have emerged as essential mechanisms for enhancing resilience against climate change impacts on water resources, particularly in vulnerable regions. In Rajasthan, India, rural communities have successfully implemented rainwater harvesting systems as a response to the recurring droughts exacerbated by climate change [70, 71]. These systems not only provide an efficient means to capture and store rainwater but also help to recharge local aquifers, ensuring a more reliable water supply throughout the dry seasons. The effectiveness of these systems in promoting water security and reducing dependency on external water sources is supported by literature on traditional water management practices [30, 80].

In Ethiopia, the utilization of indigenous knowledge plays a crucial role in managing seasonal water availability. Traditional irrigation methods, such as spate irrigation, are relied upon for effective resource management in arid regions. These methods are time-tested, having been adapted over generations to the local hydrological conditions. Research indicates that integrating indigenous knowledge with modern practices can significantly enhance water security and agricultural productivity while fostering community empowerment and self-reliance [12, 119].

The Andes region, particularly in countries like Peru, exhibits striking examples of community-led watershed restoration efforts aimed at improving water regulation and biodiversity. These restoration initiatives often involve reforesting watersheds and enhancing natural ecosystems, which in turn positively influence local climate conditions and water availability. Programs that focus on the collaborative involvement of local communities have demonstrated success in achieving sustainable outcomes while fortifying traditional practices of land and water management [30, 115].

Additionally, women's initiatives across sub-Saharan Africa exemplify innovation in water conservation and purification. Women-led projects that deploy solar-powered filtration systems highlight the intersection of gender equity and environmental sustainability. Findings suggest that empowering women in these roles not only results in enhanced water quality and access but also contributes to broader community health and economic improvements. Such initiatives affirm that addressing water scarcity can also advance gender equality in resource management [10].

Participatory monitoring of local water sources, as witnessed in the Philippines, underscores the significance of community engagement in water resource management. By involving residents in monitoring activities, communities enhance their adaptive capacity to climate variability, thereby refining local decision-making processes. Research indicates that such participatory approaches increase the recognition of water issues and improve the management of shared water resources [79, 134].

Moreover, the construction of small-scale sand dams in Kenya has been vital in providing water during the dry seasons. These manageable and cost-effective structures enable communities to store surface runoff and recharge local groundwater, effectively extending water availability for both drinking and agricultural purposes. Evidence points to the success of sand dams in enhancing resilience among pastoral and agricultural communities facing the challenges of climate-induced water scarcity [11, 32].

2.7.2 National and Regional Strategies: Lessons from Global South and North

National and regional strategies for water resource management provide essential insights into broader climate adaptation practices. The Netherlands' Delta Programme showcases a multifaceted approach that integrates structural and nature-based solutions for flood and water management. The programme emphasizes proactive flood prevention rather than reactive responses, highlighting the importance of long-term planning in an era of uncertainty driven by climate change. Evaluations reveal that the Delta Programme effectively balances ecological preservation with infrastructural development, ensuring a resilient future for water management in the nation [11, 18]. Australia's Murray-Darling Basin Plan illustrates adaptive governance necessary for managing river basins under prolonged drought conditions. By employing a collaborative governance framework, the plan facilitates stakeholder engagement in water allocation and management. Research findings indicate that such adaptive governance structures significantly increase the resilience of water systems while fostering sustainable ecosystem services and agricultural productivity in the region [112, 115].

In Rwanda, the National Adaptation Plan that integrates water resource management into national development planning exemplifies a strategic framework responsive to climate impacts. By systematically addressing water availability and quality while aligning with the country's development goals, Rwanda sets a progressive example for other emerging economies. The plan's emphasis on stakeholder involvement and cross-sectoral collaboration is crucial for effective implementation and sustainability [30]. South Africa's water licensing reforms have sought to address issues related to equity and sustainability, especially under changing climate conditions. These reforms aim to ensure that water allocation is equitable while preserving ecosystems and livelihoods. Studies highlight how these legislative changes facilitate more diversity in water use rights, enabling adaptation to climatic uncertainties and fostering socio-economic resilience [10, 79]. In Malawi, decentralized solar-powered water pumps are transforming rural communities by addressing both energy and water accessibility issues. These systems ensure year-round access to safe drinking water, significantly reducing reliance on traditional energy sources that may be less reliable. Research indicates that the incorporation of renewable energy in rural water supply systems can pave the way for sustainable solutions, especially in areas with limited infrastructure [10]. Cape Town's experience illustrates the effectiveness of integrated urban water planning in averting severe crises, such as the potential "Day Zero," when the city faced imminent water shortages. Through proactive demand management, public involvement, and the exploration of new water supply sources, the city has demonstrated that comprehensive urban water management is essential in response to climate variability and urban growth [79, 134].

In West Africa, the Senegal River Basin Development Authority (OMVS) coordinates resource-sharing and flood mitigation efforts among member countries. This initiative showcases a regional approach to developing infrastructure and shared resources that are responsive to the impacts of climate change while promoting food security and economic stability [10]. Community-managed boreholes in Nigeria provide another successful example of resilience-building measures during dry seasons. These initiatives not only enhance water access but also support sanitation efforts, contributing to the overall health and resilience of rural populations. The importance of community engagement in resource management is emphasized for achieving long-term sustainability and effective adaptation to climate variability [11, 32].

Chile's National Water Plan, which encompasses glacier monitoring, aquifer recharge, and drought insurance mechanisms, addresses the specific challenges posed by melting glaciers and declining water availability. The plan is a proactive measure designed to sustain water supply in the face of rapid environmental changes driven by climate change [11, 134]. Canada's adaptation framework incorporates climate data into water infrastructure planning effectively, underscoring the pivotal role of evidence-based decision-making. By utilizing climate projections and hydrological models, Canada is advancing adaptive measures for various water sectors, thus enhancing both public safety and resource sustainability [32, 80].

Transboundary river management initiatives are crucial in addressing climate-induced variability across national borders. The Nile Basin Initiative highlights cooperative frameworks that promote shared water use among Eastern African countries. This collaborative approach not only aims to optimize water usage but also seeks to build resilience against climate impacts. Studies have shown that establishing such cooperative strategies is vital for reducing potential conflicts over shared resources while fostering regional development [18, 112]. The Mekong River Commission addresses the challenges posed by upstream dam impacts and climate-induced flow variability among Southeast Asian nations. A collective management strategy emphasizes data sharing and joint planning to mitigate adverse effects on local ecosystems and communities. The effectiveness of this cooperation lies in its ability to adapt to varying climate conditions while fostering sustainable economic activities across the basin [12, 119].

Governance surrounding the Rhine River offers a historic precedent for integrated flood management and pollution control among European countries. Utilizing collaborative frameworks, countries within the Rhine Basin have developed comprehensive strategies to tackle water quality issues and protect flood-prone areas. Research underscores that such cooperation is critical in enhancing resilience to climate-induced changes and ensuring the continued health of river ecosystems [30, 115]. The Indus Waters Treaty, which governs water sharing between India and Pakistan, has undergone scrutiny under the strain of climate stress. Given the increasing impacts of climate change, this treaty is being tested for resilience and adaptability, highlighting the complexities of transboundary water governance in the face of environmental challenges. Evaluations indicate that maintaining cooperative relations and adaptable frameworks is essential for ensuring water security in the region [79, 134]. Zambezi River Basin action plans emphasize the importance of shared early warning systems and climate-resilient hydropower development. This collaborative effort not only aims to improve flood management and resource allocation but is also essential in addressing the broader challenges posed by climate variability on regional energy supply and water accessibility [11, 32].

Urban water resource management strategies often diverge significantly from rural approaches due to the inherent differences in population density, infrastructure, and resource availability. Singapore's NEWater initiative exemplifies innovative urban water management, reclaiming wastewater for drinking and industrial use. This initiative has transformed Singapore into one of the world's leaders in sustainable water management, demonstrating the effectiveness of advanced treatment and recycling technologies in addressing urban water security challenges [18, 112]. In New York City, the implementation of urban green infrastructure, such as green roofs and rain gardens, significantly mitigates stormwater runoff and reduces flood risk. These practices align with broader sustainability goals while improving urban aesthetics and enhancing biodiversity in metropolitan settings. Studies have shown that such infrastructure investments can yield substantial long-term economic and environmental benefits [12, 119]. Rural water adaptation practices often focus on improving efficiency and resource management, showcasing innovative solutions to enhance agricultural resilience and food security. For instance, rainwater harvesting and drip irrigation techniques implemented in Israel serve as effective methods to optimize water use and boost crop productivity in arid climates [30, 115].

2.8 Challenges in Adapting Water Systems to Climate Change

Adapting water systems to climate change presents a complex set of challenges tied to economic, technical, and governance issues. Recognizing these challenges is crucial for developing effective management strategies that enhance resilience amid evolving climate dynamics. Climate change is expected to intensify existing vulnerabilities in water supply systems, affecting both the quality and quantity of available water resources, thus highlighting the need for integrated frameworks to address these issues comprehensively. A significant barrier to efficient adaptation in many developing countries arises from economic constraints and technical limitations. These shortcomings are evident in inadequate funding for climate-resilient infrastructure and a lack of technical expertise. The Global Commission on Adaptation notes that prioritizing investments in climate-resilient infrastructure and early warning systems is essential, particularly in developing regions where such investments can yield profound benefits [46]. Further complexities arise from the necessity for water resource management to accommodate interactions between climate variability, ecological needs, and socio-political factors, as emphasized by Ochoa-Tocachi et al. [98]. This interconnectedness can result in fragmented solutions that fail to address the systemic nature of the water crisis driven by climate change [51].

Fragmented governance structures also pose a significant challenge in addressing the impacts of climate change on water resources. A coherent framework for governance is vital for aligning diverse stakeholders and integrating their actions towards common goals. Grasham et al. [41] stress that meaningful inclusion of marginalized communities in decision-making processes is critical for achieving climate resilience in water security. A lack of coordinated policies can lead to redundant efforts and inefficiencies that further complicate adaptation strategies. The connection between governance deficiencies, poverty, and environmental sustainability underscores the need for integrated governance that crosses sectoral and disciplinary lines [117].

Capacity gaps in data collection and climate-water research are further obstacles to effective adaptation. Successful climate adaptation relies on robust data-driven decision-making, necessitating reliable datasets related to climate threats and water resource conditions. Insufficient climate-water data can hinder policymakers' ability to anticipate impacts and act decisively [73]. The complexities of threats to water resources call for adaptive management strategies supported by ongoing research and real-time data [51]. Developing capacities in this area may involve investing in technology and infrastructure, as well as enhancing educational initiatives that equip local experts with necessary skills [96]. Social inequities further complicate water resource management in the context of climate change. Vulnerable communities often suffer the most significant impacts from climate-induced water scarcity yet are frequently excluded from planning and implementing resilient water solutions. Such inequities are exacerbated by socio-economic disparities affecting access to clean water [116]. Addressing these disparities requires tailored strategies that prioritize the unique needs and experiences of marginalized groups, ensuring that climate adaptation efforts are inclusive [41].

Integrating social, economic, and environmental considerations into water governance necessitates a fundamental rethinking of policy frameworks to promote equity and resilience. Bayu et al. [15] argue that sustainable governance structures must prioritize local knowledge and community participation, empowering communities in climate resilience planning. Successful traditional practices, such as collective water management models, demonstrate the potential of integrating local knowledge into broader climate adaptation strategies [133]. Exploring nature-based solutions (NbS) is another promising avenue for enhancing water system resilience. Approaches that emulate natural processes to improve water management not only mitigate climate variability risks but can also enhance biodiversity and habitat quality [66]. Together with community-based management strategies, such solutions present a pathway to systematically address climate impacts while contributing to sustainability objectives.

2.9 Future Directions for Climate Change and Water Resource Management

Climate change significantly challenges global water resource management, necessitating adaptive strategies for sustainable governance. To effectively navigate the complexities introduced by climatic variability and extreme weather events, it is crucial to enhance adaptive capacity through education and training. Various studies emphasize that integrated educational frameworks can inform stakeholders about water resource dynamics increasingly impacted by climate phenomena such as altered precipitation patterns and rising temperatures [148]. For instance, Muleta [91] highlights the importance of scenario analysis in identifying climate-related risks and opportunities essential for sustainable water resource management. Additionally, Sarker [127] advocates incorporating climatology principles into engineering education to equip practitioners with the skills needed to address the challenges posed by climate change on hydrological systems.

Furthermore, promoting cross-sectoral and international collaboration is vital in addressing the transboundary nature of water resources affected by climate change. As water scarcity intensifies in regions facing socio-economic and environmental pressures, cooperative management frameworks become essential. Findings by Al-Kalbani et al. [7] indicate that demographic and economic factors, alongside climate change, increase demands for freshwater resources, underscoring the importance of collaborative approaches that transcend national boundaries. Effective collaboration can harness diverse stakeholder perspectives, providing a more robust understanding of water management challenges in a climate-impacted world. Reports suggest that such partnerships are crucial for sharing technology, resources, and best practices that enhance resilience to climate impacts on water resources [142].

To foster these collaborations, leveraging finance and investment is paramount for enhancing water resilience in the face of climate change. Effective climate adaptation strategies require adequate funding and investments in infrastructure and technology [65]. Research shows that countries investing in smart irrigation and water management systems experience enhanced resilience to climate variability [25]. The framework proposed by Jeuland and Whittington [65] emphasizes evaluating the robustness of investment strategies to align financial flows with sustainability objectives while addressing uncertainties driven by climate change. This approach highlights the significance of private and public sector partnerships to increase financial capacity for infrastructure development and maintenance aimed at protecting and managing water resources. Another critical aspect of managing water resources amidst climate change is promoting Sustainable Development Goals (SDG 6: Clean Water and Sanitation and SDG 13: Climate Action). The interplay between these goals is evident, as SDG 13 aims to combat climate change and its impacts on the availability and quality of freshwater systems governed under SDG 6. Research by Gain et al. illustrates that climate change adaptation strategies are vital for maintaining sustainable water supply systems in the face of increasing pressures from an altered climate [38]. Effective governance frameworks can align national and regional policies with global sustainability objectives, thus creating a cohesive strategy that addresses both climate action and water resource management [28].

2.10 Conclusion

Climate change continues to exert significant and complex pressures on global water resources, manifesting through altered precipitation patterns, declining groundwater levels, glacier retreat, sea-level rise, and the intensification of extreme weather events. These impacts have far-reaching consequences for water availability, quality, and distribution, threatening critical sectors such as agriculture, energy, health, and biodiversity. The challenges posed by climate-induced water stress require a comprehensive understanding of hydrological changes, coupled with adaptive, forward-looking responses that integrate scientific insights and community-based practices. To address these emerging challenges, several strategic recommendations are essential. First, strengthening integrated water resources management should be prioritized to ensure coordinated planning across sectors and governance levels. Second, investing in climate-resilient infrastructure, including early warning systems, sustainable irrigation, and flood control mechanisms, will enhance preparedness and adaptive capacity. Third, harnessing innovative technologies such as remote sensing, smart water systems, and water reuse techniques is crucial for improving efficiency and enabling real-time decision-making. Fourth, policy frameworks must be inclusive and adaptive, allowing local participation while aligning with national and global sustainability goals. Fifth, capacity building, education, and public awareness are key to fostering resilience, especially at the community level. Lastly, enhanced research and climate-water data systems should be developed to fill knowledge gaps and inform evidence-based decision-making.

References

1. Abbaspour KC, Faramarzi M, Ghasemi SS, Yang H (2009) Assessing the impact of climate change on water resources in Iran. Water Resour Res 45(10). https://doi.org/10.1029/2008wr007615
2. Adama G, Jimoh D, Otache M (2020) Optimization of irrigation water allocation framework based on genetic algorithm approach. J Water Resour Prot 12(04):316–329. https://doi.org/10.4236/jwarp.2020.124019
3. Adelani F, Okafor E, Jacks B, Ajala O (2024) Exploring theoretical constructs of urban resilience through smart water grids: case studies in african and u.s. cities. Eng Sci Technol J 5(3):984–994. https://doi.org/10.51594/estj.v5i3.952
4. Adewumi J, Ilemobade A, Zyl V (2013) Application of a multi-criteria decision support tool in assessing the feasibility of implementing treated wastewater reuse. Int J Decis Support Syst Technol 5(1):1–23. https://doi.org/10.4018/jdsst.2013010101
5. Aghoghovwia OA, Miri FA, Izah SC (2018) Impacts of anthropogenic activities on heavy metal levels in surface water of Nun River around Gbarantoru and Tombia towns, Bayelsa State, Nigeria. Annal Ecol Environ Sci 2(2):1–8
6. Ahmed V, Saad A, Saleh H, Saboor S, Kasianov N, Alnaqbi T (2020) Implementation of water demand forecasting model to aid sustainable water supply chain management in uae. https://doi.org/10.20944/preprints202011.0205.v1
7. Al-Kalbani M, Price M, Abahussain A, Ahmed M, O'Higgins T (2014) Vulnerability assessment of environmental and climate change impacts on water resources in al jabal al akhdar, sultanate of oman. Water 6(10):3118–3135. https://doi.org/10.3390/w6103118
8. Al-Qawasmi J, Asif M, Fattah A, Babsail M (2019) Water efficiency and management in sustainable building rating systems: examining variation in criteria usage. Sustainability 11(8):2416. https://doi.org/10.3390/su11082416
9. Alahacoon N, Edirisinghe M (2021) Spatial variability of rainfall trends in sri lanka from 1989 to 2019 as an indication of climate change. ISPRS Int J Geo Inf 10(2):84. https://doi.org/10.3390/ijgi10020084
10. Ali F, Kumar J, Yadav G, Yadav V (2024) Applications of photovoltaic water pumping system in developing countries. Recent Pat Eng 19.https://doi.org/10.2174/0118722121336917241029062947
11. Alphen J (2015) The delta programme and updated flood risk management policies in the netherlands. J Flood Risk Manag 9(4):310–319. https://doi.org/10.1111/jfr3.12183
12. Anbarasu L (2024) Evolving power factor correction in solar water pumping systems through zeta converter-based bldc motor drives. jes 20(6s):2778–2785. https://doi.org/10.52783/jes.3286
13. Anderson E, Jackson S, Tharme R, Douglas M, Flotemersch J, Zwarteveen M, Arthington A (2019) Understanding rivers and their social relations: a critical step to advance environmental water management. Wiley Interdiscip Rev Water 6(6). https://doi.org/10.1002/wat2.1381
14. Aswathi K, Ajith K, Ajithkumar B, Abida P (2022) Impact of climate change on rice yield under projected scenarios in central zone of kerala. J Agrometeorol 24(3). https://doi.org/10.54386/jam.v24i3.1544
15. Bayu T, Kim H, Oki T (2020) Water governance contribution to water and sanitation access equality in developing countries. Water Resour Res 56(4). https://doi.org/10.1029/2019wr025330
16. Belete M, Deng J, Abubakar G, Teshome M, Wang K, Woldetsadik M, Gudo A (2020) Partitioning the impacts of land use/land cover change and climate variability on water supply over the source region of the blue nile basin. Land Degrad Dev 31(15):2168–2184. https://doi.org/10.1002/ldr.3589
17. Brentan B, Meirelles G, Luvizotto E, Izquierdo J (2018) Joint operation of pressure-reducing valves and pumps for improving the efficiency of water distribution systems. J Water Resour Plan Manag 144(9). https://doi.org/10.1061/(asce)wr.1943-5452.0000974

18. Buuren A (2019) The dutch delta approach, pp 201–217. https://doi.org/10.1093/oso/9780198843719.003.0011
19. Bwambale E, Abagale F, Anornu G (2023). Smart irrigation for climate change adaptation and improved food security. https://doi.org/10.5772/intechopen.106628
20. Cheng H, Hu Y (2011) Improving china's water resources management for better adaptation to climate change. Clim Change 112(2):253–282. https://doi.org/10.1007/s10584-011-0042-8
21. Cheng H, Hu Y (2011) Improving china's water resources management for better adaptation to climate change. Clim Change 112(2):253–282. https://doi.org/10.1007/s10584-011-0042-8
22. Ciampittiello M, Marchetto A, Boggero A (2024) Water resources management under climate change: a review. Sustainability 16(9):3590. https://doi.org/10.3390/su16093590
23. de Pádua EJBG, Gonçalves FB, Mendonça LA (2024) Development of a decision-making system based on the mandala of water and agricultural sustainability. J Eng Exact Sci 10(2):17200–17200. https://doi.org/10.18540/jcecvl10iss2pp17200
24. Doroszkiewicz J, Romanowicz R (2017) Guidelines for the adaptation to floods in changing climate. Acta Geophys 65(4):849–861. https://doi.org/10.1007/s11600-017-0050-9
25. Durmuş Y, Arslansoy G, Gürbüz T (2024) Impact of smart irrigation systems and water management on climate change. Bio Web of Conf 85:01057. https://doi.org/10.1051/bioconf/20248501057
26. Durmuş Y, Arslansoy G, Gürbüz T (2024) Impact of smart irrigation systems and water management on climate change. Bio Web of Conf 85:01057. https://doi.org/10.1051/bioconf/20248501057
27. Durmuş Y, Arslansoy G, Gürbüz T (2024) Impact of smart irrigation systems and water management on climate change. Bio Web of Conf 85:01057. https://doi.org/10.1051/bioconf/20248501057
28. Elfithri R, Mokhtar M, Zakaria S (2019) The need for awareness raising, advocacy, and capacity building in integrated water resources management toward sustainable development: a case study in Malaysia. World Water Policy 5(1):43–54. https://doi.org/10.1002/wwp2.12002
29. Enaregha EB, Omotete TR, Izah S, Odubo TC (2022) Comparison of total viable bacteria counts and in situ characteristics in drinking water sources in Sagbama town, Bayelsa state, Nigeria. Enaregha EB, Omotete TR, Izah SC, Odubo TC (2022) Comparison of total viable bacteria counts and in situ characteristics in drinking water sources in Sagbama town, Bayelsa state, Nigeria. Int J Pathog Res 11(3–4):79–87
30. Fabbina C, Kavitha P, Kavin K, Sakthidhasan K (2024) Improved landsman converter for pv based agriculture motor pump system, pp 69–82. https://doi.org/10.58532/v3bdee1p2ch4
31. Fayiah M, Kandeh B, Izah SC, Saccoh S (2024) Rural water management status, threats, and prospects: a case study of Sierra Leone. In: Water resources management for rural development. Elsevier, pp 263–274
32. Febriani S, Alqarana A, Wibowo M, Fahriannur A (2024) Assessing the performance of off-grid solar photovoltaic power plants in supporting aquaponics systems. Pena Teknik Jurnal Ilmiah Ilmu-Ilmu Teknik 9(1):54. https://doi.org/10.51557/pt_jiit.v9i1.2170
33. Fletcher S, Lickley M, Strzepek K (2019) Learning about climate change uncertainty enables flexible water infrastructure planning. Nat Commun 10(1). https://doi.org/10.1038/s41467-019-09677-x
34. Fragoso R, Noéme C (2018) Economic effects of climate change on the mediterranean's irrigated agriculture. Sustain Account Manag Policy J 9(2):118–138. https://doi.org/10.1108/sampj-07-2017-0078
35. Freitas C, Helbig M, Matzarakis A (2013) Hydroclimatic assessment of water resources of low pacific islands: evaluating sensitivity to climatic change and variability. Int J Climatol 34(3):881–892. https://doi.org/10.1002/joc.3731
36. Freund E, Abbaspour K, Lehmann A (2017) Water resources of the black sea catchment under future climate and landuse change projections. Water 9(8):598. https://doi.org/10.3390/w9080598

37. Fu Z, Xie Y, Zhang Y, Jiang X, Guo H, Wang S (2022) Water resource availability assessment through hydrological simulation under climate change in the huangshui watershed of the qinghai–tibet plateau. Front Earth Sci 9.https://doi.org/10.3389/feart.2021.755119
38. Gain A, Giupponi C, Renaud F (2012) Climate change adaptation and vulnerability assessment of water resources systems in developing countries: a generalized framework and a feasibility study in bangladesh. Water 4(2):345–366. https://doi.org/10.3390/w4020345
39. Garofalo P, Vonella A, Rinaldi M (2023) A simple-structured framework for decision support system in irrigation water management: tool description and preliminary testing. https://doi.org/10.21203/rs.3.rs-2904576/v1
40. Giupponi C, Gain A (2017) Integrated water resources management (iwrm) for climate change adaptation. Reg Environ Change 17(7):1865–1867. https://doi.org/10.1007/s10113-017-1173-x
41. Grasham C, Calow R, Casey V, Charles K, Wit S, Dyer E, Zaidi H (2021) Engaging with the politics of climate resilience towards clean water and sanitation for all. NPJ Clean Water 4(1). https://doi.org/10.1038/s41545-021-00133-2
42. Gulley J, Martin J, Spellman P, Moore P, Screaton E (2012) Influence of partial confinement and holocene river formation on groundwater flow and dissolution in the florida carbonate platform. Hydrol Process 28(3):705–717. https://doi.org/10.1002/hyp.9601
43. Gulley J, Mayer A, Martin J, Bedekar V (2016) Sea level rise and inundation of island interiors: assessing impacts of lake formation and evaporation on water resources in arid climates. Geophys Res Lett 43(18):9712–9719. https://doi.org/10.1002/2016gl070667
44. Hannah L, Roehrdanz P, Ikegami M, Shepard A, Shaw M, Tabor G, Hijmans R (2013) Climate change, wine, and conservation. Proc Natl Acad Sci 110(17):6907–6912. https://doi.org/10.1073/pnas.1210127110
45. Haro-Monteagudo D, Knox J, Holman I (2019) D-risk: a decision-support webtool for improving drought risk management in irrigated agriculture. Comput Electron Agric 162:855–858. https://doi.org/10.1016/j.compag.2019.05.029
46. Haron N, Sapuan N, Kumaran V, Saudi N, Ridzuan A (2022) Climate change, environmental pollution and covid-19 pandemic: the triangle impacts on mental health. Environ Manag Sustain Dev 11(4):53. https://doi.org/10.5296/emsd.v11i4.21051
47. Hellin J, Fisher E (2019) Climate-smart agriculture and non-agricultural livelihood transformation. Climate 7(4):48. https://doi.org/10.3390/cli7040048
48. Hirsch D, Abrami G, Giordano R, Liersch S, Matin N, Schlüter M (2010) Participatory research for adaptive water management in a transition country-a case study from uzbekistan. Ecol Soc 15(3). https://doi.org/10.5751/es-03549-150323
49. Hong S, Xia J, Chen J, Wan L, Ning L, Shi W (2016) Multi-object approach and its application to adaptive water management under climate change. J Geog Sci 27(3):259–274. https://doi.org/10.1007/s11442-017-1375-7
50. Hong S, Xia J, Chen J, Wan L, Ning L, Shi W (2016) Multi-object approach and its application to adaptive water management under climate change. J Geog Sci 27(3):259–274. https://doi.org/10.1007/s11442-017-1375-7
51. Howard G, Calow R, MacDonald A, Bartram J (2016) Climate change and water and sanitation: likely impacts and emerging trends for action. Annu Rev Environ Resour 41(1):253–276. https://doi.org/10.1146/annurev-environ-110615-085856
52. Howard G, Calow R, MacDonald A, Bartram J (2016) Climate change and water and sanitation: likely impacts and emerging trends for action. Annu Rev Environ Resour 41(1):253–276. https://doi.org/10.1146/annurev-environ-110615-085856
53. Hunt A, Watkiss P (2010) Climate change impacts and adaptation in cities: a review of the literature. Clim Change 104(1):13–49. https://doi.org/10.1007/s10584-010-9975-6
54. Huntjens P, Lebel L, Pahl-Wostl C, Camkin J, Schulze R, Kranz N (2012) Institutional design propositions for the governance of adaptation to climate change in the water sector. Glob Environ Chang 22(1):67–81. https://doi.org/10.1016/j.gloenvcha.2011.09.015
55. Iyiola AO, Afolabi OA, Alimi SK, Akingba OO, Izah SC, Ogwu MC (2024) Climate change and water crisis in the global south. In: Water crises and sustainable management in the Global

South. Springer Nature Singapore, Singapore, pp 111–140.https://doi.org/10.1007/978-981-97-4966-9_4
56. Izah SC, Ogwu MC (2025) Modeling solutions for microbial water contamination in the global south for public health protection. Front Microbiol 16:1504829. https://doi.org/10.3389/fmicb.2025.1504829
57. Izah SC, Richard G, Sawyer WE (2021) Distribution of Fungi density and diversity in a surface water of Epie Creek in Yenagoa Metropolis, Nigeria. Arch Epidemiol Public Health 3:1–5. https://doi.org/10.15761/AEPH.1000121Volume3:1-5
58. Izah SC, Stanley HO, Richard G, Sawyer WE, Uwaeme OR (2024a). Surface water quality: a statistical perspective on the efficacy of environmental and human health assessment tools. Water Air Soil Pollut 235(3). https://doi.org/10.1007/s11270-024-06965-1
59. Izah SC, Ngun CT, Richard G (2022) Microbial quality of groundwater in the Niger Delta region of Nigeria: health implications and effective treatment technologies. In: Current directions in water scarcity research, vol 6. Elsevier, pp 149–172. https://doi.org/10.1016/B978-0-323-91838-1.00010-5
60. Izah SC, Ngun CT, Iniaghe PO, Aigberua AO, Odubo TC (2023) Processes of decontamination and elimination of toxic metals from water and wastewaters. In: Metals in water. Elsevier, pp 239–262. https://doi.org/10.1016/B978-0-323-95919-3.00003-3
61. Izah SC, Ogwu MC, Loukas A, Hamidifar H (eds) (2024) Water crises and sustainable management in the Global South. Springer. https://doi.org/10.1007/978-981-97-4966-9_2
62. Jacob DE, Nelson IU, Izah SC (2024) Indigenous water management strategies in the global south. In: Water crises and sustainable management in the Global South. Springer Nature Singapore, Singapore, pp 487–525. https://link.springer.com/chapter/https://doi.org/10.1007/978-981-97-4966-9_16
63. Jacob DE, Nelson IU, Eniang EA, Izah SC (2024) Advancing water security and resilience in the Global South through recreational development. In: Water crises and sustainable management in the Global South. Springer Nature Singapore, Singapore, pp 527–554. https://link.springer.com/chapter/https://doi.org/10.1007/978-981-97-4966-9_17
64. Jacob DE, Nelson IU, Izah SC, Daniel KS (2024a) Rural water crises in the Global South: understanding the scope and impact. In: Izah SC, Ogwu MC, Loukas A, Hamidifar H (eds) Water crises and sustainable management in the Global South. Springer, Singapore, pp 3–44. https://doi.org/10.1007/978-981-97-4966-9_1
65. Jeuland M, Whittington D (2014) Water resources planning under climate change: assessing the robustness of real options for the blue nile. Water Resour Res 50(3):2086–2107. https://doi.org/10.1002/2013wr013705
66. Jódar J, Martos-Rosillo S, Custódio E, Mateos L, Cabello J, Casas J, Lambán L (2022) The recharge channels of the sierra nevada range (spain) and the peruvian andes as ancient nature-based solutions for the ecological transition. Water 14(19):3130. https://doi.org/10.3390/w14193130
67. Kaini S, Harrison M, Gardner T, Nepal S, Sharma A (2022) The impacts of climate change on the irrigation water demand, grain yield, and biomass yield of wheat crop in nepal. Water 14(17):2728. https://doi.org/10.3390/w14172728
68. Khamidov M, Ishchanov J, Khamidova S, Isabaev K, Altmishev A (2024) Water scarcity under global climate change: ways of addressing them in the lower reaches of the Amu Darya river in Uzbekistan. Bio Web Conf 103:00021. https://doi.org/10.1051/bioconf/202410300021
69. Kim J, Ryu J (2020) Decision-making of lid-BMPS for adaptive water management at the Boise river watershed in a changing global environment. Water 12(9):2436. https://doi.org/10.3390/w12092436
70. Kosoe EA, Osumanu IK, Ogwu MC (2025b) Reimagining rural and small-town water supply in Ghana through inclusive governance and sustainable models. J Sustain 1(2):1–21. https://doi.org/10.55845/jos-2025-1234
71. Kosoe EA, Ogwu MC, Duniah J, Ndigun FB (2025a) Sustaining indigenous water governance systems in Ayimaye and Lungni, Northern Ghana, through Hybrid Models. J Sustain 1(2). https://doi.org/10.55845/jos-2025-1235

72. Kourgialas N (2021) Editorial: could advances in geoinformatics, irrigation management and climate adaptive agronomic practices ensure the sustainability of water supply in agriculture? Water Sci Technol Water Supply 21(6):v–vii. https://doi.org/10.2166/ws.2021.244
73. Kurniawan T, Goh K, Goh H, Zhang D, Jiang M, Dai W, Meidiana C (2024) Strengthening climate resilience: urban water technologies for heat-resilient physical infrastructure in south-east Asia cities amidst extreme temperature events and el Niño challenges. Acs Es&t Water 4(9):3713–3724. https://doi.org/10.1021/acsestwater.4c00585
74. Langevin C, Zygnerski M (2012) Effect of sea-level rise on salt water intrusion near a coastal well field in southeastern florida. Ground Water 51(5):781–803. https://doi.org/10.1111/j.1745-6584.2012.01008.x
75. Lassiter A, Leonard N (2022) A systematic review of municipal smart water for climate adaptation and mitigation. Environ Plan B Urb Anal City Sci 49(5):1406–1430. https://doi.org/10.1177/23998083211072864
76. Lebeza TM, Gashaw T, Tefera GW, Mohammed JA (2023) Trend analysis of hydro-climate variables in the Jemma sub-basin of Upper Blue Nile (Abbay) Basin, Ethiopia. SN Appl Sci 5(5):129. https://doi.org/10.1007/s42452-023-05345-4
77. Lester L, Ramirez M, Kneidel A, Heckscher C (2016) Use of a florida gulf coast barrier island by spring trans-gulf migrants and the projected effects of sea level rise on habitat availability. PLoS ONE 11(3):e0148975. https://doi.org/10.1371/journal.pone.0148975
78. Li Q (2024) Exploring the interactions of climate change, freshwater supply, and environmental balance. Highlights Sci Eng Technol 116:109–113. https://doi.org/10.54097/nddpfd14
79. Lilhare AS, Kadwane SG (2024) Maximizing solar water pump efficiency: exploring MPPT strategies for optimization. AIP Adv 14(4). https://doi.org/10.1063/5.0204215
80. Longwe B, Mganga M, Sinyiza N (2019) Review of sustainable solar powered water supply system design approach by Water Mission Malawi. Water Pract Technol 14(4):749–763. https://doi.org/10.2166/wpt.2019.079
81. Lynch AJ, Taylor WW, McCright AM (2016) Stakeholder views of management and decision support tools to integrate climate change into great lakes lake Whitefish management. Fisheries 41(11):644–652. https://doi.org/10.1080/03632415.2016.1232960
82. Ma C, Pei W, Liu J, Fu G (2024) Long-Term trends and variability of hydroclimate variables and their linkages with climate indices in the Songhua river. Atmosphere 15(2):174. https://doi.org/10.3390/atmos15020174
83. Manandhar S, Pandey VP, Kazama F (2012) Hydro-climatic trends and people's perceptions: case of Kali Gandaki River Basin, Nepal. Clim Res 54(2):167–179. https://doi.org/10.3354/cr01108
84. Mannina G, Rebouças TF, Cosenza A, Sànchez-Marrè M, Gibert K (2019) Decision support systems (DSS) for wastewater treatment plants–a review of the state of the art. Biores Technol 290:121814. https://doi.org/10.1016/j.biortech.2019.121814
85. Masia S, Sušnik J, Marras S, Mereu S, Spano D, Trabucco A (2018) Assessment of irrigated agriculture vulnerability under climate change in Southern Italy. Water 10(2):209. https://doi.org/10.3390/w10020209
86. Masterson JP, Garabedian SP (2007) Effects of sea-level rise on ground water flow in a coastal aquifer system. Groundwater 45(2):209–217. 2006.00279.x
87. Mastrocicco M, Busico G, Colombani N, Vigliotti M, Ruberti D (2019) Modelling actual and future seawater intrusion in the Variconi coastal wetland (Italy) due to climate and landscape changes. Water 11(7):1502. https://doi.org/10.3390/w11071502
88. Menapace A, Avesani D (2019) Global gradient algorithm extension to distributed pressure driven pipe demand model. Water Resour Manag 33(5):1717–1736. https://doi.org/10.1007/s11269-018-2174-3
89. Menapace A, Zanfei A, Righetti M (2021) Tuning ANN hyperparameters for forecasting drinking water demand. Appl Sci 11(9):4290. https://doi.org/10.3390/app11094290
90. Moslen M, Miebaka CA, Izah SC (2024) Occurrence and management of toxicity associated with parasitic pathogens in drinking water sources in rural areas in the Niger Delta, Nigeria. In: Water resources management for rural development. Elsevier, pp 275–286

91. Muleta TN (2021) Climate change scenario analysis for Baro-Akobo river basin, Southwestern Ethiopia. Environ Syst Res 10(1):24. https://doi.org/10.1186/s40068-021-00225-5
92. Muller M (2021) Managing current climate variability can ensure water security under climate change. In: African handbook of climate change adaptation. Springer International Publishing, Cham, pp 1–27. https://doi.org/10.1007/978-3-030-45106-6_243
93. Murshid S, Singh B (2020) Single stage autonomous solar water pumping system using PMSM drive. IEEE Trans Ind Appl 56(4):3985–3994. https://doi.org/10.48175/ijarsct-18230
94. Di Nardo A, Boccelli DL, Herrera M, Creaco E, Cominola A, Sitzenfrei R, Taormina R (2021) Smart urban water networks: solutions, trends and challenges. Water 13(4):501. https://doi.org/10.3390/w13040501
95. Nasimi MN, Sagin J, Wijesekera NTS (2020) Climate and water resources variation in Afghanistan and the need for urgent adaptation measures. Int J Food Sci Agric 4(1). https://doi.org/10.26855/ijfsa.2020.03.009
96. Nijhawan A, Howard G, Poudel M, Pregnolato M, Eunice Lo YT, Ghimire A, Baidya M, Geremew A, Flint A, Mulugeta Y (2022) Assessing the climate resilience of community-managed water supplies in Ethiopia and Nepal. Water 14(8):1293. https://doi.org/10.3390/w14081293
97. Numbere AO, Kika EP, Izah SC (2024) Challenges and mitigation of water resource management in rural areas in Nigeria. In: Water crises and sustainable management in the Global South. Springer Nature Singapore, Singapore, pp 373–390. https://doi.org/10.1007/978-981-97-4966-9_12
98. Ochoa-Tocachi BF, Bardales JD, Antiporta J, Pérez K, Acosta L, Mao F, Zulkafli Z, Gil-Ríos J, Angulo O, Grainger S, Gammie G (2019) Potential contributions of pre-Inca infiltration infrastructure to Andean water security. Nat Sustain 2(7):584–593. https://doi.org/10.1038/s41893-019-0307-1
99. Ogamba EN, Charles EE, Izah SC (2021) Distributions, pollution evaluation and health risk of selected heavy metal in surface water of Taylor creek, Bayelsa State, Nigeria. Toxicol Environ Health Sci 13(2):109–121. https://doi.org/10.1007/s13530-020-00076-0
100. Ogundeji AA, Jordaan H, Groenewald J (2018) Economics of climate change adaptation: a case study of Ceres-South Africa. Clim Dev 10(4):377–384. https://doi.org/10.1080/17565529.2017.1301866
101. Ogwu MC, Kosoe EA (2025) Integrating green infrastructure into sustainable agriculture to enhance soil health, biodiversity, and microclimate resilience. Sustainability 17(9):3838. https://doi.org/10.3390/su17093838
102. Ogwu MC (2025) Sustainable development in Africa: challenges, innovations, and policy pathways for the future. Afr Res Rep 1(2):55–69. https://doi.org/10.5897/ARR2025.0005
103. Ogwu MC, Kosoe EA (2024) Place of cultural diversity in sustainable water resource management in Ghana. In: Izah SC, Ogwu MC, Loukas A, Hamidifar H (eds) Water cris sustainable management in the Global South. Springer, Singapore. https://doi.org/10.1007/978-981-97-4966-9_14
104. Ogwu MC, Izah SC, Sawyer WE, Amabie T (2025) Environmental risk assessment of trace metal pollution: a statistical perspective. Environ Geochem Health 47:94. https://doi.org/10.1007/s10653-025-02405-z
105. Okafor GC, Ogbu KN (2018) Assessment of the impact of climate change on the freshwater availability of Kaduna River basin, Nigeria. J Water Land Dev 38(1):105–114. https://doi.org/10.2478/jwld-2018-0047
106. Okello C, Tomasello B, Greggio N, Wambiji N, Antonellini M (2015) Impact of population growth and climate change on the freshwater resources of Lamu Island, Kenya. Water 7(3):1264–1290. https://doi.org/10.3390/w7031264
107. Olsson J, Arheimer B, Borris M, Donnelly C, Foster K, Nikulin G, Persson M, Perttu A, Uvo C.B, Viklander M, Yang W (2016) Hydrological climate change impact assessment at small and large scales: key messages from recent progress in Sweden. Climate 4:39
108. Opoku DR, Abubakar VF, Kobina SAL, Danso OSJ, Nyameche M (2024) Assessing the effectiveness of participatory irrigation management approaches to improve water use efficiency

and equity in Ghana. Poljoprivredna tehnika 49(2):7–16. https://doi.org/10.5937/poljteh2402007o
109. Osland MJ, Enwright N, Stagg CL (2014) Freshwater availability and coastal wetland foundation species: ecological transitions along a rainfall gradient. Ecology 95(10):2789–2802. https://doi.org/10.1890/13-1269.1
110. Parwin R, Ramadas M, Agrawal A, Atnurkar A (2023) Impacts of climate change on future crop water demand in an agricultural watershed in Mayurbhanj district of Odisha, India. J Agrometeorol 25(2):326–329. https://doi.org/10.54386/jam.v25i2.1952
111. Peres DJ, Modica R, Cancelliere A (2019) Assessing future impacts of climate change on water supply system performance: application to the Pozzillo Reservoir in Sicily, Italy. Water 11(12):2531. https://doi.org/10.3390/w11122531
112. Petersen AC, Bloemen P (2015) Planned adaptation in design and testing of critical infrastructure: the case of flood safety in The Netherlands. UCL STEaPP. https://doi.org/10.14324/000.cp.1469402
113. Phan TD, Sahin O, Trinh HD, Dao HN, Mai LS, Pham TV, Pham HQ, Do TP, Vu TM, Capon SJ (2019) An integrated approach for improved management of an island's scarce water resources under climate change and tourism development. In: 23rd International congress on modelling and simulation (MODSIM2019). Modelling and Simulation Society of Australia and New Zealand. https://doi.org/10.36334/modsim.2019.j1.phan
114. Porru MC, Arras C, Biddau R, Cidu R, Lobina F, Podda F, Wanty R, Da Pelo S (2024) Assessing recharge sources and seawater intrusion in coastal groundwater: a hydrogeological and multi-isotopic approach. Water 16(8):1106. https://doi.org/10.3390/w16081106
115. Prinsen G, Sperna Weiland F, Ruijgh E (2015) The Delta model for fresh water policy analysis in the Netherlands. Water Resour Manag 29(2):645–661.https://doi.org/10.1007/s11269-014-0880-z
116. Pullan RL, Freeman MC, Gething PW, Brooker SJ (2014) Geographical inequalities in use of improved drinking water supply and sanitation across sub-Saharan Africa: mapping and spatial analysis of cross-sectional survey data. PLoS Med 11(4):e1001626. https://doi.org/10.1371/journal.pmed.1001626
117. Pérez-Peña MDC, Jiménez-García M, Ruiz-Chico J, Peña-Sánchez AR (2021) Analysis of research on the SDGs: the relationship between climate change, poverty and inequality. Appl Sci 11(19):8947. https://doi.org/10.3390/app11198947
118. Rehana S, Mujumdar PP (2013) Regional impacts of climate change on irrigation water demands. Hydrol Process 27(20):2918–2933. https://doi.org/10.1002/hyp.9379
119. Rijke J, van Herk S, Zevenbergen C, Ashley R (2012) Room for the River: delivering integrated river basin management in the Netherlands. Int J River Basin Manag 10(4):369–382. https://doi.org/10.1080/15715124.2012.739173
120. Rocha JR, de Mello Prado R, Teixeira GCM, de Oliveira Filho ASB (2021) Si fertigation attenuates water stress in forages by modifying carbon stoichiometry, favouring physiological aspects. J Agron Crop Sci 207(4):631–643. https://doi.org/10.1111/jac.12479
121. Rosa L (2022) Adapting agriculture to climate change via sustainable irrigation: biophysical potentials and feedbacks. Environ Res Lett 17(6):063008. https://doi.org/10.1088/1748-9326/ac7408
122. Rosa L, Chiarelli DD, Sangiorgio M, Beltran-Peña AA, Rulli MC, D'Odorico P, Fung I (2020) Potential for sustainable irrigation expansion in a 3 C warmer climate. Proc Natl Acad Sci 117(47):29526–29534. https://doi.org/10.1073/pnas.2017796117
123. Royan A, Hannah DM, Reynolds SJ, Noble DG, Sadler JP (2013) Avian community responses to variability in river hydrology. PLoS ONE 8(12):e83221. https://doi.org/10.1371/journal.pone.0083221
124. SSeiyaboh EI, Youkparigha FO, Izah SC, Daniels ID (2020b) Bacteriological quality of groundwater in Imiringi town, Bayelsa State, Nigeria. J Biotechnol Biomed Sci 2(2):34–40
125. Sadr SM, Saroj DP, Mierzwa JC, McGrane SJ, Skouteris G, Farmani R, Kazos X, Aumeier B, Kouchaki S, Ouki SK (2018) A multi expert decision support tool for the evaluation of advanced wastewater treatment trains: a novel approach to improve urban sustainability. Environ Sci Policy 90:1–10. https://doi.org/10.1016/j.envsci.2018.09.006

126. Sambou MHA, Liersch S, Koch H, Vissin EW, Albergel J, Sane ML (2023) Synergies and trade-offs in water resources management in the bafing watershed under climate change. Water 15(11):2067. https://doi.org/10.3390/w15112067
127. Sarker S (2022) Fundamentals of climatology for engineers: lecture note. Eng 3(4):573–595. https://doi.org/10.3390/eng3040040
128. Schuetze T, Chelleri L (2013) Integrating decentralized rainwater management in urban planning and design: flood resilient and sustainable water management using the example of coastal cities in the Netherlands and Taiwan. Water 5(2):593–616. https://doi.org/10.3390/w5020593
129. Seebens H, Bacher S, Blackburn TM, Capinha C, Dawson W, Dullinger S, Genovesi P, Hulme PE, Van Kleunen M, Kühn I, Jeschke JM (2021) Projecting the continental accumulation of alien species through to 2050. Glob Change Biol 27(5):970–982. https://doi.org/10.1111/gcb.15333
130. Seiyaboh EI, Izah SC (2017) Review of impact of anthropogenic activities in surface water resources in the Niger Delta region of Nigeria: a case of Bayelsa state. Int J Ecotoxicol Ecobiol 2(2):61–73
131. Seiyaboh EI, Izah SC, Bokolo JE (2017) Bacteriological quality of water from river nun at Amassoma Axises, Niger Delta, Nigeria. ASIO J Microbiol Food Sci Biotechnol Innov 3(1):22–26
132. Seiyaboh EI, Youkparigha FO, Izah SC, Mientei K (2020) Assessment of bacteriological characteristics of surface water of Taylor creek in Bayelsa state, Nigeria. Noble Int J Sci Res 4(4):25–30
133. Shiono M, Ikegami K, Tsuruta T (2017) Climate resilience of collective water management in rural Japan. J Asian Rural Stud 1(2):162–171. https://doi.org/10.20956/jars.v1i2.1187
134. Shrestha R, Dipak ARDST, Sharma SM (2022) Design of digital solar water pump using microcontroller ATmega. 32. Electr Sci Eng 4(2):11–24. https://doi.org/10.30564/ese.v4i2.4945
135. Sierra-Cárdenas E, Usaquén-Perilla O, Fonseca-Molano M, Ochoa-Echeverría M, Díaz-Gómez J, del Jesus M (2022) SIE-climate: a methodological and technological tool for predicting local climate variability in managing socio-ecological systems. Int J Climatol 42(2):868–888. https://doi.org/10.1002/joc.7277
136. Storlazzi CD, Elias EP, Berkowitz P (2015) Many atolls may be uninhabitable within decades due to climate change. Sci Rep 5(1):14546. https://doi.org/10.1038/srep14546
137. Sánchez A, Broekman A, Giannakis E, Jebari S, Suhadolnik P, Adams S (2017) BEWATER–making society an active participant in water adaptation to global change–FP7 project. Impact (3):6–8. https://doi.org/10.21820/23987073.2017.3.6
138. Thomas N, James EJ, George C (2022) Water-related impacts on agriculture due to climate change: a review with reference to Kerala. Sustain Agri Food Environ Res Discontin 10.https://doi.org/10.7770/safer-v10n1-art2560
139. Tibebe D, Teferi E, Bewket W, Zeleke G (2022) Climate induced water security risks on agriculture in the Abbay river basin: a review. Front Water 4:961948. https://doi.org/10.3389/frwa.2022.961948
140. Trnka M, Feng S, Semenov MA, Olesen JE, Kersebaum KC, Rötter RP, Semerádová D, Klem K, Huang W, Ruiz-Ramos M, Hlavinka P (2019) Mitigation efforts will not fully alleviate the increase in water scarcity occurrence probability in wheat-producing areas. Sci Adv 5(9):p.eaau2406. https://doi.org/10.1126/sciadv.aau2406
141. Wang K, Sun S, Li Y, Hu X, Qi X, Wang J, Li C, Gao F (2022) Response of regional agricultural water use to the change of climate and plantation structure in the typical agricultural region of China. J Water Clim Change 13(3):1370–1388. https://doi.org/10.2166/wcc.2022.416
142. Wang HM, Qiu L, Tong JP, Li CY, Liu GF (2011) A framework for adaptive management system of water resources facing climate change—a case of Huai river basin. In ICSSSM11. IEEE, pp 1–5. https://doi.org/10.1109/icsssm.2011.5959485
143. Welch H, Brodie S, Jacox MG, Bograd SJ, Hazen EL (2020) Decision-support tools for dynamic management. Conserv Biol 34(3):589–599. https://doi.org/10.1111/cobi.13417

144. Wellens J, Traoré F, Diallo M, Tychon B (2013) A framework for the use of decision-support tools at various spatial scales for the management of irrigated agriculture in West-Africa. Agric Sci 4(08):9–15. https://doi.org/10.4236/as.2013.48a002
145. Williams BK (2011) Adaptive management of natural resources—framework and issues. J Environ Manage 92(5):1346–1353. https://doi.org/10.1016/j.jenvman.2010.10.041
146. YYoon S, Lee T (2016) Investigation of hydrological variability in the Korean Peninsula with the ENSO teleconnections. Proceed Int Assoc Hydrol Sci 374:165–173. https://doi.org/10.5194/piahs-374-165-2016
147. Zhang X (2015) Conjunctive surface water and groundwater management under climate change. Front Environ Sci 3:59. https://doi.org/10.3389/fenvs.2015.00059
148. Zhao M, Boll J (2022) Adaptation of water resources management under climate change. Front Water 4:983228. https://doi.org/10.3389/frwa.2022.983228

Chapter 3
Water Pollution and Contamination: Industrial, Agricultural, and Urban Impacts

Sylvester Chibueze Izah and Matthew Chidozie Ogwu

Abstract This Chapter explores the multifaceted issue of water pollution and contamination, focusing on the industrial, agricultural, and urban activities that contribute significantly to the degradation of water quality. The chapter synthesizes current knowledge on key pollutants, their environmental and public health impacts, and strategies for mitigation. The chapter identifies that industrial activities are major contributors to water contamination through the release of heavy metals, hydrocarbons, and chemical effluents into water bodies. Agricultural practices, particularly the overuse of fertilizers and pesticides, were linked to nutrient enrichment and chemical residues in both surface and groundwater. Urbanization was found to exacerbate microbial pollution, plastic waste accumulation, and the overall burden on already strained water infrastructure. Across all domains, the consequences of pollution included ecosystem disruption, biodiversity loss, increased waterborne diseases, and threats to food and water security. The chapter emphasizes that addressing these complex and overlapping sources of pollution requires a shift towards integrated water management strategies. Such approaches should harmonize industrial regulation, sustainable farming techniques, urban infrastructure planning, and public engagement. From the principles of sustainable development perspective, these strategies are critical for safeguarding water resources, promoting environmental justice, and achieving long-term water security in line with global targets such as Sustainable Development Goal 6.

S. C. Izah (✉)
Department of Community Medicine, Faculty of Clinical Sciences, Bayelsa Medical University, Yenagoa, Bayelsa, Nigeria
e-mail: Sylvester.izah@bmu.edu.ng

Department of Microbiology, Faculty of Science, Bayelsa Medical University, Yenagoa, Bayelsa, Nigeria

M. C. Ogwu
Goodnight Family Department of Sustainable Development, Appalachian State University, Boone, NC, USA
e-mail: ogwumc@appstate.edu

M. C. Ogwu and S. Chibueze Izah (eds.), *Water Quality and Safety in the Global South*, Springer Water, https://doi.org/10.1007/978-3-032-10602-5_3

Keywords Industrial effluents · Agricultural runoff · Urban wastewater · Environmental health · Integrated water management · Sustainable development · Ecosystem degradation

3.1 Introduction

Water pollution and contamination refer to the alteration of natural water quality [1–3], rendering it unsuitable for various purposes including drinking, swimming, fishing, or supporting wildlife. Water pollution arises from numerous sources of contamination, including physical, chemical and biological (including microbial) pollutants that significantly affect aquatic ecosystems and human health [2, 4–6]. For instance, [7] highlighted the contributions of total phosphorus (TP) and lead (Pb) in river water quality, illustrating the essential role of domestic and industrial sewage in influencing these parameters. Furthermore, pollution can occur through both point sources, such as industrial discharges, and non-point sources, such as agricultural runoff, complicating the efforts needed for effective management and remediation [8].

Clean water is essential not only for human health but also for maintaining ecosystem balance. Polluted water can transmit waterborne diseases, contributing to high mortality rates, especially in developing nations. Agarwal and Xue [9] note that water pollution is a leading cause of mortality and disease globally, affecting approximately 80% of the global population living in areas exposed to threats to water security. Moreover, the ecological importance of clean water is highlighted by its role in sustaining biodiversity. Healthy aquatic ecosystems support a wide range of organisms, and contamination can lead to biodiversity loss, destabilizing ecosystems and affecting services provided to humans, such as clean filtration and nutrient recycling [10]. One of the predominant sources of water pollution is industrial activities. Industries often release various effluents characterized by pollutants such as nutrients and heavy metals like mercury, organic chemicals, and nutrients that can lead to eutrophication [11–14]. Emphasized that industrial activities significantly contribute to pollution in river basins through diverse types of effluents. Similarly, [15] discussed the implications of untreated industrial wastewater being discharged into freshwater sources, exacerbating water quality issues and straining resources.

Agricultural activity is another substantial contributor to water pollution through the use of pesticides [16–18] and fertilizers. The runoff from agricultural lands can carry these chemicals into waterways, resulting in nutrient loading and disturbances of aquatic ecosystems [19, 20]. Noted that improper agricultural practices might lead to severe environmental impacts, emphasizing that fertilizer application is a prominent source of pollution in water bodies by enhancing nutrient levels. Such agricultural runoffs can cause issues like algal blooms that deplete oxygen in water bodies, leading to hypoxia and endangering aquatic life [8]. Urbanization also plays a critical role in water contamination. As cities expand, the increase in impermeable surfaces generates significant non-point source pollution from stormwater runoff, which can contain a variety of pollutants, from heavy metals to hydrocarbons [8, 21]. Pointed out that urban surface runoff is a growing concern for drinking water sources, highlighting how land use changes associated with urbanization threaten water quality. This shift presents additional challenges in managing and protecting water resources in urban settings, often necessitating sophisticated and multifaceted management strategies.

Moreover, urban and industrial runoff can carry numerous contaminants, including microplastics and pharmaceuticals, complicating pollution dynamics. The complexity and multiplicity of pollutants can lead to compounded adverse effects on both ecosystem health and human safety [22, 23]. The interaction between these pollution sources and their cumulative impacts on water quality demands urgent attention. Thus, the regulatory framework should effectively address pollutant interactions, especially in regions with multiple contamination sources. This includes strengthening existing laws and implementing innovative monitoring systems to ensure timely detection and remediation of polluted water sources [24].

This chapter examines the multifaceted issue of water pollution and contamination, focusing on the distinct yet interconnected impacts of industrial, agricultural, and urban activities. It explores how each sector contributes specific pollutants (ranging from chemical effluents and agricultural runoff to urban wastewater) and the resulting environmental and public health consequences. The chapter emphasizes the need for integrated management strategies to mitigate these threats and protect water resources for sustainable use.

3.2 Industrial Impacts on Water Pollution

Water resources are crucial for human survival [25], agriculture, and industrial use, yet they face growing threats from both natural and human-induced pollution [26–29]. Natural sources of pollution include geological processes, climate variability, and water–rock interactions, which can introduce minerals and other substances into water bodies. Anthropogenic activities such as solid wastes, including market wastes [30–32] agricultural runoff, industrial discharge, and sewage contribute contaminants like heavy metals, nitrates, and pathogens to both surface and groundwater [33]. These pollutants degrade water quality, disrupt ecosystems, and complicate water resource management, as illustrated in Fig. 3.1, highlighting the need to understand pollution types, sources, and pathways for effective intervention [33]. Factors like rapid urbanization, excessive agricultural inputs, poor waste management, natural gas extraction, natural disasters, agrochemical use, etc., have intensified groundwater pollution, especially in developing regions where access to clean water is already limited [34], Fig. 3.2.

The industrial sector has significantly contributed to water pollution, manifesting through various direct and indirect means [35–37]. One of the foremost sources of industrial water pollution is the discharge of untreated or partially treated effluents.

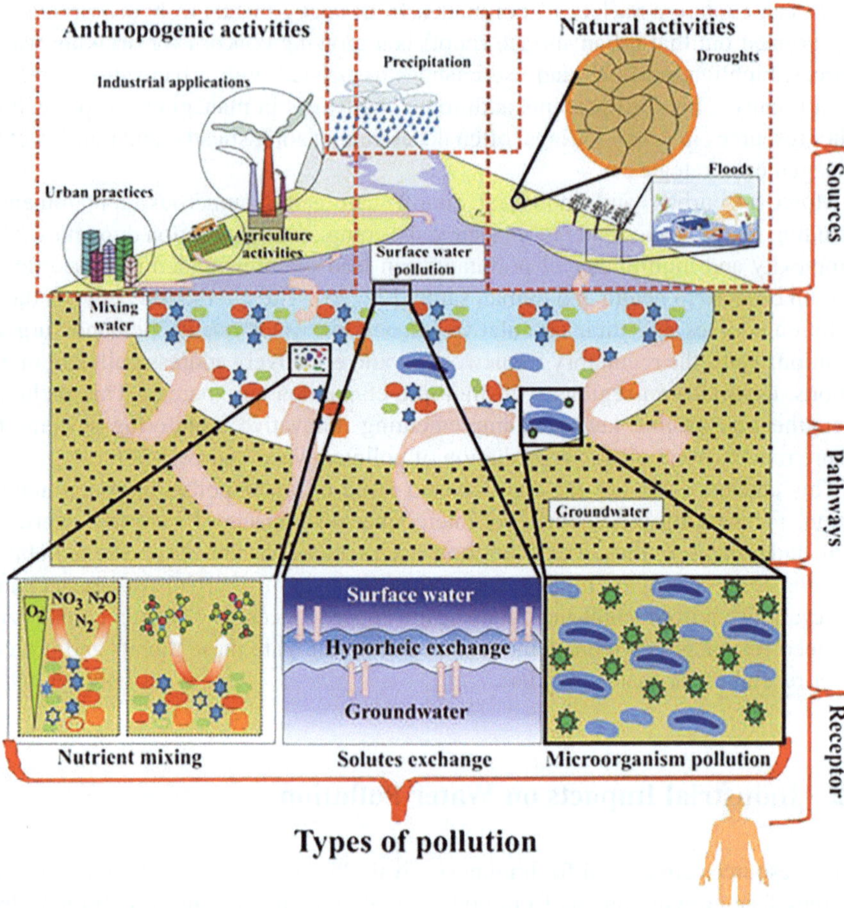

Fig. 3.1 Water contamination stems from natural and human sources, impacting ecosystems through multiple pollutant pathways [33]

Many factories, particularly in developing nations with weak environmental regulations, often release wastewater directly into waterways without adequate treatment. This wastewater may contain a range of harmful pollutants, including organic matter, acids, dyes, and various toxic substances [38]. Another critical source of industrial water pollution is the release of heavy metals, chemicals, and solvents from industries such as mining, metal plating, and battery manufacturing. Heavy metals, including mercury, lead, cadmium, and arsenic, are particularly concerning due to their persistence in the environment and potential to accumulate in the food chain. Authors have reported heavy metals in aquatic organisms [39–41], as well as sediment [42, 43].

The bioaccumulation of these metals poses significant ecological risks, as they can have deleterious effects on aquatic life and subsequently on humans who consume contaminated seafood [39, 40]. Such compounds are notorious for their toxicity and

Fig. 3.2 Some major activities leading to water contamination. *Source* [34]

can precipitate long-term environmental health crises, especially in regions heavily reliant on local waterways for sustenance and economic activities.

Thermal pollution is another alarming aspect arising from industrial processses, particularly from water used for cooling machinery. The discharge of heated water back into natural water bodies elevates water temperatures, disrupting aquatic ecosystems. Such temperature anomalies adversely affect oxygen levels in the water, making environments less hospitable for temperature-sensitive species [44]. Furthermore, this form of pollution can lead to a decrease in biodiversity, as organisms that cannot adapt to these changes face increased mortality rates. This disruption amplifies competitive pressures within ecosystems, potentially leading to population collapses in sensitive species [45].

The environmental and health effects resulting from industrial water pollution are profound. The immediate toxicity of pollutants like phenols and cyanides to aquatic organisms leads to reduced biodiversity and the collapse of fish populations in heavily polluted areas. Over time, these pollutants enter food chains through bioaccumulation, ultimately affecting top predators, including humans. For instance, long-term consumption of contaminated seafood has been associated with significant health risks, including neurological disorders and increased cases of cancer and kidney damage [46]. The vulnerability is particularly heightened among children and pregnant women, who are more susceptible to the deleterious impacts of such toxins.

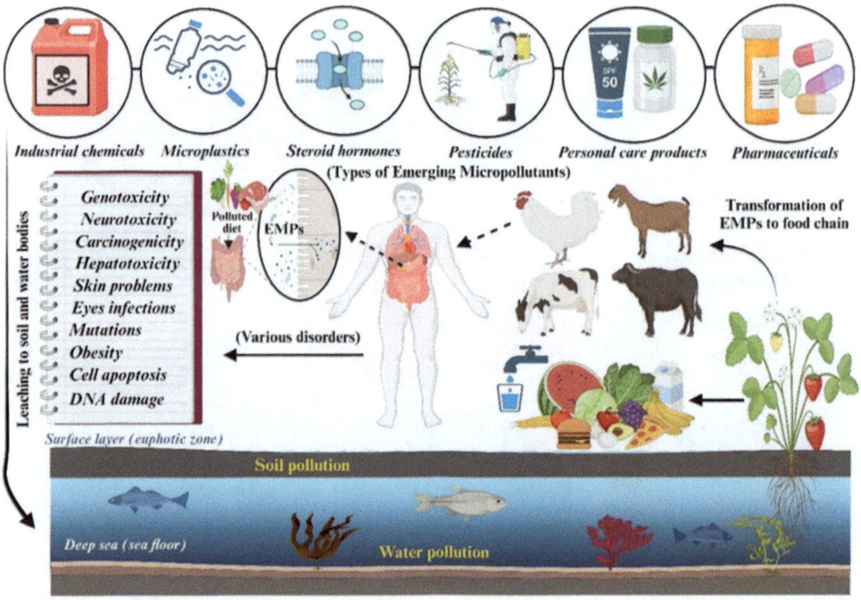

Fig. 3.3 Emerging pollutants and their negative effects on the environment particularly water. *Source* [49]

Rapid industrialization and urban expansion have led to the release of hazardous substances such as pesticides [47, 48], pharmaceuticals, personal care products and heavy metals into aquatic environments [49]. These pollutants, often classified as emerging pollutants (EPs) or micropollutants (MPs), are not fully removed by conventional water treatment and pose serious ecological and health risks through bioaccumulation and water quality degradation. Their presence disrupts ecosystems, threatens aquatic life, and raises global concerns about safe drinking water and environmental sustainability (Fig. 3.3, [49]).

Additionally, microplastics, both primary and secondary, are a growing concern in aquatic environments due to their persistence and widespread sources. Primary microplastics originate from industrial products such as synthetic clothing, plastic resins, car tires, and road paints, with fibers and particles often escaping filtration systems and entering waterways [50]. Secondary microplastics form from the degradation of larger plastic debris through environmental forces like UV radiation, mechanical abrasion, and low temperatures, especially in marine ecosystems [50]. Human activities (such as fishing, tourism, and maritime industries) further contribute to this pollution, posing significant threats to marine organisms and habitats (Fig. 3.4, [50].

In examining case studies, the Niger Delta presents a local example of industrial water pollution with severe repercussions. The region has faced extensive oil pollution from spills and gas flaring, leading to widespread contamination of rivers and

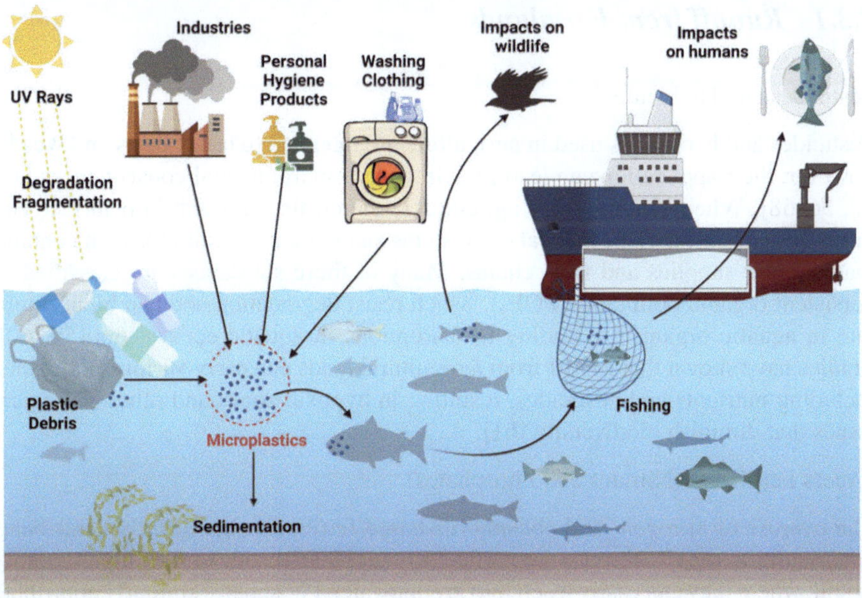

Fig. 3.4 Microplastics in aquatic environment. *Source* [50]

groundwater [51–55]. This ecological disaster has devastated traditional fishing practices and has resulted in serious health crises among local communities due to hydrocarbon exposure. In contrast, the Flint water crisis in Michigan illustrates a different aspect of industrial pollution, where changes in the water supply led to lead leaching from corroded pipes, exposing residents to toxic levels of lead. This crisis highlighted significant failures in regulatory oversight and accountability within industrial practices and brought national attention to the systemic issues surrounding industrial water pollution.

3.3 Agricultural Contributions to Water Contamination

Agricultural activities significantly contribute to water contamination, primarily through runoff from farmlands, livestock waste, and the associated management practices. Comprehending these contributions is crucial for effectively tackling the environmental and public health issues they create.

3.3.1 Runoff from Farmlands

Pesticides and Herbicides

Pesticides and herbicides used in agriculture are intended to target pests and weeds; however, their application can lead to unintended environmental consequences [17, 18, 56–58]. When it rains, these agricultural chemicals can leach into surface and groundwater, posing risks to local ecosystems and to human health through contaminated water supplies and food chains. Many of these substances are classified as persistent organic pollutants (POPs), which resist degradation and can bioaccumulate in aquatic organisms, leading to disruptions in aquatic ecosystems [59, 60]. Studies have shown that runoff from agricultural lands can carry various pollutants, including nutrients and pesticides, resulting in hypoxic zones and other ecological issues that diminish biodiversity [61].

Excess Fertilizers (Nitrates and Phosphates)

The overuse of nitrogen- and phosphorus-based fertilizers significantly contributes to agricultural runoff. Nitrates are particularly problematic due to their high mobility in soil, where they can easily leach into groundwater. Phosphates similarly contribute to nutrient loading in water bodies, leading to eutrophication, a process where excessive nutrient levels lead to algal blooms that deplete oxygen in the water and threaten aquatic life [62]. Agricultural non-point source pollution, primarily resulting from excessive fertilizer application, has been identified as a leading factor in the deterioration of water quality in various rivers and lakes [61]. This phenomenon threatens aquatic ecosystems and also affects the viability of drinking water supplies and agricultural productivity [60].

3.3.2 Livestock Waste and Antibiotics

Pathogen Transmission to Water Bodies

The management of livestock waste is crucial in assessing agricultural contributions to water contamination. Animal manure can harbor pathogens, including bacteria such as E. coli and Salmonella, which pose significant health risks when washed into water bodies. During rainfall events or due to improper waste disposal practices, these pathogens can enter rivers, streams, and groundwater, increasing the risk of waterborne diseases within rural and downstream communities. Research indicates that pathogens found in livestock waste can significantly impact community health, especially in areas with limited access to safe drinking water [63, 64]. This association highlights the need for improved livestock waste management strategies to protect water quality and public health [65].

Antibiotic Resistance Issues

A major concern related to livestock waste is the use of antibiotics in animal husbandry [35]. Residual antibiotics excreted in manure can lead to environmental contamination, fostering the proliferation of antibiotic-resistant bacteria. These pathogens can transfer their resistance genes to human pathogens, resulting in a significant public health threat [63]. Studies have shown that water sources near livestock farms often contain elevated levels of antibiotic-resistant bacteria, complicating efforts to manage water quality and protect human health [63].

3.3.3 *Impacts*

- **Eutrophication and Algal Blooms:** Excessive nutrient runoff associated with agricultural practices catalyzes ecological crises, primarily through the onset of eutrophication and harmful algal blooms (HABs). Nutrient enrichment stimulates rapid algal growth that blocks sunlight and depletes oxygen, leading to fish kills and loss of biodiversity [62]. These algal blooms can produce toxins harmful to both humans and animals, exacerbating health risks associated with contaminated water supplies [60, 61]. As such, agricultural runoff not only threatens aquatic ecosystems but raises serious concerns regarding food safety and public health [62].
- **Groundwater Contamination:** The persistent leaching of nitrates from fertilizers into aquifers has raised alarms about groundwater quality. High levels of nitrates are linked to health issues such as methemoglobinemia, commonly referred to as "blue baby syndrome," and potential carcinogenic effects, underscoring the necessity for stringent controls on agricultural runoff [62, 66]. Once groundwater is contaminated, remediation becomes both challenging and financially burdensome, emphasizing the need for proactive measures in agricultural practices [62, 66].
- **Impaired Drinking Water Sources:** The implications of agricultural runoff extend to impaired drinking water sources. Elevated nitrate and pathogen levels can exceed safe drinking water standards, placing rural communities that rely on shallow wells at greater risk [63]. Ensuring access to clean water becomes increasingly complex and costly, impacting not only public health but also the economic stability of these communities [61, 66]. The interplay between agricultural activities and water quality underscores the importance of adopting sustainable farming practices that safeguard water resources and enhance public health [62, 67].

3.3.4 Case Examples

A prime example of the consequences of agricultural runoff is observed in the Gulf of Mexico, where nutrient-rich waters from the Mississippi River Basin lead to the creation of seasonal hypoxic zones. This occurs primarily due to nutrient runoff from intensive agricultural practices across the United States [62]. These dead zones span thousands of square kilometers and threaten both marine life and the fishing industry. Similar nutrient-driven hypoxia has been reported in coastal waters of Nigeria, indicating a global challenge that requires urgent intervention [62].

In both developed and developing nations, groundwater nitrate contamination related to long-term fertilizer use is a persistent problem. In Nigeria, rural areas with significant agricultural activity could alarmingly high nitrate levels in water due to high use of pesticides in agricultural activities. Many farmers and rural inhabitants lack the resources for proper testing or treatment of contaminated groundwater, underscoring socio-economic disparities that complicate the management of agricultural runoff [63, 68, 69].

3.4 Urban Impacts on Water Quality

Urban areas significantly influence water quality through various mechanisms, primarily urban runoff and wastewater management failures. Urban surfaces, such as roads and pavements, exacerbate the problems of stormwater runoff, which accumulates pollutants and diverts them into surrounding water bodies. Pollutants commonly transported by runoff include oil, grease, brake dust, heavy metals like lead and copper, and other contaminants such as fertilizers, pesticides, and pet waste [70, 71]. These urban pollutants can severely impair the ecological balance of aquatic environments and threaten human health for communities living near these affected areas.

In addition to the contaminative effects of urban runoff, inadequate sewage systems pose a substantial risk to water quality, particularly in rapidly developing urban slums. Many of these areas lack proper wastewater treatment infrastructure, leading to direct discharges of untreated wastewater into local waterways [72]. Poorly maintained septic systems contribute further to this problem by contaminating surface and groundwater sources nearby [73, 74]. Moreover, the increasing detection of emerging pollutants, such as pharmaceuticals and hormones, in untreated waste highlights the inadequacy of existing wastewater treatment solutions [75].

The phenomenon of Combined Sewer Overflows (CSOs) is particularly problematic in older urban regions where stormwater and sanitary sewage share the same pipelines. During periods of heavy rainfall, these systems can become overwhelmed, resulting in untreated sewage being released into nearby water bodies [76]. Such overflows introduce pathogens, nutrients, and organic materials into aquatic environments, significantly degrading water quality and increasing the risk of disease

[77, 78]. Studies have shown that sewage overflows can lead to drastic spatial variations in water quality, with more severe pollution detected closer to overflow outlets [79, 80].

Urban microplastic pollution further complicates the water quality dilemma. Significant sources of microplastics are found in urban areas, including litter from single-use plastics, tire wear particles, and microbeads from personal care products [81]. The biological implications of microplastics are concerning, as they are ingested by aquatic organisms, leading to bioaccumulation within food chains. This process not only threatens aquatic ecosystems but also poses serious health risks to humans who consume contaminated seafood or who are exposed to microplastics through drinking water [74, 81].

Addressing these challenges requires a multifaceted approach to urban water management. Integrating advanced stormwater management strategies, such as constructing stormwater detention tanks and implementing sustainable drainage systems, can help mitigate the pollution caused by CSOs [82, 83]. Additionally, enhancing the maintenance and upgrading of aging sewage infrastructure is critical to prevent leaks and the release of contaminants into drinking water supplies [84, 85]. By prioritizing pollutant control and employing a combination of construction and management strategies, cities can significantly improve urban water quality and create healthier environments for both aquatic ecosystems and urban populations [81, 86].

3.5 Comparative Analysis of Water Pollution Sources

Water pollution stems from various sources, each contributing uniquely to environmental degradation and public health risks. This section of the chapter shows a comparative analysis of the key sources of water pollution, the overlaps in their impacts, and the socioeconomic and regulatory dimensions surrounding these issues as summarized in Table 3.1.

3.5.1 Comparative Analysis of Water Pollution Sources

- **Industrial Pollution:** Industrial sources of water pollution have severe implications due to the concentrated discharge of toxic chemicals, heavy metals, and organic pollutants. These discharges significantly impact large water bodies and groundwater systems, posing risks to both human health and aquatic ecosystems [87, 88]. For example, industrial activities release substantial amounts of nutrients and hazardous substances, which can lead to detrimental changes in water quality. Adverse environmental outcomes include bioaccumulation of toxins in aquatic life and significant biodiversity losses [89, 90]. In a study analyzing various sectors' impacts on water resources, it was illustrated that industries (especially those

Table 3.1 Comparative Analysis of Water Pollution Sources

Category	Explanation	Examples/Notes
Relative severity and scale of impact		
Industrial Pollution	High toxicity; concentrated discharges	Heavy metals, solvents, dyes; impact surface and groundwater
Agricultural Pollution	Widespread; diffuse sources	Nitrates, phosphates, and pesticides lead to eutrophication and groundwater contamination
Urban Runoff	Medium to high impact; episodic	Oils, plastics, metals, and pathogens from roads and domestic sources
Natural Sources	Generally less severe, but can be significant	Arsenic, salinity, and contaminants mobilized by floods or droughts
Overlaps and synergistic effects		
Industrial + Urban Runoff	Increased chemical load and toxicity	Complex pollutant mixtures are difficult to treat
Agriculture + Domestic Waste	Enhanced nutrient and microbial contamination	Promotes algal blooms, waterborne disease, and antimicrobial resistance
Natural + Anthropogenic	Climate events intensify pollution mobility	Floods disperse contaminants; droughts concentrate them
Mining + Agriculture	Compounded toxicity	Heavy metals mix with fertilizers and pesticides
Socioeconomic and regulatory dimensions		
Low- and Middle-Income Countries	Greater exposure, weaker infrastructure	Limited water treatment and poor enforcement of regulations
Cost of Inaction	Economic and health burdens	Loss of productivity, healthcare costs, and ecological damage
Regulatory Gaps	Inconsistent monitoring and enforcement	Overlapping agencies and outdated legislation
Environmental Justice	Disproportionate impact on marginalized groups	Poor communities often live near polluted water sources without alternatives
Global Disparities	Export of polluting activities	Industrialized nations shift pollution to less-regulated developing countries

engaged in chemical processing) represent substantial threats to water quality due to their extensive release of pollutants [91]. Moreover, in transitional economies, rising industrialization correlates with increased pollution exposure, necessitating remediation measures [92]. Such findings highlight the complex nature of industrial water pollution, requiring robust regulatory frameworks and infrastructure to manage and mitigate its impacts effectively [93, 94].

- **Agricultural Pollution:** Agricultural practices contribute significantly to water pollution, primarily through non-point source pollution mechanisms that are widespread and diffuse. Agriculture is responsible for substantial nutrient loading (nitrates, phosphates, and pesticides) that leads directly to eutrophication and groundwater contamination [95–97]. The OECD highlights that agricultural pollution has dire outcomes for ecosystems, affecting drinking water safety and recreational activities [98]. Additionally, agricultural runoff promotes algal blooms that deplete oxygen levels, suffocating aquatic life [99]. This dynamic demonstrates both the acute and chronic nature of agricultural pollution, posing threats to biodiversity, ecosystem functionality, and human health through contaminated water supplies.
- **Urban Runoff:** Urban runoff presents another multifaceted challenge due to the transport of pollutants such as oil, heavy metals, plastics, and pathogens from urban landscapes into adjacent water bodies [100]. The interaction of these pollutants with substances from industrial or agricultural sources compounds the water quality crisis. Reports emphasize that urban areas significantly contribute to water pollution, necessitating comprehensive urban watershed management strategies. Urban runoff particularly poses significant human health risks in less developed regions where infrastructure for managing such runoff is often inadequate [100].
- **Natural Sources:** Conversely, natural sources of water pollution, such as salinization and the mobilization of arsenic due to geological formations, are generally less severe than anthropogenic sources but still noteworthy. These factors can exacerbate existing human-induced pollution, demonstrating the need for a combined understanding of both natural and anthropogenic influences on water quality.

3.5.2 *Overlaps and Synergistic Effects*

The interactions between different sources of pollution create complex challenges for management and remediation efforts. For example, the combination of industrial waste and urban runoff can amplify the chemical load, increasing the toxicity and complexity of pollutants in water bodies [74, 101]. Conversely, untreated domestic wastewater combined with agricultural practices can enrich water bodies with nutrients while potentially introducing pathogens [97, 99]. These overlaps pose environmental and health risks that necessitate integrated approaches in policy and regulatory frameworks to manage water pollution effectively. Addressing these synergistic effects requires a robust understanding of the interactions between pollution sources and a targeted approach to mitigate their combined impacts [102, 103].

3.5.3 Socioeconomic and Regulatory Dimensions

The socioeconomic context of water pollution reveals significant disparities, particularly in low- and middle-income countries. Lack of infrastructure, inadequate regulatory frameworks, and insufficient public awareness exacerbate vulnerability to water pollution, leading to severe consequences for poorer communities [100, 104]. Studies illustrate that waterborne diseases and ecological degradation impose substantial economic burdens (both direct and indirect) on these populations [105]. Furthermore, regulatory gaps stemming from overlapping jurisdictions and outdated legislation hinder effective pollution control [104]. Environmental justice perspectives highlight that marginalized communities often bear the burden of water pollution without access to clean alternatives or legal recourse [100, 102]. Therefore, there is a pressing need for comprehensive policy reform and active engagement of local communities to enhance awareness and participation in pollution management efforts.

3.6 Strategies for Mitigation and Management of Water Pollution

3.6.1 Policy and Regulation

The enactment and enforcement of robust national and regional water quality laws are essential for safeguarding water resources from pollution. In Nigeria, the National Environmental Standards and Regulations Enforcement Agency (NESREA) regulations play a pivotal role in managing water quality and assessing environmental impacts stemming from various industries, including the hospitality sector, where improper effluent discharge poses significant threats to groundwater quality [106]. The Clean Water Act in the United States exemplifies a proactive policy that has inspired similar frameworks in other countries, emphasizing the need for comprehensive regulations that address pollutant discharge standards and thresholds [107]. By developing precise standards for effluent discharge and establishing protected water zones, regulatory bodies can significantly mitigate adverse environmental impacts, ensuring that water sources are safe for public use and wildlife habitats.

Furthermore, regular audits and sanctions for non-compliance by industries and municipalities are imperative to uphold these regulations. Effective monitoring by entities like NESREA ensures accountability among water users, thereby enhancing compliance with environmental standards [108]. Such mechanisms are crucial not only for deterring potential violations but also for reinforcing the commitment of stakeholders towards sustainable water management practices. As part of international agreements such as Sustainable Development Goal 6 (SDG 6), countries are encouraged to implement strategies that ensure the availability and sustainable management of water and sanitation [109]. Adopting transboundary water treaties, like those outlined in the Ramsar Convention, supports collaborative approaches among nations to address shared water resources and enhance compliance with international water quality benchmarks.

Reporting progress through global water quality indicators and national development plans plays a vital role in assessing the effectiveness of implemented regulations [110]. Metrics derived from local, national, and global water quality assessments help identify trends, facilitate transparent governance, and promote informed decision-making regarding water management policies. As such, a structured approach that encompasses enacted legal frameworks, continuous monitoring, and international cooperation is paramount for achieving significant improvements in water quality across diverse regions.

3.6.2 Technological Interventions

Technological advancements offer a pathway to enhance water quality through innovative wastewater treatment technologies. Modern treatment plants that employ membrane filtration, ultraviolet (UV) disinfection, and activated sludge systems represent sophisticated solutions capable of addressing complex pollution challenges [111]. The integration of these technologies in urban and rural settings not only improves the effectiveness of wastewater management but also aligns with compliance requirements outlined in national and international regulations. The adoption of decentralized treatment systems in rural regions, where infrastructure may be lacking, ensures that communities have access to safe and clean water, fostering better public health outcomes.

Moreover, retrofitting outdated infrastructure to meet contemporary environmental standards is critical. Many regions still rely on aging water systems inadequately equipped to handle current pollution levels [112]. Upgrading these systems with state-of-the-art technology can significantly decrease pollutant levels, thereby safeguarding water quality and reinforcing public trust in water supply systems. In conjunction with these technological interventions, precision agriculture techniques such as controlled irrigation and optimized fertilizer application help minimize agricultural runoff, which is a substantial contributor to water pollution [111]. By implementing bioremediation strategies that utilize plants and microbes to detoxify polluted water bodies, stakeholders can work towards restoring the ecological health of these systems [110, 113]. The promotion of organic farming combined with integrated pest management practices can mitigate the introduction of harmful chemicals into water systems. By prioritizing these sustainable agricultural techniques, it is possible to reduce the dependency on chemical fertilizers and pesticides that often contribute to non-point source pollution, thereby enhancing overall water quality.

3.6.3 Community and Stakeholder Engagement

Effective community and stakeholder engagement is critical in the pursuit of sustainable water management strategies. Public education campaigns aimed at raising awareness regarding the importance of clean water and pollution prevention are essential in cultivating a culture of water stewardship among local populations. Such initiatives are often implemented through school-based programs that encourage students to become proactive stewards of their water resources [112]. By disseminating information about water-saving techniques and creating pollution reporting hotlines, communities can directly participate in monitoring and protecting local water bodies. Citizen science programs, which involve local communities in the collection of water samples and tracking pollution levels, have proven effective in fostering environmental accountability. Engaging citizens in water monitoring empowers them, builds trust between communities and regulatory bodies, and generates valuable data that can inform decision-making. By developing open-access platforms and mobile applications for water quality reporting, collaborative networks can form among governments, non-governmental organizations, academia, and local water users. Such alliances are instrumental in ensuring that diverse stakeholders contribute to the sustainable management of water resources.

3.7 Suitable and Future Outlook

To effectively tackle water pollution, suitable approaches should involve multiple dimensions, including governance frameworks, sustainable practices, and innovative technologies. A prominent strategy is Integrated Water Resource Management (IWRM), which emphasizes a holistic approach to managing water resources while considering agriculture, urban development, and industrial needs. IWRM could promote participatory governance, allowing communities to play an active role in management decisions, which could lead to more effective pollution reductions [114]. The IWRM framework not only supports water quality improvements but also aims for equity in access to water resources, which is stressed by the need for context-specific priorities that reflect local challenges [114].

Another critical approach is the strengthening of environmental regulations and enforcement. Governments are encouraged to modernize their regulatory frameworks to incorporate stricter pollution control policies tailored to specific sectors, such as effluent standards for industries and management protocols for agricultural runoff [115]. Real-time monitoring and adaptive governance can be vital components in this regard. For instance, leveraging water quality monitoring technologies to inform regulatory actions can enhance the responsiveness of governance structures to emerging water quality issues [116]. Meanwhile, decentralized wastewater treatment systems present a pragmatic solution, especially in rural and peri-urban areas lacking centralized infrastructure. These community-based solutions can significantly reduce pollution while being cost-effective and environmentally sound [117]. Moreover, sustainable agricultural practices, such as integrated pest management and precision irrigation, are essential to minimize chemical runoff into waterways. These practices are not only beneficial for water quality but also provide economic advantages to farmers through reduced input costs [117, 118].

Urban planning incorporating green infrastructure also presents a multifaceted approach toward water pollution control. Green roofs, permeable pavements, and rain gardens are effective in managing stormwater, thereby reducing runoff and filtering pollutants before they reach water bodies [119, 120]. These infrastructure elements not only mitigate flood risks but also enhance urban biodiversity and improve public health, highlighting the interconnected benefits of green space integration [121]. Public awareness campaigns that focus on behavior change regarding water use and pollution are crucial in this context. Engaging communities in the development and management of these green infrastructures fosters a collective commitment to sustainable water use [122].

Looking towards the future, emerging technologies such as modeling [2], artificial intelligence and the Internet of Things are poised to transform water monitoring and treatment processes. These advancements allow for real-time data collection and analysis, enabling prompt pollution detection and effective response strategies [117]. Moreover, nanotechnology offers promising applications in water filtration, particularly for addressing micro-pollutants and heavy metals, thus bolstering water quality and safety [123].

Climate change adaptation will also play a critical role in shaping water pollution control strategies [23]. The increasing prevalence of droughts and floods highlights the need for resilient water infrastructure and adaptive management approaches that can withstand climate variability [124]. Furthermore, circular economy principles in water management (such as the reuse of treated wastewater and recovery of nutrients) will contribute significantly to minimizing pollution loads and conserving freshwater resources [125]. International cooperation and knowledge sharing are paramount, as water pollution transcends national boundaries. Collaborative efforts in data sharing, research, and cross-border agreements can catalyze effective responses to water contamination challenges globally [126]. Furthermore, fostering youth engagement in water management through educational initiatives empowers the next generation with the skills and knowledge necessary for innovative solutions to water sustainability.

3.8 Conclusion

Water pollution remains one of the most pressing environmental challenges of the twenty-first century, driven predominantly by industrial discharges, agricultural runoff, and urban waste. Each of these sources contributes uniquely to water contamination: industries release heavy metals, toxic chemicals, and thermal effluents; agriculture introduces excessive nutrients, pesticides, and livestock waste; while urban areas contribute untreated sewage, stormwater runoff, and plastic debris. The cumulative and often interacting effects of these pollutants severely compromise water quality, disrupt aquatic ecosystems, and pose substantial risks to human health, particularly in low- and middle-income countries where regulatory and infrastructural capacities may be limited. Addressing these interlinked sources of pollution demands an integrated water management approach—one that transcends sectoral boundaries and emphasizes coordination among stakeholders, institutions, and policy frameworks. Such an approach requires the harmonization of industrial regulations, the promotion of sustainable agricultural practices, and the modernization of urban infrastructure. It also calls for increased investment in innovative technologies, including advanced wastewater treatment, bioremediation, and precision farming, alongside the enforcement of environmental standards and monitoring mechanisms. Equally important is the role of community engagement, public education, and equitable access to safe water in fostering accountability and long-term stewardship. Embedding these strategies within the broader context of sustainable development—particularly in alignment with Sustainable Development Goal 6 (Clean Water and Sanitation)—is essential for safeguarding water resources for future generations. Ultimately, the path forward must be one of collective responsibility and systemic action, where environmental protection, public health, and sustainable growth are pursued in unison.

References

1. Ogamba EN, Seiyaboh EI, Izah SC, Ogbugo I, Demedongha FK (2015) Water quality, phytochemistry and proximate constituents of *Eichhornia crassipes* from Kolo creek, Niger Delta, Nigeria. Int J Appl Res Technol 4(9):77–84
2. Ogamba EN, Charles EE, Izah SC (2021) Distributions, pollution evaluation and health risk of selected heavy metal in surface water of Taylor creek, Bayelsa State. Nigeria. Toxicol Environ Health Sci 13(2):109–121. https://doi.org/10.1007/s13530-020-00076-0
3. Ogamba EN, Izah SC, Oribu T (2015) Water quality and proximate analysis of *Eichhornia crassipes* from River Nun, Amassoma Axis. Nigeria. Res J Phytomedicine. 1(1):43–48
4. Enaregha EB, Omotete TR, Izah SC, Odubo TC (2022) Comparison of total viable bacteria counts and in situ characteristics in drinking water sources in Sagbama town, Bayelsa state, Nigeria. Int J Pathogen Res 11(3–4):79–87
5. Seiyaboh EI, Youkparigha FO, Izah SC, Mientei K (2020) Assessment of bacteriological characteristics of surface water of Taylor creek in Bayelsa state, Nigeria. Noble Int J Sci Res 4(4):25–30
6. Seiyaboh EI, Youkparigha FO, Izah SC, Daniels ID (2020) Bacteriological Quality of Groundwater in Imiringi Town, Bayelsa State, Nigeria. J Biotechnol Biomed Sci 2(2):34–40
7. Xiao L, Zhang Q, Niu C, Wang H (2020) Spatiotemporal patterns in river water quality and pollution source apportionment in the arid Beichuan River basin of northwestern China using positive matrix factorization receptor modeling techniques. Int J Environ Res Public Health 17(14):5015. https://doi.org/10.3390/ijerph17145015
8. Huang SJ, Zhou GQ, Tang YD, Wei J (2023) Investigation on water pollution control of drinking water source in China. In: IOP conference series: earth and environmental science 1204(1), p. 012004). IOP Publishing.https://doi.org/10.1088/1755-1315/1204/1/012004

9. Agarwal A, Xue L (2020) Model-based clustering of nonparametric weighted networks with application to water pollution analysis. Technometrics 62(2):161–172
10. Dalmieda J, Kruse P (2019) Metal cation detection in drinking water. Sensors 19(23):5134. https://doi.org/10.3390/s19235134
11. Tripathi PK (2024) The impact of industrial waste on water pollution: a detailed analysis. Int J Geogr Geol Environ, 6(1), 391–395. https://doi.org/10.22271/27067483.2024.v6.i1e.245
12. Aghoghovwia OA, Umoru OD, Izah SC (2018) Physicochemical characteristics of nun river at Gbarantoru and Tombia Axis in Bayelsa State. Niger, Biosci Methods 9(1):1–11
13. Aghoghovwia OA, Miri FA, Izah SC (2018) Impacts of anthropogenic activities on heavy metal levels in surface water of nun river around Gbarantoru and Tombia Towns, Bayelsa State, Nigeria. Ann Ecol Environ Sci 2(2):1–8
14. Neidoni DG, Nicorescu V, Andres L, Negrea SC, Diaconu LA (2021) Database with pollution sources and impact assessment in Bega and Timis rivers basins. https://doi.org/10.21698/rjeec.2021.104
15. Joudah ASA, Racoviteanu G (2019) Membrane bioreactors used for treatment of food industry effluents. In E3S web of conferences (Vol. 85, p. 07013). EDP Sciences
16. Inyang IR, Izah SC, Johnson DT, Ejomarie OA (2017) Effects of Lambda cyhalothrin on some electrolytes and metabolites in organs of Parpohiocephalus obscurus. Biotechnol Res 3(1):6–10
17. Inyang IR, Thomas S, Izah SC (2016) Activities of electrolytes in kidney and liver of Clarias gariepinus exposed to fluazifop-p-butyl. J Biotechnol Res 2(9):68–72
18. Inyang IR, Okon NC, Izah SC (2016) Effect of glyphosate on some enzymes and electrolytes in Heterobranchus bidosalis (a common African catfish). Biotechnol Res 2(4):161–165
19. Tian Z, Ji S (2024) Research on pollution control and remediation of the Sangu spring area based on the evolution of the water Environment. Environ, Resour Ecol J, 8(1), 152–159. https://doi.org/10.23977/erej.2024.080119
20. Lyu Y, Gao Y, Ye H, Liu Y, Han S, Tian J, Chen L (2021) Quantifying the life cycle environmental impacts of water pollution control in a typical chemical industrial park in China. J Ind Ecol 25(6):1673–1687
21. Wang S, Islam MZ (2022) Legal institutional inefficiency and water pollution problem in Bangladesh. Resour Environ Econ, 4(2), 372–375.https://doi.org/10.25082/ree.2022.02.004
22. Rawat R, Siddiqui AR (2019) Assessment of physiochemical characteristics of drinking water quality in Allahabad Metropolitan City, India. The Oriental Anthropologist, 19(1), 121–135.https://doi.org/10.1177/0972558x19835368
23. Iyiola AO, Afolabi OA, Alimi SK, Akingba OO, Izah SC, Ogwu MC (2024) Climate change and water crisis in the global south. In: Izah SC, Ogwu MC, Loukas A, Hamidifar H (eds) Water Crises and Sustainable Management in the Global South. (pp. 111–140). Springer, Singapore. https://doi.org/10.1007/978-981-97-4966-9_4
24. Yuan Q, Peng T, Ming J (2022) Collaborative innovation mechanism of water pollution control industry chain based on complex scientific management. Discret Dyn Nat Soc 2022(1):5894496. https://doi.org/10.1155/2022/5894496
25. Seiyaboh EI, Izah SC, Bokolo JE (2017) Bacteriological quality of water from river nun at Amassoma Axises, Niger Delta, Nigeria. ASIO J Microbiol, Food Sci & Biotechnol Innov 3(1):22–26
26. Numbere AO, Kika EP, Izah SC (2024) Challenges and Mitigation of Water Resource Management in Rural Areas in Nigeria. In: Izah SC, Ogwu MC, Loukas A, Hamidifar H (eds) Water crises and sustainable management in the global south. (pp 373 – 390) Springer, Singapore. https://doi.org/10.1007/978-981-97-4966-9_12
27. Jacob DE, Nelson IU, Izah SC (2024a). Indigenous water management strategies in the global south. In: Izah SC, Ogwu MC, Loukas A, Hamidifar H (eds) Water Crises and Sustainable Management in the Global South. (pp. 487–525) Springer. https://link.springer.com/chapter/https://doi.org/10.1007/978-981-97-4966-9_16

28. Jacob DE, Nelson IU, Eniang EA, Izah SC (2024b). Advancing water security and resilience in the global south through recreational development. In: Izah SC, Ogwu MC, Loukas A, Hamidifar, H (eds) Water crises and sustainable management in the global south. (pp. 527–554)_ Springer. https://link.springer.com/chapter/https://doi.org/10.1007/978-981-97-4966-9_17
29. Jacob DE, Nelson IU, Izah SC, Daniel KS (2024c) Rural water crises in the global south: understanding the scope and impact. In: Izah SC, Ogwu MC, Loukas A, Hamidifar H (eds) Water crises and sustainable management in the global south. (pp 3–44) Springer, Singapore. https://doi.org/10.1007/978-981-97-4966-9_1
30. Ben-Eledo VN, Kigigha LT, Izah SC, Eledo BO (2017) Bacteriological quality assessment of epie creek, niger delta region of Nigeria. Int J Ecotoxicol Ecobiol. 2(3):102–108
31. Ben-Eledo VN, Kigigha LT, Izah SC, Eledo BO (2017) Water quality assessment of Epie creek in Yenagoa metropolis, Bayelsa state. Nigeria. Arch Curr Res Int 8(2):1–24
32. Seiyaboh EI, Izah SC (2017) Review of impact of anthropogenic activities in surface water resources in the niger delta region of Nigeria: A case of Bayelsa state. Int J Ecotoxicol Ecobiol. 2(2):61–73
33. Akhtar N, Syakir Ishak MI, Bhawani SA, Umar K (2021) Various natural and anthropogenic factors responsible for water quality degradation: A review. Water 13(19):2660
34. Ravindiran G, Rajamanickam S, Sivarethinamohan S, Karupaiya Sathaiah B, Ravindran G, Muniasamy SK, Hayder G (2023) A review of the status, effects, prevention, and remediation of groundwater contamination for sustainable environment. Water 15(20):3662
35. Izah SC, Nurmakhanova A, Ogwu MC, Toktarbay Z, Umirbaeva Z, Ussen K, Koibasova L, Nazarbekova S, Tynybekov B, Guo Z (2025a) Public health risks associated with antibiotic residues in poultry food products, journal of agriculture and food research, 21, 101815. https://doi.org/10.1016/j.jafr.2025.101815
36. Izah SC, Jacob DE, Nelson IU, Daniel KS, Ogwu MC (2024b). Exploring the Influence of Protected Areas on Water Crises in the Global South: A Balancing Act. In: Izah SC, Ogwu MC, Loukas A, Hamidifar H (eds) Water crises and sustainable management in the global south. (pp. 625–654) Springer. https://link.springer.com/chapter/https://doi.org/10.1007/978-981-97-4966-9_20
37. Ogidi OI, and Izah SC (2024). Water Contamination by Municipal Solid Wastes and Sustainable Management Strategies. In: Izah, S.C., Ogwu, M.C., Loukas A, Hamidifar, H. (eds) Water Crises and Sustainable Management in the Global South. (pp. 313–339). Springer. https://link.springer.com/chapter/https://doi.org/10.1007/978-981-97-4966-9_10
38. Nduka JK, Orisakwe OE (2011) Water-quality issues in the Niger Delta of Nigeria: a look at heavy metal levels and some physicochemical properties. Environ Sci Pollut Res, 18(2), 237–246.https://doi.org/10.1007/s11356-010-0366-3
39. Izah SC, Stanley HO, Richard G, Sawyer WE, Uwaeme OR (2024c). Surface water quality: a statistical perspective on the efficacy of environmental and human health assessment tools. Water, Air, & Soil Pollution. 235(3). https://doi.org/10.1007/s11270-024-06965-1
40. Aigberua AO, Izah SC, Richard G (2021) Hazard Analysis of Trace Metals in Muscle of Sarotherodon melanotheron and Chrysichthys nigrodigitatus from Okulu River, Rivers State, Nigeria. J Environ Health Sustain Dev 6(3):1340–1356
41. Izah SC, Stanley HO, Richard G, Sawyer WE, Uwaeme OR (2024) Environmental health risks of trace elements in sediment using multivariate approaches and contamination indices. Int J Environ Sci Technol. https://doi.org/10.1007/s13762-024-05974-1
42. Izah SC, Stanley HO, Richard G, Sawyer WE, Uwaeme OR, Sylva L (2024) Source and health risks of Trace Metals in Clarias batrachus and Chrysichthys nigrodigitatus from surface waters in Bayelsa State, Nigeria: A probabilistic model. Front Sustain Food Syst. 8:1419143
43. Aigberua AO, Ogbuta AA, Izah SC (2020) Selected heavy metals in sediment of Taylor creek due to anthropogenic activities in the Niger Delta region of Nigeria: geochemical spreading and evaluation of environmental risk. Biodiversity Int J 4(2):67–80. https://doi.org/10.15406/bij.2020.04.00166
44. Ewim DRE, Orikpete OF, Scott TO, Onyebuchi CN, Onukogu AO, Uzougbo CG, Onunka C (2023) Survey of wastewater issues due to oil spills and pollution in the Niger Delta area of

Nigeria: a secondary data analysis. Bull Natl Res Cent, 47(1), 116.https://doi.org/10.21203/rs.3.rs-2895102/v1
45. Liu C, Liu W (2022) Bioaccumulation of silver nanoparticles in marine copepod Tigriopus japonicus under waterborne and dietary exposure. In: IOP conference series: earth and environmental science 1087(1), p 012003). IOP Publishing
46. Wang H, Xia X, Wang Z, Liu R, Muir DC, Wang WX (2021) Contribution of dietary uptake to PAH bioaccumulation in a simplified pelagic food chain: modeling the influences of continuous vs intermittent feeding in zooplankton and fish. Environ Sci Technol 55(3):1930–1940. https://doi.org/10.1021/acs.est.0c06970
47. Inyang IR, Izah SC, Okoroba VR (2019) Effect of toluene on the organosomatic indices and electrolytes in the liver and muscle of adult clarias gariepinus. SumerZ J Biotechnol 2(5):31–34
48. Inyang IR, Izah SC, Okpogholor KD (2019) Impact of aluminum phosphide on the transferases in liver and muscle of parophiocephalus obscurus. J Plant Anim Ecol, Spec Issue 1(4):1–6
49. Ahmad HW, Bibi HA, Chandrasekaran M, Ahmad S, Kyriakopoulos GL (2024) Sustainable wastewater treatment strategies in effective abatement of emerging pollutants. Water (20734441), 16(20)
50. Ziani K, Ioniță-Mîndrican CB, Mititelu M, Neacşu SM, Negrei C, Moroşan E, Drăgănescu D, Preda OT (2023) Microplastics: a real global threat for environment and food safety: a state of the art review. Nutrients 15(3):617
51. Richard G, Izah SC, Ogwu MC (2022) Implications of artisanal crude oil refining on sustainable food production in the niger delta region of nigeria. J Environ Bioremediation Toxicol 5(2):69–77. https://doi.org/10.3389/fsufs.2024.1419143
52. Richard G, Izah SC, Morufu OR, Austin-Asomeji I (2023) Public and environmental health implications of artisanal petroleum refining and risk reduction strategies in the Niger Delta region of Nigeria. Bio-Research 21(1):1896–1910
53. Seiyaboh EI, Izah SC (2019) Impacts of soil pollution on air quality under nigerian setting. J Soil Water Sci 3(1):45–53
54. Wang H, Xia X, Liu R, Wang Z, Zhai Y, Lin H, Wen W, Li Y, Wang D, Yang Z et al (2019) Dietary uptake patterns affect bioaccumulation and biomagnification of hydrophobic organic compounds in fish. Environ Sci Technol 53(8):4274–4284. https://doi.org/10.1021/acs.est.9b00106
55. Seiyaboh EI, Izah SC (2017) A review of impacts of gas flaring on vegetation and water resources in the niger delta region of Nigeria. Int J Econ, Energy Environ. 2(4):48–55
56. Inyang IR, Kenobi A, Izah SC (2016) Effect of dimethoate on some selected metabolites in the brain, liver and muscle of Clarias lazera. Sky J Biochem Res. 5(4):63–68
57. Inyang IR, Thomas S, Izah SC (2016) Evaluation of activities of transferases and phosphatase in plasma and organs of clarias gariepinus exposed to Fluazifop-p-Butyl. J Environ Treat Tech 4(3):94–97
58. Inyang IR, Akio K, Izah SC (2016) Effect of dimethoate on lactate dehydrogenase, creatinine kinase and amylase in Clarias lazera. Biotechnol Res 2(4):155–160
59. Sampat AM, Ruiz-Mercado GJ, Zavala VM (2018) Economic and environmental analysis for advancing sustainable management of livestock waste: A Wisconsin Case Study. ACS sustainable chemistry & engineering, 6(5), 6018–6031.https://doi.org/10.1021/acssuschemeng.7b04657
60. Tao Z, Liu G (2024) Research on current situation of farmland runoff pollution and ecological treatment technology. In E3S web of conferences 520, p 01021. EDP Sciences. https://doi.org/10.1051/e3sconf/202452001021
61. Awad A, Luo W, Al-Ansari N, Elbeltagi A, El-Rawy M, Farres HN, Gabr MES (2021) Farmers' awareness in the context of climate change: An underutilized way for ensuring sustainable farmland adaptation and surface water quality. Sustain, 13(21), 11802.https://doi.org/10.3390/su132111802
62. Awad A, El-Rawy M, Abdalhi M, Al-Ansari N (2022) Evaluation of the DRAINMOD model's performance using different time steps in evapotranspiration computations. Hydrology, 9(2), 40.https://doi.org/10.3390/hydrology9020040

63. Sa'diyah SN, Kholik K, Munawaroh M, Aprianti ATD, Rahmawati SE, Riwu KHP (2023) Antibiotic Resistance in Escherichia ColiBacteria Isolated from Water Sources and Waste Disposal in Livestock Farms in East Lombok. Jurnal Sain Peternakan Indonesia 18(4), 235. https://doi.org/10.31186/jspi.id.18.4.235-241
64. Wardani I (2018) Developing a characterization factor framework for microbial contamination. In: E3s web of conferences, 74, 11002. https://doi.org/10.1051/e3sconf/201874 11002
65. Bhattacharjee M, Mehta S, Singh P, Singh M, Pongener N (2023) Diversification of livelihood from animal waste and by-product for small and marginal farmers in India. Asian J Dairy Food Res, (4), 564–569.https://doi.org/10.18805/ajdfr.b-5042
66. Pericherla S, Karnena MK, Saritha V (2022) Controlling technologies adopted for the minimization of non-point source pollution from the agriculture runoff. Pollution Research, 210–217. https://doi.org/10.53550/pr.2022.v41i01.030
67. Newton RJ, McClary JS (2019) The flux and impact of wastewater infrastructure microorganisms on human and ecosystem health. Curr Opin Biotechnol 57:145–150. https://doi.org/10.1016/j.copbio.2019.03.015
68. Kosoe EA, Osumanu IK, Ogwu MC (2025b) Reimagining rural and small-town water supply in Ghana through inclusive governance and sustainable models. J Sustain, 1(2), 1–21. https://doi.org/10.55845/jos-2025-1234
69. Kosoe EA, Ogwu MC, Duniah J, Ndigun FB (2025a) Sustaining indigenous water governance systems in ayimaye and lungni, Northern Ghana, through Hybrid Models. J Sustain, 1(2). https://doi.org/10.55845/jos-2025-1235
70. Atauzzaman M, Ali MA (2022) Effects of combined sewer overflow on water quality: a case study of Hatirjheel Lake in Dhaka. SN Appl Sci, 4(11), 303.https://doi.org/10.1007/s42452-022-05187-6
71. Gasperi J, Zgheib S, Cladière M, Rocher V, Moilleron R, Chebbo G (2012) Priority pollutants in urban stormwater: Part 2–Case of combined sewers. Water Res 46(20):6693–6703. https://doi.org/10.1016/j.watres.2011.09.041
72. Pereira Souza F, Leite Costa ME, Koide S (2019) Hydrological modelling and evaluation of detention ponds to improve urban drainage system and water quality. Water 11(8):1547. https://doi.org/10.3390/w11081547
73. Lee S, Lee J, Kim M (2012) The influence of storm-water sewer overflows on stream water quality and source tracking of fecal contamination. KSCE J Civ Eng 16(1):39–44. https://doi.org/10.1007/s12205-012-1198-0
74. Yang L, Zhou H, Dai L (2021). nvestigation and Water quality Analysis of Agricultural nonpoint Source pollution in Tanxia Town. In E3S web of conferences. 329, p 01027) EDP Sciences. https://doi.org/10.1051/e3sconf/202132901027
75. Phillips PJ, Chalmers AT, Gray JL, Kolpin DW, Foreman WT, Wall GR (2012) Combined sewer overflows: an environmental source of hormones and wastewater micropollutants. Environ Sci Technol 46(10):5336–5343. https://doi.org/10.1021/es3001294
76. Liu T, Su X, Prigiobbe V (2018) Groundwater-sewer interaction in urban coastal areas. Water 10(12):1774. https://doi.org/10.3390/w10121774
77. Mutzner L, Bohren C, Mangold S, Bloem S, Ort C (2020) Spatial differences among micropollutants in sewer overflows: a multisite analysis using passive samplers. Environ Sci Technol 54(11):6584–6593. https://doi.org/10.1021/acs.est.9b05148
78. Wang J, Guo Y (2018) An analytical stochastic approach for evaluating the performance of combined sewer overflow tanks. Water Resour Res 54(5):3357–3375. https://doi.org/10.1029/2017wr022286
79. Kapoor V, Pitkänen T, Ryu H, Elk M, Wendell D, Santo Domingo JW (2015) Distribution of human-specific bacteroidales and fecal indicator bacteria in an urban watershed impacted by sewage pollution, determined using RNA-and DNA-based quantitative PCR assays. Applied and environmental microbiology, 81(1), 91–99.https://doi.org/10.1128/aem.02446-14
80. Morgan D, Xiao L, McNabola A (2017) Evaluation of combined sewer overflow assessment methods: case study of Cork City. Ireland. Water Environ J 31(2):202–208. https://doi.org/10.1111/wej.12239

81. Ding Z, Hu Z, Sun L, Wu C, Luo K (2022) Systematic treatment of water ecological environment in plain river network city: a case study of Jiangyin. In proceedings of the 3rd international conference on green energy, environment and sustainable development (GEESD2022) (pp. 1113–1125). https://doi.org/10.3233/atde220390
82. Romei M, Lucertini M, Esposito Renzoni E, Baldrighi E, Grilli F, Manini E, Marini M, Iagnemma L (2021) A detention reservoir reduced combined sewer overflows and bathing water contamination due to intense rainfall. Water, 13(23), p.3425.https://doi.org/10.3390/w13233425
83. Andrés-Doménech I, Montanari A, Marco JB (2012) Efficiency of storm detention tanks for urban drainage systems under climate variability. J Water Resour Plan Manag 138(1):36–46
84. Boënne W, Desmet N, Van Looy S, Seuntjens P (2014) Use of online water quality monitoring for assessing the effects of WWTP overflows in rivers. Environ Sci Process Impacts 16(6):1510–1518
85. Broadhead AT, Horn R, Lerner DN (2015) Finding lost streams and springs captured in combined sewers: a multiple lines of evidence approach. Water Environ J, 29(2), 288–297.https://doi.org/10.1111/wej.12104
86. Sola KJ, Bjerkholt JT, Lindholm OG, Ratnaweera H (2021) What effect does rehabilitation of wastewater pipelines have on the share of infiltration and inflow water (I/I-water)? Water 13(14):1934. https://doi.org/10.3390/w13141934
87. Li H, Li Y, Lee MK, Liu Z, Miao C (2015) Spatiotemporal analysis of heavy metal water pollution in transitional China. Sustainability 7(7):9067–9087
88. Salehi F, Kussainova M (2023) Assessment of the water quality state in the downstream portion of Syr-Darya using the water quality index arithmetic method. Eng J Satbayev Univ, 145(6), 30–34.. https://doi.org/10.51301/ejsu.2023.i6.05
89. Zowail ME, Elezaby MM, Abdel-kareim AM, Yousef GA, Mahmoud AA (2010) Cytogenetic effect of industrial and agricultural wastes on Tilapia zilli fish. Egypt Acad J Biol Sci, B Zool, 2(1), 9–16. https://doi.org/10.21608/eajbsz.2010.15903
90. Ogwu MC, Izah SC, Sawyer WE, Amabie T (2025) Environmental risk assessment of trace metal pollution: a statistical perspective. Environ Geochem Health 47, 94 (2025). https://doi.org/10.1007/s10653-025-02405-z
91. Jing S (2024) Quantitative methodology for industrial activities and water resources: an analysis on healthcare, materials, and consumer discretionary sectors and implications for chula vista. Adv Econ Manag Polit Sci, 129(1), 210–219. https://doi.org/10.54254/2754-1169/2024.18370
92. Xu H, Tan X, Liang J, Cui Y, Gao Q (2022) Impact of agricultural non-point source pollution on river water quality: evidence from China. Front Ecol Evol 10:858822. https://doi.org/10.3389/fevo.2022.858822
93. Ogwu MC, Kosoe EA (2024) Place of Cultural Diversity in Sustainable Water Resource Management in Ghana. In: Izah SC, Ogwu MC, Loukas A, Hamidifar H (eds) Water crises and sustainable management in the global south. Springer, Singapore. https://doi.org/10.1007/978-981-97-4966-9_14
94. Ogwu MC, Kosoe EA (2025) Integrating green infrastructure into sustainable agriculture to enhance soil health, biodiversity, and microclimate resilience. Sustain 17(9):3838. https://doi.org/10.3390/su17093838
95. Ma Y, Chen E (2024) Effects of agricultural insurance on water pollution in china. Journal of Global Economy Business and Finance, 6(10), 12–19. https://doi.org/10.53469/jgebf.2024.06(10).04
96. Madjar RM, Vasile Scăețeanu G, Sandu MA (2024) Nutrient water pollution from unsustainable patterns of agricultural systems, effects and measures of integrated farming. Water 16(21):3146
97. Shortle JS, Ribaudo M, Horan RD, Blandford D (2012) Reforming agricultural nonpoint pollution policy in an increasingly budget-constrained environment. Environ Sci Technol 46(3):1316–1325. https://doi.org/10.1021/es2020499

98. Parris K (2011) Impact of agriculture on water pollution in oecd countries: recent trends and future prospects. Int J Water Resour Dev 27(1):33–52. https://doi.org/10.1080/07900627.2010.531898
99. Stephenson K, Shabman L, Shortle J, Easton Z (2022) Confronting our agricultural nonpoint source control policy problem. JAWRA J Am Water Resour Assoc, 58(4), 496–501.https://doi.org/10.1111/1752-1688.13010
100. Jinadasa KBSN, Wijewardena SKI, Zhang DQ, Gersberg RM, Kalpage CS, Tan SK, Wang JY, Ng WJ (2012) Socio-environmental impact of water pollution on the mid-canal (Meda Ela), Sri Lanka. J Water Resour Prot, 4(7), pp.451–459.https://doi.org/10.4236/jwarp.2012.47052
101. Stoyanova Z, Harizanova H (2019) Impact of agriculture on water pollution. AGROFOR Int J, 4(1), 111–118.https://doi.org/10.7251/agreng1901111s
102. Abeygunawardane AWGN, Dayawansa NDK, Pathmarajha S (2011) Socioeconomic implications of water pollution in an urban environment a case study in Meda Ela catchment, Kandy, Sri Lanka. Trop Agric Res, 22(4). https://doi.org/10.4038/tar.v22i4.3788
103. Farzin YH, Grogan KA (2013) Socioeconomic factors and water quality in California. Environ Econ Policy Stud 15(1):1–37.https://doi.org/10.2139/ssrn.1890587
104. Henegama HP, Dayawansa NDK, De Silva S (2015) An assessment of social and environmental implications of agricultural water pollution in Nuwara Eliya. Trop Agric Res, 24(4). https://doi.org/10.4038/tar.v24i4.8016
105. Abayomi O, Olayemi TE, Ogungbade T (2021) Environmental pollution and its ecological consequences on the Niger Delta: A review of the literature. Afr J Environ Nat Sci Res, 4, 27–42. https://doi.org/10.52589/ajensr-bjggacsv
106. Imam AS, Ojochenemi I (2024) The effects of hotels effluent discharge on groundwater quality in Lokoja, Kogi State, Nigeria. J Geogr Environ Earth Sci Int 28(7). https://doi.org/10.9734/jgeesi/2024/v28i7791
107. Aga TA, Anyadike CC, Mbajiorgu CC, Ogwo V (2020) Assessment of the effect of cement industry effluent discharge on water quality of Ngo River in Benue. Nigeria. Niger J Technol 39(3):918–924. https://doi.org/10.4314/njt.v39i3.34
108. Okechukwu N (2024) (nesrea) and the challenges of environmental regulation in nigeria. Br J Mass Commun Media Res, 4(1), 1–11. https://doi.org/10.52589/bjmcmr-fljqlr8s
109. Lawal JA, Sanni I, Jimoh I (2025) Physicochemical analysis of stormwater runoff from a cement manufacturing plant in Nigeria. J Chem Soc Niger, 50(2), 303–309.https://doi.org/10.46602/jcsn.v50i2.1055
110. Ogwu MC (2025) Sustainable development in africa: challenges, innovations, and policy pathways for the future (2025) Afr Res Rep, 1(2), 55–69. https://doi.org/10.5897/ARR2025.0005
111. Ma Y, Cao H, Ma Y, Wu S (2020) Does technological innovation reduce water pollution intensity in the context of informal environmental regulation? Asia-Pac J Chem Eng 15:e2493
112. Suleiman HA, Hanafiah MM (2021) Spatiotemporal assessment of effects of leachate from landfills on groundwater in Kano Metropolis, Nigeria. In: IOP conference series: earth and environmental science 880(1), p. 012051). IOP Publishing. https://doi.org/10.1088/1755-1315/880/1/012051
113. Ojeih CA, Ogidan O, Oluwajobi YF, Adebayo BO (2024) Biotechnology regulatory conundrum: balancing innovation and oversight. J Sustain Dev Law Policy (The), 15(3), 120–144.https://doi.org/10.4314/jsdlp.v15i3.5
114. Katusiime J, Schütt B (2020) Integrated water resources management approaches to improve water resources governance. Water, 12(12), 3424. Katusiime, J, Schütt, B
115. Ignatieva M (2021) Evolution of the approaches to planting design of parks and gardens as main greenspaces of green infrastructure. In Urban services to ecosystems: Green infrastructure benefits from the landscape to the urban scale (pp. 435–452). Cham: Springer International Publishing.https://doi.org/10.1007/978-3-030-75929-2_23
116. Ferreyra C, De Loe RC, Kreutzwiser RD (2008) Imagined communities, contested watersheds: Challenges to integrated water resources management in agricultural areas. J Rural Stud 24(3):304–321

117. Mishra BK, Kumar P, Saraswat C, Chakraborty S, Gautam A (2021) Water security in a changing environment: Concept, challenges and solutions. Water, 13(4), 490.https://doi.org/10.3390/w13040490
118. Esposito K, Tsuchihashi R, Anderson J, Selstrom J (2005) The role of water reclamation in water resources management in the 21st century. Proc Water Environ Fed 2005(6):8621–8633. https://doi.org/10.2175/193864705783812792
119. Demuzere M, Orru K, Heidrich O, Olazabal E, Geneletti D, Orru H, Bhave AG, Mittal N, Feliú E, Faehnle M (2014) Mitigating and adapting to climate change: Multi-functional and multi-scale assessment of green urban infrastructure. J Environ Manag, 146, pp.107–115.https://doi.org/10.1016/j.jenvman.2014.07.025
120. Pugh TA, MacKenzie AR, Whyatt JD, Hewitt CN (2012) Effectiveness of green infrastructure for improvement of air quality in urban street canyons. Environ Sci Technol 46(14):7692–7699. https://doi.org/10.1021/es300826w
121. Gelan E, Girma Y (2021) Sustainable urban green infrastructure development and management system in rapidly urbanized cities of Ethiopia. Technol, 9(3), 66.https://doi.org/10.3390/technologies9030066
122. Jiménez A, Saikia P, Giné R, Avello P, Leten J, Liss Lymer B, Schneider K, Ward R (2020) Unpacking water governance: A framework for practitioners. Water, 12(3), p.827.https://doi.org/10.3390/w12030827
123. Hsu KW, Chao JC (2021) Study on the value model of urban green infrastructure development—A case study of the central district of Taichung city. Sustainability, 13(13), 7402. https://doi.org/10.3390/su13137402, https://doi.org/10.1007/s11069-010-9529-2, https://doi.org/10.1016/j.jrurstud.2007.11.001, https://doi.org/10.1051/e3sconf/20198507013, https://doi.org/10.1061/(asce)wr.1943-5452.0000144, https://doi.org/10.1080/00401706.2019.1623076, https://doi.org/10.1088/1755-1315/1087/1/012003, https://doi.org/10.1111/jiec.13149, https://doi.org/10.3390/su7079067
124. Zhao M, Boll J (2022) Adaptation of water resources management under climate change. Front Water, 4. https://doi.org/10.3389/frwa.2022.983228
125. Geng Y, Yi J (2006) Integrated water resource management at the industrial park level: A case of the Tianjin Economic Development Area. Int J Sustain Dev World Ecol, 13(1), 37–50.https://doi.org/10.1080/13504500609469660
126. Grison C, Koop S, Eisenreich S, Hofman J, Chang IS, Wu J, Savic D, Van Leeuwen K (2023) Integrated water resources management in cities in the world: global challenges. Water Resour Manage 37(6):2787–2803. https://doi.org/10.1007/s11269-023-03475-3

[illegible]

Chapter 4
Waterborne Diseases and Public Health: Risks and Interventions

Kingsley Excel Dunga

Abstract Waterborne diseases remain a pressing global health issue, accounting for approximately 485,000 deaths annually from diarrheal illnesses and affecting over 2 billion people who lack access to safe drinking water. These figures highlight the critical relationship between water quality, sanitation, and public health outcomes, especially in low-resource settings. This chapter focuses on identifying long-term, sustainable interventions for preventing and managing waterborne diseases through a multifaceted, integrated public health approach. The findings emphasize that effective disease control requires combining infrastructure development with household-level water treatment, improved sanitation, and widespread hygiene education. Advanced surveillance techniques, including the use of Geographic Information Systems, play a vital role in mapping high-risk areas and informing targeted interventions. Moreover, the chapter explores how environmental changes, seasonal variability, and the rise of antibiotic-resistant pathogens complicate waterborne disease dynamics, necessitating adaptive strategies. It stresses the importance of international collaboration, policy alignment, and stakeholder engagement in supporting resilient water systems. Community-centered initiatives, culturally informed health campaigns, and ongoing monitoring contribute to long-term success. The chapter also presents case studies that demonstrate the effectiveness of integrated approaches, particularly when backed by robust governance and cross-sectoral cooperation. Therefore, safeguarding public health from waterborne diseases demands a comprehensive framework that bridges infrastructure, behavior change, innovation, and policy.

Keywords Waterborne diseases · Public health · Sanitation · Safe drinking water · GIS surveillance · Climate change · Disease prevention · Health policy · Community engagement

K. E. Dunga (✉)
Department of Medical Laboratory Science, Kingsley Ozumba University, Ogboko, Ideato, Imo State, Nigeria
e-mail: kingsleydunga1@gmail.com

M. C. Ogwu and S. Chibueze Izah (eds.), *Water Quality and Safety in the Global South*, Springer Water, https://doi.org/10.1007/978-3-032-10602-5_4

4.1 Introduction

Water quality and pollution remain critical concerns for public health, particularly in regions like the Niger Delta, where anthropogenic activities significantly influence surface and groundwater conditions [1–3]. Studies have consistently shown elevated levels of heavy metals and microbial contaminants in local water bodies, such as the Nun River and Taylor Creek, largely due to industrial discharge, agricultural runoff, and poor waste management [4–6]. These pollutants pose significant health risks, including gastrointestinal diseases, neurological disorders, and long-term carcinogenic effects. Research by [7] and Seiyaboh et al. [8, 9] stresses the alarming bacteriological loads in drinking water sources, often exceeding WHO limits, thereby compromising water safety and increasing the vulnerability of local populations to waterborne diseases. Climate change and urban expansion further exacerbate these challenges by altering water availability and quality [5, 10]. The health burden is magnified in rural areas where infrastructural deficiencies persist, and access to safe water remains limited [11, 12]. Integrated strategies involving water resource management, decontamination technologies, and environmental monitoring are therefore essential for mitigating health risks [13, 14]. Additionally, adopting culturally relevant water governance and community-based surveillance mechanisms can enhance early detection and promote public health resilience [12, 15–17].

Waterborne diseases represent a significant public health challenge worldwide, often resulting from the consumption of contaminated water that harbors pathogens. According to the World Health Organization, approximately 1.7 billion individuals utilize drinking water sources afflicted by fecal contamination, putting them at risk for ailments associated with gastrointestinal infections, including diarrheal disease, which in 2016 caused an estimated 1.6 million deaths, particularly impacting vulnerable populations such as children under five years old [18]. Understanding the definition and scope of waterborne diseases is essential: these diseases generally encompass infectious illnesses transmitted through water, primarily due to contamination by fecal matter, viruses, bacteria, and protozoa [19].

The significance of water safety in public health cannot be overstated, as clean drinking water is vital for reducing the incidence of waterborne diseases. Practices intended to ensure water safety not only protect individuals but also reduce the overall burden on public health systems. For instance, estimates suggest that annual cases of waterborne illnesses in the United States range from 12 million to 19.5 million [20], while globally, diseases linked to inadequate water, sanitation, and hygiene (WASH) systems remain key public health priorities [21]. The challenges linked to WASH, including population growth and climate change, exacerbate these issues, especially in low- and middle-income countries, where water quality and access remain insufficient [21].

Global statistics illustrate the burden of waterborne illnesses. In emergency situations, such as extreme weather events, outbreaks of diseases like cholera and leptospirosis can significantly escalate [22]. The burden varies significantly by region, with countries in Sub-Saharan Africa experiencing high prevalence rates of waterborne diseases, often driven by inadequate infrastructure and environmental factors [23, 24]. Therefore, ongoing assessments and interventions are paramount in addressing these public health concerns. Regular monitoring of water quality and implementing community-based interventions can markedly reduce the incidence of waterborne diseases, as evidenced by studies conducted in various global settings [25].

This chapter explores the significant public health challenges posed by waterborne diseases, particularly in low-resource and vulnerable settings. It examines the types, transmission routes, and health impacts of these diseases, while highlighting the socio-environmental and infrastructural risk factors [26, 27]. The chapter emphasizes the importance of public health interventions such as WASH programs, surveillance systems, and innovative technologies in mitigating outbreaks. Through case studies and future-oriented strategies, it presents a roadmap for sustainable waterborne disease prevention and control.

4.2 Types of Waterborne Diseases

Waterborne diseases are a significant public health concern globally, caused primarily by pathogenic microorganisms transmitted through contaminated water. These diseases can be classified into three major categories: bacterial, viral, and parasitic infections (Fig. 4.1), each manifesting unique symptoms and requiring different preventive measures.

Fig. 4.1 Major water borne diseases

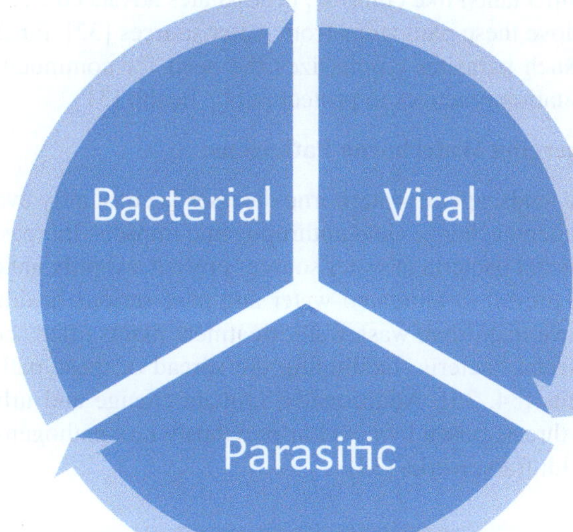

Bacterial Infections:

Bacterial pathogens such as *Vibrio cholerae* (which causes cholera), *Salmonella typhi* (responsible for typhoid fever), and *Shigella spp.* (linked to dysentery) represent a considerable fraction of waterborne diseases. Cholera, in particular, is a severe diarrheal disease that can lead to death from dehydration if untreated. Studies have documented the emergence of antibiotic-resistant strains, complicating treatment and control efforts in affected areas, particularly in regions with poor sanitation [28, 29]. The increased resistance to common antibiotics, such as ampicillin, has been reported in *Vibrio* spp. isolated from various environments, including sea turtles, indicative of how natural ecosystems can harbor resistant bacteria [29, 30].

Viral Infections:

Viral waterborne diseases include hepatitis A, norovirus, and rotavirus infections. These typically manifest as gastrointestinal symptoms, with rotavirus being a leading cause of severe diarrhea in children [31]. The transmission of hepatitis A, particularly in regions experiencing poor sanitary conditions, underscores the necessity for improved water quality management and public health interventions [31]. The role of water sources in facilitating the spread of these viruses has been documented, highlighting the importance of rigorous water treatment processes to mitigate the risks [30, 31].

Parasitic Infections:

Parasitic pathogens such as *Giardia lamblia*, *Cryptosporidium spp.*, and *Entamoeba histolytica* also contribute to the burden of waterborne diseases. *Cryptosporidium*, in particular, has emerged as a major threat due to its significant prevalence in recreational and drinking water supplies [32]. Its resilience against conventional water disinfectants, like chlorine, necessitates advanced treatment strategies to effectively remove these pathogens from water sources [32]. Furthermore, the rising incidence of such parasites emphasizes the need for continuous surveillance and improved sanitation practices to protect public health [31].

Emerging Waterborne Pathogens:

The landscape of waterborne diseases is constantly evolving, primarily due to environmental changes and anthropogenic impacts. Increasing populations of antibiotic-resistant bacteria in water sources present a significant challenge, as these pathogens can prevail in untreated water and pose serious health risks [33, 34]. Studies have shown that urban wastewater treatment plants often serve as hotspots for antibiotic-resistant bacteria, facilitating the spread of these pathogens into broader environments [34, 35]. Additionally, climate change and urbanization further exacerbate the threats posed by novel or previously rare pathogens, challenging existing public health frameworks [33].

Classification by Transmission Routes:

The primary mode of transmission for most waterborne diseases remains the fecal–oral route, which is particularly prevalent in regions suffering from inadequate sanitation infrastructure [36]. Diseases such as cholera and typhoid fever commonly spread this way, stressing the need for improved sanitation and clean drinking water distributions in vulnerable populations [31, 33, 37, 38]. In contrast, infections such as Legionnaires' disease result from inhalation of aerosolized water containing the bacteria *Legionella*, representing a different route of transmission that can also be linked to water quality [33]. Dermal contact or mucous membrane exposure is another transmission pathway observed with parasitic infections such as schistosomiasis, further complicating the public health response to waterborne diseases [36].

Other Considerations

Vulnerable populations, including children, the elderly, and immunocompromised individuals, are particularly at risk of severe outcomes from these infections [33, 39]. Moreover, geographical and seasonal variations influence the prevalence and types of waterborne diseases, necessitating localized public health strategies and commitments [35]. The challenges in diagnosing these diseases further exacerbate the situation, as symptoms can overlap significantly across different infections, complicating timely and accurate treatment [33, 40].

4.3 Transmission Pathways of Waterborne Diseases

Waterborne diseases are a significant public health concern globally, primarily transmitted through contaminated water sources. Understanding the transmission pathways of these diseases is crucial to implementing effective control measures. The main transmission pathways can be categorized into several categories, including contaminated drinking water, poor sanitation practices, unsafe food handling, recreational water use, and the influence of climate change (Table 4.1).

Contaminated Drinking Water and Poor Sanitation

One of the predominant pathways for the transmission of waterborne diseases is the consumption of drinking water contaminated with fecal matter. Pathogens such as *E. coli*, *Vibrio cholerae*, and *Giardia* can be present in water sources due to inadequate treatment and sanitation systems. Contaminated drinking water is particularly an issue in informal settlements or rural areas where access to safe water is limited, and infrastructural deficiencies lead to cross-contamination between potable and wastewater lines [22, 41]. Additionally, practices such as open defecation exacerbate environmental contamination and increase the risk of waterborne diseases [22, 42]. Recent studies have indicated that extreme weather events, including heavy rainfall, can overwhelm sanitation systems, leading to increased pathogens in water sources

Table 4.1 Transmission pathways of waterborne diseases

Transmission pathway	Mechanism of disease spread	Public health implications
Contaminated drinking water & poor sanitation	Ingestion of water contaminated with fecal pathogens- Poor infrastructure and lack of safe water access	Outbreaks of cholera, typhoid, and diarrhea- Increased child morbidity and mortality
Unsafe food handling & cross-contamination	Use of unsafe water in food prep- Improper hygiene in food handling	Foodborne illness outbreaks- Burden on food safety systems and regulatory bodies
Wastewater exposure & recreational water use	Contact with polluted rivers/ lakes during swimming or bathing- Use of wastewater for irrigation	Spread of Hepatitis A, schistosomiasis, and skin infections- Need for water quality monitoring
Climate change, flooding & disasters	Floodwaters spreading fecal contaminants- Disruption of sanitation systems	Surge in waterborne infections post-disaster- Urgent need for disaster-resilient systems
Urbanization & overpopulation	Overburdened sanitation systems- Informal settlements lacking infrastructure	Increased transmission in densely populated areas- Demand for integrated urban planning
Inadequate waste disposal practices	Open dumping of waste near water sources- Lack of wastewater treatment	Long-term contamination of drinking water sources- Calls for improved waste management policies

[22, 43]. This highlights the need for improved water infrastructure and sanitation facilities as essential components in reducing disease transmission.

Unsafe Food Handling and Cross-Contamination

Another critical pathway for transmitting waterborne diseases involves the improper use of contaminated water in food preparation processes, particularly in settings with poor hygiene practices. Utilizing unsafe water for washing fruits and vegetables or preparing food can lead to pathogen transmission, especially in street food environments common in many regions [44]. Food vendors often lack adequate methods for maintaining hygiene, resulting in cross-contamination from unwashed hands, utensils, or contaminated surfaces [45, 46]. Studies have shown a correlation between the consumption of food prepared with contaminated water and the incidence rates of gastrointestinal infections, indicating significant public health implications that warrant attention [47].

Wastewater Exposure and Recreational Water Use

Recreational water practices also contribute significantly to the spread of waterborne diseases. Direct exposure to contaminated surface water through activities such as swimming and bathing poses risks, notably for children who often engage in play in polluted aquatic environments [48, 49]. The presence of pathogens like norovirus and adenoviruses in recreational waters has been documented, causing a range of illnesses from gastrointestinal infections to respiratory diseases [50, 51]. Furthermore, the calibration of environmental monitoring efforts that assess the quality of recreational waters for fecal indicator bacteria can help predict risks associated with recreational exposure [48, 52]. Unregulated discharge of untreated wastewater into lakes and rivers has been repeatedly shown to contribute to the high microbial load in these environments, emphasizing the need for public health policies that address these risks [22, 53].

4.4 Influence of Climate Change, Flooding, and Disasters

Climate change has emerged as a significant factor affecting the transmission dynamics of waterborne diseases. Increased rainfall and flooding can disrupt sanitation infrastructure and lead to extensive water contamination [22, 43]. Moreover, shifts in temperature can promote the survival and growth of pathogens like *Cryptosporidium* and *Legionella*, further complicating disease management efforts [54]. The relation between extreme precipitation events and increased waterborne disease outbreaks has been demonstrated, showing that climate variability acts as a risk multiplier for vulnerable populations already lacking adequate sanitation and clean water [54, 55]. As the impact of climate change exacerbates existing vulnerabilities, it is essential for health authorities to implement adaptive management strategies to mitigate the risks of waterborne diseases.

4.5 Epidemiological Trends in Waterborne Diseases

Waterborne diseases are a significant public health challenge, contributing to an estimated 2 billion cases of diarrheal diseases annually, particularly in developing regions such as Sub-Saharan Africa and South Asia. The epidemiological trends indicate a stark increase in incidence rates in these areas, largely due to deficient water infrastructure and sanitation facilities [56]. Communities experiencing inadequate sanitation often exhibit high prevalence rates of pathogens that lead to diseases such as cholera and dysentery [19]. In contrast, developed nations may witness sporadic outbreaks linked to localized contamination of recreational or municipal water sources, reflecting variability in exposure and control measures across different regions [57]. Emerging economies face unique challenges, juggling the legacy of historical water issues with pressures from urbanization, this dual burden exacerbates the critical necessity for robust water quality management strategies [58, 59].

High-risk populations are particularly vulnerable to the adverse effects of waterborne diseases. Children under the age of five represent the most affected demographic, as diarrheal diseases remain among the leading causes of mortality within this age group, driven by inadequate access to clean water and sanitation resources

[60, 61]. Other vulnerable groups include pregnant women, elderly individuals, and those with compromised immune systems, such as HIV/AIDS patients, who are more susceptible to severe complications arising from infections [62]. Furthermore, marginalized populations living in refugee camps, slums, or areas affected by disasters face heightened risks due to overcrowding and inadequate sanitation infrastructure. Socioeconomic status closely correlates with disease exposure; those in low-income brackets often struggle to obtain safe drinking water, perpetuating the cycle of poverty and disease [63].

Seasonal and geographic factors intricately influence the dynamics of waterborne disease outbreaks. Research indicates that outbreaks typically surge during rainy seasons, as flooding can contaminate water supplies with sewage and waste [64]. In tropical climates, warmer temperatures can enhance microbial growth rates, thus increasing the likelihood of outbreaks [65]. Geographic features such as river deltas and floodplains are more susceptible to waterborne disease transmission due to their unique hydrological and ecological characteristics. With climate change altering precipitation patterns and increasing the frequency of extreme weather events, the seasonal dynamics of waterborne diseases are becoming increasingly unpredictable, threatening regions previously considered low-risk [66, 67].

4.5.1 Health Impacts of Waterborne Diseases

Waterborne diseases are a significant public health challenge, contributing to an estimated 2 billion cases of diarrheal diseases annually, particularly in developing regions such as Sub-Saharan Africa and South Asia (Table 4.2). The epidemiological trends indicate a stark increase in incidence rates in these areas, largely due to deficient water infrastructure and sanitation facilities [56]. Communities experiencing inadequate sanitation often exhibit high prevalence rates of pathogens that lead to diseases such as cholera and dysentery [19]. In contrast, developed nations may witness sporadic outbreaks linked to localized contamination of recreational or municipal water sources, reflecting variability in exposure and control measures across different regions [57, 68]. Emerging economies face unique challenges, juggling the legacy of historical water issues with pressures from urbanization, this dual burden exacerbates the critical necessity for robust water quality management strategies [58, 59].

High-risk populations are particularly vulnerable to the adverse effects of waterborne diseases. Children under the age of five represent the most affected demographic, as diarrheal diseases remain among the leading causes of mortality within this age group, driven by inadequate access to clean water and sanitation resources [60, 61]. Other vulnerable groups include pregnant women, elderly individuals, and those with compromised immune systems, such as HIV/AIDS patients, who are more susceptible to severe complications arising from infections [62]. Furthermore, marginalized populations living in refugee camps, slums, or areas affected by disasters face heightened risks due to overcrowding and inadequate sanitation infrastructure. Socioeconomic status closely correlates with disease exposure; those in

Table 4.2 Health impacts of waterborne diseases

Health impact area	Details of health effects	Public health focus
Short- and long-term health consequences	Acute symptoms: diarrhea, vomiting, dehydration, fever- Chronic issues: malnutrition, stunted growth, liver damage, post-infectious syndromes	Early detection and treatment- Public health campaigns on hygiene and sanitation
Impact on children	Leading cause of death in under-five children in low-resource settings- Repeated infections impair growth and development	Child-focused sanitation programs- Nutrition support and immunization efforts
Impact on elderly	Increased severity due to age-related immunity decline- Greater risk of complications and hospitalization	Targeted education and preventive care for elderly populations
Impact on immunocompromised individuals	Higher susceptibility and prolonged illness- Severe complications from common waterborne pathogens	Tailored intervention strategies- Water safety education and access to treatment
Impact on pregnant women	Risk of miscarriage, preterm birth, and congenital infections	Prenatal health programs- Access to clean water for maternal care
Socioeconomic burden	Lost productivity, school/work absenteeism- Financial burden on families for treatment	Investment in water infrastructure and preventive healthcare
Healthcare costs	Increased demand for medical services and hospitalization- Strain on public healthcare systems in outbreak situations	Strengthening healthcare systems for outbreak response
Public health resource allocation	Government spending on water treatment, sanitation, and outbreak control- Need for long-term public health infrastructure investment	Prioritizing water safety in public health policy and budgeting
Community-level effects	Disruption of daily life, especially in vulnerable and remote communities- Stigma and social exclusion during outbreaks	Community engagement and localized public health interventions

low-income brackets often struggle to obtain safe drinking water, perpetuating the cycle of poverty and disease [63].

Seasonal and geographic factors intricately influence the dynamics of waterborne disease outbreaks. Research indicates that outbreaks typically surge during rainy seasons, as flooding can contaminate water supplies with sewage and waste [64]. In tropical climates, warmer temperatures can enhance microbial growth rates, thus increasing the likelihood of outbreaks [65]. Geographic features such as river deltas and floodplains are more susceptible to waterborne disease transmission due to their unique hydrological and ecological characteristics. With climate change altering precipitation patterns and increasing the frequency of extreme weather events, the seasonal dynamics of waterborne diseases are becoming increasingly unpredictable, threatening regions previously considered low-risk [66, 67].

4.5.2 Risk Factors for Waterborne Diseases

Waterborne diseases remain a critical public health issue globally, exacerbated by a myriad of risk factors including inadequate infrastructure, poverty, conflict, displacement, and various behavioral and cultural considerations (Fig. 4.2). The complex interplay between these factors significantly influences the transmission and impact of waterborne pathogens.

Inadequate infrastructure, especially in developing regions, correlates with a heightened incidence of waterborne diseases. Poorly maintained water supply systems and sanitation facilities facilitate the spread of pathogens. Research indicates that flooding, a consequence of climate change and infrastructure failure, drastically increases the risk of infectious diseases as floodwaters contaminate drinking water

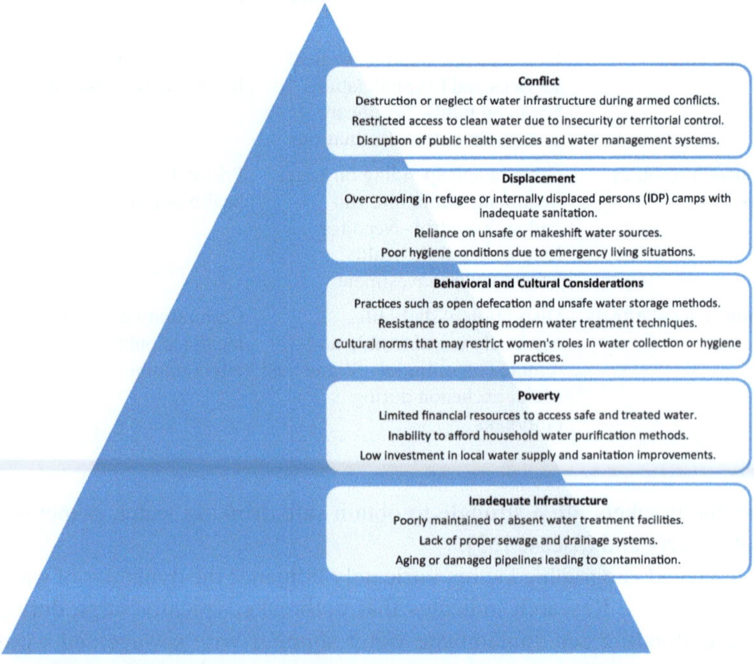

Fig. 4.2 Interplay of factors leading to poor water quality and increased incidence of waterborne diseases

supplies with pathogens such as bacteria, viruses, and parasites [69]. This contamination is particularly troubling in regions where heavy rains overwhelm drainage systems, leading to fecal–oral disease outbreaks such as cholera and typhoid fever [70]. As depicted in findings from Nigeria and other African nations, this risk is compounded where there is persistent reliance on surface water sources that are frequently contaminated during such events [60].

Further exacerbating the risk landscape is the reality of poverty and limited access to clean water, which directly contributes to the prevalence of waterborne diseases. The World Health Organization estimates that approximately 2 billion people lack access to safely managed drinking water, a situation that heightens the vulnerability of these populations to diseases [71]. Countries grappling with high poverty rates often struggle to implement effective public health interventions, leaving communities without the necessary resources for clean water and sanitation, which results in disease outbreaks [72]. For example, studies in sub-Saharan Africa have demonstrated a strong link between poor water quality and chronic illnesses, showing that populations without reliable access to clean water face severe health risks [73].

Conflict and displacement also pose significant risks, as seen in areas experiencing humanitarian crises. Displaced populations often reside in overcrowded conditions where sanitation and access to clean water are severely compromised. One study highlights that mass displacement into camps leads to increased dependency on precarious external aid, further facilitating the spread of waterborne diseases [74]. Complex emergencies often see a surge in outbreaks due to compromised water and sanitation infrastructure, highlighting the urgent need for targeted interventions in these crisis settings [75].

Also, behavioral and cultural factors, such as local beliefs regarding water treatment and hygiene practices, affect the adoption of effective interventions. Evidence from outbreaks in Malawi shows that despite the provision of resources for water treatment and hygiene, skepticism about transmission mechanisms often hampers community engagement in health-promoting behaviors [76, 77]. Cultural practices surrounding water use can lead to transmission risks, for example, inadequate handwashing before handling food or water remains a significant barrier to controlling the spread of pathogens [62]. The challenge is not only to supply clean water but also to motivate behavioral change through education and community efforts tailored to the cultural context.

4.6 Public Health Interventions for Waterborne Diseases

Waterborne diseases continue to pose significant public health challenges worldwide, particularly in developing regions, where lack of access to clean water and sanitation exacerbates the problem. To mitigate the incidence and transmission of these diseases, comprehensive public health interventions, primarily focusing on Water, Sanitation, and Hygiene (WASH) programs, are imperative (Table 4.3).

Table 4.3 Public health management and prevention of waterborne diseases

Public health intervention	Summary	Public health focus
Safe water supply, sanitation, and hygiene (WASH)	Provision of clean water, proper sanitation, and hygiene promotion through infrastructure and education	Prevents contamination and transmission of pathogens via water; reduces disease incidence
Household water treatment	Methods such as boiling, chlorination, and filtration used at the household level to purify drinking water	Empowers households to take immediate action to reduce exposure to waterborne pathogens
Sanitation facilities	Construction and maintenance of toilets, sewage systems, and wastewater management	Minimizes environmental contamination and prevents fecal–oral disease transmission
Hygiene Education Programs	Campaigns and school-based initiatives to promote handwashing and hygiene practices	Builds long-term behavioral change to reduce disease transmission, especially among children
Community health education	Health promotion through local communication channels and culturally sensitive messaging	Enhances awareness, empowers communities, and encourages proactive health behaviors
Training of health workers	Capacity building for public health professionals to deliver WASH and disease prevention interventions	Strengthens health systems and ensures sustainable implementation of public health strategies
Vaccination campaigns	Immunization efforts against diseases like cholera, typhoid, and hepatitis A in vulnerable regions	Reduces disease incidence and severity; useful during outbreaks or in endemic areas
National policy and regulation	Enforcement of water safety standards, sanitation guidelines, and waste disposal regulations	Creates a legal framework supporting public health goals and environmental safety
Disease surveillance systems	Monitoring and early detection of waterborne disease outbreaks to enable timely response	Facilitates targeted interventions, resource allocation, and minimizes outbreak spread
International collaboration and funding	Partnerships with WHO, UNICEF, and NGOs for resources, expertise, and coordination	Enhances capacity for large-scale public health programs and accelerates progress in disease reduction

The development and maintenance of infrastructure for providing access to safe drinking water are foundational elements of effective WASH programs. In rural areas, where access to improved water sources is often limited, heightened efforts are needed to reduce the prevalence of waterborne illnesses such as diarrhea, which can be traced directly to unsafe water and poor sanitation practices [78–80]. Evidence demonstrates that enhancing water supply infrastructure can reduce diarrheal disease by up to one-third [80]. Moreover, positive trends in water supply improvements coincide with decreases in the rates of waterborne diseases, affirming the connection between improved water access and better health outcomes [81].

Household water treatment methods, such as chlorination, boiling, and filtration, play a vital role in safeguarding communities against waterborne pathogens. Studies indicate that communities employing regular household water treatment exhibit lower instances of disease [82, 83]. However, the sustainability of such practices is often challenged, as reports reveal that many households discontinue water treatment over time, which diminishes health benefits [82]. Therefore, promotion of these methods, alongside ensuring community-wide adoption, is necessary.

Another key intervention is enhancing sanitation facilities to eliminate practices such as open defecation, which directly contaminates water sources and heightens health risks [84]. Improving sanitation infrastructure not only curtails disease transmission but also fosters a healthier local environment, reducing the burden of waterborne diseases significantly [85]. Hygiene education further complements these efforts by emphasizing essential practices like handwashing with soap at critical times, thereby limiting pathogen spread [84, 86]. Integrating WASH initiatives within schools and healthcare facilities is also a strategic approach to ingraining these health practices in communities, reducing disease transmission rates among vulnerable populations [85, 87].

Behavior change strategies are equally crucial for the long-term success of public health interventions. Community-based campaigns can effectively raise awareness about the origins and preventive measures against waterborne diseases, utilizing culturally relevant materials to foster understanding [84, 88]. Training local health workers to communicate consistent health messages can also enhance community engagement [88, 89]. Active involvement from local leaders and influencers can mobilize communities around behavior change, making preventive practices more acceptable [23, 81]. Additionally, continuous evaluation and monitoring of these behavior change programs ensure responsiveness and improvements over time [84].

Vaccination efforts address specific waterborne diseases like cholera, typhoid, and hepatitis A. Implementing national immunization programs targeted at high-risk populations is crucial, particularly in regions prone to outbreaks [88, 90]. The recent success of cholera vaccination campaigns has demonstrated significant improvements in community health practices, including increased adherence to safe water treatment methods [91]. Furthermore, stringent policies regulating water safety and sanitation standards are imperative to establish a robust framework for preventing waterborne diseases [86]. Strengthening public health systems for rapid detection and response to waterborne disease outbreaks is essential. Collaborations with international organizations can provide the necessary funding and technical assistance needed to support these efforts [90, 92].

4.7 Surveillance Systems for Waterborne Diseases

The complex landscape of surveillance systems for waterborne diseases (WBDs) involves multifaceted approaches essential for effective disease monitoring and outbreak detection. The continuous tracking of incidence rates for key waterborne illnesses such as cholera, typhoid, and hepatitis A is crucial, as evidenced by historical analyses indicating that most outbreaks emerge from issues related to water supply and quality [69]. These epidemiologic trends illustrate the need for robust data collection from hospitals, clinics, and laboratories to enable timely reports of abnormal spikes in disease occurrences [93]. Moreover, syndromic surveillance systems play an important role by capturing early warning signs based on reported symptoms prior to laboratory confirmation, allowing for a proactive public health response to potential outbreaks [94]. Strengthening community-based reporting mechanisms, particularly in remote or underserved regions, enhances the ability to detect early signs of an outbreak, emphasizing the need for localized data in emergency health situations [95].

The integration of Geographic Information Systems (GIS) and remote sensing technologies further augments the capabilities of surveillance systems for waterborne diseases. Mapping high-risk regions using environmental and socioeconomic indicators improves the understanding of disease hotspots [96]. Studies utilizing satellite imagery have demonstrated effective monitoring of water quality, sanitation infrastructure, and climatic conditions—factors significantly contributing to the emergence of waterborne diseases [97]. By enhancing spatial–temporal analysis, stakeholders can predict outbreak occurrences and plan preventive strategies in vulnerable areas effectively. Furthermore, the application of GIS data supports targeted interventions like vaccination campaigns and improvements in water, sanitation, and hygiene (WASH) facilities, proving invaluable in outbreak preparedness and response [98].

Integrating surveillance systems into national health infrastructure is paramount for unified reporting and response planning regarding waterborne diseases. Coordination between public health entities and local health departments ensures that collected surveillance data serves broader epidemiological goals, facilitating the establishment of standardized protocols for data sharing across multiple sectors, including public health and water management [90]. The interoperability with global networks, such as the World Health Organization's Global Outbreak Alert and Response Network (GOARN), allows for comprehensive data-sharing frameworks that bolster outbreak responses on an international scale [95]. Moreover, incorporating surveillance outcomes into national disease control policies and emergency preparedness frameworks fortifies the response mechanism against potential waterborne disease outbreaks, reflecting an integrated approach to public health [99].

4.8 Case Studies and Regional Interventions in Waterborne Disease Control

The increasing prevalence of waterborne diseases necessitates the implementation of effective control programs, evidenced by recent case studies from regions such as Bangladesh and Haiti. In Bangladesh, the Arsenic Mitigation Program has played a crucial role in addressing the issue of arsenic contamination, primarily through the introduction of deep tube wells as alternative water sources. These wells are vital in regions heavily afflicted by arsenic, significantly contributing to public health by minimizing exposure to contaminated water. The program's success is underscored by the consideration of various interventions, as the choice of mitigation strategy can also inadvertently increase the burden of other water-related infectious diseases if not carefully managed [90, 100].

In Haiti, the Cholera Elimination Plan illustrates the importance of a comprehensive approach combining Water, Sanitation, and Hygiene (WASH) strategies with vaccination efforts. Following the devastating earthquake in 2010 and the subsequent choleral outbreak, authorities implemented a focused response incorporating oral cholera vaccination campaigns alongside improved sanitation and water quality monitoring. Findings indicate that such comprehensive plans are essential in reducing cholera incidence and preventing its recurrence, demonstrating the effectiveness of well-coordinated health strategies, particularly in post-disaster scenarios where health systems are vulnerable [101–103].

Moreover, innovation in water supply solutions has been showcased through initiatives like the Safe Water Network in Ghana, which established small water enterprises to ensure sustainable access to clean water in rural communities. This model empowers local populations by engaging them in the management and operation of these water systems, significantly improving their resilience to waterborne diseases. Similarly, Kenya's Water Safety Plans reflect a proactive community-driven approach, involving local stakeholders in risk assessment and management of water quality, which has been associated with lower rates of diarrheal diseases [102].

Despite these successes, several challenges persist that undermine the effectiveness of waterborne disease control programs. Inadequate infrastructure and maintenance of water supply systems, coupled with limited funding and inconsistent governmental support, create barriers to sustainable health interventions [90, 104]. Additionally, sociocultural resistance to behavioral changes concerning sanitation practices can hinder the adoption of recommended health measures [90, 103]. Moreover, climatic disruptions resulting from climate change continue to threaten water safety and supply reliability, exacerbating existing vulnerabilities [105].

On the other hand, numerous success factors contribute to the positive outcomes of waterborne disease interventions. Strong collaboration among stakeholders, including governments, non-governmental organizations, and local communities, has been crucial in sustaining these health initiatives. For instance, continuous public education and behavior change campaigns have strengthened community engagement [101, 103]. Implementing real-time water quality monitoring and ensuring the integration of disease surveillance with water quality testing are also crucial to effectively preempting and responding to outbreaks [105, 106].

Community-led initiatives, such as the Community-Led Total Sanitation (CLTS) programs, show significant promise in eliminating open defecation and fostering hygienic practices at the grassroots level. Concurrently, government-led strategies, including the establishment of national drinking water quality standards and public–private partnerships, provide the regulatory framework necessary for supporting these initiatives [101, 104]. The integration of waterborne disease control strategies into broader health and disaster preparedness frameworks is essential for ensuring long-term sustainability and resilience against future outbreaks, particularly in water-scarce and disaster-prone areas [107, 108].

4.9 Innovations and Future Directions in Waterborne Disease Management

Waterborne diseases remain a significant public health challenge, particularly in developing regions. Innovations in water management, including advanced purification technologies and real-time monitoring systems, are crucial for addressing this issue. Several technological solutions are currently being deployed to enhance water quality and ensure safe drinking water, thereby reducing the incidence of waterborne illnesses (Table 4.4).

One promising technological advancement is the development of low-cost water filtration systems that utilize nanotechnology and bio-sand filters. These systems have been shown to enhance the removal of pathogens and contaminants from water, making it safer for consumption [109]. Such innovations are critical in areas lacking access to centralized water treatment facilities. Additionally, the optimization of chlorine injection in water distribution networks has been demonstrated to ensure that adequate residual chlorine reaches even the most remote consumers [110, 111]. Automating this process can significantly reduce the risk of disease outbreaks tied to insufficient disinfection in distributed water systems [112].

Moreover, the integration of Internet of Things (IoT) technologies allows for real-time monitoring of water quality across both rural and urban settings. IoT-based devices enable continuous assessment of multiple water quality parameters like temperature, pH, and turbidity, thus ensuring prompt responses to contamination risks [113]. The use of remote sensors, combined with cloud computing technologies, facilitates widespread monitoring and timely public health interventions [114,

Table 4.4 Innovations and future directions in waterborne disease management

Innovation area	Application	Public health focus
Technological solutions for water purification and monitoring	Nanotech and bio-sand filtration systems- IoT-enabled real-time water quality sensors- Smart chlorination infrastructure- Solar-powered purification for remote areas	Ensures safe drinking water, reduces exposure to pathogens, and prevents outbreaks in vulnerable populations
Mobile tools for public health data collection	Mobile health (mHealth) apps for outbreak alerts- GPS tools for mapping contaminated zones- Crowdsourced water safety platforms- Mobile diagnostic and telemedicine kits	Facilitates rapid disease surveillance, improves early response times, and enhances community reporting
Multi-sectoral and interdisciplinary approaches	Cross-sector collaboration (health, water, environment, education)- Local government and community co-designed water safety plans- Use of behavioral science and One Health approach	Promotes sustainable prevention, empowers local action, and addresses the root causes of waterborne diseases

115]. Such systems can inform local authorities about potential pollution sources and enable proactive measures to mitigate health risks [116].

Another avenue of innovation lies in smart infrastructure, specifically automated chlorination systems that can adjust chemical dosages based on real-time water quality data. This technology is essential for maintaining water safety across extensive distribution networks, particularly in regions where manual monitoring may be inefficient or insufficient [117]. Furthermore, solar-powered purification units represent a sustainable solution for off-grid communities, allowing for localized water treatment without dependency on electrified infrastructure [109]. Mobile tools for public health data collection are also gaining traction in the management of waterborne diseases. The use of mHealth applications allows for the rapid reporting of disease outbreaks related to water quality [109]. GPS-enabled devices play a key role in tracing contaminated water sources and identifying areas at heightened risk for disease, thus facilitating targeted public health responses [118]. Crowdsourcing platforms further enhance community engagement in data collection regarding water safety, empowering residents to contribute to the monitoring process [116].

Multi-sectoral and interdisciplinary approaches are increasingly being recognized as vital for effective waterborne disease management. Collaboration across health, environmental, and educational sectors can lead to more holistic strategies for disease prevention and water safety [116, 119]. Integrating insights from behavioral science, climate modeling, and environmental engineering into policy frameworks enhances the adaptability and efficacy of public health interventions [109, 120]. Furthermore, a

One Health approach—linking the health of humans, animals, and the environment—provides a comprehensive strategy for addressing the multifaceted risks associated with waterborne diseases [119].

4.10 Conclusion

The persistent burden of waterborne diseases highlights the urgent need for sustainable, long-term solutions that address the root causes of unsafe water access and poor sanitation. Epidemiological data show that billions of people remain at risk due to inadequate water infrastructure, especially in low-income and climate-vulnerable regions. Key lessons from global health efforts emphasize the importance of integrated strategies—combining disease surveillance, behavior change, sanitation improvement, and resilient infrastructure development. Climate change, antimicrobial resistance, and population displacement further complicate water safety, underscoring the necessity of multi-sectoral responses grounded in strong public health systems. Preventing waterborne diseases is not just a technical challenge—it requires inclusive community engagement and policy innovation. The path to resilient, health-secure communities lies in global cooperation and aligned policy frameworks that prioritize water quality as a fundamental health determinant. A future-ready vision involves leveraging technologies such as GIS-based surveillance, digital health platforms, and eco-sensitive water treatment innovations. These must be supported by international funding mechanisms, research collaborations, and policy harmonization to close infrastructure gaps and promote equitable access. The establishment of comprehensive waterborne disease surveillance systems, alongside continued capacity building and advocacy, can transform how the world confronts these threats. Therefore, safeguarding public health from waterborne diseases is both a moral and strategic imperative, thus demanding coordinated, sustainable, and inclusive action at every level.

References

1. Izah SC, Richard G, Sawyer WE (2021) Distribution of Fungi density and diversity in a Surface water of Epie Creek in Yenagoa Metropolis. Archives of Epidemiology and Public Health, Nigeria. https://doi.org/10.15761/AEPH.1000121Volume3:1-5
2. Moslen M, Miebaka CA, Izah SC (2024) Occurrence and management of toxicity associated with parasitic pathogens in drinking water sources in rural areas in the Niger Delta, Nigeria. In: Madhav S, Izah SC, van Hullebusch ED, Srivastav AL (eds) Water Resources Management for Rural Development. Elsevier, pp 275–286
3. Numbere AO, Kika EP, Izah SC (2024) Challenges and mitigation of water resource management in rural areas in Nigeria. In: Izah SC, Ogwu MC, Loukas A, Hamidifar H (eds) Water crises and sustainable management in the global south. Springer, Singapore, pp 373–390. https://doi.org/10.1007/978-981-97-4966-9_12

4. Aghoghovwia OA, Miri FA, Izah SC (2018) Impacts of anthropogenic activities on heavy metal levels in surface water of nun river around Gbarantoru and Tombia Towns, Bayelsa State, Nigeria. Ann Ecol Environ Sci 2(2):1–8
5. Izah SC, Stanley HO, Richard G, Sawyer WE, Uwaeme OR (2024a) Surface water quality: a statistical perspective on the efficacy of environmental and human health assessment tools. Water, Air, Soil Pollut 235(3). https://doi.org/10.1007/s11270-024-06965-1
6. Ogamba EN, Charles EE, Izah SC (2021) Distributions, pollution evaluation and health risk of selected heavy metal in surface water of Taylor creek, Bayelsa State Nigeria. Toxicol Environ Health Sci 13(2):109–121. https://doi.org/10.1007/s13530-020-00076-0
7. Enaregha EB, Omotete TR, Izah SC, Odubo TC (2022) Comparison of total viable bacteria counts and in situ characteristics in drinking water sources in Sagbama town, Bayelsa state Nigeria. Int J Pathog Res 11(3–4):79–87
8. Seiyaboh EI, Youkparigha FO, Izah SC, Mientei K (2020) Assessment of bacteriological characteristics of surface water of Taylor creek in Bayelsa state, Nigeria. Noble Int J Sci Res 4(4):25–30
9. Seiyaboh EI, Youkparigha FO, Izah SC, Daniels ID (2020) Bacteriological quality of groundwater in Imiringi Town, Bayelsa State, Nigeria. J Biotechnol Biomed Sci 2(2):34–40
10. Iyiola AO, Afolabi OA, Alimi SK, Akingba OO, Izah SC, Ogwu MC (2024) Climate change and water crisis in the global South. In: Izah SC., Ogwu MC, Loukas A, Hamidifar H (eds) Water crises and sustainable management in the global South. Springer, Singapore, pp 111–140. https://doi.org/10.1007/978-981-97-4966-9_4
11. Fayiah M, Kandeh B, Izah SC, Saccoh S (2024) Rural water management status, threats, and prospects: a case study of Sierra Leone. In: Madhav S, Izah SC, van Hullebusch ED, Srivastav AL (eds) Water Resources Management for Rural Development. Elsevier, pp 263–274
12. Jacob DE, Nelson IU, Izah SC, Daniel KS (2024a) Rural water crises in the global south: understanding the scope and impact. In: Izah SC, Ogwu MC, Loukas A, Hamidifar H (eds) Water crises and sustainable management in the global south. Springer, Singapore, pp 3–44. https://doi.org/10.1007/978-981-97-4966-9_1
13. Izah SC, Etim NG, Alaka OO, Nwachukwu BC, Joshua G (2024c) Public health interventions for cholera control and prevention: water, sanitation, and hygiene (WASH) perspective. Greener J Epidemiol Public Health 12(1):1–15. https://doi.org/10.15580/gjeph.2024.1.102024145
14. Izah SC, Ngun CT, Iniaghe PO, Aigberua AO, Odubo TC (2023) Processes of decontamination and elimination of toxic metals from water and wastewaters. In: Shukla SK, Kumar S, Madhav S, Mishra PK (eds) Metals in water: global sources, significance, and treatment. Woodhead Publishing Advances in Pollution Research. Elsevier, pp 239–262. https://doi.org/10.1016/B978-0-323-95919-3.00003-3
15. Izah SC, Ngun CT, Richard G (2022). Microbial quality of groundwater in the Niger Delta region of Nigeria: health implications and effective treatment technologies. In: Srivastav AL, Madhav S, Bhardwaj AK, Valsami-Jones E (eds) Urban water crisis and management: strategies for sustainable development. Current directions in water scarcity research. Elsevier, pp 149–172. https://doi.org/10.1016/B978-0-323-91838-1.00010-5
16. Jacob DE, Nelson IU, Izah SC (2024c) Indigenous water management strategies in the global south. In: Izah SC, Ogwu MC, Loukas A, Hamidifar H (eds) Water crises and sustainable management in the global south. Springer, pp 487–525. https://link.springer.com/chapter/10.1007/978-981-97-4966-9_16
17. Jacob DE, Nelson IU, Eniang EA, Izah SC (2024b) Advancing water security and resilience in the global south through recreational development. In: Izah SC, Ogwu MC, Loukas A, Hamidifar H (eds) Water crises and sustainable management in the global south. Springer, pp 527–554. https://link.springer.com/chapter/10.1007/978-981-97-4966-9_17
18. Antunez EE, Mahon CS, Tong Z, Voelcker NH, Müllner M (2020) A regenerable biosensing platform for bacterial toxins. Biomacromol 22(2):441–453
19. Ramírez-Castillo FY, Loera-Muro A, Jacques M, Garneau P, Avelar-González FJ, Harel J, Guerrero-Barrera AL (2015) Waterborne pathogens: detection methods and challenges. Pathogens 4(2):307–334. https://doi.org/10.3390/pathogens4020307

20. Collier SA, Deng L, Adam EA, Benedict KM, Beshearse EM, Blackstock AJ, Bruce BB, Derado G, Edens C, Fullerton KE et al (2021) Estimate of burden and direct healthcare cost of infectious waterborne disease in the United States. Emerg Infect Dis 27(1):140. https://doi.org/10.3201/eid2701.190676
21. Machado A, Amorim E, Bordalo AA (2022) Spatial and seasonal drinking water quality assessment in a Sub-Saharan Country (Guinea-Bissau). Water 14(13):1987. https://doi.org/10.3390/w14131987
22. Cann KF, Thomas DR, Salmon RL, Wyn-Jones AP, Kay D (2013) Extreme water-related weather events and waterborne disease. Epidemiol Infect 141(4):671–686
23. Degbey C, Houessionon E, de Brouwer C (2025) Prevalence and factors associated with waterborne diseases in Couffo, southwestern Benin: the case of Aplahoué. Int J Environ Res Public Health 22(1):58
24. Jacob B, Kazaura M (2021) Access to safe water, sanitation, and hygiene: a cross-sectional study among the maasai in tanzania. Am J Trop Med Hyg 104(4):1535–1539. https://doi.org/10.4269/ajtmh.20-0134
25. Mehboob G, Assad Hassan MA, Shah S. An assessment of community intervention for reduction of waterborne illnesses in Peshawar, Khyber Pakhtunkhwa, Pakistan. J Dev Policy, Res Pract 106. https://doi.org/10.59926/jodprp.vol02/05
26. Ogwu MC, Kosoe EA (2024) Place of Cultural Diversity in Sustainable Water Resource Management in Ghana. In: Izah SC, Ogwu MC, Loukas A, Hamidifar H (eds) Water crises and sustainable management in the global South. Springer, Singapore. https://doi.org/10.1007/978-981-97-4966-9_14
27. Ogwu MC, Kosoe EA (2025) Integrating green infrastructure into sustainable agriculture to enhance soil health, biodiversity, and microclimate resilience. Sustainability 17(9):3838. https://doi.org/10.3390/su17093838
28. Stange C, Sidhu JPS, Tiehm A, Toze S (2016) Antibiotic resistance and virulence genes in coliform water isolates. Int J Hyg Environ Health 219(8):823–831. https://doi.org/10.1016/j.ijheh.2016.07.015
29. Zavala-Norzagaray AA, Aguirre AA, Velazquez-Roman J, Flores-Villaseñor H, León-Sicairos N, Ley-Quiñonez CP, Hernández-Díaz LDJ, Canizalez-Roman A (2015) Isolation, characterization, and antibiotic resistance of Vibrio spp. in sea turtles from Northwestern Mexico. Front Microbiol 6:635. https://doi.org/10.3389/fmicb.2015.00635
30. Alduina R, Gambino D, Presentato A, Gentile A, Sucato A, Savoca D, Filippello S, Visconti G, Caracappa G, Vicari D, Arculeo M (2020) Is Caretta caretta a carrier of antibiotic resistance in the Mediterranean Sea? Antibiotics 9(3):116
31. Lupo A, Coyne S, Berendonk TU (2012) Origin and evolution of antibiotic resistance: the common mechanisms of emergence and spread in water bodies. Front Microbiol 3:18. https://doi.org/10.3389/fmicb.2012.00018
32. Labana RV (2019) Cryptosporidium in the Philippines. Int Ann Sci 6(1):18–27. https://doi.org/10.21467/ias.6.1.18-27
33. Aslam B, Wang W, Arshad MI, Khurshid M, Muzammil S, Rasool MH, Nisar MA, Alvi RF, Aslam MA, Qamar MU, Salamat MKF (2018) Antibiotic resistance: a rundown of a global crisis. Infect Drug Resist 1645–1658
34. Rizzo L, Manaia C, Merlin C, Schwartz T, Dagot C, Ploy MC, Michael I, Fatta-Kassinos D (2013) Urban wastewater treatment plants as hotspots for antibiotic resistant bacteria and genes spread into the environment: a review. Sci Total Environ 447:345–360. https://doi.org/10.1016/j.scitotenv.2013.01.032
35. Hassan NL, Yern KK, Zain NAM (2020) Isolation of antibiotic resistant bacteria from rivers in Kelantan, Malaysia. Int J Life Sci Biotechnol 3(3):291–307. https://doi.org/10.38001/ijlsb.712542
36. Chuen-Im T, Sawetsuwannakun K, Neesanant P, Kitkumthorn N (2021) Antibiotic-resistant bacteria in green turtle (Chelonia mydas) rearing seawater. Animals 11(6):1841. https://doi.org/10.3390/ani11061841

37. Alaka OO, Etim NG, Joshua G, Udensi CI, Izah SC, Madu PC (2024) Microbial Pathogenesis of Vibrio cholerae and Antimicrobial Resistance Trends. ES General 6:1358. https://doi.org/10.30919/esg1358
38. Etim NG, Joshua G, Izah SC, Alaka OO, Udensi CI, Etim EN (2024) Cholera vaccine development: progress, efficacy, and public health strategies. Greener J Biomed Health Sci 7(1):47–60. https://doi.org/10.15580/gjbhs.2024.1.102024146
39. Suzuki S, Ogo M, Miller TW, Shimizu A, Takada H, Siringan MAT (2013) Who possesses drug resistance genes in the aquatic environment?: sulfamethoxazole (SMX) resistance genes among the bacterial community in water environment of Metro-Manila, Philippines. Front Microbiol 4:102. https://doi.org/10.3389/fmicb.2013.00102
40. Mahmood AR, Al-Haideri HH, Hassan FM (2019) Detection of antibiotics in drinking water treatment plants in Baghdad City, Iraq. Adv Public Health 2019(1):7851354. https://doi.org/10.1155/2019/7851354
41. Adegoke AA, Amoah ID, Stenström TA, Verbyla ME, Mihelcic JR (2018) Epidemiological evidence and health risks associated with agricultural reuse of partially treated and untreated wastewater: a review. Front Public Health 6:337. https://doi.org/10.3389/fpubh.2018.00337
42. Dorevitch S, Pratap P, Wroblewski M, Hryhorczuk DO, Li H, Liu LC, Scheff PA (2012) Health risks of limited-contact water recreation. Environ Health Perspect 120(2):192–197. https://doi.org/10.1289/ehp.1103934
43. Guzman Herrador BR, De Blasio BF, MacDonald E, Nichols G, Sudre B, Vold L, Semenza JC, Nygård K (2015) Analytical studies assessing the association between extreme precipitation or temperature and drinking water-related waterborne infections: a review. Environ Health 14(1): 29. https://doi.org/10.1186/s12940-015-0014-y
44. Ferrer A, Nguyen-Viet H, Zinsstag J (2012) Quantification of diarrhea risk related to wastewater contact in thailand. EcoHealth 9(1):49–59. https://doi.org/10.1007/s10393-012-0746-x
45. Hughes B, Beale DJ, Dennis PG, Cook S, Ahmed W (2017) Cross-comparison of human wastewater-associated molecular markers in relation to fecal indicator bacteria and enteric viruses in recreational beach waters. Appl Environ Microbiol 83(8):e00028-e117. https://doi.org/10.1128/aem.00028-17
46. Nour I, Hanif A, Ryan M, Eifan S (2021) Insights into gastrointestinal Virome: etiology and public exposure. Water 2021(13):2794. https://doi.org/10.3390/w13192794
47. Farrer D, Counter M, Hillwig R, Cude C (2015) Health-based cyanotoxin guideline values allow for cyanotoxin-based monitoring and efficient public health response to cyanobacterial blooms. Toxins 7(2):457–477
48. Lamparelli CC, Pogreba-Brown K, Verhougstraete M, Sato MIZ, de Castro Bruni A, Wade TJ, Eisenberg JN (2015) Are fecal indicator bacteria appropriate measures of recreational water risks in the tropics: a cohort study of beach goers in Brazil? Water Res 87:59–68. https://doi.org/10.1016/j.watres.2015.09.001
49. Schijven JF, Blaak H, Schets FM, de Roda Husman AM (2015) Fate of extended-spectrum β-lactamase-producing Escherichia coli from faecal sources in surface water and probability of human exposure through swimming. Environ Sci Technol 49(19):11825–11833. https://doi.org/10.1021/acs.est.5b01888
50. McClary JS, Sassoubre LM, Boehm AB (2017) Staphylococcus aureus strain newman photoinactivation and cellular response to sunlight exposure. Appl Environ Microbiol 83(17):e01052–17. https://doi.org/10.1128/aem.01052-17
51. Wyn-Jones AP, Carducci A, Cook N, D'agostino M, Divizia M, Fleischer J, Gantzer C, Gawler A, Girones R, Höller C, de Roda Husman AM (2011) Surveillance of adenoviruses and noroviruses in European recreational waters. Water Res 45(3):1025–1038. https://doi.org/10.1016/j.watres.2010.10.015
52. Wade TJ, Sams E, Brenner KP, Haugland R, Chern E, Beach M, Wymer L, Rankin CC, Love D, Li Q, Noble R (2010) Rapidly measured indicators of recreational water quality and swimming-associated illness at marine beaches: a prospective cohort study. Environ Health 9(1):66. https://doi.org/10.1186/1476-069x-9-66

53. Jørgensen SB, Søraas AV, Arnesen LS, Leegaard TM, Sundsfjord A, Jenum PA (2017) A comparison of extended spectrum β-lactamase producing Escherichia coli from clinical, recreational water and wastewater samples associated in time and location. PLoS ONE 12(10):e0186576. https://doi.org/10.1371/journal.pone.0186576
54. Semenza JC, Suk JE, Estevez V, Ebi KL, Lindgren E (2012) Mapping climate change vulnerabilities to infectious diseases in Europe. Environ Health Perspect 120(3):385–392. https://doi.org/10.1289/ehp.1103805
55. Herrador BG, Lund V, Fonahn W, Hisdal H, Hygen HO, Hyllestad S, Nordeng Z, Skaland RG, Sunde LS, Vold L, White R (2021) Heavy weather events, water quality and gastroenteritis in Norway. One Health 13:100297. https://doi.org/10.1016/j.onehlt.2021.100297
56. Khabo-Mmekoa CMN, Momba MNB (2019) The impact of social disparities on microbiological quality of drinking water supply in Ugu District Municipality of Kwazulu-Natal Province, South Africa. Int J Environ Res Public Health 16(16):2972. https://doi.org/10.3390/ijerph16162972
57. Pons W, Young I, Truong J, Jones-Bitton A, McEwen S, Pintar K, Papadopoulos A (2015) A systematic review of waterborne disease outbreaks associated with small non-community drinking water systems in Canada and the United States. PLoS ONE 10(10):e0141646. https://doi.org/10.1371/journal.pone.0141646
58. Levy K, Woster AP, Goldstein RS, Carlton EJ (2016) Untangling the impacts of climate change on waterborne diseases: a systematic review of relationships between diarrheal diseases and temperature, rainfall, flooding, and drought. Environ Sci Technol 50(10):4905–4922. https://doi.org/10.1021/acs.est.5b06186
59. McIver L, Kim R, Woodward A, Hales S, Spickett J, Katscherian D, Hashizume M, Honda Y, Kim H, Iddings S, Naicker J (2016) Health impacts of climate change in Pacific Island countries: a regional assessment of vulnerabilities and adaptation priorities. Environ Health Perspect 124(11):1707–1714. https://doi.org/10.1289/ehp.1509756
60. Kombo Mpindou GOM, Escuder Bueno I, Chordà Ramón E (2021) Review on emerging waterborne pathogens in Africa: the case of Cryptosporidium. Water 13(21):2966. https://doi.org/10.3390/w13212966
61. Naganathan T, O'Connor A, Sargeant JM, Shapiro K, Totton S, Winder C, Greer AL (2022) The prevalence of Cyclospora cayetanensis in water: a systematic review and meta-analysis. Epidemiol Infect 150:e15. https://doi.org/10.1017/s0950268821002521
62. Sukami JM, Mufungizi I, Lombo JB, Kadima AU, Ngunga DY, Akilimali A (2025) Spatiotemporal analysis of the distribution of waterborne diseases in children under 5 years of age from 2018 to 2022 in the lemba health zone in kinshasa, dr congo: a retrospective and observational analysis. Health Sci Rep 8(4). https://doi.org/10.1002/hsr2.70605
63. Trovão N, Neves-Silva P, Pinheiro LC, Peixoto SV, Heller L (2023) Changes in access to water and incidence of waterborne diseases after the Vale dam collapse in Brumadinho (MG), Brazil. Revista Brasileira de Epidemiologia 26:e230010. https://doi.org/10.1590/1980-549720230010
64. Sadik NJ, Uprety S, Nalweyiso A, Kiggundu N, Banadda NE, Shisler JL, Nguyen TH (2017) Quantification of multiple waterborne pathogens in drinking water, drainage channels, and surface water in Kampala, Uganda, during seasonal variation. GeoHealth 1(6):258–269. https://doi.org/10.1002/2017gh000081
65. Radi MFM, Hashim JH, Jaafar MH, Hod R, Ahmad N, Nawi AM, Baloch GM, Ismail R, Ayub NIF (2018) Leptospirosis outbreak after the 2014 major flooding event in Kelantan, Malaysia: a spatial-temporal analysis. Am J Trop Med Hyg 98(5):1281. https://doi.org/10.4269/ajtmh.16-0922
66. Hedlund C, Blomstedt Y, Schumann B (2014) Association of climatic factors with infectious diseases in the Arctic and subarctic region–a systematic review. Glob Health Action 7(1):24161. https://doi.org/10.3402/gha.v7.24161
67. Mellor JE, Levy K, Zimmerman J, Elliott M, Bartram J, Carlton E, Clasen T, Dillingham R, Eisenberg J, Guerrant R, Lantagne D (2016) Planning for climate change: the need for mechanistic systems-based approaches to study climate change impacts on diarrheal diseases. Sci Total Environ 548:82–90. https://doi.org/10.1016/j.scitotenv.2015.12.087

68. Ogwu MC (2025) Sustainable development in Africa: challenges, innovations, and policy pathways for the future. African Research Reports 1(2):55–69. https://doi.org/10.5897/ARR 2025.0005
69. Okaka FO, Odhiambo BD (2018) Relationship between flooding and out break of infectious diseasesin Kenya: a review of the literature. J Environ Public Health 2018(1):5452938.https://doi.org/10.22437/jop.v9i1.28689
70. Hutton G, Chase C (2016) The knowledge base for achieving the sustainable development goal targets on water supply, sanitation and hygiene. Int J Environ Res Public Health 13(6):536. https://doi.org/10.3390/ijerph13060536
71. Gürsu H (2024) Assessment of theoretical and test performance considerations of concentrated solar water purification system "parabosol" in underserved regions. Sustainability 16(15):6611. https://doi.org/10.3390/su16156611
72. Adelodun B, Ajibade FO, Ighalo JO, Odey G, Ibrahim RG, Kareem KY, Bakare HO, Tiamiyu AO, Ajibade TF, Abdulkadir TS et al (2021) Assessment of socioeconomic inequality based on virus-contaminated water usage in developing countries: a review. Environ Res 192:110309. https://doi.org/10.1016/j.envres.2020.110309
73. Sente C, Erume J, Naigaga I, Mulindwa J, Ochwo S, Magambo PK, Namara BG, Kato CD, Sebyatika G, Muwonge K, Ocaido M (2016) Prevalence of pathogenic free-living amoeba and other protozoa in natural and communal piped tap water from Queen Elizabeth protected area Uganda. Infect Dis Poverty 5(04):84–97. https://doi.org/10.3390/w15112121
74. Hammer CC, Brainard J, Hunter PR (2018) Risk factors and risk factor cascades for communicable disease outbreaks in complex humanitarian emergencies: a qualitative systematic review. BMJ Glob Health 3(4)
75. Tricou V, Bouscaillou J, Laghoe-Nguembe GL, Béré A, Konamna X, Sélékon B, Nakouné E, Kazanji M, Komas NP (2020) Hepatitis E virus outbreak associated with rainfall in the Central African Republic in 2008–2009. BMC Infect Dis 20(1):260. https://doi.org/10.1186/s12879-020-04961-4
76. Bennett SD, Lowther SA, Chingoli F, Chilima B, Kabuluzi S, Ayers TL, Warne TA, Mintz E (2018) Assessment of water, sanitation and hygiene interventions in response to an outbreak of typhoid fever in Neno District Malawi. PloS one 13(2):e0193348
77. Izah SC, Ogwu MC (2025) Modeling solutions for microbial water contamination in the global south for public health protection. Front Microbiol 16:1504829. https://doi.org/10.3389/fmicb.2025.1504829
78. Kosoe EA, Osumanu IK, Ogwu MC (2025b) Reimagining rural and small-town water supply in Ghana through inclusive governance and sustainable models. J Sustain 1(2):1–21. https://doi.org/10.55845/jos-2025-1234
79. Kosoe EA, Ogwu MC, Duniah J, Ndigun FB (2025a) Sustaining indigenous water governance systems in Ayimaye and Lungni, Northern Ghana, through Hybrid Models. J Sustain 1(2). https://doi.org/10.55845/jos-2025-1235
80. Simelane MS, Shongwe MC, Vermaak K, Zwane E (2020) Determinants of households' access to improved drinking water sources: a secondary analysis of eswatini 2010 and 2014 multiple indicator cluster surveys. Adv Public Health 2020(1):6758513. https://doi.org/10.1155/2020/6758513
81. Taylor DL, Kahawita TM, Cairncross S, Ensink JH (2015) The impact of water, sanitation and hygiene interventions to control cholera: a systematic review. PLoS ONE 10(8):e0135676. https://doi.org/10.1371/journal.pone.0135676
82. Moropeng RC, Momba MNB (2020) Assessing the sustainability and acceptance rate of cost-effective household water treatment systems in rural communities of Makwane Village South Africa. Crystals 10(10):872. https://doi.org/10.3390/cryst10100872
83. Wright JA, Yang H, Rivett U, Gundry SW (2012) Public perception of drinking water safety in South Africa 2002–2009: a repeated cross-sectional study. BMC Public Health 12(1):556. https://doi.org/10.1186/1471-2458-12-556
84. Meki CD, Ncube EJ, Voyi K (2022) Community-level interventions for mitigating the risk of waterborne diarrheal diseases: a systematic review. Syst Rev 11(1):73. https://doi.org/10.1186/s13643-022-01947-y

85. Collins OC, Duffy KJ (2018) Analysis and optimal control intervention strategies of a waterborne disease model: a realistic case study. J Appl Math 2018(1):2528513. https://doi.org/10.1155/2018/2528513
86. Collier SA, Stockman LJ, Hicks LA, Garrison LE, Zhou FJ, Beach MJ (2012) Direct healthcare costs of selected diseases primarily or partially transmitted by water. Epidemiol Infect 140(11):2003–2013. https://doi.org/10.1017/s0950268811002858
87. Saxena D, Raheja L, Tamma RR, Jain PK, Takhelchangbam N, TAMMA RR, Jain P (2024) Assessment of safe drinking water handling practices in households of Northern India: a cross-sectional study. Cureus 16(3). https://doi.org/10.7759/cureus.55888
88. Ivers LC, Teng JE, Lascher J, Raymond M, Weigel J, Victor N, Jerome JG, Hilaire IJ, Almazor CP, Ternier R, Cadet J (2013) Use of oral cholera vaccine in Haiti: a rural demonstration project. Am J Trop Med Hyg 89(4):617. https://doi.org/10.4269/ajtmh.13-0183
89. Rosa G, Clasen T (2010) Estimating the scope of household water treatment in low-and medium-income countries. Am J Trop Med Hyg 82(2):289. https://doi.org/10.4269/ajtmh.2010.09-0382
90. Hossain M, Rahman SN, Bhattacharya P, Jacks G, Saha R, Rahman M (2015) Sustainability of arsenic mitigation interventions—an evaluation of different alternative safe drinking water options provided in MATLAB, an arsenic hot spot in Bangladesh. Front Environ Sci 3:30. https://doi.org/10.3389/fenvs.2015.00030
91. Aibana O, Franke MF, Teng JE, Hilaire J, Raymond M, Ivers LC (2013) Correction: cholera vaccination campaign contributes to improved knowledge regarding cholera and improved practice relevant to waterborne disease in rural Haiti. PLoS Negl Trop Dis 7(12):10–1371
92. Collins OC, Govinder KS (2016) Stability analysis and optimal vaccination of a waterborne disease model with multiple water sources. Nat Resour Model 29(3):426–447. https://doi.org/10.1111/nrm.12095
93. Beer KD, Gargano JW, Roberts VA, Hill VR, Garrison LE, Kutty PK, et al (2015) Surveillance for waterborne disease outbreaks associated with drinking water—united states, 2011–2012. MMWR Morb Mortal Wkly Rep 64(31):842–848. https://doi.org/10.15585/mmwr.mm6431a2
94. Bounoure F, Mouly D, Beaudeau P, Bentayeb M, Chesneau J, Jones G, Skiba M, Lahiani-Skiba M, Galey C (2020) Syndromic surveillance of acute gastroenteritis using the french health insurance database: discriminatory algorithm and drug prescription practices evaluations. Int J Environ Res Public Health 17(12):4301
95. Atif M, Ayub G, Shakoor F, Farooq M, Iqbal M, Zaman Q, Ilyas M (2024) Evolution of waterborne diseases: a case study of Khyber Pakhtunkhwa Pakistan. SAGE Open Med 12:20503121241263030
96. Gatto M, Mari L, Bertuzzo E, Casagrandi R, Righetto L, Rodriguez-Iturbe I, Rinaldo A (2012) Generalized reproduction numbers and the prediction of patterns in waterborne disease. Proc Natl Acad Sci 109(48):19703–19708. https://doi.org/10.1073/pnas.1217567109
97. Yomwan P, Cao C, Rakwatin P, Suphamitmongkol W, Tian R, Saokarn A (2015) A study of waterborne diseases during flooding using Radarsat-2 imagery and a back propagation neural network algorithm. GeomatS, Nat Hazards Risk 6(4):289–307.https://doi.org/10.1080/19475705.2013.853325
98. Haar R, Abdelrahman S, Muzzall E, Aladhan I, Shobak K, Kawas MR, Aksh A, Blair AH, Reingold A, AlMhawish N (2023) Water and war: the effect of functioning chlorinated water stations in reducing waterborne diseases during conflict in northwest Syria, 2017–2021. PLOS Global Public Health 3(12):e0002696. https://doi.org/10.1371/journal.pgph.0002696
99. Tien JH, Earn DJ (2010) Multiple transmission pathways and disease dynamics in a waterborne pathogen model. Bull Math Biol 72(6):1506–1533. https://doi.org/10.1007/s11538-010-9507-6
100. Nahar N (2009) Impacts of arsenic contamination in groundwater: case study of some villages in Bangladesh. Environ Dev Sustain 11(3):571–588. https://doi.org/10.1007/s10668-007-9130-3

101. Lee EC, Chao DL, Lemaitre JC, Matrajt L, Pasetto D, Perez-Saez J, Finger F, Rinaldo A, Sugimoto JD, Halloran ME, Longini IM (2020) Achieving coordinated national immunity and cholera elimination in Haiti through vaccination: a modelling study. Lancet Glob Health 8(8):e1081–e1089. https://doi.org/10.1016/s2214-109x(20)30310-7
102. Rebaudet S, Bulit G, Gaudart J, Michel E, Gazin P, Evers C, Beaulieu S, Abedi AA, Osei L, Barrais R, Pierre K (2019) The case-area targeted rapid response strategy to control cholera in Haiti: a four-year implementation study. PLoS NeglEd Trop Dis 13(4):e0007263. https://doi.org/10.1371/journal.pntd.0007263
103. Rebaudet S, Dély P, Boncy J, Henrys JH, Piarroux R (2021) Toward cholera elimination, Haiti. Emerg Infect Dis 27(11):2932. https://doi.org/10.3201/eid2711.203372
104. Rahman M, Tondel M, Ahmad SA, Axelson O (1998) Diabetes mellitus associated with arsenic exposure in Bangladesh. Am J Epidemiol 148(2):198–203. https://doi.org/10.1093/oxfordjournals.aje.a009624
105. Bashir MT, Alrowais R, Daniyal M (2021) Arsenic poisoning in Bangladesh, impact on health and ecology and mitigation strategies. Int J Eng Res Technol 10(1):595–601
106. Andrews JR, Basu S (2011) Transmission dynamics and control of cholera in Haiti: an epidemic model. The Lancet 377(9773):1248–1255
107. Alam MT, Weppelmann TA, Weber CD, Johnson JA, Rashid MH, Birch CS, Brumback BA, de Rochars VEMB, Glenn J, Ali A (2014) Monitoring water sources for environmental reservoirs of toxigenic Vibrio cholerae O1 Haiti. Emerg Infect Dis 20(3):356
108. Kirpich A, Weppelmann TA, Yang Y, Morris JG Jr, Longini IM Jr (2017) Controlling cholera in the Ouest Department of Haiti using oral vaccines. PLoS Negl Trop Dis 11(4):e0005482. https://doi.org/10.1371/journal.pntd.0005482
109. Latif S, Alim MA, Rahman A, Haque MM (2023) A review on chlorination of harvested rainwater. Water 15(15):2816. https://doi.org/10.3390/w15152816
110. Frederick FD, Marlim MS, Kang D (2024) Optimization of chlorine injection schedule in water distribution networks using water age and breadth-first search algorithm. Water 16(3):486. https://doi.org/10.3390/w16030486
111. Goyal RV, Patel HM (2015) Analysis of residual chlorine in simple drinking water distribution system with intermittent water supply. Appl Water Sci 5(3):311–319. https://doi.org/10.1007/s13201-014-0193-7
112. Siomara G, Wilber Q, Doris E, Torrico-Irahola R (2024) On-site chlorine: a promising technology in drinking water treatment in Santa Cruz Bolivia. Water 16(12):1738. https://doi.org/10.3390/w16121738
113. Ly YS, Hairom NHH, Harun NH, Fhong SC (2022) A LoRa-cloud based water pH and air temperature sensor hub: IoT water quality monitoring sensor hub. Emerg Adv Integr Technol 3(1):15–24. https://doi.org/10.30880/emait.2022.03.01.003
114. Fan Y, Zhu Q, Liu Y (2018) Cloud/fog computing system architecture and key technologies for south-north water transfer project safety. Wirel Commun Mob Comput 2018(1):7172045. https://doi.org/10.1155/2018/7172045
115. Kamienski C, Soininen JP, Taumberger M, Dantas R, Toscano A, Salmon Cinotti T, Filev Maia R, Torre Neto A (2019) Smart water management platform: IoT-based precision irrigation for agriculture. Sensors 19(2):276. https://doi.org/10.3390/s19020276
116. Коваленко А (2024) Application of iot protocols in surface water pollution monitoring systems. Cybersecur Educ Sci Tech 3(23):85–96. https://doi.org/10.28925/2663-4023.2024.23.8596
117. Yoo DG, Lee SM, Lee HM, Choi YH, Kim JH (2018) Optimizing re-chlorination injection points for water supply networks using harmony search algorithm. Water 10(5):547.https://doi.org/10.3390/w10050547
118. Abdurrafi A, Maulana D, Kurniadi NT (2023) Optimization water conservation through IoT sensor implementation at Smartneasy Nusantara. J Appl Intell Syst 8(3):432–441. https://doi.org/10.33633/jais.v8i3.9475
119. Stolecka K (2019) Analysis of the accidental release of chlorine. Syst Saf: Hum-Tech Facil-Environ 1(1). https://doi.org/10.2478/czoto-2019-0019

120. Ogwu MC, Izah SC, Sawyer WE, Amabie T (2025) Environmental risk assessment of trace metal pollution: a statistical perspective. Environ Geochem Health 47:94. https://doi.org/10.1007/s10653-025-02405-z
121. Izah SC, Jacob DE, Nelson IU, Avez S (2024b) Urban water crisiș in the global south. In: Izah SC, Ogwu MC, Loukas A, Hamidifar H (eds) Water crises and sustainable management in the global south. Springer, Singapore, pp 45–83. https://doi.org/10.1007/978-981-97-4966-9_2
122. Okaka FO, Odhiambo BD (2018) Relationship between flooding and out break of infectious diseasesin Kenya: a review of the literature. J Environ Public Health 2018(1):5452938. https://doi.org/10.1155/2018/5452938
123. Seiyaboh EI, Izah SC, Oweibi S (2017) Assessment of Water quality from Sagbama Creek, Niger Delta Nigeria. Biotechnol Res 3(1):20–24
124. Seiyaboh EI, Izah SC (2017) Review of Impact of anthropogenic activities in surface water resources in the Niger delta region of Nigeria: a case of Bayelsa state. Int J Ecotoxicol Ecobiology 2(2):61–73
125. Seiyaboh EI, Izah SC, Bokolo JE (2017) Bacteriological quality of water from river nun at Amassoma Axises, Niger Delta, Nigeria. ASIO J Microbiol, Food Sci Biotechnol Innov 3(1):22–26
126. Sente C, Erume J, Naigaga I, Mulindwa J, Ochwo S, Magambo PK, Namara BG, Kato CD, Sebyatika G, Muwonge K, Ocaido M (2016) Prevalence of pathogenic free-living amoeba and other protozoa in natural and communal piped tap water from Queen Elizabeth protected area, Uganda. Infect Dis Poverty 5(04):84–97. https://doi.org/10.1186/s40249-016-0162-5

Chapter 5
Low-Cost Water Filtration Technologies for the Purification of Water

Nand Kumar Kashyap, Amit Kumar Chaturwedi, Harishankar Dewangan, Ashish Tiwari, Kiran Thakur, and Milan Hait

Abstract For underserved communities around the world, access to clean and safe drinking water continues to be a major challenge. Conventional water treatment techniques are costly, and many communities lack access to such systems, creating a demand for inexpensive, effective, and sustainable filtration technologies. Some other low-cost water purification technologies exist to solve this global problem; this chapter aims to describe some of them targeted to low-income and rural populations. We also describe the potential of specific innovative filtration approaches, including ceramic filters, bio sand filtration, activated carbon and solar disinfection (SODIS), to be both effective and affordable. We also highlight emerging technologies such as nanotechnology-based filtration, membrane filtration with locally available materials, and bio-inspired methods of purification. We particularly focus on community-based methods as well as decentralized water purification systems that enhance the local populace through capacity building and sustainable management practices. It discusses social enterprises and NGOs, government initiatives that support the scale up of these technologies. Well documented case studies from diverse geographical settings are documented data indicating the degree of reduction in occurrence rate of the waterborne diseases as a result of implementing the above solutions. This chapter addresses the technology gap in three aspects of filtration technology affordability, adaptability, and environmental sustainability by showing how innovative but low-cost technologies can help to close the water access gap. The results highlight the importance of interdisciplinary engagement and policy support to further refine and scale these interventions. Ultimately, this study can motive further

N. K. Kashyap · A. K. Chaturwedi · M. Hait (✉)
Department of Chemistry, Dr. C. V. Raman University, Kota, Bilaspur, Chhattisgarh, India
e-mail: haitmilan@gmail.com

H. Dewangan
Department of Chemistry, Government Naveen College, Janjgir-Champa, Baloda, Chhattisgarh, India

A. Tiwari
Department of Chemistry, Government College Dumariya, Jarhi, Surajpur, Chhattisgarh, India

K. Thakur
Department of Chemistry, Government Bilasa Girls P.G. College, Bilaspur, Chhattisgarh, India

M. C. Ogwu and S. Chibueze Izah (eds.), *Water Quality and Safety in the Global South*, Springer Water, https://doi.org/10.1007/978-3-032-10602-5_5

research and investment in scalable, cost-effective water-filtration technologies for vulnerable groups.

Keywords Sustainable technology · Low-cost filtration · Renewable energy · Water treatment · Innovative water purification solution

Abbreviations

As	Arsenic
CDs	Carbon dots
CO_2	Carbon dioxide
DBPs	Disinfection Byproducts
DNA	Deoxyribonucleic acid
EPA	Environmental Protection Agency
HWISE	Household Water Insecurity Experiences
IONPs	Inorganic nanoparticles
LED	Light Emitting Diode
LNGs	Liquefied natural gases
MCL	Maximum Contaminant Level
MWCNTs	Multi Walled Carbon Nanotubes
NGOs	Non-Government Organizations
NO_2	Nitrogen dioxide
NPs	Nanoparticles
ONPs	Organic nanoparticles
PONPs	Polymeric nanoparticles
POU	Point-of-Use
PPPs	Public-Private Partnerships
RGV	Rio Grande Valley
RO	Reverse Osmosis
SO_2	Sulfar dioxide
SODIS	Solar Disinfection
SWCNTs	Single Walled Carbon Nanotubes
TCEQ	Texas Commission on Environmental Quality
USEPA	United States Environmental Protection Agency
UV	Ultraviolet
WHO	World Health Organization

5.1 Introduction

Clean drinking water remains elusive globally [1–3] especially in rural backwaters where fancy water purifiers are woefully unaffordable or altogether absent. Sustainable potable water solutions are offered by low-cost filtration technologies ensuring access for people in economically disadvantaged remote regions. Innovative filtration methods prioritize ease of use and efficiency quite practically for household implementation and some community-level applications too. Solutions like bio-sand filters and solar disinfection purify water by removing bacteria viruses and heavy metals without expensive infrastructure or electricity. Advancements in nanotechnology and membrane filtration have vastly improved effectiveness of low-cost solutions remarkably under various operating conditions [4]. These filtration methods leveraging locally sourced stuff empower communities pretty effectively to take charge of their water security. Social enterprises and shrewd government policies play a crucial role in scaling these innovative technologies up rather quickly nationwide [5]. This chapter explores various low-cost water purification techniques, their benefits, and the challenges and opportunities in making safe drinking water accessible to all, especially in resource-limited settings. The ongoing drinking water crisis in the United States starkly highlights a pressing failure in infrastructure and governance, leaving over two million Americans without reliable access to running water or basic sanitation. This alarming situation is exacerbated by communities grappling with inadequate water connections, often resulting in the costly necessity of resorting to private water sources. The root causes of these challenges are deeply entrenched, stemming from a combination of careless development practices, political discord, insufficient regulatory enforcement, and daunting environmental obstacles. As the nation grapples with these multifaceted issues, it becomes increasingly clear that addressing the drinking water crisis is not merely an act of humanitarian concern but an essential step towards ensuring equity and sustainability for all citizens [6]. The pervasive issue of arsenic contamination in Texas water systems underscores a pressing public health crisis that demands urgent attention. With 65 community water systems exceeding the Safe Drinking Water Act standards, it is alarming to note that approximately 51,000 residents across 34 communities have been unwittingly exposed to this toxic substance for over a decade. This situation is particularly dire along the Texas-Mexico border, where the impacts of arsenic-laden water not only threaten individual health but also strain community resources and erode trust in local governance. As the state grapples with these challenges, it becomes increasingly vital to prioritize remediation efforts and implement stringent regulations to safeguard the health of its residents and ensure access to safe drinking water for all [7]. The presence of arsenic in U.S. drinking water poses a significant public health risk, as evidenced by its classification by the U.S. Environmental Protection Agency as a Group A human carcinogen. Historically, the maximum contaminant level (MCL) was set at 50 μg/L; however, recent assessments have shown that this standard is no longer adequate to safeguard human health. The growing body of evidence highlights the urgent need for stricter regulations and lower limits on arsenic levels in drinking water to protect communities and prevent the serious health consequences associated with long-term exposure. Addressing this issue is not merely a regulatory challenge but a moral imperative, as ensuring safe drinking water is fundamental to the well-being of all citizens. Communities must come together to advocate for these necessary changes, raising awareness about the risks of arsenic contamination and supporting policymakers in implementing more stringent standards. By prioritizing public health over industrial convenience, we can create a safer environment for future generations [8]. The pervasive threat of arsenic exposure, with levels ranging from 50 to 10 μg/L, unveils a serious public health crisis that cannot be ignored. Individuals suffering from arsenic poisoning endure a range of debilitating symptoms, including headache, diarrhea, vomiting, and muscle cramps, which significantly diminish their quality of life. In addition to these immediate problems, arsenicosis is linked to serious health issues like chronic bronchitis and diabetes, as well as a higher chance of developing cancers of the skin, lungs, and urinary bladder. As more evidence shows how harmful arsenic is to human health, it becomes more important to raise awareness and take action to reduce exposure and protect public health. Such action includes implementing stricter regulations on industrial emissions, promoting the use of safe drinking water sources, and educating communities about the dangers of arsenic contamination. By prioritizing these measures, we can work toward a healthier future for those at risk and ultimately reduce the incidence of arsenicosis and its associated health complications. Basu et al. [9] conducted a study in West Bengal, India, using a two-step oxidation and adsorption technique to remove arsenic (As) from groundwater. The results of the water treatment study underscore the effectiveness of utilizing standard plastic shower heads in enhancing dissolved oxygen levels to above 4 mg/L, a crucial factor for promoting aquatic life and improving water quality. The comprehensive design of the treatment plant—including a storage tank, oxidation station, and fortified water supply pipes—demonstrates a robust approach to ensuring safe and sustainable water management. Notably, the achievement of arsenic levels below the World Health Organization's recommended standard of 10 mg/dL not only highlights the plant's commitment to public health but also reinforces its role in safeguarding environmental integrity. Together, these findings illustrate the potential for innovative treatment solutions that can significantly improve both human and ecological well-being [9]. The potential of plants to generate potable water stands as a promising solution to one of the world's most pressing challenges. By harnessing natural processes, these systems provide clean drinking water and do so without producing harmful sludge, thereby minimizing environmental impact. Furthermore, the astonishingly low daily operational cost of just one dollar makes this approach not only sustainable but also accessible for communities struggling with water scarcity. As we continue to explore innovative solutions for water generation, the role of plant-based systems could become integral in ensuring that clean water is within reach for everyone [10]. The Biosand Filter stands out as an innovative and economically viable solution for addressing the critical issue of arsenic contamination in drinking water. Constructed from readily available materials such as nails and sand, this filter not only boasts a remarkably low cost of just five US dollars but also achieves an impressive 98% effectiveness in arsenic removal. Rural, low-income communities, often disproportionately affected by waterborne contaminants, find it particularly well-suited due to its affordability and efficiency [11–13]. By implementing this simple yet powerful technology, we can take significant strides toward ensuring safe drinking water for vulnerable populations, ultimately fostering improved health outcomes and enhancing quality of life in these areas. The ongoing water crisis along the Texas-Mexico border highlights a complex web of challenges stemming from unreliable water quality and service interruptions, which together contribute to pervasive water shortages in the region. Many communities face limited access to safe running water, intensifying the already difficult living conditions. Furthermore, the alarming presence of arsenic contamination reveals not only environmental neglect but also significant public health risks for residents. This multifaceted issue calls for immediate attention and concerted efforts from both local authorities and government agencies to ensure safe, reliable water sources, ultimately safeguarding the health and well-being of affected populations while aiming for long-term sustainability in this critical border region [14]. The imperative for clean water access underscores the critical nature of ongoing research into filtration systems designed to eliminate harmful contaminants, such as arsenic. As studies reveal significant gaps in our understanding and application of these technologies, they illuminate the urgent need for practical interventions that can be implemented in communities most affected by water quality issues [15]. Such advancements promise to enhance public health and serve as a catalyst for health equity and environmental justice initiatives, ensuring that every individual has their fundamental right to safe drinking water upheld. By addressing these challenges head-on, we can foster a more just society where clean water is not a privilege but a universal standard [16].

The innovative water filtration methods applied by Barnaby et al. [17] reveal a multifaceted approach to addressing global water quality challenges. By using activated alumina and iron coatings, these methods improve the ways water particles stick together and settle, making them useful in different situations. The direct treatment of water within aquifers promotes efficiency and harnesses the natural sedimentation process, allowing for cleaner water through passive means [10]. Additionally, solar oxidation provides an eco-friendly option that leverages renewable energy for purification. The spectrum of solutions, from commercial filtration systems to home multi-stage filtration, illustrates the versatility necessary to meet diverse needs, ensuring that clean water access can be tailored to both large-scale and individual contexts. Ultimately, these methods represent a significant step forward in sustainable water management strategies crucial for the health of communities globally [18]. These methods have shown promising results in improving water quality by effectively removing contaminants. The comprehensive approach taken by Barnaby et al. emphasizes the need for diverse strategies to tackle varying pollution challenges in different environments. The innovative Mesopaper, crafted from bamboo fibers by Nicomel et al. [19], stands as a beacon of hope for the communities along the Texas-Mexico border, where water scarcity has been a pressing concern since the 1980s. This groundbreaking filter not only offers a sustainable solution to a long-standing crisis but also exemplifies how resourcefulness and environmental stewardship can converge to create practical applications that benefit vulnerable populations. By harnessing the potential of bamboo—a rapidly renewable resource—Mesopaper underscores the importance of addressing critical issues with ingenuity and community-focused solutions. As these border communities continue to grapple with water accessibility challenges, innovations like Mesopaper pave the way for a more resilient future, demonstrating that effective change is achievable through creativity and collaboration [19]. The HWISE Network Initiative stands at the forefront of addressing the pressing issue of water insecurity by championing innovative solutions such as water filters made from bamboo charcoal, which exemplify a harmonious blend of scientific research and practical application [20]. By fostering a collaborative approach that unites researchers, communities, and stakeholders, the initiative not only aims to tackle immediate challenges but also seeks to establish sustainable practices for the long term. This commitment to informed research and community-driven strategies underscores a proactive stance toward improving public health and overall well-being. Ultimately, the HWISE Network Initiative embodies a vision where resilient water solutions pave the way for healthier communities and a more secure future [21]. The pressing issue of arsenic contamination, primarily stemming from agricultural pesticide use and mining activities, underscores the intricate relationship between environmental health and global water scarcity. With approximately 1.2 billion people facing the dire consequences of inadequate access to clean water, the presence of arsenic in groundwater exacerbates health risks and highlights the urgent need for sustainable practices in agriculture and industry [22]. In regions like the Rio Grande Valley, where water pollution is rampant, addressing these challenges becomes essential for ensuring safe drinking water and protecting vulnerable populations. As we confront the dual crises of arsenic contamination and water scarcity, a concerted effort toward innovative solutions and stricter regulations will be vital to safeguarding our most precious resource—clean water—for generations to come [23].

5.1.1 Global Water Shortages and Their Severely Impact

Addressing the challenge of waterborne diseases is imperative not only for improving public health but also for fostering equitable opportunities within communities. The staggering mortality rates from illnesses like cholera and typhoid underscore the urgency of tackling this issue, especially as women and children bear the brunt of water collection responsibilities, which hampers their educational and economic prospects [24]. The high costs associated with effective filtration technology further exacerbate existing health disparities, leaving vulnerable populations at greater risk. However, the emergence of innovative, low-cost water purification methods offers a beacon of hope, paving the way for sustainable solutions that empower communities to take charge of their health and well-being. By prioritizing these accessible technologies, society can create a healthier future while dismantling the barriers that perpetuate inequality [25]. This collaborative approach enhances public health and

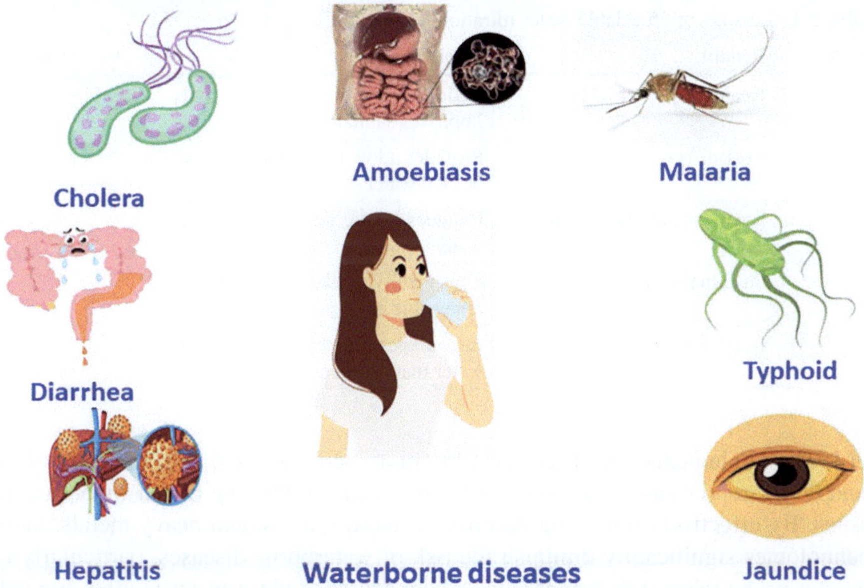

Fig. 5.1 Waterborne diseases that affect humans

fosters economic development, as clean water is essential for productivity and growth. Ultimately, investing in these technologies can lead to a ripple effect, improving quality of life and promoting social equity across diverse communities. Water borne diseases that affect human life are given in Fig. 5.1.

5.1.2 Significance of Relatively Cheap Technologies for Filtering Water

Millions face severe contamination issues in underserved communities lacking safe drinking water essential for human health and wellbeing ordinarily. Sustainable solutions emerge from fairly affordable water filtration tech enabling communities purify water sans reliance on ridiculously pricey infrastructure overnight. Low-cost filtration methods like ceramic filters and bio sand filters effectively remove bacteria and viruses and heavy metals reducing waterborne disease risk significantly [24]. Technologies like these prove especially beneficial in disaster-prone regions and rural backwaters lacking conventional facilities for water treatment. These filtration methods help slash healthcare costs and boost child survival rates by making clean water ridiculously affordable for people. Implementing community-based filtration solutions promotes self-sufficiency and resilience, ensuring long-term sustainability

Table 5.1 Benefits of affordable water filtration techniques

S. N	Benefit	Application	References
1	Health improvement	Reduces waterborne diseases and improves sanitation	[26]
2	Cost-effective	Provides a low-cost alternative to bottled water	
3	Environmentally friendly	Reduces plastic waste from bottled water	
4	Sustainability	Uses locally available materials and renewable energy	
5	Community empowerment	Encourages self-reliance and local water management	

[26]. The implementation of low-cost filtration methods, such as ceramic and bio-sand filters, represents a transformative approach to addressing global water quality issues. By effectively removing harmful bacteria, viruses, and heavy metals, these technologies significantly diminish the risk of waterborne diseases, particularly in disaster-prone regions and rural areas where access to clean water is often limited. The positive impact on healthcare costs and child survival rates further underscores the importance of these filtration systems; they not only enhance public health but also foster self-sufficiency and resilience within vulnerable communities. As we continue to explore sustainable solutions for clean water access, the adoption of these simple yet effective filtration methods stands out as a crucial step toward improving the quality of life for countless individuals around the world. The benefits of affordable water filtration techniques are given in Table 5.1.

5.2 Traditional and Low-Cost Water Filtration Methods

5.2.1 Ceramic Filters and Biosand Filtration

Ceramic filters and bio sand filtration are prominent water purification methods. Ceramic filters are made from porous ceramic clay that traps pathogen and particles. Some models incorporated silver for enhanced antibacterial action, an often-affordable anti microband for supply as affordable and straight forward. Archicly, bio sand filtration used bio sand filtration as a sustainable effective and sustainable. The exploration of water purification methods reveals the remarkable efficacy and sustainability offered by ceramic filters and biosand filtration. Ceramic filters serve as a frontline defense against pathogens and particulate matter, ensuring that water remains safe for consumption. Notably, certain models incorporate silver to boost their antibacterial properties, further enhancing their protective capabilities. On the other hand, biosand filtration stands out as an environmentally friendly solution that

Table 5.2 Ceramic filters and biosand filtration techniques

Technique	Principles: a summary	References
Ceramic filters	Ceramic filters of porous pathogens and particles. Some models incorporated silver for antibacterial action	[28]
Bio sand filtration	Water passes through a bed of sand as a biological layer on surface for ends, pathogens removed	[29]

purifies water and promotes sustainability through its use of natural materials [27]. The path toward ensuring global water quality hinges on the adoption of innovative approaches that marry tradition with modern technology, fostering healthier communities and paving the way for a sustainable future. As we look ahead, the potential integration of smart sensors promises to revolutionize real-time water quality monitoring, enabling swift responses to contamination issues and enhancing public safety. By combining time-tested methods with cutting-edge innovations, we can create resilient water supply systems capable of adapting to the challenges posed by climate change and population growth. This holistic approach safeguards our most vital resource and empowers communities to thrive in a world where clean water is no longer a luxury but a fundamental right [16]. Some filters are highly effective and sustainable. These filters provide excellent purification and minimize environmental impact, making them a responsible choice for consumers. Their design and materials contribute to long-term use without compromising performance. This implies that they not only improve water quality but also align with eco-friendly practices, thereby encouraging individuals to make thoughtful decisions. As awareness of environmental issues grows, such filters are becoming increasingly popular among households and businesses alike. Ceramic filters and biosand filtration techniques and their principle descriptions are given in Table 5.2 and Fig. 5.2 respectively.

5.2.2 *Activated Carbon Filtration*

Activated carbon filtration is an innovative, low-cost water purification technology widely used for removing contaminants such as chlorine, pesticides, and volatile organic compounds. Derived from coconut shells or coal, activated carbon has a highly porous surface that enables effective adsorption of impurities. This method enhances taste and odor while maintaining essential minerals in drinking water. Its affordability, ease of use, and reusability make it ideal for underserved and rural communities. Activated carbon filters are often integrated into household and portable water systems, contributing significantly to public health by reducing exposure to harmful pollutants [30]. Innovation means continuously improving efficiency markedly alongside sustainability very slowly over time. The remarkable capabilities of activated carbon filtration underscore its significance as a transformative solution in water purification. By utilizing this innovative and cost-effective technology, we can effectively eliminate a range of contaminants, including chlorine, pesticides,

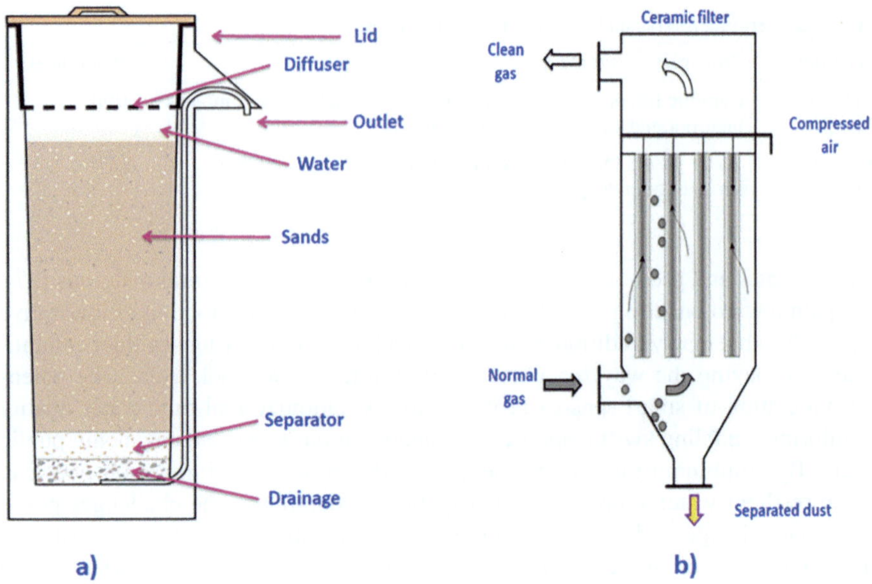

Fig. 5.2 Filtration techniques **a** biosand techniques; **b** ceramic techniques

and volatile organic compounds that threaten water quality. Sourced from environmentally friendly materials like coconut shells or coal, activated carbon boasts a highly porous surface that maximizes impurity adsorption, ensuring clean and safe drinking water. Beyond just purification, this method enhances the taste and odor of water while preserving essential minerals that contribute to overall health. As we seek sustainable solutions for clean water access, activated carbon filtration stands out as an essential tool for improving our water systems and promoting better public health outcomes (Fig. 5.3).

5.2.3 *Charcoal-Based Filtration*

Charcoal-based filtration serves as an effective low-cost method for purifying water widely used in rural communities and some urban areas. Activated charcoal, derived mostly from coconut shells or sometimes wood, has a remarkably high surface area that adsorbs impurities efficiently. Filtration techniques enhance taste and odor, remarkably improving overall water quality greatly beneath the surface [31]. Charcoal-based systems are ideally suited for underserved populations owing largely to their dirt-cheap cost and pretty straightforward reusability. Sustainable charcoal production innovations and quirky compact filter designs help scale this tech globally with considerable alacrity in various regions. Charcoal filtration emerges as

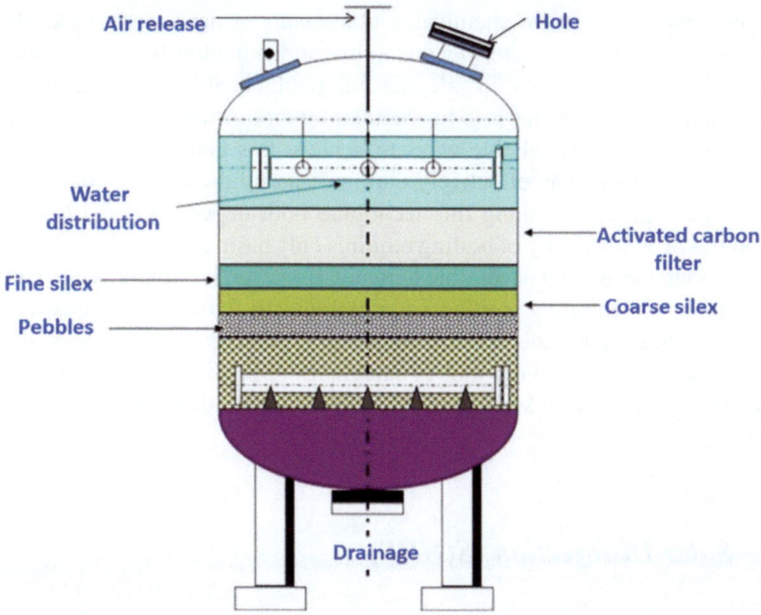

Fig. 5.3 Activated carbon filtration

a promising fix for safe drinking water needing minimal upkeep and zero electricity [32]. The implementation of charcoal-based filtration systems represents a transformative approach to water purification, particularly in both rural and urban settings where access to clean water can be limited. By utilizing activated charcoal sourced from sustainable materials like coconut shells and wood, these filters effectively adsorb a wide range of impurities, making them a low-cost yet powerful solution. Beyond merely removing contaminants, the techniques employed in this filtration process also enhance the taste and odor of water, significantly improving its overall quality. As communities increasingly recognize the efficacy and affordability of charcoal-based systems, the potential for widespread adoption could lead to healthier populations and a more sustainable future for global water resources.

5.2.4 Boiling Filtration

Boiling water is a ridiculously simple, super effective purification method used globally, especially in areas with scarce resources. Harmful pathogens like bacteria and viruses get killed pretty effectively by bringing water to a rolling boil for roughly sixty seconds, making it safe for drinking. Boiling needs pretty basic gear like a heat source and container, making it rather suitable for nasty emergency situations or super remote communities. It significantly reduces the risk of waterborne diseases, although it does not remove chemical contaminants or necessarily make the taste somewhat better. Boiling's utter dependability and ridiculously low expense make it an indispensable instrument in safeguarding public health pretty effectively [33]. Ensuring safe drinking water is a fundamental public health measure, and boiling remains one of the most reliable ways to achieve this goal. By heating water for about 60 s, individuals can effectively eliminate harmful pathogens that contribute to waterborne diseases, making this technique both dependable and low-cost for communities. The simplicity of boiling requires only basic equipment—a heat source and a container—making it accessible to many. However, it is important to recognize its limitations; while boiling purifies water biologically, it does not address chemical contaminants or enhance the taste of the water. Despite these drawbacks, the practice of boiling serves as a vital tool in safeguarding public health and underscores the importance of accessible methods for ensuring safe drinking water in various settings.

5.2.5 Solar Disinfection (SODIS)

SODIS an innovative low-cost method purifies water rather effectively by harnessing sunlight and subsequently eliminating most pathogens lurking in it. SODIS involves filling transparent plastic bottles with contaminated water and putting them under direct sunlight for several hours slowly. UV-A rays coupled with elevated temperature pretty effectively annihilate various bacteria viruses and protozoa [34]. SODIS employs an eerily straightforward method extremely sustainable and eco-friendly requiring neither noxious chemicals nor massive amounts of electricity. It remains highly effective in improving microbiological water quality although failing utterly to remove nasty chemical pollutants and toxic heavy metals. Such a method proves especially valuable in arid regions beneath intense sunlight and with scarce access to pristine water sources [35]. The SODIS method stands out as an ingenious and accessible solution for water purification, particularly in regions lacking advanced infrastructure. By harnessing the power of sunlight, this low-cost technique transforms contaminated water into a safe drinking source. The process is straightforward: filling transparent plastic bottles with water and leaving them exposed to sunlight allows UV-A rays and increased temperatures to effectively eradicate harmful bacteria, viruses, and protozoa. This innovative approach promotes health and empowerment by providing a sustainable means of ensuring access to clean water. Thus, SODIS exemplifies how simple yet effective strategies can address critical public health challenges (Fig. 5.4 and Table 5.3).

Fig. 5.4 Solar water disinfections (SODIS) techniques

Table 5.3 Advantages and limitations of SODIS

Aspect	Advantages	Limitations	References
Cost	Very low-cost; uses freely available sunlight	Limited to sunny regions and requires consistent sunlight	[36]
Simplicity	Easy to implement; requires only transparent plastic bottles	Not effective on cloudy or rainy days	
Energy requirement	No external energy source needed	Requires 6–48 h of exposure depending on weather conditions	
Effectiveness	Kills bacteria, viruses, and some protozoa through UV-A and heat	Does not remove turbidity, chemicals, or heavy metals	
Materials needed	Reuses PET bottles; easily accessible	PET bottles degrade over time and need replacement	
Environmental impact	Eco-friendly with no carbon emissions	Risk of microplastic release from prolonged PET bottle usage	
Scalability	Ideal for individual or household use	Not suitable for large-scale or community-level water needs	

5.2.6 UV-Purification

UV purification is a modern, innovative, low-cost water treatment method that utilizes ultraviolet light to disinfect water. It works by exposing contaminated water to UV-C light, which penetrates microorganisms and disrupts their DNA, rendering them inactive and unable to reproduce. This technology is particularly effective against bacteria, viruses, and protozoa, making it an excellent option for ensuring microbiological safety in drinking water [37]. UV systems are compact, easy to install, and require low maintenance. However, UV purification does not remove heavy metals, sediments, or chemical contaminants, so it is often used alongside filtration systems. It requires electricity or solar power to operate, making solar-powered UV purifiers ideal for off-grid communities. With advancements in LED-based UV systems, these devices are becoming more affordable and accessible in developing regions [38]. The transformative power of UV purification stands out as a modern and cost-effective solution in the quest for safe drinking water. By harnessing the energy of ultraviolet light, this method effectively disinfects contaminated sources, targeting harmful microorganisms with precision. The exposure to UV-C light disrupts the DNA of bacteria, viruses, and protozoa, rendering them inactive and ensuring microbiological safety for consumers. As global awareness surrounding water quality continues to grow, the adoption of UV purification systems not only addresses immediate health concerns but also represents a sustainable approach to water treatment that prioritizes both efficacy and environmental responsibility. Embracing this advanced technology paves the way for healthier communities and a cleaner future (Fig. 5.5 and Table 5.4).

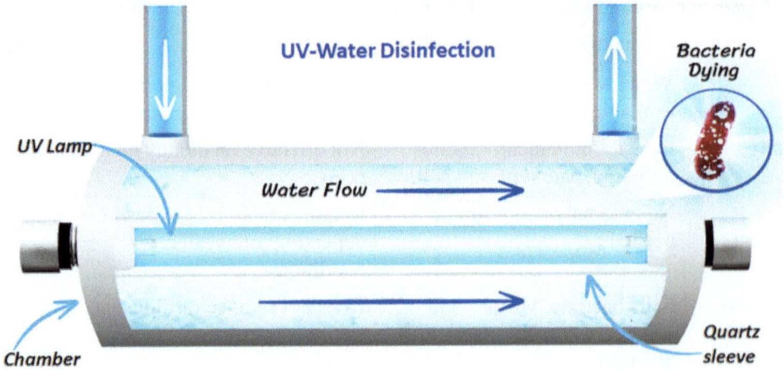

Fig. 5.5 UV-water disinfection techniques

Table 5.4 Benefits and limitations of UV purification

Aspect	Details	References
Effectiveness	High against pathogens (bacteria, viruses, protozoa)	[39, 40]
Cost	Moderate; decreasing with LED technology	
Power requirement	Requires electricity or solar energy	
Maintenance	Low; periodic cleaning and lamp replacement	
Limitation	Does not remove chemical pollutants or turbidity	
Best use	In combination with filtration for comprehensive water purification	

5.3 Innovative Low-Cost Water Purification Technologies

Nanotechnology-based filtration signifies a notable upward mobility; however, numerous other forward-thinking and cost-effective water purification technologies are being developed and implemented worldwide, especially in areas confronting limited water supplies and affordability issues. These innovations not only aim to improve access to clean water but also seek to enhance sustainability by reducing waste and energy consumption. As communities adopt these technologies, the potential for improved public health and economic growth becomes increasingly promising.

5.3.1 Nanotechnology-Based Water Filtration

Filtration utilizing nanotechnology employs materials meticulously designed at the nanoscale (one billionth of a meter) to produce filters that are both highly efficient and selectively effective. The filters offer significant advantages over regular filtration methods because of the unique features of nanomaterials, such as their large surface area compared to their volume and the ability to change the size of their pores. The various types of nanoparticles have been employed in water remediations, like ONPs, IONPs, and PONPs [41]. The remarkable advancements in nanoscale filters have revolutionized the field of water remediation, showcasing their unparalleled efficiency and selectivity in removing contaminants. With their large surface area and the ability to adjust pore sizes, these filters are adept at targeting specific pollutants while preserving essential water properties. Adding organic, inorganic, and polymer nanoparticles makes these filters even better, providing flexible options for different water quality problems. As we continue to confront global water scarcity and pollution issues, the potential of nanoscale filtration technologies stands out as a beacon of hope for sustainable water management practices that can secure clean drinking water for future generations.

5.3.1.1 Organic Nanoparticles (ONPs)-Based Water Filtration

The remarkable properties of organic nanomaterials position them at the forefront of modern science and technology. Their unique electrical, chemical, and physical characteristics unlock a vast array of applications, revolutionizing fields such as electronics, materials science, and biomedicine. Researchers are always finding new ways to use important nanostructures like single-walled carbon nanotubes (SWCNTs), multi-walled carbon nanotubes (MWCNTs), graphene, carbon dots, and fullerenes like C60 and C540 for creative solutions. As we advance our understanding of these organic nanomaterials, their potential to drive technological progress becomes increasingly apparent, paving the way for a future where their exceptional capabilities can be fully realized [42]. The remarkable attributes of carbon dots position them at the forefront of nanomaterial innovation, capturing attention in various fields due to their biocompatibility and low toxicity. Sourced from abundant raw materials, these quasi-spherical nanoparticles, with diameters under 10 nm, offer a cost-effective solution for numerous applications. Their great surface features improve chemical stability and water solubility, and their natural ability to glow creates new opportunities in imaging and sensing technologies. As research continues to unlock their full potential, carbon dots are not just a trend but a transformative force that promises to redefine material science and its practical applications [4]. The potential of carbon nanomaterials, such as single-walled nanotubes (SWNT), multi-walled nanotubes (MWNT), and graphene, in the realm of decontamination cannot be overstated. Their remarkable adsorption characteristics enable them to effectively capture both organic and inorganic pollutants from contaminated water, making them invaluable in environmental remediation efforts. However, the efficacy of these materials hinges on appropriate surface modification and activation techniques, which enhance their interaction with various contaminants. As research continues to advance in this field, the integration of carbon nanomaterials into decontamination strategies promises not only to improve water quality but also to pave the way for innovative solutions to environmental challenges [43, 44]. Photocatalysis involves using light to accelerate a chemical reaction that breaks down pollutants in water, making it a powerful method for purification. This process complements adsorption techniques by providing an alternative means to remove contaminants. By utilizing light to enhance chemical reactions, photocatalysis offers an efficient way to degrade harmful substances, thereby improving overall water quality. This method serves as a valuable addition to existing purification technologies, broadening the options available for effective contamination removal [42].

5.3.1.2 Inorganic Nanoparticles (IONPs)-Based Water Filtration

The exploration of inorganic nanoparticles in water remediation has unveiled a promising pathway to address the pressing issues of heavy metal contamination and organic pollutants. Key materials in this area, like titanium dioxide (TiO_2) nanoparticles, silver nanoparticles (Ag-NPs), iron and iron oxide nanoparticles (Fe and FeO-NPs), copper oxide nanoparticles (CuO-NPs), and silica nanoparticles, have shown great success in trapping harmful substances from water. Their unique properties enhance pollutant removal and highlight the potential for innovative applications in environmental cleanup technologies. By harnessing the capabilities of these multifunctional nanomaterials, we can pave the way for more sustainable solutions to safeguard our water resources, ensuring a cleaner future for generations to come [45]. Mesoporous silica nanoparticles, with their remarkable properties, are poised to revolutionize the field of decontamination. Their rapid kinetics and impressive adsorption capacity underscore their effectiveness in removing pollutants, while their inherent elasticity allows for versatility in various applications. The ability to change the size of the pores and modify the surface makes them even more attractive, allowing for customized solutions to address specific pollution problems. Additionally, the presence of -OH groups is important for changing the surface, which boosts the nanoparticles' ability to find and remove different types of pollutants. These features not only show that mesoporous silica nanoparticles can be an effective tool for cleaning up pollution but also open up possibilities for new methods in environmental cleanup [46]. The remarkable properties of mesoporous silica nanoparticles position them as a transformative solution in the field of decontamination. Their rapid kinetics and exceptional adsorption capacity, combined with their inherent elasticity, make them highly effective in addressing a wide range of contaminants. The adjustable pore diameter allows for tailored approaches to specific pollutants, while surface functionalization enhances their efficacy even further. Particularly noteworthy is the role of -OH groups, which facilitate targeted surface modifications that optimize the interaction with various pollutants. As research continues to unlock the full potential of these nanoparticles, it becomes increasingly clear that they offer a versatile and powerful tool in the ongoing quest for cleaner environments and improved public health [47].

5.3.1.3 Polymeric Nanoparticles (PONPs)-Based Water Filtration

The quest for enhancing nanoparticle stability reveals a promising pathway through the innovative use of polymeric hosts. While traditional nanoparticles often face challenges such as aggregation that diminishes their adsorption capacity, integrating them into a polymeric matrix offers a viable solution. This approach retains the nanoparticles effectively and employs stabilizing agents, surface-modification ligands, emulsifiers, and surfactants that work synergistically to enhance performance and prevent instability. As researchers keep studying these hybrid systems, it's becoming clearer that they could lead to better use of nanoparticles in technology and medicine [48]. This could result in improved drug delivery systems, more efficient catalysts, and advanced materials with tailored properties. The ongoing exploration of these innovative combinations promises to unlock new applications that were previously unattainable with conventional methods. The advancements in membrane technologies represent a significant leap forward in addressing the critical challenges of contamination detection and removal. By effectively identifying and eliminating heavy metals, organic pollutants, gases, and microbes, these innovative solutions play a vital role in safeguarding environmental and public health. Some of the most exciting advancements include nanoparticle and nanofiber membranes that improve how well we filter things, aligned nanotube membranes that are very accurate; self-assembled two-dimensional layer materials that boost performance; and composite membranes that mix different useful features. Collectively, these technologies offer effective means for contamination management and pave the way for a cleaner, safer future as we strive to protect our ecosystems from the pervasive threats of pollution [49]. The advancements in water purification technologies, particularly through the use of nanoparticle and nanofiber membranes, represent a significant leap toward

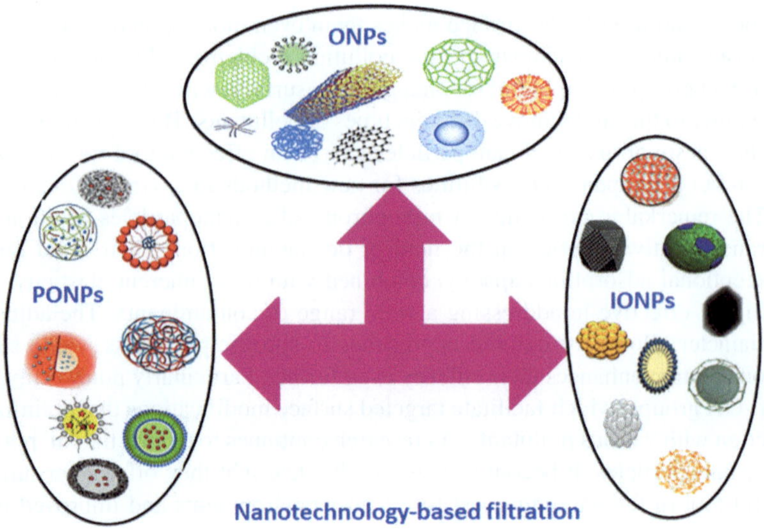

Fig. 5.6 Nanotechnology-based water filtration

efficient filtration processes that improve water quality and address pressing environmental safety concerns. By harnessing self-assembled two-dimensional layer materials, these innovations pave the way for sustainable purification systems that prioritize ecological health. As we strive to protect our ecosystems and ensure clean water access for future generations, it is clear that these cutting-edge solutions are integral to creating a resilient and safe environment. Ultimately, the fusion of technology and sustainability in water purification holds the promise of a brighter future where clean water is not just a privilege but a fundamental right for all. The integration of these technologies into existing frameworks could revolutionize our approach to contamination challenges, marking a significant step toward both technological advancement and ecological responsibility. The nanotechnology based water filtrations are given in Fig. 5.6.

5.3.2 Solar-Powered Water Purification Techniques

It uses battery solutions to provide sufficient energy storage for a purification unit's continuous operation, especially during periods of low solar radiation. A place's average solar power supply and the amount of electricity the filter unit uses every day determine how well the electricity storage system works [50]. The integration of solar electricity with filtration systems is garnering increasing interest from both academics and businesses due to the substantial improvements it brings to efficiency and sustainability [51].

To add solar electricity to water filter technologies, one must combine electrical and mechanical control systems. Pumps control the water extraction process in a filtration system that uses solar energy. This system comprises sedimentation filters, charcoal filters, reverse osmosis membranes, and ultraviolet light sterilization units. In different working situations, a power management system makes sure that the solar power moves smoothly from the cells to the batteries and filter unit. Using this technique, filtration units may continue to operate at peak efficiency even when solar power output is minimal. Security measures are put in place to prevent battery overcharging, which will ensure the system lasts for a long time [52]. Novel solar energy collecting methods and filtering modalities are the subject of intense research efforts across a wide range of environmental regimes. Solar water purification systems choose filtration technologies according to their efficiency, energy consumption, and capacity to target certain contaminants. It is fairly uncommon for research and commercial applications to employ many filtering processes concurrently. Activated carbon filtration efficiently removes contaminants and enhances flavor and aroma, whereas mechanical purification physically separates particles from liquid by use of a barrier. Effectively eliminating microorganisms, viruses, and inorganic compounds from liquids is the goal of membrane filtration, which includes microfiltration, ultrafiltration, and nanofiltration [52]. Solar-powered water purification techniques are given in Fig. 5.7.

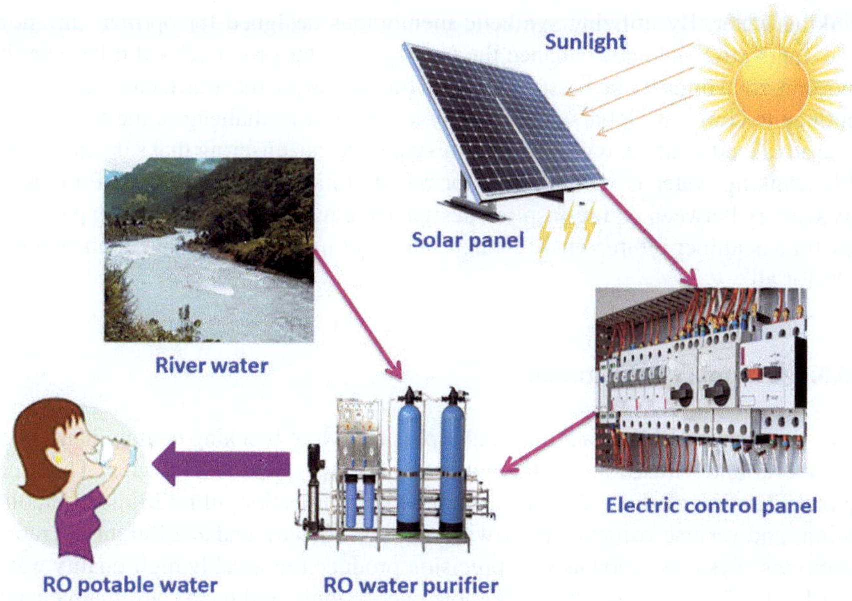

Fig. 5.7 Solar panels provide power, which is then used to operate RO water purifier that move water through a multi-stage filtration process

The materials used heavily influence the performance and durability of a system or an item over time. For membrane-based filtering procedures, materials with high strength and resilience, such as polyamide or polysulfone, are often used. To withstand high radiation for an extended period of time, UV reactors usually use UV-resistant polymers or stainless steel. The materials used in purification units in outlying areas must be able to withstand the long-term effects of the extreme weather and not corrode [53].

5.3.3 *Membrane and Bio-Inspired Purification Techniques*

Membrane and bio-inspired purification techniques signify a substantial leap forward in water treatment technology, offering unusually high efficiency and considerable sustainability. Synthetic membranes or natural processes are harnessed rather effectively for filtration purposes, making them super suitable for clean drinking water production [54]. These advancements reduce energy consumption and minimize the reliance on harmful chemicals, promoting a healthier environment. As research continues to evolve, the potential for these technologies to transform water access in underserved regions becomes increasingly promising. The integration of membrane technology and bio-inspired approaches in water treatment represents a significant advancement toward achieving high efficiency and sustainability in the quest for clean drinking water. By utilizing synthetic membranes designed for optimal filtration, these innovative methods enhance the quality of water produced and minimize the environmental impact often associated with traditional purification techniques. As we continue to confront global water scarcity and pollution challenges, the adoption of these advanced systems stands as a promising solution, ensuring that safe and accessible drinking water is within reach for communities worldwide [55]. Embracing this synergy between nature-inspired design and cutting-edge engineering paves the way for a healthier future, where clean water is not just a privilege but a fundamental right for all.

5.3.3.1 Membrane Filtration

Semi-permeable barriers facilitate water passage while blocking contaminants such as bacteria and viruses and salts and heavy metals effectively always. There are several filtration methods available, such as microfiltration, ultrafiltration, nanofiltration, and reverse osmosis, each with varying efficacy under different operating conditions. Systems renowned for precision produce remarkably high-quality water effortlessly. They may need considerable energy input and haphazard maintenance pretty regularly or else fouling occurs and clogging gets pretty bad. This can lead to reduced flow rates and compromised water quality, necessitating timely interventions and careful monitoring. Ultimately, the choice of filtration system should

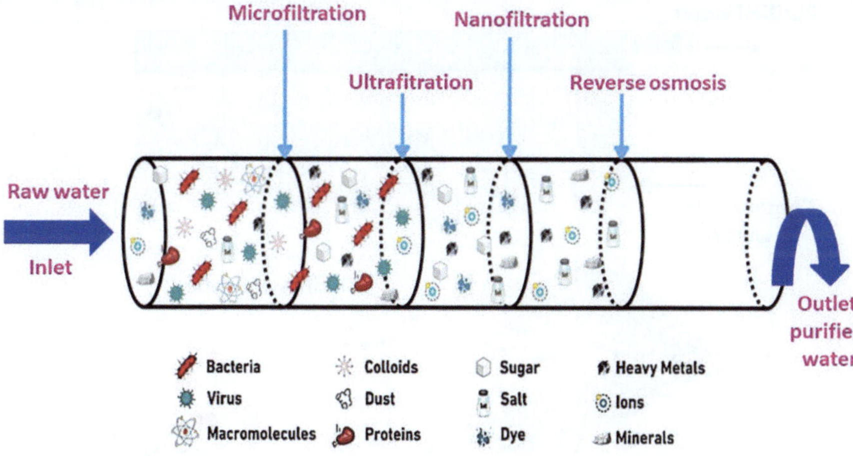

Fig. 5.8 Schematic representation of microfiltration, ultrafiltration, nanofiltration, and reverse osmosis

balance efficiency, cost, and maintenance requirements to ensure optimal performance over time [56]. A delicate balance between performance, cost, and upkeep determines the effectiveness of filtration systems. Precision reverse osmosis systems stand out for their ability to deliver high-quality water, making them a popular choice for many. However, this efficiency comes at the cost of energy consumption and the need for regular maintenance to ensure optimal functioning. Moreover, challenges such as fouling and clogging can undermine water quality if not addressed promptly. Therefore, selecting the right filtration system involves careful consideration of these factors to achieve a sustainable and efficient solution that meets both immediate needs and long-term goals in water quality management. Membrane filtration process and their mechanism are given in Fig. 5.8.

5.3.3.2 Bio-Inspired Filtration

Bio-inspired filtration techniques garner inspiration from nature, such as the filtering action of plant roots or the gills of fish swimming in murky waters. Aquaporin-based membranes that eerily mimic water channels found in living cells somehow enable ultra-selective purification with remarkable energy efficiency. Rural water treatment settings explore biofilms and natural coagulants like Moringa seeds as low-cost filtration options. These innovative approaches enhance water quality and promote sustainability by utilizing readily available natural resources. As researchers continue to study and refine these methods, the potential for scalable solutions to global water scarcity issues becomes increasingly promising [57]. The innovative use of aquaporin-based membranes signifies a substantial advancement in the pursuit of ultra-selective water purification, closely resembling the natural water channels

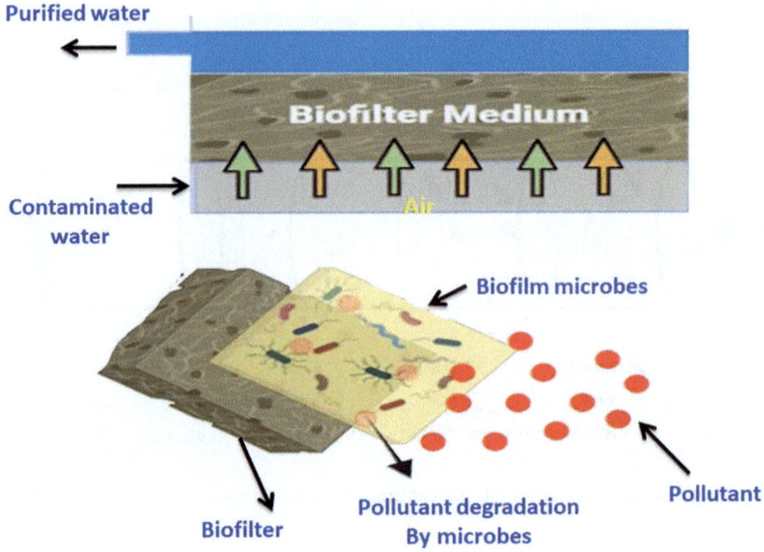

Fig. 5.9 Bioinspired technique by removal pollutant from water

present in living cells. This technology not only demonstrates exceptional energy efficiency but also paves the way for the incorporation of biofilms and natural coagulants like Moringa seeds, which offer promising cost-effective filtration solutions. By improving water quality through these cutting-edge methods, we progress toward a more sustainable future, tackling pressing issues of global water scarcity. Ultimately, the potential for scalability of these solutions underscores a transformative approach to securing clean water access for communities around the globe, emphasizing the urgent necessity to invest in and embrace such groundbreaking technologies. Bioinspired technique by removal pollutant from water is given Fig. 5.9.

Differences and Benefits of Membrane Filtration and Bio-Inspired Techniques

Membrane filtration and bio-inspired techniques represent quite innovative water purification methods with rather unique mechanisms and varying application areas. Synthetic semi-permeable membranes physically block contaminants making membrane filtration surprisingly effective for removing nasty pathogens and various chemicals. It gets used pretty ubiquitously in industrial setups and household rigs alike nowadays [58]. Bio-inspired techniques mimic natural processes such as aquaporins or plant filtration removing impurities effectively underwater occasionally. Such methods frequently necessitate fewer energetic resources and prove relatively more ecofriendly especially within environments plagued by meager infrastructural provisions. Membrane systems offer decent precision and pretty high speed but may involve higher maintenance costs over time somewhat inexplicably. Bio-inspired

Table 5.5 Differences and benefits of membrane filtration and bio inspired technique

Feature	Membrane filtration	Bio-inspired techniques	References
Mechanism	Physical separation using synthetic membranes	Mimics natural biological filtration	[60]
Efficiency	High, especially for pathogens and heavy metals	Moderate to high, depending on the design	[61]
Sustainability	Moderate (energy and material intensive)	High (eco-friendly, often locally sourced)	[62]
Cost	Higher initial and maintenance cost	Lower cost, suitable for rural areas	[63]
Application	Industrial, urban, household		[60]

methods promote sustainability somewhat effectively using locally sourced materials or structures that vaguely mimic nature's quirky designs. Both play a vital role in advancing low-cost and sustainable water purification [59] (Table 5.5).

5.3.4 *Locally Sourced Filtration*

Locally sourced filtration materials offer a sustainable and cost-effective approach to water purification, especially in underserved and remote communities. These materials, often readily available in the local environment, can be repurposed into effective filters that remove contaminants from drinking water. Common locally sourced filtration media include sand, gravel, charcoal, clay, and natural fibers like coconut husks or rice husks [64]. By utilizing materials that are both accessible and environmentally friendly, communities can enhance their water quality without incurring significant expenses. This approach not only addresses immediate health concerns but also promotes local resource management and sustainability. This modern approach of repurposing local materials for water filtration not only enhances water quality but also champions sustainability and resource management within communities. By harnessing readily available resources such as sand, gravel, charcoal, clay, and natural fibers, this method offers a cost-effective solution to pressing health concerns associated with contaminated water. As communities implement these strategies, they improve their immediate health outcomes, foster a deeper connection to their environment, and promote a culture of sustainability. Ultimately, this grassroots initiative empowers local populations to take charge of their health while making prudent use of the materials at hand. For example, charcoal made from coconut shells or clay mixed with sawdust can serve as efficient filtration components. These natural materials are not only affordable but also biodegradable and environmentally friendly. They can remove particulates, improve taste, reduce turbidity, and in some cases, eliminate microbial contaminants [65].

Using local materials reduces the dependency on imported or expensive technologies and promotes community participation in water management. Moreover, it encourages innovation and traditional knowledge application, strengthening community resilience. When combined with simple construction techniques and basic hygiene education, these systems can be scaled across rural settings for long-term use. However, quality control, periodic maintenance, and proper training are essential to ensure their effectiveness and safety. Overall, locally sourced filtration materials represent an empowering, sustainable strategy to improve water accessibility in resource-limited settings [66]. The pathway to effective water management lies in fostering community participation, which not only sparks innovation but also honors traditional knowledge, creating a rich tapestry of local solutions. By strengthening resilience within communities, this approach empowers individuals to adapt and thrive amid challenges. The simplicity of construction techniques and hygiene education makes these initiatives scalable, ensuring they can be tailored to fit diverse contexts. However, rigorous quality control, ongoing maintenance, and comprehensive training programs must underpin such sustainable efforts. Making clean water a reality for more people is also made possible by using locally sourced filtration materials, which improve accessibility in environments with limited resources. Together, these elements weave a robust framework that champions both community engagement and sustainable practices in water management.

5.4 Community-Driven Water Filtration Solutions

Community-driven water filtration solutions involve strategies in which the local community actively participates in the establishment, management, and maintenance of water filtration systems to guarantee access to safe drinking water. These solutions are frequently employed in regions where centralized water treatment infrastructure is insufficient, inconsistent, or cost-prohibitive. By engaging community members in the process, these initiatives not only empower individuals but also foster a sense of ownership and responsibility regarding their water resources. This collaborative approach often leads to more sustainable and effective solutions tailored to the specific needs and challenges of the community.

5.4.1 Decentralized Water Treatment System

Decentralized water treatment systems become a much cheaper, sustainable, and still managed locally way of finding clean water for the underserved communities. In contrast to centralized treatment facilities, those systems work at the level of the household, community, or small community, thus being an easily manageable alternative to provide safe drinking water where there is no access to it. The application of affordable filtration technologies like bio sand filters, ceramic filters, activated carbon, and also membrane filtration has turned these small-scale systems into a reliable way of getting clean water. De-centralized systems can even go further through the use of solar disinfection (SODIS), rainwater harvesting, and ultraviolet (UV) purification to be completely off-grid and thus, be autonomously suitable for different locations [67]. The virtues of decentralized water treatment can be seen in lower infrastructure costs, minimal maintenance, and ease of adaptation to local conditions. At the same time, these systems have shown to be less reliable and lacking in connection with the typically state wide water systems, which thus are not infrequently prone to failure or even non-existing in the regions of the country where the rural or disaster-affected areas have been located [68]. In addition, these systems are empowering community participation by facilitating arrangements with local organizations, NGOs, and social enterprises, which are the main participants in the process of organizing, maintaining and sensitizing the residents to proper usage [69]. The self-management of water resources in the community is the main aspect that is emphasized by the decentralized treatment systems that implies long-term sustainable and resilient methods against water scarcity, as well as the virus of contamination [70]. By managing to work that out in a larger number, these systems can have a considerable impact both in improving water security and people's readiness to take their health risks as well as in raising living standards of vulnerable communities across the globe [71].

5.4.2 Capacity Building and Community Engagement

Capacity building and community engagement are essential for the success and sustainability of community-driven water filtration solutions. Empowering local communities with knowledge, skills, and resources ensures long-term management and maintenance of low-cost water filtration technologies. Training programs on water quality testing, filtration system maintenance, and hygiene practices enable residents to take ownership of their water sources. Involving local leaders, schools, and women's groups helps drive awareness and encourages behavioral change toward safe water practices. Community engagement fosters collaboration with NGOs, government agencies, and private organizations, ensuring financial and technical support. By integrating local knowledge and traditional water purification methods with modern filtration systems, solutions become culturally appropriate and widely accepted. Encouraging community participation in decision-making strengthens trust and accountability, leading to better adoption and long-term sustainability. Ultimately, investing in capacity building enhances water security, health, and economic stability in underserved regions. The journey toward sustainable water access hinges on the active involvement and empowerment of local communities. By equipping individuals with the knowledge, skills, and resources necessary for effective water filtration, we lay the groundwork for long-term success and resilience. Training programs focused on water quality testing, system maintenance, and hygiene practices not only enhance community capabilities but also foster a culture of awareness and responsibility. Engaging local leaders, educational institutions, and women's groups further amplifies this impact, driving behavioral change and ensuring that safe drinking water becomes a shared priority. Ultimately, these community-driven solutions create a ripple effect that transforms not just individual lives but entire communities, paving the way for healthier futures.

5.4.3 Case Studies of Successful Implementations

Several communities have been set up with the objective of using them to purify water projects around the world and at this stage, they are very successful affordable, and sustainable which have also become an alternative to make it. One example is the Naandi Foundation's Safe Water Project in India, which has established community-run water purification plants in rural villages. These systems use reverse osmosis and ultraviolet (UV) filtration technologies, ensuring safe drinking water at low costs. Project training empowers local women economically as plant operators enhancing community participation markedly amidst radical social changes somehow. Rural Kenyan households now utilize bio sand filtration systems under WASH program initiatives heavily promoted across vast swaths of Africa recently. These filters, made from locally available materials, effectively remove bacteria, viruses, and sediment, providing clean drinking water without the need for electricity. Through education programs, the initiative has improved hygiene practices and reduced waterborne diseases [72]. Another successful model is the Aguajira Water Project in Colombia, which introduced solar-powered desalination units in indigenous Wayuu communities. This project ensures clean water access in arid regions, reducing health risks from contaminated sources. These case studies demonstrate that community engagement, local leadership, and innovative technologies are key to scaling up sustainable water filtration solutions and improving public health in underserved regions worldwide [73, 74]. Community driven water filtration solutions are given in Fig. 5.10.

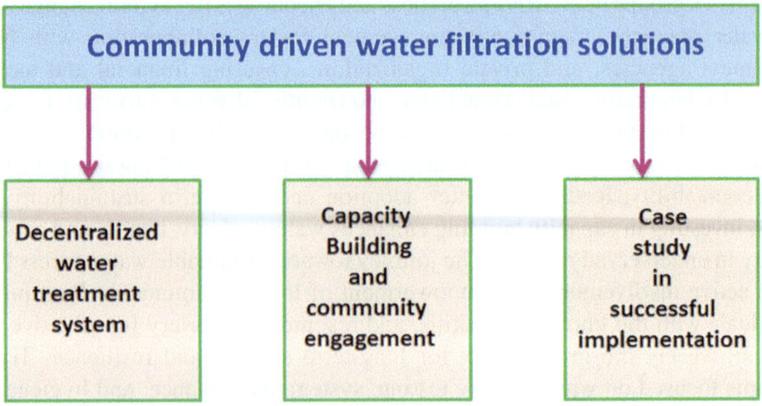

Fig. 5.10 Community driven water filtration solutions

5.5 Scaling up Affordable Water Technologies

The implementation of water technologies that are both inexpensive and scalable is essential for tackling the global water problem and making sure that everyone has access to water supplies that are both safe and sustainable. By investing in innovative solutions, communities can enhance their resilience to water scarcity and improve public health outcomes. Furthermore, collaboration between governments, the private sector, and non-profit organizations will be crucial in driving these initiatives forward. Addressing the intricate web of global water challenges demands a multifaceted approach that integrates affordable and scalable technologies with a steadfast commitment to ensuring safe and sustainable water supplies. By investing in innovative solutions, we can enhance resilience against climate variability and improve public health outcomes for communities worldwide. Moreover, fostering collaboration between governments, the private sector, and non-profit organizations will amplify our collective impact and create a robust framework for sustainable water management. Together, these strategies pave the way for a future where clean water is accessible to all, transforming lives and fortifying ecosystems in the process.

5.5.1 Role of Social Enterprises and NGOs

Social enterprises and NGOs are very important in the distribution of low-cost water filtration technologies to new locations, which guarantees that the people in the local areas have access to clean water because it is affordable. By forming the link between idea creation and its realization, they enable financial support, community engagement, and advocacy for regulatory instruments. The NGOs are very committed and have dedicated their resources to capacity building, education, and programs that raise awareness. The local people are not only empowered but also made to be the main drivers in the uptake and the maintenance of the local solutions that are sustainable. Social enterprises make use of more innovative, market-based methods and are building such that scalable, affordable, and impact-driven models ensure accessibility to clean water over a longer time [75, 76]. In their approach, partnering is key as they work with the governments, the scientific community, and the organizations in the private sector to set up and run low-cost filtration facilities, facilitate public–private interactions, and foster and manage micro-finance schemes. Thanks to them, this approach is community-based, which means that the benefits will not only be sustainable but they will also perfectly fit the cultural and economic aspects, and they will be environmentally friendly. One of the effective ways to uplift the very important role of social enterprises and NGOs is to facilitate the quick uptake of affordable water filtration technologies. Thus, health, hygiene, and in the long run, life quality of the people in disadvantaged communities.

5.5.2 Government Initiatives and Policy Support

Government initiatives and policy support are vital to boost low-cost water filtration technologies and provide the safe and cost-effective water to communities that are not served. The encouragement of the development and application of effective filtration systems is done by the policies that provide research funding, subsidies, and tax incentives for water purification innovations [77]. The role that governments can play in regulatory frameworks to guarantee the quality and cost effectiveness of accessible water technologies is significant. The implementation of public–private partnerships (PPPs) results in collaborating between various institutions, private sector innovators, and NGOs and hence, it leads to mass utilizations [78]. The departments in charge of infrastructure development can through projects like community water kiosks and autonomous purification plants increase accessibility. Moreover, the national clean water programs can realize the sustainability and far-reaching impact of water filtration technologies [66]. Governments, by adopting the water security policies, financing the innovative solutions, and enhancing the implementation strategies, can greatly accelerate the wide-ranging use of affordable filtration technologies to enable the population to access clean water and those infected with waterborne diseases to decrease [79].

5.5.3 Challenges and Opportunities for Large-Scale Adoption

The large-scale adoption of low-cost water filtration technologies presents both challenges and opportunities in ensuring clean water access for underserved communities. The large-scale adoption of low-cost water filtration technologies stands at the intersection of significant challenges and remarkable opportunities, ultimately shaping the future of clean water access for underserved communities. By addressing financial constraints and promoting innovative solutions, these technologies can transform the landscape of public health and environmental sustainability. As we navigate the complexities of implementation, collaboration among governments, NGOs, and local stakeholders becomes crucial to overcoming barriers and maximizing impact. The potential to provide safe drinking water to countless individuals not only fosters healthier communities but also empowers them with the independence to thrive. Embracing this movement is not merely a technological shift; it represents a collective commitment to equity and dignity in one of life's most essential resources. Some challenges and opportunities are discussed below:

Challenges:

- High Initial Costs: Despite being low-cost in the long run, initial investments for production, distribution, and maintenance can be a barrier.
- Infrastructure Limitations: Many remote areas lack proper water supply systems to integrate filtration technologies.

- Public Awareness and Acceptance: Communities may be reluctant to adopt new water purification methods due to a lack of education or trust.
- Regulatory Hurdles: Compliance with government policies and quality standards can slow deployment [80].

Opportunities:

- Innovations in Technology: Advancements in nanomaterials, solar purification, and biofiltration can improve efficiency and affordability.
- Public–Private Partnerships: Collaboration between governments, NGOs, and social enterprises can drive large-scale implementation [81].
- Decentralized Solutions: Community-based filtration models can offer self-sustaining, localized access to clean water [82].

By addressing these challenges, affordable water filtration technologies can be effectively scaled to provide safe drinking water globally. The journey toward ensuring safe drinking water for all hinges on the advancement and accessibility of affordable water filtration technologies. By tackling the multifaceted challenges associated with global water safety, these innovations play a crucial role in transforming contaminated sources into reliable supplies. As communities around the world face increasing threats from pollution and climate change, scaling effective solutions becomes imperative. The integration of low-cost filtration systems not only empowers individuals and families but also fosters resilience in vulnerable populations, paving the way for a healthier future. Ultimately, by prioritizing these technologies, we can cultivate an environment where clean water is not a privilege but a universal right, thus safeguarding public health and promoting sustainable development across the globe. The challenges and opportunities for large scale adoption of affordable water technology are given in Table 5.6.

Table 5.6 Challenges and opportunities for large-scale adoption of affordable water technologies

Challenges	Opportunities	References
High initial investment	Funding and government subsidies	[83, 84]
Lack of infrastructure	Decentralized community solutions	
Regulatory barriers	Policy reforms for faster approval	
Public resistance and trust	Awareness campaigns and education	
Scalability issues	Technological advancements in filtration	

5.6 Environmental and Economic Considerations

The environmental and economic considerations of community-driven water filtration solutions and the expansion of cost-effective water technologies are interrelated and essential for their lasting effectiveness and sustainability. Ultimately, the implementation of community-driven water filtration solutions stands as a powerful testament to the symbiotic relationship between environmental stewardship and economic viability. By engaging local populations in the development and maintenance of these systems, not only are we addressing immediate health concerns, but we are also fostering a sense of ownership and responsibility that fuels long-term sustainability. The effective integration of sustainable practices with economic strategies ensures that communities can thrive while preserving their vital natural resources. As these solutions take root, they not only cleanse contaminated water sources but also empower residents to become active participants in shaping a healthier future—one where clean water is not just a privilege but a shared right for all.

5.6.1 Filtration Techniques for Environment and Sustainability

Inexpensive water filtration technologies must be sustainable and environmentally friendly ultimately minimizing damage and providing long-term clean water access. Different filtration methods exhibit varying energy usage and generate disparate amounts of waste with some having substantial environmental impact [84]. Delving beneath the surface of filtration methods reveals intricate environmental and economic considerations that extend far beyond their immediate function of water purification. A truly sustainable approach requires a profound understanding of the entire life cycle of these technologies, from raw material extraction to disposal or recycling. This comprehensive perspective ensures that the chosen filtration methods not only effectively purify water but also minimize negative impacts on the environment and promote resource efficiency throughout their lifespan. By considering these factors, stakeholders can make informed decisions that align with sustainability goals (Table 5.7).

Table 5.7 Environmental impact of different water filtration methods

Filtration method	Energy use	Waste generation	sustainability level	References
Bio-sand filtration	Low	Minimal	High	[85–89]
Reverse osmosis (RO)	High	High (wastewater)	Moderate	
Ceramic filters	Low	Minimal	High	
Activated carbon	Moderate	Low	High	
UV purification	High	None	Moderate	

Sustainability Considerations:

- Energy Efficiency: Some filtration methods require electricity, while others operate passively (e.g., gravity-based systems).
- Waste Generation: Filtration processes, such as reverse osmosis (RO), generate wastewater, whereas bio-sand filters produce minimal waste.
- Material Sustainability: Use of natural and biodegradable materials, such as activated carbon and plant-based filters, reduces environmental impact.
- Chemical Use: Some methods involve chemical treatment, which can lead to secondary pollution.

5.6.2 Long-Term Viability and Cost-Effectiveness

The prosperity of low-cost water filtration technologies hinges heavily on reasonable costs and feasibility for a somewhat uncertain future. Communities lacking adequate supply should be given the chance to decide their fate and lift themselves out of such dire situations entirely on their own. It is not only important to supply them with clean water but also to empower them with the knowledge and tools necessary to maintain these systems sustainably. By fostering local engagement and investment, we can create resilient communities that thrive in the face of environmental challenges [90] (Table 5.8).

Cost Considerations:

- **Initial Investment:** Some technologies, like reverse osmosis (RO) and UV purification, require higher upfront costs, while ceramic filters and bio-sand filters are more affordable.
- **Maintenance Costs:** Technologies that need frequent filter replacements or energy consumption can become costly over time.
- **Lifespan of Filtration Systems:** Long-lasting filters (e.g., ceramic and bio-sand) provide better long-term savings compared to disposable filters.

Table 5.8 Cost-effectiveness of different water filtration methods

Filtration method	Initial cost	Maintenance cost	Lifespan	Overall feasibility	References
Bio-sand filter	Low	Low	High	High	[91]
Reverse osmosis (RO)	High	High	Moderate	Moderate	[92]
Ceramic filters	Low	Low	High	High	[93]
Activated carbon	Moderate	Moderate	Moderate	High	[94]
UV purification	High	Low	High	Moderate	[95]

Long-Term Feasibility Factors:

- **Availability of Materials:** Filters using locally available and renewable materials lower production and maintenance costs.
- **Community Involvement:** Training local communities in maintenance and repair ensures sustainability.
- **Scalability and Adaptability:** Technologies that can be easily adapted to different water sources are more feasible in the long run.

By prioritizing low-cost, sustainable filtration technologies, communities can achieve affordable and long-term access to clean water.

5.7 Future Perspectives and Challenges

Technologies that treat water using nanoparticles show enormous potential in eliminating microbiological contaminants and enhancing water quality. However, to fully realize their potential and ensure sustainable implementation, they must remove several significant obstacles. These obstacles include regulatory challenges, scalability of production processes, and public acceptance of nanoparticle-based solutions. Addressing these issues will be crucial in advancing the adoption of such innovative technologies in water treatment systems worldwide [96]. The successful integration of nanoparticle-based water treatment technologies hinges on overcoming several significant challenges that currently hinder their widespread adoption. Regulatory hurdles must be navigated to ensure safety and efficacy, while the scalability of production processes requires further enhancement to meet global demand effectively. Additionally, fostering public acceptance is vital, as community trust can significantly influence the implementation of these innovative solutions. As future research endeavors concentrate on these critical areas, there lies a promising potential for nanoparticle-based treatments to revolutionize water purification practices and address the growing concerns surrounding water quality worldwide [97].

5.7.1 Emerging Trends and Potential Breakthroughs

Affordable water purification technologies are transforming access to safe water, especially in low-income communities. Novel technologies like nanomaterial filters, biomimetic membranes, and solar-driven purification offer a path toward more efficient and cost-effective solutions [98]. Future implications of this research should include scalability and sustainability, as well as integration with smart monitoring systems to provide safe, accessible, and cost-effective water solutions on a worldwide scale. Additionally, fostering partnerships between governments, NGOs, and the private sector will be crucial in implementing these technologies effectively. By ensuring that these innovations reach the communities that need them most, we can create a more equitable future where clean water is a basic right for everyone [99]. The path forward in addressing global water challenges hinges on effective scalability and sustainability, ensuring that innovative solutions can be adapted to diverse contexts. By integrating advanced smart monitoring systems, we can enhance our ability to manage water resources efficiently and responsively. Collaborative efforts between governments, non-governmental organizations, and the private sector are essential in fostering partnerships that leverage strengths across sectors, leading to more impactful outcomes. Ultimately, these collective initiatives pave the way for a future where access to clean water is recognized as a fundamental right for all people, transcending borders and promoting health and prosperity on a global scale. Through dedicated action and cooperation, we can transform the vision of equitable water access into a tangible reality [100].

5.7.2 *Recommendations for Further Research and Investment*

Dirty dirties are against innovative, sustainable, scalable, and community-oriented low-cost water filtration technologies. Solar-based purification and further innovations in nanotechnology and biofiltration can improve cost and effectiveness [101]. However, without increased funding, policy support, and collaboration across sectors, these efforts alone will not lead to long-lasting impact, accessibility, or the provision of clean water for underserved populations globally [102]. The path toward ensuring clean water access for underserved populations around the globe is illuminated by advancements in low-cost water filtration technologies, particularly through solar-based purification and innovative nanotechnology [103]. These breakthroughs enhance the effectiveness of water treatment and significantly reduce costs, making them more accessible to those in need. However, the journey does not end with technological innovation, it requires a concerted effort involving increased funding, supportive policies, and collaboration across sectors to create sustainable solutions. By combining these elements, we can foster a future where clean water is not a privilege but a fundamental right enjoyed by all, ultimately transforming lives and communities worldwide.

5.8 Conclusion and Future Directions

The study emphasizes the importance of affordable household water purification methods for developing regions through techniques such as cloth filtration combined with boiling water and using chlorination and solar disinfection. While each treatment approach provides cost-effective simplicity users face issues with recontamination improper dosage and material limitations. Ceramic water filters constructed from materials found locally present themselves as a sustainable point-of-use (POU) treatment solution that maintains effectiveness. Ongoing research seeks to enhance the performance and durability of ceramic filters while maintaining their cost-effective water purification capability and microbial removal efficiency. The field of nanotechnology provides promising methods to improve water purification systems but requires more research to decrease toxicity levels while enhancing stability and making these solutions affordable for broader use. The journey toward ensuring safe drinking water in developing countries relies heavily on accessible and effective low-cost household water treatment techniques such as cloth filtering, boiling, chlorination, and solar water disinfection. Each method plays a vital role in inactivating harmful microorganisms and contaminants that threaten public health. The choice of cloth material can significantly enhance the efficacy of filtering high-turbidity water, demonstrating the importance of tailored approaches to local conditions. However, challenges persist; boiling may be hindered by the availability of clean fuels and risks of recontamination, while chlorination must navigate the complexities of maintaining minimum residual levels and dealing with disinfection byproducts at the point of use. By understanding these intricacies and addressing their limitations, communities can harness these techniques to improve water quality effectively, fostering a healthier future for everyone. Ceramic filters have the potential to treat household water, especially for liquefied natural gas (LNG) applications, and present a compelling case for innovation in water purification. Crafted from locally sourced materials and clay, these filters offer a sustainable and cost-effective point-of-use solution that not only meets the pressing need for clean drinking water but also supports local economies. Their low maintenance requirements and affordability make them accessible to a wide range of households, promoting health and well-being. However, addressing challenges like removing contaminants and improving bacterial filtration efficiency is crucial to maximizing their effectiveness. As research continues to refine these technologies, ceramic filters stand poised to play a vital role in delivering safe water solutions that align with both environmental sustainability and community needs. The potential of low-cost ceramic filters as point-of-use (POU) devices for water treatment is undeniably promising, particularly when considering the significant enhancements that can be achieved through modifications. By integrating ceramic filtration with complementary techniques, these filters can effectively tackle multicomponent contaminants, ensuring cleaner water for communities in need. As the field evolves, future research will play a critical role in developing smart ceramic filters that not only improve efficiency but also streamline monitoring and maintenance processes. Ultimately, the success of these innovative solutions hinges on proper installation and timely replacement, underscoring the importance of a comprehensive approach to safeguarding against water pollution and enhancing public health.

Nanotechnology can cure harmful chemicals in water, which is essential for sustainable development. Although further large-scale investigations are required, solutions based on nanotechnology show promise for being both competent and cost-effective. The broad usage of nanoparticles in water filtration relies on their low manufacturing costs. The interaction mechanisms involved in water treatment, as well as ways to increase economic efficiency and stability and decrease toxicity, should be the subject of future research. Affordable water filtration technologies serve as essential tools for delivering safe drinking water to populations with limited access. Future research and financial support must target improvements in filtration efficiency together with affordability and sustainability. Nanomaterials and biofiltration advancements along with solar-powered purification methods show potential yet demand additional efforts to achieve large-scale production capabilities while enhancing product longevity and incorporating smart monitoring functionalities. Scientists working with government officials and business executives must join forces to create affordable filtration systems that protect the environment. To guarantee that these technologies achieve wide usage and lasting benefits governments and organizations need to boost their financial support and policy guidance. Upcoming projects must focus on involving local communities in water treatment activities and educational initiatives and implement decentralized systems to enable people to manage their own water solutions. Investment in innovative filtration technologies that are scalable and accessible will provide clean water to every community while minimizing health risks and fostering sustainable development.

References

1. Jacob DE, Nelson IU, Izah SC (2024c) Indigenous water management strategies in the global south. In: Water crises and sustainable management in the global south, pp 487–525. Singapore: Springer Nature Singapore. https://link.springer.com/chapter/10.1007/978-981-97-4966-9_16
2. Jacob, D. E., Nelson, I. U., Eniang, E. A., & Izah, S. C. (2024b). Advancing Water Security and Resilience in the Global South Through Recreational Development. In Water Crises and Sustainable Management in the Global South (pp. 527–554). Singapore: Springer Nature Singapore. https://link.springer.com/chapter/10.1007/978-981-97-4966-9_17
3. Jacob DE, Nelson IU, Izah SC, Daniel KS (2024a) Rural water crises in the global south: understanding the scope and impact. In: Izah SC, Ogwu MC, Loukas A, Hamidifar H (eds) Water crises and sustainable management in the global south, pp 3–44. Springer, Singapore. https://doi.org/10.1007/978-981-97-4966-9_1
4. Etefa HF, Tessema AA, Dejene FB (2024) Carbon dots for future prospects: synthesis, characterizations and recent applications: a review (2019–2023). C J Carbon Res 10(3):60. https://doi.org/10.3390/c10030060
5. Afreen MY, Suma MS, Sri G, Rohit J, Raju DA (2024) Low-cost water purifier using natural and waste material. J Phys Conf Ser 2779:012091. https://doi.org/10.1088/1742-6596/2779/1/012091
6. Levin R, Villanueva CM, Beene D, Cradock AL, Donat-Vargas C, Lewis J, Martinez-Morata I, Minovi D, Nigra AE, Olson ED, Schaider LA, Ward MH, Deziel NC (2024) US drinking water quality: exposure risk profiles for seven legacy and emerging contaminants. J Expos Sci Environ Epidemiol 34:3–22. https://doi.org/10.1038/s41370-023-00597-z
7. Foster SA, Pennino MJ, Compton JE, Leibowitz SG, Kile ML (2019) Arsenic drinking water violations decreased across the united states following revision of the maximum contaminant level. Environ Sci Technol 53:11478–11485. https://doi.org/10.1021/acs.est.9b02358
8. Hait M, Gir ND, Kashyap NK, Dey A, Chaturwedi AK (2019) Bioremoval of water pollutants from the effluents of Tifra industrial area in Bilaspur (C.G.). Int J Curr Eng Sci Res 6:399–401
9. Basu S, Dutta A, Mukherjee SK, Hossain ST (2021) Exploration of green technology for arsenic removal from groundwater by oxidation and adsorption using arsenic-oxidizing

bacteria and metal nanoparticles. In: New trends in removal of heavy metals from industrial wastewater, pp 177–211. https://doi.org/10.1016/B978-0-12-822965-1.00009-X
10. Hait M, Kashyap NK, Giri AK (2024) Synthesis of metal NPs using plant flowers and fruits. In: Bhardwaj AK, Srivastav AL, Rai S (eds) Biogenic wastes-enabled nanomaterial synthesis, pp 195–237. Springer, Cham. https://doi.org/10.1007/978-3-031-59083-2_8
11. Chaturwedi AK, Kashyap NK, Gupta P, Biswas S, Shriwas SK, Vaishnav MM, Hait M (2024) Industrial effluents and their impact on water pollution- an overview. ES Gen 5:1203. https://doi.org/10.30919/esg1203
12. Izah SC, Ogwu MC (2025) Modeling solutions for microbial water contamination in the global south for public health protection. Front Microbiol 16:1504829. https://doi.org/10.3389/fmicb.2025.1504829
13. Nitzsche KS, Lan VM, Trang PT, Viet PH, Berg M, Voegelin A, Planer-Friedrich B, Zahoransky J, Müller SK, Byrne JM, Schröder C, Behrens S, Kappler A (2015) Arsenic removal from drinking water by a household sand filter in Vietnam–effect of filter usage practices on arsenic removal efficiency and microbiological water quality. Sci Total Environ 502:526–536. https://doi.org/10.1016/j.scitotenv.2014.09.055
14. Smith K, Li Z, Chen B, Liang H, Zhang X, Xu R, Li Z, Dai H, Wei C, Liu S (2017) Comparison of sand-based water filters for point-of-use arsenic removal in China. Chemosphere 168:155–162. https://doi.org/10.1016/j.chemosphere.2016.10.021
15. Gupta P, Tamrakar G, Biswas S, Dhundhel RS, Soni VK, Das HS, Sahu P, Chaturwedi AK, Kashyap NK, Akitsu T, Hait H (2024) Quantum dots as a fluorescent sensor for detecting heavy metal ions in aqueous environment: an overview. ES Chem Sustain 1:1359. https://doi.org/10.30919/escs1359
16. Patel D, Chaturwedi AK, Kashyap N K, Roymahapatra G, Hait M (2024) Quality control and standardization of nanomaterials. In Thakur A, Kumar A (eds) Sustainability, safety, and applications of nanomaterials-based corrosion inhibitors, IGI Global, USA, pp 297–325. https://doi.org/10.4018/979-8-3693-7640-9.ch012
17. Barnaby R, Liefeld A, Jackson BP, Hampton TH, Stanton BA (2017) Effectiveness of table top water pitcher filters to remove arsenic from drinking water. Environ Res 158:610–615. https://doi.org/10.1016/j.envres.2017.07.018
18. Hait M, Kashyap NK, Sahu P, Jana U (2024) Fabrication of metal NPS from plant root and tuber. In: Bhardwaj AK, Srivastav AL, Rai S (eds) Biogenic wastes-enabled nanomaterial synthesis, pp 29–71. Springer, Cham. https://doi.org/10.1007/978-3-031-59083-2
19. Nicomel NR, Leus K, Folens K, Van Der Voort P, Du Laing G (2015) Technologies for Arsenic removal from water: current status and future perspectives. Int J Environ Res Pub Health 13:13010062. https://doi.org/10.3390/ijerph13010062
20. Kashyap NK, Hait M, Bhardwaj AK (2024) Planktons as a sustainable biomonitoring tool of aquatic ecosystem. In: Izah SC, Ogwu MC, Hamidifar H (eds) Biomonitoring of pollutants in the global south. Springer, Singapore, pp 275–319. https://doi.org/10.1007/978-981-97-1658-6_8
21. Kimbrell E (2019) Wendy Jepson engages stakeholders to address global water insecurity. Leshner Leadership Institute for Public Engagement with Science. https://www.aaas.org/news/wendy-jepson-engages-stakeholders-address-global-water-insecurity
22. Khosravi-Darani K, Rehman Y, Katsoyiannis IA, Kokkinos E, Zouboulis AI (2022) Arsenic exposure via contaminated water and food sources. Water 14:1884. https://doi.org/10.3390/w14121884
23. Shankar S, Shanker U, Shikha (2014) Arsenic contamination of groundwater: a review of sources, prevalence, health risks, and strategies for mitigation. Sci World J 2014: 304524. https://doi.org/10.1155/2014/304524
24. Hait M, Kashyap NK, Chaturwedi AK (2018) Physicochemical investigation of water bodies of Arpa River at Bilaspur urban areas. Int J Chem Stud 6:1773–1777
25. Anonymous (2025) Imminent risk of a global water crisis, warns the UN World Water Development Report 2023. 221st session of the Executive Board, 2–17 April 2025. https://www.unesco.org/en/articles/imminent-risk-global-water-crisis-warns-un-world-water-development-report-2023.

26. Venkatasubramanian G, Das S, Pandey S (2021) Community-scale water purification technologies for sustainable rural development. Environ Sci Pollut Res 28:15742–15756. https://doi.org/10.1007/s11356-020-11894-9
27. Shepard ZJ, Zhang Y, Anaya NM, Cardace D, Oyanedel-Craver V (2020) Development of ceramic water filter clay selection criteria. Water 12:1657. https://doi.org/10.3390/w12061657
28. Mahlangu O, Mamba B, Momba M (2012) Efficiency of Silver Impregnated Porous Pot (SIPP) filters for production of clean potable water. Int J Environ Res Pub Health 9:3014–3029. https://doi.org/10.3390/ijerph9093014
29. Tellen V, Nkeng G, Dentel S (2010) Improved filtration technology for pathogen reduction in rural water supplies. Water 2:285–306. https://doi.org/10.3390/w2020285
30. Muttil N, Jagadeesan S, Chanda A, Duke M, Singh SK (2023) Production, types, and applications of activated carbon derived from waste tyres: an overview. Appl Sci 13:257. https://doi.org/10.3390/app13010257
31. Chew TW, H'Ng PS, Luqman Chuah Abdullah BCTG, Chin KL, Lee CL, Mohd Nor Hafizuddin BMS, TaungMai L (2023) A review of bio-based activated carbon properties produced from different activating chemicals during chemicals activation process on biomass and its potential for Malaysia. Materials 16:7365. https://doi.org/10.3390/ma16237365
32. Ahmed AS, Alsultan M, Hameed RT, Assim YF, Swiegers GF (2022) High surface area activated charcoal for water purification. J Composites Sci 6:311. https://doi.org/10.3390/jcs6100311
33. Rosa G, Miller L, Clasen T (2010) Microbiological effectiveness of disinfecting water by boiling in rural Guatemala. Am J Trop Med Hygiene 82:473–477. https://doi.org/10.4269/ajtmh.2010.09-0320
34. Sales-Lérida D, Grosso J, Martínez-Jiménez PM, Manzano M (2023) A low cost and eco-sustainable device to determine the end of the disinfection process in SODIS. Sensors 23:575. https://doi.org/10.3390/s23020575
35. Schmid P, Kohler M, Meierhofer R, Luzi S, Wegelin M (2008) Does the reuse of PET bottles during solar water disinfection pose a health risk due to the migration of plasticisers and other chemicals into the water? Water Res 42:5054–5060. https://doi.org/10.1016/j.watres.2008.09.025
36. Onubogu NO, Chong KK, Tan MH (2021) Review of active and passive daylighting technologies for sustainable building. Int J Photoenergy. https://doi.org/10.1155/2021/8802691
37. Tchonkouang RD, Lima AR, Quintino AC, Cristofoli NL, Vieira MC (2023) UV-C Light: a promising preservation technology for vegetable-based nonsolid food products. Foods 12:3227. https://doi.org/10.3390/foods12173227
38. Kim HJ, Yoon HW, Lee MA, Kim YH, Lee CJ (2023) Impact of UV-C irradiation on bacterial disinfection in a drinking water purification system. J Microbiol Biotechnol 33:106–113. https://doi.org/10.4014/jmb.2211.11027
39. Elkhatat AM (2022) Recent trends of microbial decontamination for occupational, industrial and domestic applications. Bull Natl Res Cent 46:75. https://doi.org/10.1186/s42269-022-00761-9
40. Zulkifli SN, Rahim HA, Lau WJ (2018) Detection of contaminants in water supply: a review on state-of-the-art monitoring technologies and their applications. Sens Actuator B Chem 255:2657–2689. https://doi.org/10.1016/j.snb.2017.09.078
41. Joudeh N, Linke D (2022) Nanoparticle classification, physicochemical properties, characterization, and applications: a comprehensive review for biologists. J Nanobiotechnol 20:262. https://doi.org/10.1186/s12951-022-01477-8
42. Ajith MP, Rajamani P (2021) Nanotechnology for water purification–current trends and challenges. J Nanotechnol Nanomater 2:88–91. https://doi.org/10.33696/Nanotechnol.2.025
43. Arora B, Attri P (2020) Carbon nanotubes (CNTs): a potential nanomaterial for water purification. J Composites Sci 4:135. https://doi.org/10.3390/jcs4030135
44. Ogwu, M.C., Izah, S.C., Sawyer, W.E., and Amabie, T. (2025). Environmental risk assessment of trace metal pollution: a statistical perspective. Environ Geochem Health 47:94. https://doi.org/10.1007/s10653-025-02405-z

45. Alhalili Z (2023) Metal oxides nanoparticles: general structural description, chemical, physical, and biological synthesis methods, role in pesticides and heavy metal removal through wastewater treatment. Molecules (Basel, Switzerland) 28:3086. https://doi.org/10.3390/molecules28073086
46. Dikshit PK, Kumar J, Das AK, Sadhu S, Sharma S, Singh S, Gupta PK, Kim BS (2021) Green synthesis of metallic nanoparticles: applications and limitations. Catalysts 11:902. https://doi.org/10.3390/catal11080902
47. Ali A, Zafar H, Zia M, Ul Haq I, Phull AR, Ali JS, Hussain A (2016) Synthesis, characterization, applications, and challenges of iron oxide nanoparticles. Nanotechnol Sci Appl 9:49–67. https://doi.org/10.2147/NSA.S99986
48. Kulkarni D, Sherkar R, Shirsathe C, Sonwane R, Varpe N, Shelke S, More MP, Pardeshi SR, Dhaneshwar G, Junnuthula V, Dyawanapelly S (2023) Biofabrication of nanoparticles: sources, synthesis, and biomedical applications. Front Bioengin Biotechnol 11:1159193. https://doi.org/10.3389/fbioe.2023.1159193
49. Wang L, Hu C, Shao L (2017) The antimicrobial activity of nanoparticles: present situation and prospects for the future. Int J Nanomed 12:1227–1249. https://doi.org/10.2147/IJN.S121956
50. Hasan MM, Hossain S, Mofijur M, Kabir Z, Badruddin IA, Khan YTM, Jassim E (2023) Harnessing solar power: a review of photovoltaic innovations, solar thermal systems, and the dawn of energy storage solutions. Energies 16:6456. https://doi.org/10.3390/en16186456
51. Bosnjakovic M, Santa R, Crnac Z, Bosnjakovic T (2023) Environmental impact of PV power systems. Sustainability 15:11888. https://doi.org/10.3390/su151511888
52. Anjimoon S, Anandhi RJ, Singh N, Parmar A, Sharma N, Mohammad Q (2024) Comprehensive Review on low-cost, solar-powered water purification technologies for remote areas. E3S Web of Conferences 505:02005. https://doi.org/10.1051/e3sconf/202450502005
53. Freeland B, McCarthy E, Balakrishnan R, Fahy S, Boland A, Rochfort KD, Dabros M, Marti R, Kelleher SM, Gaughran J (2022) A review of Polylactic acid as a replacement material for single-use laboratory components. Materials 15:2989. https://doi.org/10.3390/ma15092989
54. Tripathy DB, Gupta A (2023) Nanomembranes-affiliated water remediation: chronology, properties, classification, challenges and future prospects. Membranes 13:713. https://doi.org/10.3390/membranes13080713
55. Obotey Ezugbe E, Rathilal S (2020) Membrane technologies in wastewater treatment: a review. Membranes 10:89. https://doi.org/10.3390/membranes10050089
56. Othman NH, Alias NH, Fuzil NS, Marpani F, Shahruddin MZ, Chew CM, David Ng KM, Lau WJ, Ismail AF (2021) A review on the use of membrane technology systems in developing countries. Membranes 12:30. https://doi.org/10.3390/membranes12010030
57. Ede SR, Yu H, Sung CS, Kisailus D (2024) Bio-inspired functional materials for environmental applications. Adv Biomimetic Mater 8:2301227. https://doi.org/10.1002/smtd.202301227
58. Parveen S, Malviya A (2022) Analytical review on membrane water filter using different materials to prevent microbial activities. J Pure Appl Microbiol 16:2352–2362. https://doi.org/10.22207/JPAM.16.4.68
59. El Batouti M, Al-Harby NF, Elewa MM (2021) A review on promising membrane technology approaches for heavy metal removal from water and wastewater to solve water crisis. Water 13:3241. https://doi.org/10.3390/w13223241
60. Goel G, Helix-Nielsen C, Upadhyaya HM (2021) Goel S (2021) A bibliometric study on biomimetic and bioinspired membranes for water filtration. NPJ Clean Water 4:41. https://doi.org/10.1038/s41545-021-00131-4
61. Abdelrasoul A, Doan H, Lohi A (2017) Fabrication of biomimetic and bioinspired membranes. INTECH. https://doi.org/10.5772/intechopen.71718
62. Chen T, Wei X, Chen Z, Morin D, Alvarez SV, Yoon Y, Huang Y (2022) Designing energy-efficient separation membranes: knowledge from nature for a sustainable future. Adv Membr 2:100031. https://doi.org/10.1016/j.advmem.2022.100031
63. Rahmah W, Khoiruddin K, Wenten IG, Kawi S (2024) Advancing carbon capture with bio-inspired membrane materials: a review. Carbon Capt Sci Technol 13:100318. https://doi.org/10.1016/j.ccst.2024.100318

64. Ciawi Y, Khoiruddin K (2024) Low-cost antibacterial ceramic water filters for decentralized water treatment: advances and practical applications. ACS Omega 9:12457–12477. https://doi.org/10.1021/acsomega.3c09311
65. Samdin S, Lim HP, Marzuki M (2013) Investigation of coconut shells activated carbon as the cost-effective absorbent in drinking water filter. J Teknol 77:22. https://doi.org/10.11113/jt.v77.6656
66. George-Williams HEM, Hunt DVL, Rogers CDF (2024) Sustainable water infrastructure: visions and options for Sub-Saharan Africa. Sustainability 16:1592. https://doi.org/10.3390/su16041592
67. Sha C, Shen S, Zhang J, Zhou C, Lu X, Zhang H (2024) A review of strategies and technologies for sustainable decentralized wastewater treatment. Water 16:3003. https://doi.org/10.3390/w16203003
68. Swan KH, Surinkul N, Visvanathan C (2023) Investigation of decentralized wastewater treatment system in urban wastewater management: case study in Yangon. Myanmar. Sustainability 15:16756. https://doi.org/10.3390/su152416756
69. Whaley L, Cleaver F (2017) Can 'functionality' save the community management model of rural water supply? Water Resour Rural Dev 9:56–66. https://doi.org/10.1016/j.wrr.2017.04.001
70. Annonymous (2022) Open Resources for Nursing (Open RN). In: Ernstmeyer K, Christman E, (eds) Nursing: mental health and community concepts [Internet]. Eau Claire (WI): Chippewa Valley Technical College; 2022. Chapter 17 Vulnerable Populations. https://www.ncbi.nlm.nih.gov/books/NBK590046/
71. Kruk ME, Gage AD, Arsenault C, Jordan K, Leslie HH, Roder-DeWan S, Adeyi O, Barker P, Daelmans B, Doubova SV, English M, García-Elorrio E, Guanais F, Gureje O, Hirschhorn LR, Jiang L, Kelley E, Lemango ET, Liljestrand J, Malata A, Pate M (2018) High-quality health systems in the sustainable development goals era: time for a revolution. Lancet Global health 6:e1196–e1252. https://doi.org/10.1016/S2214-109X(18)30386-3
72. Ogwu MC (2025) Sustainable development in Africa: challenges, innovations, and policy pathways for the future. Afr Res Rep 1(2):55–69. https://doi.org/10.5897/ARR2025.0005
73. Ogwu MC, Kosoe EA (2024) Place of cultural diversity in sustainable water resource management in Ghana. In: Izah SC, Ogwu MC, Loukas A, Hamidifar H (eds) Water crises and sustainable management in the global south. Springer, Singapore. https://doi.org/10.1007/978-981-97-4966-9_14
74. Ogwu MC, Kosoe EA (2025) Integrating green infrastructure into sustainable agriculture to enhance soil health, biodiversity, and microclimate resilience. Sustainability 17(9):3838. https://doi.org/10.3390/su17093838
75. Kosoe EA, Osumanu IK, Ogwu MC (2025b) Reimagining rural and small-town water supply in Ghana through inclusive governance and sustainable models. J Sustain 1(2):1–21. https://doi.org/10.55845/jos-2025-1234
76. Kosoe EA, Ogwu MC, Duniah J, Ndigun FB (2025a) Sustaining indigenous water governance systems in Ayimaye and Lungni, northern Ghana, through hybrid models. J Sustain 1(2). https://doi.org/10.55845/jos-2025-1235
77. Hutton G, Chase C (2017) Water supply, sanitation, and hygiene. In: Mock CN, Nugent R, Kobusingye O (eds) Injury prevention and environmental health, 3rd edon. Washington (DC): The International Bank for Reconstruction and Development/The World Bank
78. Ssekyanzi G, Ahmad MJ, Choi KS (2024) Sustainable solutions for mitigating water scarcity in developing countries: a comprehensive review of innovative rainwater storage systems. Water 16:2394. https://doi.org/10.3390/w16172394
79. Mondejar ME, Avtar R, Lellani H, Diaz B, Dubey RK, Esteban J, Gómez-Morales A, Hallam B, Mbungu NT, Okolo CC, Prasad AK, She Q, Garcia-Segura S (2021) Digitalization to achieve sustainable development goals: steps towards a smart green planet. Sci Total Environ 794:148539. https://doi.org/10.1016/j.scitotenv.2021.148539
80. Shemer H, Wald S, Semiat R (2023) Challenges and solutions for global water scarcity. Membranes 13:612. https://doi.org/10.3390/membranes13060612

81. Prescott MF, Dobbie MF, Ramirez-Lovering D (2021) Green infrastructure for sanitation in settlements in the Global South: a narrative review of socio-technical systems. Sustainability 13:2071. https://doi.org/10.3390/su13042071
82. Newton PW, Rogers BC (2020) Transforming built environments: towards carbon neutral and blue-green cities. Sustainability 12:4745. https://doi.org/10.3390/su12114745
83. Mhlongo S, Mbatha K, Ramatsetse B, Dlamini R (2023) Challenges, opportunities, and prospects of adopting and using smart digital technologies in learning environments: an iterative review. Heliyon 9:e16348. https://doi.org/10.1016/j.heliyon.2023.e16348
84. Mulugeta A, Bekele T, Abebe M (2020) Low-cost water filtration systems for rural households: a review of sustainable approaches. Sustain Water Resour Manag 6:89. https://doi.org/10.1007/s40899-020-00431-1
85. Aziz S, Mazhar AR, Ubaid A, Shah SMH, Riaz Y, Talha T, Jung DW (2024) A comprehensive review of membrane-based water filtration techniques. Appl Water Sci 14:169. https://doi.org/10.1007/s13201-024-02226-y
86. Hakami MW, Alkhudhiri A, Al-Batty S, Zacharof MP, Maddy J, Hilal N (2020) Ceramic microfiltration membranes in wastewater treatment: filtration behavior. Fouling Prevention Membr 10:248. https://doi.org/10.3390/membranes10090248
87. Moezzi SA, Rastgar S, Faghani M, Ghiasvand Z, Khoei AJ (2025) Optimization of carbon membrane performance in reverse osmosis systems for reducing salinity, nitrates, phosphates, and ammonia in aquaculture wastewater. Chemosphere 376:144304. https://doi.org/10.1016/j.chemosphere.2025.144304
88. Osman AI, Chen Z, Elgarahy AM, Farghali M, Mohamed IMA, Priya AK, Hawash HB, Yap PS (2024) Membrane technology for energy saving: principles, techniques, applications, challenges, and prospects. Adv Energy Sustain Res 5:2400011. https://doi.org/10.1002/aesr.202400011
89. Ulusoy A, Atılgan A, Rolbiecki R, Jagosz B, Rolbiecki S (2024) Innovative approaches for sustainable wastewater resource management. Agriculture 14:2111. https://doi.org/10.3390/agriculture14122111
90. Mdee OJ, Lwasa A, Sadiki N (2024) Overview of the low-cost technologies for household water treatment in developing countries. Water Pract Technol 19:3051–3071. https://doi.org/10.2166/wpt.2024.177
91. Duran Romero DA, de Almeida Silva MC, Chauque BJ, Benetti A (2020) Biosand filter as a point-of-use water treatment technology: influence of turbidity on microorganism removal efficiency. Water 12:2302. https://doi.org/10.3390/w12082302
92. George CM, Smith AH, Kalman DA, Steinmaus CM (2006) Reverse osmosis filter use and high arsenic levels in private well water. Arch Environ Occup Health 61:171–175. https://doi.org/10.3200/AEOH.61.4.171-175
93. Bulta AL, Micheal GAW (2010) Evaluation of the efficiency of ceramic filters for water treatment in Kambata Tabaro zone, southern Ethiopia. Environ Syst Res 8:1. https://doi.org/10.1186/s40068-018-0129-6
94. Madani N, Moulefera I, Boumad S, Cazorla-Amorós D, Gandía FJV, Cherifi O, Bouchenafa-Saib N (2022) Activated carbon from *Stipa tenacissima* for the adsorption of atenolol. C J Carbon Res 8:66. https://doi.org/10.3390/c8040066
95. Matsumoto T, Tatsuno I, Hasegawa T (2019) Instantaneous water purification by deep ultraviolet light in water waveguide: *Escherichia Coli* bacteria disinfection. Water 11:968. https://doi.org/10.3390/w11050968
96. Tripathy J, Mishra A, Pandey M, Thakur RR, Chand S, Rout PR, Shahid MK (2024) Advances in nanoparticles and nanocomposites for water and wastewater treatment: a review. Water 16:1481. https://doi.org/10.3390/w16111481
97. Nasrollahzadeh M, Sajjadi M, Iravani S, Varma RS (2021) Carbon-based sustainable nanomaterials for water treatment: state-of-art and future perspectives. Chemosphere 263:128005. https://doi.org/10.1016/j.chemosphere.2020.128005
98. Gehrke I, Geiser A, Somborn-Schulz A (2015) Innovations in nanotechnology for water treatment. Nanotechnol Sci Appl 8:1–17. https://doi.org/10.2147/NSA.S43773

99. Lazarenko NS, Golovakhin VV, Shestakov AA, Lapekin NI, Bannov AG (2022) Recent advances on membranes for water purification based on carbon nanomaterials. Membranes 12:915. https://doi.org/10.3390/membranes12100915
100. Elhenawy S, Khraisheh M, AlMomani F, Al-Ghouti M, Selvaraj R, Al-Muhtaseb A (2024) Emerging nanomaterials for drinking water purification: a new era of water treatment technology. Nanomaterials 14:1707. https://doi.org/10.3390/nano14211707
101. Kotia A, Yadav A, Raj RT, Gertrud Keischgens M, Rathore H, Sarris IE (2020) Carbon nanoparticles as sources for a cost-effective water purification method: a comprehensive review. Fluids 5:230. https://doi.org/10.3390/fluids5040230
102. Pachaiappan R, Cornejo-Ponce L, Rajendran R, Manavalan K, Femilaa Rajan V, Awad F (2022) A review on biofiltration techniques: recent advancements in the removal of volatile organic compounds and heavy metals in the treatment of polluted water. Bioengineered 13:8432–8477. https://doi.org/10.1080/21655979.2022.2050538
103. Cao JJ, Huang Y, Zhang Q (2021) Ambient air purification by nanotechnologies: from theory to application. Catalysts 11:1276. https://doi.org/10.3390/catal11111276
104. Annonymous 1(980) National research council (US) safe drinking water committee. Drinking Water and Health: Volume 2. Washington (DC): National Academies Press (US); 1980. IV, An Evaluation of Activated Carbon for Drinking Water Treatment. https://www.ncbi.nlm.nih.gov/books/NBK234593/
105. Hait M, Gir ND, Dey A, Chaturwedi AK Kashyap NK (2018a) Removal of water pollutants from the effluent of Tifra industrial area in Bilaspur (C.G.) using bio- adsorbents, Pharm Innov J 6(3):15–20
106. Hait M, Kashyap, NK, Bhardwaj AK (2024) Emerging pollutants from urbanization and industrialization in the global south. In: Izah SC, Ogwu MC, Hamidifar H (eds) Biomonitoring of pollutants in the global south, pp 39–87. Springer, Singapore. https://doi.org/10.1007/978-981-97-1658-6_2
107. Institute of medicine (US) roundtable on environmental health sciences, research, and medicine. Global environmental health: research gaps and barriers for providing sustainable water, sanitation, and hygiene services: workshop summary. Washington (DC): National Academies Press (US); 2009. 5, Achieving Water and Sanitation Services for Health in Developing Countries. https://www.ncbi.nlm.nih.gov/books/NBK50770/

Chapter 6
Rainwater Harvesting and Community-Led Solutions for Sustainable Water Management Practices

Amit Kumar Chaturvedi, Nand Kumar Kashyap, Asish Tiwari, Sant Ram Kamlesh, Supriya Biswas, and Milan Hait

Abstract Urban and rural regions alike may benefit from rainwater collection, a sustainable and efficient method of water management that helps alleviate water shortages. Communities can reduce their reliance on groundwater and ensure access to potable, agricultural, and sanitary water by collecting, storing, and utilizing rainwater. The successful implementation of these systems relies heavily on community-led solutions, which guarantee local involvement, cost-effectiveness, and sustainability in the long run. Various rainwater harvesting techniques, including rooftop collection, surface runoff harvesting, and underground storage, have been successfully adopted worldwide. Grassroots initiatives, supported by non-governmental organizations, local governments, and social enterprises, help empower communities to take charge of their water security. Additionally, policy frameworks and technological advancements enhance the efficiency and scalability of these solutions. This chapter explores the importance, methods, benefits, and challenges of rainwater harvesting, along with case studies showcasing successful community-driven projects. By integrating traditional wisdom with modern innovation, rainwater harvesting can become a key strategy in achieving sustainable water management and climate resilience.

A. K. Chaturvedi · N. K. Kashyap · M. Hait (✉)
Department of Chemistry, Dr. C. V. Raman University, Kota, Bilaspur, India
e-mail: haitmilan@gmail.com

A. Tiwari
Department of Chemistry, Govt. College Dumariya, Jarhi, Chhattisgarh, India

S. R. Kamlesh
Govt. Bilasa Girls P.G. College, Bilaspur, Chhattisgarh, India

S. Biswas
Department of Applied Chemistry, Shri Shankaracharya Institute of Professional Management and Technology, Raipur, Chhattisgarh, India

M. C. Ogwu and S. Chibueze Izah (eds.), *Water Quality and Safety in the Global South*, Springer Water, https://doi.org/10.1007/978-3-032-10602-5_6

Keywords Rainwater · Harvesting · Water scarcity · Community-led solutions · Sustainable water management · Groundwater conservation · Policy frameworks

Abbreviations

BC	Before Christ
CC	Climate change
DBPs	Disinfection byproducts
GAC	Granular activated carbon
H2O2	Hydrogen peroxide
IoT	Internate of thing
K2FeO4	Pottasium ferrate (VI)
km	Kilometers
KMnO4	Pottasium per mangate
NGOs	Non-governmental organizations
RWH	Rainwater harvesting
RWHS	Rainwater harvesting systems
SDWHs	Sustainable domestic water harvesting system
UV	Ultraviolet
UV/PS	Ultraviolet/Persulfate

6.1 Introduction

The practice of rainwater harvesting stands as a testament to humankind's ability to innovate in the face of growing environmental challenges. By collecting and storing rainwater, individuals and communities can tap into a vital resource that not only alleviates pressure on existing water supplies but also promotes sustainability. The methods employed, whether they be as straightforward as rain barrels or as intricate as urban collection systems, cater to diverse needs and settings [17]. This versatility allows harvested rainwater to serve various purposes, from nourishing gardens through irrigation to providing essential potable water and enabling efficient toilet flushing. As societies increasingly recognize the value of this practice, rainwater harvesting emerges not just as an alternative source of water but as a crucial component in fostering resilience against climate change and ensuring water security for future generations. Rainwater harvesting (RWH) is not a new concept; it has been practiced for thousands of years across different cultures and civilizations [30]. Ancient communities relied on rainwater to sustain their agricultural practices and domestic needs, building sophisticated systems to capture and store this precious resource. The revival of rainwater harvesting in modern urban settings is a testament to its enduring utility and relevance. The practice of rainwater harvesting presents a sustainable solution for water collection and conservation, emphasizing the importance of utilizing natural resources efficiently [31]. By deliberately capturing rainwater from catchment surfaces and storing it in various physical structures or soil profiles, communities can access a source of water that is not only pure and soft but also free from harmful purging chemicals, salts, and minerals. This pristine quality enhances its value, especially in agricultural settings where plants thrive in submerged systems. As awareness of environmental issues grows, embracing rainwater harvesting could serve as a vital step toward fostering ecological balance, promoting sustainable agriculture, and ensuring clean water availability for future generations [1]. The evolution of water storage practices reveals not only the ingenuity of ancient civilizations but also underscores the critical role water management has played throughout history. Archeological findings indicate that as far back as 4,000 years ago, societies were developing sophisticated methods to store and manage this vital resource, with evidence from China suggesting that the practice may extend even further back to 6,000 years. In Israel, remnants from 2000 B. C. point to early attempts at managing overflow for both domestic and private use, highlighting a continuous human endeavor to harness water's life-sustaining properties. Collectively, these historical insights illustrate how essential effective water storage has been in supporting community development and ensuring survival in various environments, emphasizing its lasting impact on modern practices and our approach to sustainability today [29]. Ultimately, the practice of rainwater harvesting presents a sustainable solution for addressing water scarcity by efficiently capturing and storing rainwater from diverse sources. By utilizing collection vessels strategically placed in open separations, communities can transform precipitation into a valuable resource for human consumption and other essential uses. This innovative approach reduces dependency on traditional water supply systems and promotes environmental stewardship by maximizing natural resources. As awareness of the importance of water conservation continues to grow, the integration of rainwater harvesting systems into everyday life may very well pave the way for a more resilient and self-sufficient future [2]. The ingenuity of ancient Roman civilization in harnessing rainwater illustrates a remarkable understanding of resource management that not only served immediate needs but also laid the groundwork for sustainable practices. With rooftop catchment systems and other innovations dating back 6,000 years, Romans demonstrated a forward-thinking approach by utilizing rainwater for drinking and residential purposes in their manors and bustling urban areas. This legacy echoes through history, as seen in the development of tanks in the Negev desert around 2000 BC, which facilitated habitation and growth in regions like Philistine. Such advancements not only highlight the necessity of adapting to environmental conditions but also reflect a broader commitment to innovation that resonates even today as we confront modern water challenges [34]. The lessons gleaned from ancient Rome's rainwater harvesting strategies continue to inspire contemporary solutions for sustainable living. The rich history of rainwater collection in Southeast Asia reveals a remarkable integration of traditional practices and sustainable resource management. From the early methods employed in Thailand, where communities ingeniously collected rainwater from roof troughs and canals, to the advanced systems seen in South Asia dating back to the ninth or tenth century, this ancient practice has evolved while maintaining its core purpose—efficiently harnessing nature's bounty [34]. Evidence of hometop rainwater collection and simple brush dam developments illustrates not only the ingenuity of these early societies but also their understanding of environmental stewardship long before modern sustainability became a global focus. As contemporary challenges regarding water scarcity intensify, revisiting these time-honored methods could provide valuable insights into creating resilient water management systems that honor both heritage and innovation [2]. The practice of rainwater harvesting, rooted in the necessity of managing short and sporadic rainfall, stands out as a sustainable solution for contemporary water challenges. As communities increasingly adapt to modern recovery techniques, the efficiency and quality of harvested rainwater shine through, often surpassing that of contaminated groundwater sources. This method not only addresses immediate water needs but also integrates seamlessly into flexible water systems, providing resilience against climate variability. Embracing rainwater harvesting not only enhances water security but also fosters a deeper connection between communities and their natural resources, paving the way for a more sustainable for upcoming years [5]. The practice of water harvesting in urban areas emerges as a versatile solution tailored to meet the unique ecological demands of each locality. In regions like India, where urban infrastructure often grapples with water scarcity, implementing rooftop rainwater harvesting systems has proven essential for civil structures. By capturing and utilizing rainwater, cities can enhance their resilience to water shortages while promoting sustainable practices that benefit both the environment and local communities. As urbanization continues to accelerate, embracing innovative water management strategies such as these will be crucial in ensuring a reliable and flexible water supply for generations to come [5]. The significance of water in our daily lives and its increasingly precarious status cannot be overstated. As urban populations in developing countries turn to costly water merchants—spending more than they would on bottled water—the fundamental right to access clean and affordable water becomes a pressing issue. Meanwhile, the staggering amount of precipitation required for basic food production in semi-arid regions underscores the urgent need for sustainable practices and efficient resource management. With projections indicating a six-fold increase in the number of individuals dependent on water by 2025, it is imperative that nations prioritize innovative solutions to address this looming crisis. The future of billions rests on our ability to recognize water not just as a commodity but as a vital resource that requires collective stewardship and responsible consumption [34]. The stark realities of the global water crisis illuminate a pressing and often overlooked humanitarian challenge. In places like the Kibeira ghetto in Nairobi, individuals must traverse an arduous 6 km (km) to collect just 20 kg (kg) of water, highlighting the immense physical toll on communities already burdened by poverty. This struggle is compounded by the fact that the cost of accessing a single liter of water can surpass that of Americans, indicating that water scarcity is not merely a geographical issue but a profound economic one. With an alarming 1.1 billion people worldwide lacking safe water access and 470 million residing in areas grappling with extreme deficiencies, it becomes evident that urgent action is required to address this crisis. As we reflect on these statistics, it is imperative to recognize our shared responsibility in advocating for sustainable solutions and equitable access to this vital resource, ensuring that everyone has the opportunity to thrive [23].

6.1.1 Brief Overview of Rain Water Harvesting

Throughout history, the ingenuity of civilizations in harnessing water has played a crucial role in their survival and prosperity. The Romans, with their advanced roof basin systems dating back to 2000 B.C., exemplified how urban environments could effectively utilize rainwater for drinking and domestic needs, thereby enhancing public health and sanitation. Similarly, the ingenious use of tanks in Israel's Negev Desert demonstrates the timeless importance of water harvesting techniques in arid regions, where storing runoff supports agriculture and sustains residential life. These diverse practices underscore a fundamental truth: the ability to collect and manage water resources has been a cornerstone of civilization development, influcncing everything from public infrastructure to agricultural productivity across different landscapes and eras [72]. As societies continue to evolve, the innovation in water management will remain crucial, particularly as climate change challenges traditional sources. By investing in sustainable methods and technologies, communities can ensure a resilient future that prioritizes both environmental conservation and the well-being of their populations [23]. The rich tradition of rainwater harvesting in Asia underscores the region's deep-rooted relationship with water conservation, showcasing techniques that have evolved over nearly two millennia. As evidenced by historical practices in Thailand and the continued relevance of similar methods across Africa and Asia, this sustainable approach to water collection has been integral to human survival and agricultural success. The Yerebatan Cistern in Istanbul epitomizes the engineering prowess achieved in this field, standing as a monumental testament to the importance of capturing and storing rainwater. Together, these elements highlight not only the historical significance but also the enduring necessity of innovative water management strategies as societies confront modern challenges related to climate change and resource scarcity. As urban populations continue to grow and environmental pressures mount, the lessons learned from ancient practices can inform contemporary solutions [71, 74]. By blending traditional wisdom with modern technology, communities worldwide can develop resilient systems that ensure access to clean water for future generations. The evolution of rainwater harvesting, from the ingenious aqueducts of ancient Rome to the refined systems of 13th-century Venice, illustrates humanity's enduring quest to optimize water resources. These historical advancements laid the groundwork for contemporary techniques that harness modern materials and technologies, such as advanced pumps and reinforced concrete tanks [50]. Today, rainwater harvesting not only reflects a continuation of these age-old practices but also embodies a commitment to sustainable water management in an era increasingly challenged by water scarcity. As we look ahead, the integration of historical wisdom with modern innovation will be crucial in ensuring that communities can effectively capture and utilize this precious resource for generations to come. This synergy between past and present offers a roadmap for developing resilient infrastructure that can withstand the pressures of climate changes. By prioritizing education and community involvement, we can further enhance the effectiveness of rainwater harvesting systems, ensuring they meet the diverse needs of urban and rural populations [4]. The intricate relationship between the Ottoman Empire (1669–1898) and its water culture reveals a profound understanding of the significance of water in both daily life and spiritual practice. The emphasis placed on water within Muslim tradition not only inspired the construction of ornate fountains and public baths but also catalyzed the development of advanced hydraulic systems that supported large cisterns across the empire. These cisterns became vital lifelines in remote regions, ensuring access to this precious resource where other water systems were absent. Additionally, the adaptation from rectangular Byzantine designs to circular cisterns in rural areas highlights a unique architectural evolution tailored to meet local needs. Ultimately, this sophisticated approach to water management not only enhanced public health and hygiene but also fostered a sense of community and cultural identity throughout the Ottoman Empire [63]. The historical and ongoing significance of rainwater harvesting underscores its vital role in human survival and sustainability. From ancient civilizations that relied on this method for irrigation and livestock care to modern-day initiatives addressing water scarcity, the practice of collecting and storing rainwater has proven essential across various contexts. As a primary source of both potable and non-potable water, it not only fulfills fundamental domestic needs but also supports agricultural practices in arid regions where conventional water sources are scarce. Embracing rainwater harvesting as a sustainable solution is crucial for ensuring access to clean water, particularly in areas facing increasing environmental challenges. Ultimately, the lessons drawn from our ancestors highlight the enduring importance of this resource in fostering resilience and promoting a sustainable future for communities worldwide. By integrating modern technology with traditional methods, communities can enhance their rainwater harvesting systems, maximizing efficiency and effectiveness. This approach conserves water and empowers individuals to take an active role in managing their resources, fostering a sense of stewardship that can lead to broader environmental benefits [3].

6.1.2 Classical Concept of Rainwater Harvesting

The practice of rainwater harvesting emerges as a vital strategy in addressing the pressing challenges of water scarcity, particularly in arid and semi-arid regions. By efficiently collecting and storing rainwater, whether on the surface or underground, communities can significantly reduce their dependence on dwindling non-renewable groundwater sources. Innovative techniques for capturing surface runoff not only promote sustainable water management but also enhance overall water use efficiency and productivity. As we continue to confront the realities of climate change and population growth, embracing rainwater harvesting can pave the way for a more resilient future, ensuring that this precious resource is harnessed effectively for generations to come [80]. Ultimately, the classification of water catchment systems reveals a complex interplay between various agricultural and environmental needs. Microsystem classifications such as farm and roof systems illustrate how localized strategies can effectively capture and manage water resources, promoting sustainability at the community level. On the other hand, macrosystem classifications like valley floor and non-valley systems focus on larger environmental factors that go beyond single properties to include regional water flow and land use. By understanding these diverse approaches to water management, we can better appreciate the vital role these systems play in preserving our natural resources while addressing the challenges posed by climate change and population growth. Embracing both microsystems and macrosystems will be crucial in developing integrated strategies that ensure water security for future generations [91]. The multifaceted approach to rainwater harvesting highlights its adaptability to varying water demands and climatic conditions, making it a pivotal strategy for sustainable water management. Among the various systems, roof rainwater harvesting stands out as the most prevalent method, efficiently channeling water through gutters into storage systems. The choice between underground storage options, such as cisterns or soil infiltration, and surface storage solutions like tanks and ponds, further emphasizes the versatility of these systems in catering to diverse needs [60]. By harnessing this natural resource effectively, communities can not only alleviate pressure on existing water supplies but also promote environmental stewardship and resilience in the face of changing weather patterns. Ultimately, rainwater harvesting emerges as a crucial tool in the quest for sustainable living and resource conservation. As awareness of its benefits grows, more communities are adopting rainwater harvesting practices, integrating them into urban planning and agricultural strategies. This shift not only enhances water security but also fosters a culture of sustainability that encourages individuals and organizations to take proactive steps in managing their water resources responsibly [40].

6.1.3 Advantages of Rain Harvesting

The practice of rainwater harvesting stands as a remarkable solution to the pressing challenges of water scarcity and resource management. By enabling communities to capture and treat rainwater for safe drinking, it fosters self-sufficiency while safeguarding vital water resources. This sustainable approach extends its benefits to agriculture, providing an efficient irrigation system that supports healthy landscaping and crop cultivation. Furthermore, by reducing runoff into surface water bodies, rainwater harvesting contributes to environmental conservation efforts. With its straightforward technology and low operational expenses, this method ensures that individuals can maintain their systems with ease, ultimately decreasing reliance on conventional surface and groundwater sources [79]. Embracing rainwater harvesting enhances local resilience and promotes a more sustainable future for our planet's precious water supply. Additionally, community education and outreach programs can further amplify the benefits of rainwater harvesting, encouraging wider adoption and innovative uses. By fostering a culture of conservation, we can collectively work toward safeguarding our water resources for generations to come. The multifaceted advantages of rainwater harvesting present a compelling case for its adoption in both urban and rural settings. By effectively managing storm water, it plays a crucial role in flood control during intense rainfall events, mitigating the risks associated with sudden downpours. Additionally, it serves as a vital alternative water source for households that struggle with access to clean water, addressing a pressing societal need. Compared to traditional dams and piped distribution systems, rainwater harvesting boasts a lower environmental footprint, making it a more sustainable choice for communities [42]. As a cost-effective solution, it ensures the availability of high-quality water and allows for surplus water to be redirected toward groundwater replenishment, promoting long-term ecological health. Embracing rainwater harvesting is not merely an innovative approach to resource management,it reflects a commitment to sustainability and resilience in the face of growing environmental challenges. By integrating rainwater harvesting into urban planning and community practices, we can create a more adaptable infrastructure that responds effectively to climate variability. This forward-thinking strategy conserves vital water resources and fosters community stewardship over local ecosystems [25].

6.1.3.1 Environmental Advantage of Rainwater Harvesting

Embracing rainwater harvesting offers a multitude of environmental and practical advantages that cannot be overlooked. By effectively capturing and utilizing rainwater, we can significantly reduce storm water runoff, which in turn decreases peak flow volume and velocity, thus minimizing the risk of stream bank erosion. This practice aligns with municipal storm water management requirements and presents an excellent, chemical-free water source for nurturing plants and landscaping. As communities strive for sustainability, integrating rainwater harvesting into our water management strategies emerges as an essential step toward preserving natural resources and promoting ecological balance. Moreover, public awareness and education about the benefits of rainwater harvesting can further enhance community participation and support. By fostering a culture of conservation, we empower individuals to take action in their households, ultimately contributing to a collective effort that benefits the environment and local ecosystems [15].

6.1.3.2 Economic Advantage of Rainwater Harvesting

The pressing challenges of rising costs associated with supplying treated water to communities highlight the urgent need for innovative solutions in water management.

With expenses stemming from extensive infrastructure—encompassing dams, pipes, treatment plants, and ongoing maintenance—ratepayers are increasingly feeling the weight of these financial demands. However, by embracing rainwater harvesting as a viable alternative, communities can significantly reduce their overall water demand while also benefiting from more affordable potable water services [69]. This sustainable approach not only alleviates the burden on existing water infrastructure but also fosters a more responsible relationship with our precious natural resources, paving the way for a more resilient and economically sound future in water supply. This shift toward rainwater harvesting encourages a proactive mind set regarding water conservation, ultimately leading to enhanced sustainability. By prioritizing such initiatives, communities can better prepare for future challenges related to water scarcity and infrastructure strain. The integration of rainwater collection systems into our communities fosters sustainable practices and opens the door to a myriad of employment opportunities within the burgeoning green infrastructure movement [83]. By harnessing the power of onsite rainwater storage, we can bolster wildfire protection efforts, safeguarding both properties and lives in vulnerable areas. Moreover, the financial incentives, such as potential insurance discounts for those who implement these systems, serve to further enhance their appeal. As we continue to prioritize environmental responsibility and resilience, the role of rainwater collection will undoubtedly expand, paving the way for a more sustainable and secure future that benefits individuals and communities alike.

6.1.4 The Need for Community-Led Water Solutions

Ensuring access to clean water is not merely a matter of infrastructure; it is a fundamental human right that underscores the dignity and well-being of individuals and communities alike. However, traditional top-down approaches often fail to address the intricate challenges faced by local populations, resulting in solutions that do not align with their actual needs. By prioritizing community engagement and adopting more inclusive strategies, we can create sustainable frameworks that empower individuals and acknowledge their unique circumstances. Ultimately, bridging the gap between policy-making and grassroots realities will be essential in overcoming water access challenges, fostering resilience, and promoting equity for all [28]. In contrast, community-led water solutions empower residents to identify issues, design appropriate interventions, and sustainably manage their water resources. One major advantage of community-led initiatives is that they leverage local knowledge. Community members understand their environment, seasonal water patterns, and cultural practices better than external agencies. This insight leads to more relevant and effective solutions. For instance, small-scale rainwater harvesting or localized filtration systems often have greater long-term success when planned and maintained by the community [48]. Moreover, involving communities in water projects increases accountability and ownership, when people contribute to building and maintaining their own water systems, they are more likely to ensure their proper use and upkeep

[70]. This reduces dependency on external aid and fosters self-reliance. Community-led water projects also enhance social cohesion and resilience. By coming together to solve shared problems, communities strengthen trust and collective responsibility [7]. In times of drought, contamination, or infrastructure failure, these networks are crucial for rapid and adaptive responses. In a world facing growing water stress due to climate change and population growth, it is more important than ever to invest in participatory water management. Community-led solutions are not just about water—they are about dignity, empowerment, and sustainable development from the ground up [26, 71].

6.1.5 Climate Change and Rainwater Harvesting (RWH)

As the intricate web of climate change (CC) continues to unravel, its profound impact on water demand and supply becomes increasingly evident. The confluence of rapid population growth, industrialization, and urbanization has already strained our water resources, but the looming specter of global warming complicates this challenge further. Shifts in temperature affect evapotranspiration rates and reshape atmospheric water storage, creating an unpredictable landscape for water availability [22]. Moreover, alterations in rainfall patterns—manifested through changes in magnitude, intensity, and frequency—underscore the urgent need for adaptive measures. Without a concerted effort to address these interlinked issues, we risk exacerbating an already critical situation that threatens both human livelihoods and ecological health [37]. To mitigate these challenges, it is imperative that we invest in sustainable water management practices and innovative technologies. Collaboration across sectors and communities will play a vital role in developing resilient systems that can withstand the changing climate and ensure equitable access to water for all [89]. The intricate relationship between rainfall patterns, the reliability of Rainwater Harvesting Systems (RWHS), and the looming effects of climate change underscores the necessity for adaptive strategies in water resource management. As shifting climate dynamics introduce variability in precipitation, traditional Water Balance Model designs may fall short if they do not account for these unpredictable changes. This inadequacy raises critical questions about the sustainability of RWHS, particularly during prolonged dry spells or unexpected rainfall events. Thus, embracing a more flexible approach that integrates climate projections into RWHS planning becomes essential to ensure reliable water supply and resilience against future uncertainties [14]. By acknowledging and addressing these challenges, communities can better safeguard their water resources in an increasingly uncertain climate landscape. This proactive strategy enhances the effectiveness of RWHS systems and fosters community engagement in water conservation efforts. Ultimately, a collaborative approach that combines innovative technology with traditional practices will strengthen the resilience of water management systems and promote sustainable usage for generations to come [88]. As the effects of climate change continue to unfold, the reliability of Rainwater Harvesting Systems (RWHS) across the United States faces a complex and nuanced future. While some regions may experience a marginal decline in RWHS reliability, with runoff capture potentially plummeting to as low as 12%, others could benefit from an unexpected boost in water supply reliability, reaching increases of up to 22%. However, on average, communities may confront a decrease in water supply reliability by as much as 18% [6]. This duality underscores the need for adaptive management strategies that account for regional variations and changing weather patterns. By understanding these dynamics, stakeholders can better prepare for the challenges ahead and harness the potential benefits of RWHS in an era increasingly defined by climate uncertainty. This proactive approach enhances resilience and promotes sustainable water use practices. As communities implement these strategies, they can foster collaboration among local governments, organizations, and residents to ensure equitable access to water resources for all [67]. The intricate relationship between climate change and regional water security in China reveals a pressing challenge that demands immediate attention. As dry periods intensify, resulting in a staggering 27% reduction in water security, the stark reality of shifting weather patterns—where the already arid regions become drier while wetter areas experience increased rainfall—emerges. This duality underscores the necessity for adaptive strategies tailored to local conditions, such as expanding tank sizes in arid zones to enhance water storage capacity [92]. Ultimately, addressing the diverse and location-specific impacts of climate change is essential for fortifying water security across the nation, ensuring that vulnerable regions are not left to grapple with the consequences alone. By fostering collaboration between governments, communities, and scientific experts, we can develop innovative solutions that mitigate the effects of climate change and promote resilience in the face of future challenges. Empowering local populations through education and resource management will be keys to creating sustainable practices that safeguard water resources for the next generations [51]. Rainfall patterns and system design delicately balance each other for the effectiveness of rainwater harvesting systems. By prioritizing rainfall as the key variable, designers can tailor RWHS to match specific climatic conditions, ensuring optimal performance. A critical aspect of this design process is determining the ideal tank size; striking the right balance is essential—too large a tank leads to unnecessary resource expenditure, while too small a tank can result in inadequate water supply for users. Ultimately, achieving efficiency in RWHS requires careful consideration of these elements, demonstrating that thoughtful design can harness nature's bounty while promoting sustainability and resource conservation. In addition to tank size, factors such as local rainfall patterns and potential evaporation rates must also be evaluated to create a truly effective system. By integrating these considerations into the design, communities can enhance their resilience to climate variability and secure a more reliable water source for future generations. Therefore, when designing an RWHS, it is essential to consider water demand, intended uses, and geographical characteristics [84]. Table 6.1 presents additional seminal research related to climate change and RWHS.

Table 6.1 Climate change and RWHS

Authors and year	Study area	Methods and models	Summary
[45]	Nepal	CLIM software	Rooftop rainwater collecting, artificial groundwater recharge and grey water recycling can supply water needs until 2050
[55]	Africa	Continental scale modelling	Rainwater collection devices might overcome 40% of water shortage-related production deficiencies in present circumstances and 30% in expected future climates by 2050 during the major maize growing season
[59]	Sri Lanka	Long Ashton research station weather generator	By the period of 2080–2099, climate change (CC) is expected to lead to increased rainfall, which will have a greater impact on residential rainwater collection systems compared to non-residential systems
[42]	Bangladesh	Simulation model	Individual-based rainwater collection systems have 35–40% volumetric dependability under historical rainfall patterns and 48–50% under expected future rainfall
[37]	Australia	Water-balance simulation model	CC is expected to harm rainwater gathering systems
[93]	China	Downscaling technique	Future rainfall increases water-saving performance, but larger tank sizes are required for efficient storm water capture

6.1.6 Water Qualities and Rain Water Harvestings (RWHs)

The journey of rainwater from the sky to our homes is fraught with challenges that can compromise its quality. While it begins as a pure resource, factors such as harvesting methods, storage conditions, and household usage can introduce various contaminants. Wind-blown soil, animal feces, insects, leaves, and decaying organic matter all have the potential to taint this precious water supply. Consequently, the risks associated with consuming polluted rainwater from storage tanks cannot be overlooked. Ensuring that rainwater remains safe for use requires diligent attention to collection and maintenance practices, highlighting the importance of safeguarding this vital resource for both health and sustainability [18]. The pressing issue of health concerns related to drinking water underscores the critical need for effective management and design strategies. Contaminants from storage tanks and point-of-use systems can pose serious health risks, necessitating a proactive approach to safeguard public health. Implementing scientifically informed practices can significantly mitigate these risks, particularly through the adoption of rainwater harvesting systems that feature clean catchments and well-maintained storage tanks. However, it is essential to recognize that poorly designed systems can exacerbate health hazards, highlighting the importance of careful planning and ongoing maintenance [32]. By prioritizing these measures, communities can ensure safer drinking water while promoting overall well-being and environmental sustainability. Additionally, public education initiatives are crucial in raising awareness about the benefits and potential challenges of rainwater harvesting. Engaging local stakeholders in training sessions can foster a sense of ownership and responsibility, ultimately leading to more effective and sustainable water management practices [65]. The intricate relationship between rainwater quality and various influencing factors underscores the complexity of managing this vital resource. The presence of contaminants such as heavy metals and trace organic pollutants in roof runoff highlights the need for careful consideration of the materials used in roofing, as well as the surrounding environmental conditions. Additionally, physical and chemical quality is not only dictated by rainfall intensity but also by the specific sources of pollution that may accumulate during dry periods. Furthermore, microbiological quality is intricately linked to both roof materials and the duration of preceding dry spells, emphasizing how environmental and structural choices can significantly affect water safety. Collectively, these insights reveal that achieving optimal rainwater quality requires a holistic approach that addresses both material selection and environmental management practices [64]. The persistent presence of trace organics and heavy metals, as well as the limitations of conventional cleaning methods, underscore the challenge of effectively removing them from roof runoff. The characteristics of roofs significantly affect the concentration levels of organic pollutants in runoff, especially when combined with their unique chemical properties. To tackle this issue, the initial phase involves straining out larger contaminant particles, setting the stage for a more thorough filtration process [66]. Utilizing materials such as sponge, charcoal, coarse sand, and fine sand ensures that finer contaminants are effectively captured. Moreover, incorporating chlorination plays a critical role in neutralizing any bacteria that may evade earlier treatment stages and enhancing overall water safety. Together, these strategies create a comprehensive approach to managing trace organics in roof runoff while highlighting the complexity of environmental remediation efforts [49]. The exploration of rainwater disinfection methods reveals a diverse array of techniques, each offering unique advantages in ensuring safe water for consumption. Chlorination and chloramination stand out for their effectiveness in eliminating pathogens while simultaneously reducing carbonaceous disinfection by-products, a vital consideration in water treatment. Complementing these methods, UV/H_2O_2 disinfection and UV/PS disinfection harness the power of ultraviolet light coupled with advanced oxidants, providing an environmentally friendly alternative that minimizes chemical residues. Furthermore, pre-oxidation with $KMnO_4$ and K_2FeO_4 emerges as a critical step to enhance the overall efficiency of the treatment process, particularly in reducing disinfection by-products [58]. Together, these methodologies not only improve the safety and quality of rainwater but also showcase the importance of integrating innovative approaches to achieve optimal results in water purification systems. As communities increasingly turn to rainwater harvesting as a sustainable resource, understanding and implementing these disinfection techniques will be paramount in safeguarding public health while promoting environmental stewardship [77]. The implementation of Granular Activated Carbon (GAC) post-treatment has proven to be a transformative approach in the realm of water purification. By significantly lowering the concentrations of disinfection byproducts (DBPs), GAC enhances water safety and mitigates the associated toxic effects that can compromise public health. This dual benefit underscores the importance of integrating advanced filtration technologies in water treatment processes. As communities increasingly prioritize safe drinking water, the adoption of GAC post-treatment emerges as a crucial strategy, demonstrating that innovative solutions can effectively address both environmental and health concerns [27]. Table 6.2 provides a summary of studies on water quality related to rainwater harvesting systems.

6.2 Rainwater Harvesting Techniques

The integration of rainwater harvesting into urban planning offers a transformative approach to water management that enhances resilience and fosters sustainable living. By significantly reducing reliance on surface and groundwater sources, communities can mitigate the risks associated with water scarcity. The strategic capture of rainwater plays a crucial role in preventing urban flooding, minimizing soil erosion, and alleviating waterlogging, thereby creating safer and more stable environments. Techniques such as percolation pits and recharge wells further contribute to the replenishment of aquifers, ensuring a reliable supply even during periods of drought or dry spells. Ultimately, embracing rainwater harvesting not only safeguards vital water resources but also empowers communities to adapt to changing climate conditions, making it an indispensable component in the pursuit of environmental sustainability. Emphasizing the vital role of rainwater harvesting, it becomes clear that this practice is not merely a sustainable approach but an essential strategy for both agricultural and urban landscapes. By capturing and utilizing rainfall, RWH significantly enhances water availability for crops and livestock, thereby promoting agricultural productivity and food security. In urban areas, it alleviates the strain on municipal water supply systems, presenting a practical solution to growing water demands. Furthermore, as unpredictable weather patterns become more common, these techniques foster resilience against erratic rainfall, ensuring that communities can adapt effectively to changing conditions. Ultimately, rainwater harvesting not only strengthens water security but also supports diverse ecosystems and fortifies our collective response to climate challenges, paving the way for a sustainable future. Future initiatives should focus on expanding awareness and accessibility of rainwater harvesting systems, particularly in regions facing severe water scarcity. By investing in education and infrastructure, communities can harness this valuable resource, empowering individuals to take an active role in water conservation and environmental stewardship. Some important techniques are discussed below by systematically.

Table 6.2 Water quality and RWHS

Authors and year	Study area	Methods and models	Summary results
[90]	Developing countries	Meta-analysis	Potable water obtained in different situations loses bacteriological integrity
[64]	India	Review paper	Rooftop gathered rainwater must be microbiologically analysed to ensure its purity
[47]	South Africa	Exploratory analysis	Waterborne infections pose concerns beyond the expense of installing, maintaining, and using a Sustainable Domestic Water Harvesting system (SDWHs)
[13]	Uganda	Logistic regression	The possibility of biological degradation in collected water was recognized by 85% of responders
[42]	Bangladesh	Ac–Vc relation method	Up to three months without treatment, Dhaka rainwater is safe to drink
[21]	Nepal	Free listing and household survey	Effective execution and management are essential in gathering excellent water, since water quality tests usually reveal positive findings
[54]	Nigeria	Sampling-cum-laboratory analysis	Filtered and chlorinated rainwater stored in an underground tank for a month is drinkable
[24]	Mexico	Descriptive analysis	Before drinking, harvested water must be treated
[62]	USA	Descriptive analysis	Over 80% of tests exceeded international microbiological water quality criteria and were within US metal restrictions
[78]	Philippines	Process analysis	Rainwater may be used for more than drinking
[27]	Saudi Arabia	Tried disinfection technique	Municipal rainwater is contaminated with sediments, microorganisms, toxic substances, and organic materials and must be treated

6.2.1 Rooftop Rainwater Collection

As the world grapples with increasing water scarcity, rooftop rainwater harvesting emerges as a beacon of hope, offering a practical and sustainable solution to this pressing challenge. This innovative approach not only captures and utilizes

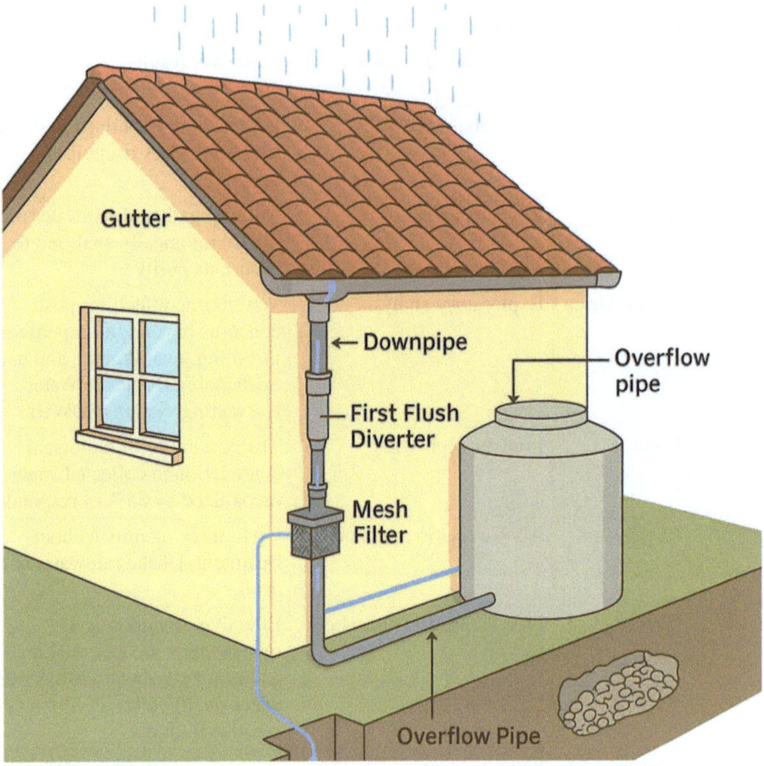

Fig. 6.1 Rooftop rainwater harvesting (RWH) system

precious rainfall but also significantly benefits regions struggling with limited reliable water sources. By transforming rooftops into functional reservoirs, communities can enhance their resilience against droughts and water shortages, ultimately fostering a more sustainable relationship with our environment. The adoption of rooftop rainwater harvesting can pave the way for more self-sufficient neighborhoods, empowering individuals to play an active role in conserving one of our planet's most vital resources. This method involves capturing and storing rainwater from the roofs of buildings, effectively turning an abundant natural resource into a viable supply of freshwater for various uses [46]. Figure 6.1 show the rooftop rainwater harvesting (RWH) system.

6.2.1.1 Rooftop Rainwater Harvesting Procedure

Collection System

The first step involves installing a system of gutters and downspouts to channel rainwater from the rooftop into a storage tank. The design must ensure an effective and efficient flow of water while minimizing contamination.

Filtration and Treatment

As rainwater can carry debris and pollutants, filtration systems are essential. A simple mesh screen can prevent larger particles from entering the tank, while additional filtration may be needed for potable water use.

Storage Facilities

Rainwater can be stored in tanks made from materials such as plastic, concrete, or metal. The size of the tank should be determined based on the rainfall patterns and water needs of the community.

Distribution System

A network of pipes can distribute water from the storage tank to various points of use, such as gardens, toilets, or kitchens. Gravity systems or pumps can be utilized to facilitate this distribution. Community-led initiatives often play a vital role in the implementation and maintenance of these systems. By engaging local residents, promoting education about RWH benefits, and fostering ownership, communities can significantly enhance their resilience to water scarcity while promoting sustainable practices. Thus, rooftop rainwater harvesting emerges as a key strategy for sustainable water management in urban and rural areas alike.

6.2.2 Surface Runoff Harvesting

Harnessing the potential of surface runoff harvesting emerges as a pivotal strategy for sustainable water management. By capturing excess rainwater that would otherwise be lost to evaporation or runoff, this method not only efficiently directs vital resources into useful storage systems but also plays a crucial role in preventing soil erosion. As communities increasingly face the challenges posed by climate change and water scarcity, adopting surface runoff harvesting can lead to innovative solutions that optimize water utilization while preserving the integrity of our landscapes.

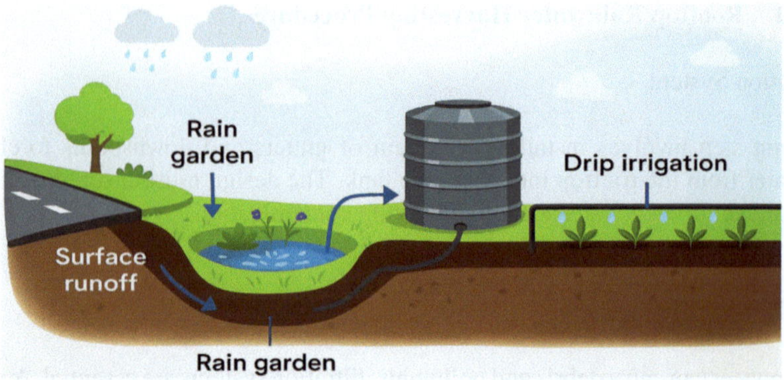

Fig. 6.2 Surface runoff harvesting

Ultimately, this approach supports ecological balance and empowers societies to thrive in harmony with their natural environment. As these practices gain traction, they encourage greater community involvement and awareness of sustainable water management. By fostering a culture of conservation, communities can adapt to changing conditions and ensure a resilient future for both people and ecosystems. The surface runoff harvesting water system is given in Fig. 6.2

6.2.2.1 Surface Runoff Harvesting Procedure

1Catchment Design

The successful implementation of runoff harvesting begins with the design of catchment areas. These can include sloped surfaces, bunds, or contour trenches that collect rainwater flowing downhill, directing it toward designated storage sites.

Storage Systems

Once captured, surface runoff can be stored in various structures such as ponds, reservoirs, or underground tanks. The size and type of the storage facility should align with the volume of runoff expected and the community's water needs.

Diversion Channels

Constructing diversion channels or swales can help redirect the flow of runoff to storage areas. These channels reduce soil erosion and ensure that water is efficiently harvested, particularly during heavy rainfall events.

Infiltration Pits

Incorporating infiltration pits can enhance groundwater recharge. These pits allow collected runoff to percolate back into the ground, improving the local aquifer levels and ensuring the availability of groundwater for future needs. Community-led initiatives are crucial in promoting awareness and education about surface runoff harvesting techniques. By involving local stakeholders in the design and maintenance of harvest systems, communities foster a collective responsibility that enhances sustainability. This grassroots approach not only addresses immediate water scarcity but also empowers communities to advocate for sustainable practices, creating resilience against climate challenges. Thus, surface runoff harvesting becomes an integral component of sustainable water management strategies.

6.2.3 Underground Storage Systems

The implementation of underground storage systems for rainwater harvesting emerges as a vital solution in addressing the pressing issue of water scarcity, particularly in arid regions where every drop counts. By capturing and storing rainwater below the surface, these systems significantly reduce evaporation losses that often plague traditional above-ground methods, ensuring that valuable water resources are preserved for future use. This innovative approach enhances water accessibility and promotes sustainable practices that can help mitigate the impacts of drought and climate change. As communities increasingly adopt these systems, they pave the way for a more resilient and water-secure future, demonstrating that effective management of natural resources is essential for long-term sustainability [85]. Underground water storage system is given in Fig. 6.3.

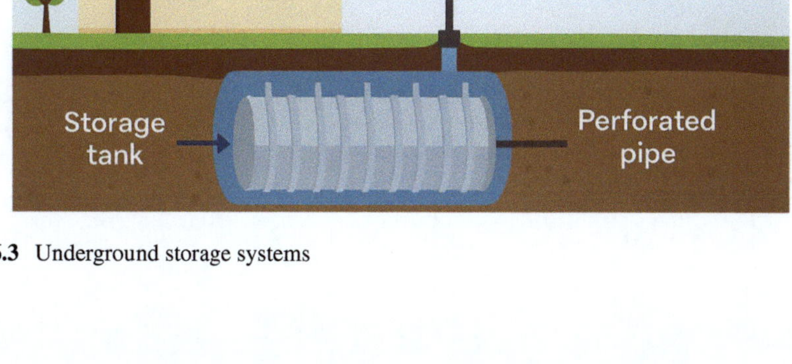

Fig. 6.3 Underground storage systems

6.2.3.1 Underground Storage Systems Procedure

Recharge Pits

Construction of recharge pits allows rainwater to infiltrate into the ground, replenishing local aquifers. These pits are strategically placed in areas with high runoff potential, maximizing the amount of water diverted into the ground.

Cisterns

Underground cisterns are engineered to collect and store rainwater from rooftops and other surfaces. These storage tanks can be fitted with filtration systems to ensure the stored water is clean and usable for irrigation, household use, or other needs.

Percolation Tanks

Designing percolation tanks helps manage storm water effectively. These structures capture runoff and facilitate natural filtration through soil layers, enhancing groundwater recharge.

Sustainable Materials

Using eco-friendly materials for the construction of underground storage systems ensures longevity and reduces environmental impact. Materials such as recycled concrete or bio-composite containers lend to sustainable practices. Community-led initiatives play a vital role in the successful implementation and maintenance of these systems. By engaging local residents in training sessions, workshops, and collaborative design processes, grassroots movements foster a sense of ownership and responsibility toward water management practices. Communities are empowered to identify their specific water needs, enabling customized solutions that reflect their unique socio-cultural contexts. Ultimately, underground storage systems not only address immediate water scarcity but also contribute to a sustainable, resilient future for communities facing climate variability.

6.3 Benefits and Challenges

6.3.1 Environmental and Economic Benefits

Rainwater harvesting (RWH) and community-led solutions present compelling environmental and economic benefits in sustainable water management practices. These initiatives not only mitigate water scarcity but also enhance community resilience and promote sustainable development.

6.3.1.1 Environmental Benefits

1Water Conservation

RWH systems help capture and store rainwater, reducing reliance on groundwater and surface water sources, which can become depleted.

Ecosystem Preservation

By mitigating runoff, RWH can reduce soil erosion and minimize pollution entering local water bodies, preserving aquatic ecosystems.

3Biodiversity Enhancement

Improved water availability encourages biodiversity by supporting local flora and fauna.

6.3.1.2 Economic Benefits

Cost Savings

By utilizing harvested rainwater, communities can lower their water bills and reduce the expenses associated with municipal water supply.

Job Creation

Installing and maintaining RWH systems can create local employment opportunities, fostering economic growth within communities.

Table 6.3 Benefits of rainwater harvesting

Benefit	Summary	Reference
Reduces water bills	Collecting rainwater lowers dependence on municipal supply	[44]
Eases groundwater pressure	Reduces over-extraction of groundwater, allowing natural recharge	[61]
Flood mitigation	Harvesting slows runoff, reducing urban flooding risk	[11]
Water for non-potable uses	Ideal for irrigation, flushing, and cleaning, easing demand on treated water	[44]
Promotes self-sufficiency	Empowers communities and households with a reliable water source	[10]
Improves water quality	Reduces surface runoff, lowering pollution in water bodies	[12]

Resilience to Climate Change

With increasingly unpredictable rainfall patterns, RWH provides a buffer against drought, enhancing food security and economic stability. Benefits of rainwater harvestings are given in Table 6.3.

6.3.2 Challenges in Implementation

While rainwater harvesting (RWH) and community-led solutions offer many benefits for sustainable water management, their implementation faces several challenges that can impede their effectiveness.

6.3.2.1 Initial Capital Investment

The cost of installing RWH systems can be significant, posing a barrier for many low-income communities.

6.3.2.2 Technical Challenges

Technical challenges—including system design, water quality issues, and maintenance requirements—can impede the implementation of rainwater harvesting. Overcoming these challenges necessitates technical expertise, continuous research, and the establishment of standardized practices. Effective RWH requires understanding and skills that may not be readily available in all communities, necessitating training and education initiatives.

6.3.2.3 Regulatory Issues

In some areas, regulations may restrict RWH practices, complicating implementation efforts and discouraging community participation.

6.3.2.4 Maintenance and Sustainability

Maintaining rainwater harvesting systems is important to keep them working well and lasting a long time. To achieve this, it's necessary to regularly check the systems, clean the areas where rainwater is collected, and take care of the storage and transportation parts of the system. This helps prevent problems like contamination and system breakdowns. Additionally, giving people the training and tools they need for maintenance can help address these issues effectively. Ensuring the long-term maintenance of RWH systems can be challenging, requiring ongoing community engagement and resources.

6.3.2.5 Economic and Social Barriers

Economic and social barriers, such as the initial installation costs and public resistance, can hinder the adoption of rainwater harvesting. Offering financial incentives, showcasing long-term cost savings, and engaging communities in the planning process can help overcome these obstacles. Challenges of rainwater harvesting implementation are given in Table 6.4.

Overall, while RWH and community-driven initiatives hold great potential for addressing water scarcity, these barriers must be acknowledged and addressed to enhance their adoption and success. By fostering collaboration with governments,

Table 6.4 Challenges of rainwater harvesting implementation

Challenge	Application	Reference
High initial cost	Installation of tanks, filters, and piping can be expensive	[44]
Maintenance issues	Systems require regular cleaning and upkeep to ensure efficiency	[10]
Limited storage capacity	Storage may not meet long-term water demand during dry periods	[44]
Water quality concerns	Risk of contamination if not properly filtered or maintained	[12]
Public awareness and acceptance	Lack of knowledge or misconceptions can hinder adoption	[11]
Policy and regulatory barriers	Absence of supportive laws and incentives in some regions	[61]

NGOs, and local stakeholders, solutions can be developed that empower communities and ensure sustainable water management practices are widely implemented.

6.4 Case Studies of Successful Initiatives

6.4.1 Community-Driven Rainwater Projects Worldwide

Around the globe, numerous case studies highlight the success of community-driven rainwater harvesting (RWH) initiatives in addressing water scarcity through sustainable practices. In Rajasthan, India, the "Jal Sahelis" program exemplifies grassroots mobilization for water conservation. Local women, trained in RWH techniques, enhance community resilience by constructing rooftop rainwater harvesting systems. This initiative has led to significant improvements in water availability and has empowered women, making them pivotal in managing local water resources. In Australia, the city of Brisbane implemented a citywide rainwater harvesting program aimed at reducing dependency on potable water. By incentivizing residents to install rainwater tanks and promoting the integration of RWH systems in new developments, the initiative has effectively decreased water demand, thereby establishing a model for urban water management. In South Africa, the community of Makhanda faced severe water shortages due to drought. Its residents implemented a community-led RWH project that constructed large collection tanks and trained locals in maintenance and usage. This approach not only improved water access but also fostered a sense of ownership and responsibility among community members, leading to long-term sustainability. These initiatives demonstrate that RWH, coupled with strong community engagement, can provide effective solutions for water scarcity. By leveraging local knowledge, fostering collaboration, and emphasizing empowerment, these successful case studies serve as blueprints for scaling sustainable water management practices worldwide, highlighting the importance of community-led efforts in climate adaptation and water security.

6.4.2 Lessons Learned

Case studies of successful rainwater harvesting (RWH) initiatives demonstrate the efficacy of community-led solutions in addressing water scarcity. In India, the village of Kothapally implemented a comprehensive RWH system that empowered residents to manage their water resources collaboratively. By involving local farmers in design and implementation, the project fostered greater ownership and sustainability. A key lesson learned was that leveraging local expertise and traditional practices led to enhanced community engagement and increased trust in the initiative. In Brazil, the "Cisterns for Life" program provided rural families with rainwater storage tanks.

Community participation was vital; training sessions educated residents on maintenance and usage. The program highlighted the importance of inclusive planning, ensuring that marginalized groups were also considered, thereby promoting equity in water access. In South Africa, the "Rainwater Harvesting for Resilience" project focused on schools, teaching students about water conservation. Engagement through education proved essential—students became ambassadors for change within their communities, fostering a culture of sustainability. These case studies underline the significance of community involvement, education, and adaptability in RWH initiatives. By learning from these grassroots efforts, regions facing water scarcity can adopt tailored, effective solutions that leverage local knowledge and foster long-term resilience.

6.4.3 Case Studies of Successful Rainwater Harvesting

Cities worldwide have successfully implemented rainwater harvesting systems. For instance, Singapore's Marina Barrage is a comprehensive water management project that includes rainwater harvesting to support the city's water supply. Similarly, India's Chennai has mandated rainwater harvesting for all buildings, significantly improving groundwater levels [86]. Urban rainwater harvesting projects, such as the Green Infrastructure Initiative in Portland, Oregon, demonstrate the potential of integrating rainwater harvesting into urban planning. These projects showcase how rainwater harvesting can be scaled up to address city-wide water management challenges. As cities continue to expand at an unprecedented rate, the convergence of rapid urbanization, population growth, and climate change exacerbates the looming urban water crisis. This multifaceted challenge necessitates innovative and sustainable solutions that prioritize not only immediate access to clean water but also long-term resilience against future shortages. Among these strategies, rainwater harvesting has emerged as a promising approach, harnessing natural precipitation to supplement dwindling water supplies and empower communities [16]. By embracing such sustainable practices, we can pave the way for a more secure and equitable water future, ensuring that urban areas are equipped to thrive despite the pressures of a changing environment. Furthermore, integrating education and community involvement into these initiatives can enhance their effectiveness, fostering a sense of ownership and responsibility among residents. As communities become more engaged in the management of their water resources, they can develop tailored solutions that reflect their unique needs and challenges [39]. The successful urban rainwater harvesting projects are given in Fig. 6.4.

Fig. 6.4 Location of rainwater harvesting system

6.5 Future Perspectives

The potential of rainwater harvesting systems to address pressing water scarcity issues becomes increasingly evident. Their integration into urban planning promotes sustainable water management and fosters community resilience against climate change. Advances in technology and innovative design can enhance the efficiency and effectiveness of these systems, making them more accessible to a wider range of users. Furthermore, public awareness and education will play a crucial role in encouraging adoption and investment in these solutions. Ultimately, embracing rainwater harvesting conserves precious resources and paves the way for a more sustainable future, where communities thrive in harmony with their environment. This harmonious relationship not only mitigates the impacts of climate change but also fosters biodiversity and enhances local ecosystems [8].

6.5.1 Policy Support and Technological Innovations

The integration of rainwater harvesting into community solutions stands as a vital strategy for addressing the pressing challenges of sustainable water management amid climate change and burgeoning populations. By capturing and utilizing this precious resource, communities can significantly reduce their reliance on traditional water supplies, enhancing resilience against the unpredictability of climate impacts.

Looking ahead, the advent of IoT-based monitoring systems promises to revolutionize rainwater harvesting practices, allowing for more efficient collection and storage tailored to specific local needs. This innovative approach maximizes water conservation efforts and fosters a sense of collective responsibility among community members. Together, these elements underscore the transformative potential of rainwater harvesting as a sustainable solution for future generations [75]. Governments can play a pivotal role by creating incentives for installing RWH systems, providing funding for community projects, and incorporating RWH into urban planning frameworks. Policy recommendations for advancing rainwater harvesting include the establishment of clear regulations, the provision of financial incentives, and the encouragement of public participation. Governments can play a vital role in fostering the adoption of rainwater harvesting through supportive policies and funding initiatives [81]. Grassroots initiatives promote local participation, ensuring that solutions are tailored to community needs. Educating communities about RWH benefits can foster ownership and encourage sustainable practices. Policies that support decentralized management empower residents to take charge of their water supply, enhancing resilience against droughts. Innovative materials, such as permeable pavements and green roofs, can be promoted to augment RWH efforts in urban areas. Furthermore, awareness campaigns can emphasize the importance of water conservation, complementing technological solutions. The intersection of policy and technology presents a powerful opportunity to combat water scarcity, fostering a sustainable future for communities. By establishing robust policy frameworks and encouraging community-led initiatives, we can create an environment where effective water management thrives. Simultaneously, the rise of technological innovations in rainwater harvesting not only enhances efficiency but also makes these systems more user-friendly and accessible to urban populations. As these advancements gain traction, they transform urban rainwater harvesting into an appealing solution that addresses both immediate water needs and long-term sustainability goals. Together, these elements forge a path toward integrated water resource management that is both viable and transformative for our communities [56].

6.5.2 Scaling up Sustainable Water Management Solutions

Scaling up sustainable water management solutions, particularly through rainwater harvesting (RWH) and community-led initiatives, is essential in addressing global water scarcity issues [80]. Future perspectives focus on enhancing collaborations between governments, NGOs, and local communities to implement RWH systems effectively. Policymakers can facilitate this process by integrating RWH into national water strategies, offering financial incentives, and supporting infrastructural development that prioritizes efficient water storage and distribution [82]. Technological innovation will play a vital role in maximizing the potential of RWH. Advances such as rainwater sensors, data analytics, and eco-friendly materials can streamline collection methods and ensure optimal use [9]. Community-led solutions, emphasizing local ownership, ensure that initiatives cater to specific regional needs, fostering resilience against climate variability [85]. Education and outreach programs are crucial for raising awareness and encouraging participation in RWH practices. By empowering communities with knowledge and skills, grassroots movements can drive sustainable water management from the ground up [57]. As these efforts gain momentum, scaling up successful models can inspire broader adoption across diverse regions. Ultimately, a collective approach combining innovative technology, supportive policies, and empowered communities will create a sustainable framework to combat water scarcity in the coming decades [68].

6.6 Recommendations

Rainwater harvesting offers a sustainable solution to urban water crises, delivering numerous environmental, economic, and social benefits. By capturing and utilizing rainwater, cities can lessen their reliance on traditional water sources, alleviate flooding, boost groundwater recharge, and foster resilient urban environments. Effective implementation of rainwater harvesting necessitates supportive policies, innovative technologies, community involvement, and continuous education. Adopting this ancient but increasingly significant practice can help ensure a sustainable water future for urban areas globally. The multifaceted approach to addressing water scarcity demonstrates not only the necessity of innovative strategies but also the potential for sustainable change. By capturing and storing rainwater, communities can significantly reduce their reliance on conventional water supplies, ensuring that they have a more resilient and self-sufficient water system. This practice not only contributes to the recharge of groundwater aquifers but also enhances overall water availability for vital domestic, agricultural, and sanitation needs. Furthermore, empowering local populations to take ownership of their water resources fosters a sense of responsibility and stewardship that is crucial for long-term sustainability. Together, these strategies create a holistic framework that not only mitigates the immediate challenges posed by water scarcity but also lays the groundwork for a more equitable and resource-conscious future. This collaborative approach encourages innovation and the sharing of best practices, enabling communities to adapt to changing environmental conditions. By integrating traditional knowledge with modern techniques, we can ensure that water management becomes a shared priority, benefiting both present and future generations. The integration of community involvement in the development and implementation of systems proves to be a transformative approach, as it not only tailors solutions to the unique contexts of local environments but also significantly enhances both success rates and sustainability. By embracing the wisdom embedded in traditional knowledge while simultaneously incorporating modern techniques, communities can create systems that are not only effective but also deeply rooted in cultural relevance. This collaborative strategy fosters a sense of ownership among community members, ensuring that these systems are not merely imposed from above but are nurtured from within, leading to enduring impacts that resonate for generations. Ultimately, harnessing the collective strengths of a community paves the way for innovative solutions that honor the past while building a resilient future. Future initiatives can benefit from this approach by emphasizing local knowledge and traditions, which can inspire new ideas and practices. By engaging diverse voices and perspectives, communities can adapt to challenges more effectively while preserving their unique identities.

- **Policy support**: Governments should incentivize rainwater harvesting through policies, subsidies, and mandates in building codes.
- **Capacity building**: Training programs must be provided to educate communities on system design, maintenance, and water quality monitoring.
- **Technology integration**: Simple, cost-effective innovations should be promoted to increase the effectiveness of rainwater collection and storage.
- **Partnerships**: Collaborations between NGOs, local authorities, and private sectors can amplify impact and scalability.
- **Research**: Continued evaluation of long-term outcomes will help refine models and foster innovation in community-led water management.

6.7 Conclusions

As the world grapples with increasing water scarcity, the shift toward alternative water sources has become not just a necessity but a focal point of innovation and sustainability. Current water projects illuminate the complex interplay of socio-economic and environmental factors, highlighting the pressing need for solutions that address both human needs and ecological balance. The rising global interest in greywater reuse, desalination, and rainwater harvesting systems reflects a proactive approach to resource management. Among these options, rainwater harvesting is notable for being environmentally friendly, needing less treatment, and helping to reduce flooding—showing that good water management can support both nature and community strength. Embracing these technologies will enhance water security and foster a more sustainable relationship with our planet's most precious resource. As communities adopt these innovative practices, they not only improve their resilience to climate change but also promote a culture of conservation and responsibility. By prioritizing sustainable water management, we can inspire future generations to value and protect our environment. The effective integration of rainwater harvesting systems (RWHS) presents a promising avenue toward enhancing sustainability, particularly when grounded in strategic implementation. By critically addressing financial viability, reliability analysis, water quality, and the impacts of climate change, these systems can become both efficient and resilient. With average payback periods falling below 15 years—thanks in part to insights from the Water Balance Model—investors and communities alike find these initiatives increasingly attractive. Furthermore, by adopting prudent measures throughout the harvesting and purification processes, health risks associated with consuming collected rainwater can be substantially mitigated. Thus, as we navigate the challenges of water scarcity and environmental change, rainwater harvesting emerges not just as a practical solution but as a vital component of a sustainable future. This transition encourages innovative practices and technologies that can further enhance water conservation efforts. Ultimately, fostering a culture of sustainability will empower communities to take ownership of their water resources, ensuring access for generations to come. The Rainwater Harvesting System (RWHS) for potable water supply emerges as a promising solution to the pressing global water crisis, effectively leveraging the Balance Model that incorporates monthly rainfall data to optimize water collection. Its adaptability makes it suitable for both developed and developing nations, demonstrating that economic disparities should not preclude access to sustainable water solutions. However, the ongoing debate regarding the benefits of individual versus community-owned systems highlights the need for further research into their respective efficiencies and impacts. Moreover, the scarcity of studies evaluating the economic viability of small-scale RWHS in developing countries underscores a critical gap that must be addressed to fully harness this technology's potential. As we look toward a future where equitable access to clean water is paramount, prioritizing investment in RWHS could serve as a vital step toward achieving sustainable development goals across diverse socioeconomic landscapes. Ensuring the safety of rainwater for human consumption is a multifaceted endeavor that hinges on effective treatment protocols tailored to the specific geographic context of each system. By utilizing biofilm to diminish metal contaminants, we can enhance the overall quality of harvested rainwater. Additionally, maximizing sun exposure and maintaining clean rooftop surfaces play crucial roles in minimizing microbial contamination and further safeguarding public health. Regular maintenance and the strategic use of disinfectants are essential to prevent any degradation of treated water over time. Collectively, these strategies underscore the importance of a comprehensive approach to rainwater management, ultimately promoting sustainable water practices that prioritize human safety and environmental integrity. The exploration of rainwater quality and maintenance serves as a vital foundation for fostering both consumer confidence and community adoption of rainwater harvesting systems. By investigating best practices in system upkeep, stakeholders can ensure that these sustainable solutions meet health standards and operate efficiently over time. However, the current landscape reveals a significant gap in empirical data, which is essential for refining rainwater harvesting system modeling and enhancing overall effectiveness. Moreover, aligning institutional frameworks with socio-political support is crucial for creating an environment conducive to widespread acceptance and successful implementation of these systems. Collectively, these elements highlight the importance of a comprehensive approach that not only prioritizes technological advancements but also addresses the social dynamics that influence the adoption of rainwater harvesting practices. As the challenges of urban water demand and flooding continue to escalate, Rainwater Harvesting Systems (RWHS) emerge as a sustainable solution that promises long-term benefits for communities. However, their successful implementation hinges on comprehensive research to refine these systems and maximize their effectiveness. To facilitate this transition, it is essential to invest in education and outreach through training sessions, workshops, and public lectures that engage community members and stakeholders alike. Additionally, pilot initiatives can serve as valuable learning experiences that demonstrate the practicality of RWHS in real-world settings. By prioritizing these strategies today, we address current water issues and lay the groundwork for a resilient water provision system that will benefit future generations. Future generations will not only inherit a more sustainable water management system but also gain the knowledge and skills necessary to innovate and adapt in the face of evolving environmental challenges. Ultimately, fostering a culture of collaboration and continuous improvement will ensure that our water resources are managed wisely and equitably in the coming years.

References

1. Adekalu KO, Baloguna JA, Alukoa OB, Okunadeb DA, Gowing JW, Faborode MO (2009) Runoff water harvesting of dry spell mitigation for cowpea in the Savannah Belt of Nigeria, Agric Water Manag 96:1502–1508. https://doi.org/10.1016/j.agwat.2009.06.005
2. Adham A, Riksen M, Ouessar M, Ritsema CJ (2016) A methodology to assess and evaluate rainwater harvesting techniques in (semi-) arid regions. Water 8(5):198. https://doi.org/10.3390/w8050198
3. Ahmed AT, El Gohary F, Tzanakakis VA, Angelakis AN (2020) Egyptian and Greek water cultures and hydro-technologies in ancient times. Sustainability 12:9760. https://doi.org/10.3390/su1222976
4. Ahrestani Z (2023) An overview of atmospheric water harvesting methods, the inevitable path of the future in water supply. RSC Adv 13:10273–10307. https://doi.org/10.1039/d2ra07733g
5. Al-Adamat R, Rawajfih Z, Easter M, Paustian K, Coleman K, Milne E, Falloon P, Powlson DS, Batjes NH (2007) Predicted soil organic carbon stocks and changes in Jordan between 2000 and 2030 made using the GEFSOC modelling system. J Agric Ecosyst Environ 122:35–45. https://doi.org/10.1016/j.agee.2007.01.006
6. Alamdari N, Sample DJ, Liu J, Ross A (2018) Assessing climate change impacts on the reliability of rainwater harvesting systems. Resour Conserv Recycl 132:178–189. https://doi.org/10.1016/j.resconrec.2017.12.013
7. Ananga E, Naiga R, Agong G, Njoh AJ, Vickers HP (2021) Examining the contribution of community participation in water resource production and management: perspectives from developing countries. SN Soc Sci 1:37. https://doi.org/10.1007/s43545-020-00050-0
8. Angelakis AN, Capodaglio AG, Kumar R, Valipour M, Ahmed AT, Baba A, Güngör EB, Mandi L, Tzanakakis VA, Kourgialas NN, Dercas N (2025) Water supply systems: past, present challenges, and future sustainability prospects. Land 14:619. https://doi.org/10.3390/land14030619
9. Anindita M, Alim MA, Rahman A (2024) Rainwater harvesting for drinking water production: teaching sustainable practices to school students in Bangladesh. Int Conf Adv Eng Educ 1:26. https://doi.org/10.71427/iCAEED2025/26
10. Anonymous (2009) World health organization calcium and magnesium in drinking-water: public health significance. In: Cotruvo J, Bartram J (eds.), World Health Organization Press, Geneva, Switzerland
11. Anonymous (2016) UN habitat world cities report 2016: urbanization and development – emerging futures, p. 262. https://unhabitat.org/world-cities-report-2016
12. Anonymous (2020) US environmental protection agency (EPA) Green infrastructure: managing weather with a storm water management approach. EPA Office of Water, Washington, DC

13. Baguma D, Loiskandl W (2010) Water management, rainwater harvesting and predictive variables in rural households. Water Resour Manag 24:3333–3348. https://doi.org/10.1007/s11269-010-9609-9
14. Basinger M, Montalto F, Lall U (2010) A rainwater harvesting system reliability model based on nonparametric stochastic rainfall generator. J Hydrol 392:105–118. https://doi.org/10.1016/j.jhydrol.2010.07.039
15. Bose D, Bhattacharya R, Kaur T, Banerjee R, Bhatia T, Ray A, Batra B, Mondal A, Ghosh P, Mondal S (2024) Overcoming water, sanitation, and hygiene challenges in critical regions of the global community. Water-Energy Nexus 7:277–296. https://doi.org/10.1016/j.wen.2024.11.003
16. Cattani L, Figoni R, Cattani P, Magrini A (2025) Integrated atmospheric water generators for building sustainability: a simulation-based approach. Energies 18:1839. https://doi.org/10.3390/en18071839
17. Chaturwedi AK, Kashyap NK, Gupta P, Biswas S, Shriwas SK, Vaishnav MM, Hait M (2024) Industrial effluents and their impact on water pollution- an overview. ES Gen 5:1203. https://doi.org/10.30919/esg1203
18. Chidamba L, Korsten L (2015) Pyrosequencing analysis of roof-harvested rainwater and river water used for domestic purposes in Luthengele village in the Eastern Cape Province of South Africa. Environ Monit Assess 187:41. https://doi.org/10.1007/s10661-014-4237-0
19. Crosson C, Tong D, Zhang Y, Zhong Q (2021) Rainwater as a renewable resource to achieve net zero urban water in water stressed cities. Resour Conserv Recycl 164:105203. https://doi.org/10.1016/j.resconrec.2020.105203
20. Davis AP, Shokouhian M, Ni S (2001) Loading estimates of lead, copper, cadmium, and zinc in urban runoff from specific sources. Chemosphere 44:997–1009. https://doi.org/10.1016/s0045-6535(00)00561-0
21. Domènech L, Heijnen H, Souri D (2012) Rainwater harvesting for human consumption and livelihood improvement in rural Nepal: benefits and risks. Water Environ J 26:465–472. https://doi.org/10.1111/j.1747-6593.2011.00305.x
22. Elmahdi A (2009) Global warming-water shortage and food security future planning interaction, system analysis approach WRM2009. Water Resour Manag 9:11
23. Ertop H, Kocięcka J, Atilgan A, Liberacki D, Niemiec M, Rolbiecki R (2023) The importance of rainwater harvesting and its usage possibilities: antalya example (Turkey). Water 15:2194. https://doi.org/10.3390/w15122194
24. Fuentes-Galván ML, Ortiz Medel J, Arias Hernández LA (2018) Roof rainwater harvesting in central Mexico: uses, benefits, and factors of adoption. Water 10:116. https://doi.org/10.3390/w10020116
25. García-Ávila F, Guanoquiza-Suárez M, Guzmán-Galarza J, Cabello-Torres R, Valdiviezo-Gonzales L (2023) Rainwater harvesting and storage systems for domestic supply: an overview of research for water scarcity management in rural areas. Results Eng 18:101153. https://doi.org/10.1016/j.rineng.2023.101153
26. George-Williams HEM (2024) Sustainable water infrastructure: visions and options for Sub-Saharan Africa. Sustainability 16:1592. https://doi.org/10.3390/su16041592
27. Ghernaout D, Elboughdiri N (2020) Controlling disinfection by-products formation in rainwater: technologies and trends. Open Access Lib J 7:1–12. https://doi.org/10.4236/oalib.1106162
28. Gomez M (2019) Socioeconomic factors affecting water access in rural areas of low- and middle-income countries. Water 11:202. https://doi.org/10.3390/w11020202
29. Gould J, Nissen-Peterson E (1999) Rainwater catchment systems for domestic supply. IT Publications Ltd, London
30. Gupta P, Tamrakar G, Biswas S, Dhundhel RS, Soni VK, Das HS, Sahu P, Chaturwedi AK, Kashyap NK, Akitsu T, Hait H (2024) Quantum dots as a fluorescent sensor for detecting heavy metal ions in aqueous environment: an overview. ES Chem Sustain 1:1359. https://doi.org/10.30919/escs1359

31. Hait M, Gir ND, Dey A, Chaturwedi AK Kashyap NK (2018) Removal of water pollutants from the effluent of Tifra industrial area in Bilaspur (C.G.) using bio- adsorbents. Pharm Innov J 6(3):15–20
32. Hait M, Gir ND, Kashyap NK, Dey A, Chaturwedi AK (2019) Bioremoval of water pollutants from the effluents of Tifra industrial area in Bilaspur (C.G.). Int J Curr Eng Sci Res 6:399–401
33. Hait M, Kashyap NK, Chaturwedi AK (2018) Physicochemical investigation of water bodies of Arpa River at Bilaspur urban areas. Int J Chem Stud 6:1773–1777
34. Hait M, Kashyap NK, Giri AK (2024) Synthesis of metal NPs using plant flowers and fruits. In: Bhardwaj AK, Srivastav AL, Rai S (eds) Biogenic wastes-enabled nanomaterial synthesis. Springer, Cham, pp 195–237. https://doi.org/10.1007/978-3-031-59083-2_8
35. Hait M, Kashyap NK, Sahu P, Jana U (2024) Fabrication of metal NPS from plant root and tuber. In: Bhardwaj AK, Srivastav AL, Rai S (eds) Biogenic wastes-enabled nanomaterial synthesis. Springer, Cham, pp 29–71. https://doi.org/10.1007/978-3-031-59083-2
36. Hait M, Kashyap, NK, Bhardwaj AK (2024) Emerging pollutants from urbanization and industrialization in the global south. In: Izah SC, Ogwu MC, Hamidifar H (eds) Biomonitoring of pollutants in the global south. Springer, Singapore, pp 39–87. https://doi.org/10.1007/978-981-97-1658-6_2
37. Haque MM, Rahman A, Hagare D, Kibria G, Karim F (2015) Estimation of catchment yield and associated uncertainties due to climate change in a mountainous catchment in Australia. Hydrol Process 29:4339–4349. https://doi.org/10.1002/hyp.10492
38. Haque MM, Rahman A, Samali B (2016) Evaluation of climate change impacts on rainwater harvesting. J Clean Prod 137:60–69. https://doi.org/10.1016/j.jclepro.2016.07.038
39. Huang J, Li S, Dalagah S, Stone S, Ahles D, Mouawad J, Sharbatmaleki M (2025) Closed-circuit reverse osmosis performance evaluation in a high-recovery-rate water reclamation pilot project. Sustainability 17:3388. https://doi.org/10.3390/su17083388
40. Ingrao C, Strippoli R, Lagioia G, Huisingh D (2023) Water scarcity in agriculture: an overview of causes, impacts and approaches for reducing the risks. Heliyon 9:e18507. https://doi.org/10.1016/j.heliyon.2023.e18507
41. Islam MM, Afrin S, Redwan AM (2015) Impact of climate change on reliability of rainwater harvesting system: a case study in Mongla, Bangladesh. J Mod Trends Sci Technol 3:220–230
42. Islam MM, Kabir MR, Chou FNF (2010) Quality assessment of stored rainwater and its utilization in Dhaka City, Bangladesh. Water Pract Technol 5:wpt2010059. https://doi.org/10.2166/wpt.2010.059
43. Izah SC, Ogwu MC (2025) Modeling solutions for microbial water contamination in the global south for public health protection. Front Microbiol 16:1504829. https://doi.org/10.3389/fmicb.2025.1504829
44. James CL, Paolo B, Warren GF, Julie EG, Karyn LH, Lorenz RR, Jane S, Gerard S, Glen VDK, Amy LW (2012) Critical comments on the WHO-UNEP state of the science of endocrine disrupting chemicals – 2012. Regul Toxicol Pharmacol 69:22–40. https://doi.org/10.1016/j.yrtph.2014.02.002
45. Jha PK (2012) Climate change: impact, adaptation and vulnerability in the water supply of Kathmandu Valley. WIT Trans Ecol Environ 155:563–574
46. Judeh T (2022) Smart rainwater harvesting for sustainable potable water supply in arid and semi-arid areas. Sustainability 14:9271. https://doi.org/10.3390/su14159271
47. Kahinda JMM (2007) Domestic rainwater harvesting to improve water supply in rural South Africa. Phys Chem Earth Parts A/B/C 32:1050–1057
48. Kalpana Balasooriya BMJ, Rajapakse J, Gallage C (2023) A review of drinking water quality issues in remote and indigenous communities in rich nations with special emphasis on Australia. Sci Total Environ 903:166559. https://doi.org/10.1016/j.scitotenv.2023.166559
49. Karim MR (2010) Assessment of rainwater harvesting for drinking water supply in Bangladesh. Water Sci Technol: Water Supply 10:243–249
50. Kashyap NK, Hait M, Bhardwaj AK (2024) Planktons as a sustainable biomonitoring tool of aquatic ecosystem. In: Izah SC, Ogwu MC, Hamidifar H (eds) Biomonitoring of pollutants in the global south. Springer, Singapore, pp 275–319. https://doi.org/10.1007/978-981-97-1658-6_8

51. Kisakye V, Akurut M, Van der Bruggen B (2018) Effect of climate change on reliability of rainwater harvesting systems for Kabarole District, Uganda. Water 10:71. https://doi.org/10.3390/w10010071
52. Kosoe EA, Osumanu IK, Ogwu MC (2025) Reimagining rural and small-town water supply in Ghana through inclusive governance and sustainable models. J Sustain 1(2):1–21. https://doi.org/10.55845/jos-2025-1234
53. Kosoe EA, Ogwu MC, Duniah J, Ndigun FB (2025) Sustaining indigenous water governance systems in Ayimaye and Lungni, Northern Ghana, through hybrid models. J Sustain 1(2). https://doi.org/10.55845/jos-2025-1235
54. Lade O, Okunlola O (2017) Rainwater quality in Ibadan (Nigeria): effect of short-term storage in surface and underground tanks. World J Res Rev 5:6–13
55. Lebel S, Fleskens L, Forster PM, Jackson LS, Lorenz S (2015) Evaluation of in situ rainwater harvesting as an adaptation strategy to climate change for maize production in Rainfed Africa. Water Resour Manag 29:4803–4816. https://doi.org/10.1007/s11269-015-1091-y
56. Lebu S, Lee A, Salzberg A et al (2024) Adaptive strategies to enhance water security and resilience in low- and middle-income countries: a critical review. Sci Total Environ 925:171520. https://doi.org/10.1016/j.scitotenv.2024.171520
57. Lepcha R, Patra SK, Ray R et al (2024) Rooftop rainwater harvesting a solution to water scarcity: a review. Groundw Sustain Dev 26:101305. https://doi.org/10.1016/j.gsd.2024.101305
58. Liu Z, Lin YL, Chu WH, Xu B, Zhang TY, Hu CY, Cao TC, Gao NY, Dong CD (2020) Comparison of different disinfection processes for controlling disinfection by-product formation in rainwater. J Hazard Mater 385:121618. https://doi.org/10.1016/j.jhazmat.2019.121618
59. Lo KFA, Koralegedara SB (2015) Effects of climate change on urban rainwater harvesting in Colombo City, Sri Lanka. Environments 2:105–124. https://doi.org/10.3390/environments2010105
60. Mamun MA, Sakib N, Gozal D, Bhuiyan AKMI, Hossain S, Bodrud-Doza M, Mamun FA, Hosen I, Safiq MB, Abdullah AH, Sarker MA, Rayhan I, Sikder MT, Muhit M, Lin CY, Grifiths MD, Pakpour AH (2021) The COVID-19 pandamic and serious psychological consequence in Bangladesh: a population based nation-wide study. J Affect Disord 279:462–472. https://doi.org/10.1016/j.jad.2020.10.036
61. Masoner JR, Kolpin DW, Cozzarelli IM, Barber LB, Burden DS, Foreman WT, Forshay KJ, Furlong ET, Groves JF, Hladik ML, Hopton ME, Jaeschke JB, Keefe SH, Krabbenhoft DP, Lowrance R, Romanok KM, Rus DL, Selbig WR, Williams BH, Bradley PM (2019) Urban stormwater: an overlooked pathway of extensive mixed contaminants to surface and groundwaters in the United States. Environ Sci Technol 53:10070–10081. https://doi.org/10.1021/acs.est.9b02867
62. Mattos K, King E, Lucas C, Snyder EH, Dotson A, Linden K (2019) Rainwater catchments in rural Alaska have the potential to produce high-quality water and high quantities of water for household use. J Water Health 17:788–800. https://doi.org/10.2166/wh.2019.238
63. Mays L (2013) History of water cisterns: legacies and lessons. Water 5:1916–1940. https://doi.org/10.3390/w5041916
64. Meera V, Ahammed MM (2006) Water quality of rooftop rainwater harvesting systems: a review. J Water Supply: Res Technol Aqua 55:257–268. https://doi.org/10.2166/aqua.2006.0010
65. Mendez CB, Klenzendorf JB, Afshar BR, Simmons MT, Barrett ME, Kinney KA, Kirisits MJ (2011) The effect of roofing material on the quality of harvested rainwater. Water Res 45:2049–2059. https://doi.org/10.1016/j.watres.2010.12.015
66. Moilleron R, Gonzalez A, Chebbo G, Thevenot DR (2002) Determination of aliphatic hydrocarbons in urban runoff samples from the "Le Marais" experimental catchment in Paris centre. Water Res 36:1275–1285
67. Musayev S, Burgess E, Mellor J (2018) A global performance assessment of rainwater harvesting under climate change. Resour Conserv Recycl 132:62–70. https://doi.org/10.1016/j.resconrec.2018.01.023

68. Mutiro J, Makurira H, Senzanje A, Mul ML (2006) Water productivity analysis for smallholder rainfed systems: a case study of Makanya catchment, Tanzania. Phys Chem Earth Parts A/B/C 31:901–909. https://doi.org/10.1016/j.pce.2006.08.019
69. Naiga R, Penker M, Hogle K (2015) Challenging pathways to safe water access in rural Uganda: from supply to demand-driven water governance. Int J Commons 9:237–260. https://doi.org/10.18352/ijc.480
70. Nelson S, Drabarek D, Jenkins A, Negin J, Abimbola S (2021) How community participation in water and sanitation interventions impacts human health, WASH infrastructure and service longevity in low-income and middle-income countries: a realist review. BMJ Open 12:e053320. https://doi.org/10.1136/bmjopen-2021-053320
71. Ogwu MC (2025) Sustainable development in Africa: challenges, innovations, and policy pathways for the future. Afr Res Rep 1(2):55–69. https://doi.org/10.5897/ARR2025.0005
72. Ogwu MC, Kosoe EA (2024) Place of cultural diversity in sustainable water resource management in Ghana. In: Izah SC, Ogwu MC, Loukas A, Hamidifar H (eds) Water crises and sustainable management in the global south. Springer, Singapore. https://doi.org/10.1007/978-981-97-4966-9_14
73. Ogwu MC, Kosoe EA (2025) Integrating green infrastructure into sustainable agriculture to enhance soil health, biodiversity, and microclimate resilience. Sustainability 17(9):3838. https://doi.org/10.3390/su17093838
74. Ogwu MC, Izah SC, Sawyer WE, Amabie T (2025) Environmental risk assessment of trace metal pollution: a statistical perspective. Environ Geochem Health 47:94. https://doi.org/10.1007/s10653-025-02405-z
75. Olley J, Cvitanovic M, Ginige T et al (2024) A systematic literature review of sustainable water management in South Africa. Sustain Water Resour Manag 10:162. https://doi.org/10.1007/s40899-024-01135-x
76. Pachpute JS (2010) A package of water management practices for sustainable growth and improved production of vegetable crop in labour and water scarce Sub-Saharan Africa. Agric Water Manag 97:1251–1258. https://doi.org/10.1016/j.agwat.2009.11.009
77. Patel D, Chaturwedi AK, Kashyap N K, Roymahapatra G, Hait M (2024) Quality control and standardization of nanomaterials. In: Thakur A, Kumar A (eds) Sustainability, safety, and applications of nanomaterials-based corrosion inhibitors, IGI Global, USA, pp 297–325. https://doi.org/10.4018/979-8-3693-7640-9.ch012
78. Pitao RL (2019) Design for harvesting and treatment of rainwater in Naval, Biliran. Int J Trend Sci Res Devlop 3:770–776. https://doi.org/10.31142/ijtsrd23897
79. Rahman S, Khan MT, Akib S et al (2014) Sustainability of rainwater harvesting system in terms of water quality. Sci World J 2014:721357. https://doi.org/10.1155/2014/721357
80. Raimondi A, Quinn R, Abhijith GR et al (2023) Rainwater harvesting and treatment: state of the art and perspectives. Water 15:1518. https://doi.org/10.3390/w15081518
81. Richards S, Rao L, Connelly S, Raj A, Raveendran L, Shirin S, Jamwal P, Helliwell R (2021) Sustainable water resources through harvesting rainwater and the effectiveness of a low-cost water treatment. J Environ Manag 286:112223. https://doi.org/10.1016/j.jenvman.2021.112223
82. Shemer H (2023) Challenges and solutions for global water scarcity. Membranes 13:612. https://doi.org/10.3390/membranes13060612
83. Shunglu R, Kopke S, Kanoi L, Nissanka TS, Withanachchi CR, Gamage DU, Dissanayake HR, Kibaroglu A, Ünver O, Withanachchi SS (2022) Barriers in participative water governance: a critical analysis of community development approaches. Water 14:762. https://doi.org/10.3390/w14050762
84. Silva CM (2015) Evaluation of rainwater harvesting in Portugal: application to single-family residences. Resour Conserv Recycl 94:21–34
85. Ssekyanzi G (2024) Sustainable solutions for mitigating water scarcity in developing countries: a comprehensive review of innovative rainwater storage systems. Water 16:2394. https://doi.org/10.3390/w16172394

86. Suni S, Firdaous MH, Zailani FF, Gödeke S, Raffi RM, Abas PE (2025) Urban water management and public acceptance of rainwater harvesting systems: insights from young and educated respondents in Muslim communities. Sustainability 17:3046. https://doi.org/10.3390/su17073046
87. Ullah S, Iqbal M, Waseem M, Abbas A, Masood M, Nabi G, Tariq MAUR, Sadam M (2024) Potential sites for rainwater harvesting focusing on the sustainable development goals using remote sensing and geographical information system. Sustainability 16:9266. https://doi.org/10.3390/su16219266
88. Wallace CD, Bailey RT, Arabi M (2015) Rainwater catchment system design using simulated future climate data. J Hydrol 529:1798–1809. https://doi.org/10.1016/j.jhydrol.2015.08.006
89. Wang XJ, Zhang JY, Yang ZF, Shahid S, He R, Xia X, Liu H (2015) Historic water consumptions and future management strategies for Haihe River basin of Northern China. Mitig Adapt Strateg Glob Chang 20:371–387. https://doi.org/10.1007/s11027-013-9496-5
90. Wright J (2004) Household drinking water in developing countries: a systematic review of microbiological contamination between source and point-of-use. Trop Med Int Health 9:106–117. https://doi.org/10.1046/j.1365-3156.2003.01160.x
91. Zabidi HA, Goh HW, Chang CK, Chan NW, Zakaria NA (2020) A review of roof and pond rainwater harvesting systems for water security: the design, performance and way forward. Water 12:3163. https://doi.org/10.3390/w12113163
92. Zhang S, Zhang J, Jing X, Wang Y, Wang Y, Yue T (2018) Water saving efficiency and reliability of rainwater harvesting systems in the context of climate change. J Clean Prod 196:1341–1355. https://doi.org/10.1016/j.jclepro.2018.06.133
93. Zhang S, Zhang J, Yue T, Jing X (2019) Impacts of climate change on urban rainwater harvesting systems. Sci Total Environ 665:262–274. https://doi.org/10.1016/j.scitotenv.2019.02.135

Chapter 7
Water Quality and Safety: Policy Interventions and International Support

Godgift Nabebe, Wisdom Richard Poyeri, Wasiu Olanrewaju Opadokun, Adams Ovie Iyiola, Odangowei Inetiminebi Ogidi, and Saoban Sunkanmi Salimon

Abstract Ensuring access to safe and quality water remains a pressing global challenge, particularly in regions vulnerable to pollution, climate change, and infrastructural deficits. National governments and global institutions have developed various frameworks to improve water safety and address contamination issues. This chapter focuses on policy interventions and international support systems that promote sustainable water quality management and inclusive access in vulnerable regions. It highlights the role of organizations such as the World Health Organization and the World Bank in supporting national water programs. Studies from Rwanda, Nigeria, Ghana, South Africa, Uganda, and other countries illustrate effective policy implementation, community engagement, and multisectoral collaboration. These examples show that with political will and technical support, significant progress in water safety can be achieved. Despite these efforts, several barriers continue to

G. Nabebe
Department of Biological Sciences, Faculty of Science, Niger Delta University, Wilberforce Island, Nigeria

W. R. Poyeri
Department of Microbiology, Faculty of Science, Bayelsa Medical University, Yenagoa, Bayelsa State, Nigeria

W. O. Opadokun
Department of Biological Sciences, Faculty of Natural and Applied Sciences, Al-Hikmah University, Ilorin, Nigeria

A. O. Iyiola (✉)
Department of Fisheries and Aquatic Resources Management, Faculty of Renewable Natural Resources Management, College of Agriculture and Renewable Natural Resources, Ejigbo Campus, Osun State University, Ejigbo, Osun State, Nigeria
e-mail: adams.iyiola@uniosun.edu.ng

O. I. Ogidi
Department of Biochemistry, Faculty of Basic Medical Sciences, Bayelsa Medical University, Yenagoa, Bayelsa State,, Nigeria

S. S. Salimon
School of Biological and Medical Physics, Moscow Institute of Physics and Technology, Dolgoprudny, Moscow Region, Russian Federation

M. C. Ogwu and S. Chibueze Izah (eds.), *Water Quality and Safety in the Global South*, Springer Water, https://doi.org/10.1007/978-3-032-10602-5_7

limit success. Important challenges include weak governance structures, overlapping institutional mandates, limited funding, and a lack of technical expertise at the local level. The chapter also explores how emerging technologies, such as remote sensing, mobile water testing, and real-time monitoring tools, are transforming water quality management. Strategic recommendations emphasize the importance of integrated water policies, inter-agency coordination, capacity building, and performance-based funding. Additionally, there is a growing need to strengthen cross-border cooperation and climate resilience in water safety planning. Overall, the chapter provides a comprehensive review of global water governance strategies and highlights actionable insights for improving water quality and safety. It advocates for a holistic and evidence-based approach that brings together policy, science, and community involvement for long-term impact.

Keywords Water quality · Water safety · Policy intervention · International support · Water governance · Contamination · Monitoring · Global health

7.1 Introduction

Water quality and safety are critical components of public health and environmental sustainability [2]. These issues are challenged by contamination from various pollutants, increased demand due to population growth, and the effects of climate change [41]. Contaminated water sources are responsible for a substantial burden of disease worldwide, particularly in developing countries, where access to clean water remains a persistent challenge [3, 4]. Addressing these challenges requires a multifaceted and integrated approach to water resource management, as highlighted in the literature [10, 74, 77].

Water contaminants, including heavy metals, biological pathogens, and synthetic chemicals, pose significant threats to both ecosystem integrity and human health [7, 8, 14, 15, 39, 43, 45–47, 86]. Heavy metals such as arsenic and cadmium are particularly alarming due to their acute toxicity and potential for bioaccumulation [1, 6, 80–82]. Arsenic contamination can occur through natural processes or anthropogenic activities, leading to chronic exposure that may cause health issues, including cancer and cardiovascular diseases [49, 94]. Similarly, cadmium has well-documented severe health impacts, including cellular metabolism disruptions and damage to major organ systems upon long-term exposure, stressing the need for urgent intervention strategies [114].

Challenges in ensuring water quality are compounded by urbanization and industrialization, which exacerbate pressures on water resources [50, 52–54]. Urban stormwater runoff, characterized by pollutant accumulation from impervious surfaces, adversely affects downstream water quality, leading to ecological degradation and health hazards for communities [58, 109]. Such industrial activities diminish the quantity of available clean water and increase competition among various sectors, particularly agriculture, energy, and municipal needs. This interaction is relevant to the Water-Energy-Food nexus, which emphasizes the interconnectedness of water management with energy production and food security [9, 84].

Recognizing water as a finite and vulnerable resource, the importance of water sustainability in development planning cannot be overstated. The United Nations' Sustainable Development Goals (SDGs), especially SDG 6, emphasize the necessity of ensuring access to clean water and sanitation. Achieving such goals requires a concerted effort toward integrated water resources management (IWRM), which harmonizes ethical, social, and political considerations to ensure equitable access and sustainability [33, 98]. This approach involves assessing trade-offs across sectors, addressing current inequities, and promoting environmentally sound practices.

Policies that emphasize community involvement and stakeholder engagement are central to successful water management interventions [9, 10]. Inclusive governance structures can enhance resilience among communities by improving adaptive capacity in the face of ongoing climate variability and resource scarcity. Effective localized water management systems have proven beneficial in both urban and rural contexts, leading to the development of green infrastructure and rainwater harvesting systems that promote conservation and sustainable practices [58, 78].

To mitigate the impacts of water contamination and improve human health, investing in technological innovations and educational campaigns is essential. Advanced water treatment technologies, such as biosensors for detecting heavy metals and water quality parameters, facilitate rapid and accurate assessment of contaminants, improving public health responses [90, 109, 112]. Public awareness programs regarding water quality, sustainable practices, and safe consumption can empower communities to undertake proactive measures in safeguarding their water sources.

Governmental and international organizations should acknowledge the interdependence of water quality and socioeconomic status, advocating for policies that ensure affordable access to clean water. Such measures are vital not only for health outcomes but also for overall economic development, as unsustainable water practices can lead to escalated healthcare costs and lost economic productivity due to water-related diseases [71]. Insights from programs aimed at rural revitalization and sustainable water utilization offer valuable lessons applicable to various geographical contexts [55, 62, 63, 105].

Water sustainability is inherently linked to environmental integrity [44], and the degradation of aquatic ecosystems can profoundly affect societal well-being. Poor wastewater management and industrial pollution disrupt aquatic habitats, compromise water quality, and endanger the myriad species that thrive in these ecosystems [40, 98]. Therefore, integrating environmental preservation into all aspects of water governance is essential to ensure the long-term viability of water resources and ecosystems [42, 43, 85].

This chapter explores the critical role of policy interventions and international support in ensuring water quality and safety, particularly in low- and middle-income countries. It examines national and global frameworks, regulatory strategies, and the contributions of organizations such as WHO and UNICEF in addressing water contamination and access challenges. The chapter also highlights case studies, implementation barriers, and future directions for sustainable and inclusive water safety governance.

7.2 Major Policy Frameworks for Water Quality Management

The management of water quality is increasingly framed within robust policy frameworks that emphasize comprehensive governance systems, as demonstrated by National Water Safety Plans (WSPs) developed by the World Health Organization (WHO). These plans represent a systematic approach that encompasses the entire water supply chain from source to consumer, prioritizing hazard identification and risk assessment. Regular monitoring and emergency response protocols are necessary to safeguard public health and ensure sustainable water management [56, 76]. Countries have adapted these frameworks to their specific infrastructure and national contexts, facilitating the establishment of minimal water quality guidelines and permissible contaminant levels [76, 79].

Successful examples of national water policies illustrate the effectiveness of these frameworks. Nigeria's National Water Resources Policy, for example, promotes Integrated Water Resources Management (IWRM) to protect water sources through public–private partnerships [79]. Similarly, South Africa's National Water Act (1998) enshrines equitable access to water as a human right while promoting sustainable practices through regulatory agencies [76]. India has also introduced measures aimed at groundwater contamination control as outlined in its National Water Policy (2012), emphasizing conservation and efficient water use [79]. Moreover, countries like Chile and Australia have implemented strong legal frameworks coupled with innovative water pricing models to enhance accountability in water resource management [72, 76, 91].

The integration of environmental, health, and agricultural policies is crucial for effective water quality management. Frameworks that recognize the interplay between water quality, sanitation, food safety, and ecosystem health promote the "One Health" concept. These frameworks advocate for collaborative management strategies that align environmental protection laws with controls on agricultural runoff, facilitating coordinated efforts among agencies responsible for health, environment, and rural development [20, 38]. Multi-sectoral platforms are vital for policy development, encouraging participation across various governmental tiers and including diverse stakeholder perspectives [20].

Stakeholder engagement is essential in creating effective water governance frameworks. Successful management relies on the active participation of stakeholders spanning governmental bodies, civil society, and private enterprises [35]. Collaboration fosters knowledge sharing and collective decision-making that support long-term sustainability initiatives. Research indicates that stakeholder participation can significantly improve water governance outcomes, reflecting a positive correlation with the performance of governance systems [67]. However, issues surrounding governance dynamics, such as power imbalances among stakeholders and challenges in information sharing, should be addressed to cultivate effective collaborative structures [20, 36].

Fostering transparency and accountability in water governance is imperative. Effective governance frameworks include clear engagement processes that enhance stakeholders' understanding of how their inputs can influence decision-making [76, 95]. To achieve robust collaborative governance, a culture of transparency should be fostered, including the dissemination of information regarding water quality management practices and policies, thereby building trust among stakeholders [36, 79]. Such measures can encourage more active participation and collective responsibility in protecting water resources.

Advocating for a decentralized approach to water management is increasingly recognized as an effective strategy. Decentralization empowers local authorities and communities to craft solutions tailored to regional water quality challenges [56, 72]. This strategy can improve resource allocation, enhance service delivery, and strengthen community engagement [79, 108]. Nonetheless, for decentralization to succeed, robust institutional frameworks and capacity-building initiatives should be established to support local governance [72, 79].

The significance of aligning water quality management policies with broader socioeconomic objectives is critical. This alignment necessitates an understanding of how water quality impacts public health and economic productivity, requiring an interconnected policy framework. Effectively addressing water-related challenges mandates an agricultural paradigm that incorporates pollution control measures alongside strategies for water use efficiency. Lessons learned from successful case studies highlight the necessity for integrated approaches that consider both productive and environmental aspects of water management [83].

Moreover, continuous evaluation and improvement of water governance policies are essential. Rigorous monitoring and performance evaluation practices are vital for assessing the actual impact of implemented policies on water quality. By adopting a learning mindset within governance processes, administrative bodies can adapt and refine strategies based on real-world outcomes and stakeholder feedback, ensuring that water quality management remains dynamic and responsive to emerging challenges.

7.3 International Support and Multilateral Partnerships on Water Quality

The role of international organizations in addressing water quality and safety is paramount, as they provide essential guidelines and frameworks that facilitate nations' efforts to ensure sustainable and safe water access. The World Health Organization (WHO) develops comprehensive international water safety guidelines, notably the Guidelines for Drinking-water Quality, which serve as a benchmark for public health policies concerning safe drinking water worldwide [31]. Furthermore, WHO aids countries in implementing Water Safety Plans (WSPs), vital tools that proactively manage water quality and reduce health risks. This technical support extends to capacity-building initiatives focusing on waterborne disease surveillance and control, critical for minimizing health impacts from contaminated water sources [103].

Similarly, the United Nations Children's Fund (UNICEF) plays a significant role in promoting water quality and safety, particularly among vulnerable populations such as children in emergency settings. UNICEF leads global Water, Sanitation, and Hygiene (WASH) programs, ensuring that children in schools and communities have access to clean water and sanitation facilities [23]. These programs are designed to foster behavior change regarding hygiene education, which is essential in preventing waterborne diseases. By responding effectively in humanitarian crises, UNICEF further emphasizes the need for reliable access to clean water, showcasing the crucial intersection of health, education, and water quality.

Another critical player is the World Bank, which provides financial resources and technical expertise to enhance water quality and infrastructure in developing countries. By offering low-interest loans and facilitating public–private partnerships, the World Bank ensures sustainable water delivery systems that can endure the test of time. Their focus on policy design and water governance underscores the importance of institutional strengthening as a mechanism for improving water quality and safety [28].

UN-Water functions as a coordinating body for various UN agencies involved in water issues, ensuring that efforts to achieve global water security are streamlined and effective. It plays a significant role in tracking progress toward Sustainable Development Goal 6 (SDG 6), which focuses on ensuring availability and sustainable management of water and sanitation for all [116]. Furthermore, UN-Water promotes Integrated Water Resources Management (IWRM), which is essential for maintaining water quality amidst competing demands [21].

Global initiatives such as SDG 6 are crucial in establishing universal access to safe and affordable drinking water while aiming to improve water quality by reducing pollution [68, 69]. Addressing this goal necessitates well-implemented WASH programs that target health care facilities, schools, and communities, substantially reducing disease burdens and enhancing public health outcomes [48]. The alignment of WASH initiatives with SDG 6 ensures a comprehensive approach to water quality, sanitation, and hygiene.

Bilateral and multilateral donors significantly contribute to expanding water infrastructure through large-scale projects that emphasize the construction of treatment plants and rural water supply systems [87–89]. By incorporating monitoring and evaluation frameworks into these initiatives, donors ensure project sustainability and a measurable impact on water quality. Financial support ties closely to performance indicators, fostering a culture of governance transparency and community engagement, which is vital for sustained improvements in water safety [60].

Collaboration in research and innovation within the water sector is also emerging as a key area for enhancing water quality management. International research consortia are pivotal in developing low-cost testing kits and real-time water quality monitoring systems that provide communities with the tools to effectively manage their water resources. Such innovations can greatly enhance local responses to water quality issues and contribute to global knowledge on best practices for water safety [113].

Capacity building is essential for enhancing local expertise in managing water resources sustainably. International agencies conduct training for water professionals and local engineers, fostering strong regulatory institutions and enforcing water safety standards. Programs aimed at strengthening institutional capacities are crucial for ensuring that water quality management aligns with public health goals [56].

The interconnection of water quality with broader issues of health, sanitation, and socioeconomic development reinforces the need for a multi-faceted approach to water governance. The growing recognition of integrated frameworks underscores the importance of cross-sectoral collaboration among health, sanitation, and environmental agencies, which has proven effective in water quality management programs [66].

7.4 Water Quality Monitoring and Regulation

Water quality monitoring and regulation are critical components in safeguarding public health and ecosystem integrity. Techniques for water quality assessment can broadly be categorized into chemical, biological, and physical parameters, each serving distinct yet interconnected roles in evaluating water safety [12, 32]. Chemical parameters involve the measurement of pH, dissolved oxygen (DO), and various contaminants, including heavy metals and nitrates, which are crucial for identifying pollution sources and understanding their potential health impacts [61, 65]. For instance, heavy metals like lead and mercury can have profound effects on human health, necessitating stringent monitoring.

Detecting chemical pollutants stemming from industrial discharge and agricultural runoff highlights the necessity of stringent regulations and effective monitoring practices [57, 102]. Waterborne pathogens such as Escherichia coli and Salmonella are also critical biological parameters that require systematic testing, especially in regions with documented waterborne disease outbreaks [26]. Bioindicators, which include certain macroinvertebrates and algae, serve as invaluable tools for gauging ecological health, linking the presence of these organisms to the broader water quality status [22].

Physical parameters encompass turbidity, temperature, and total suspended solids (TSS), all of which contribute to the aesthetic qualities of water bodies and affect aquatic life [107, 115]. Advanced monitoring techniques, including the use of portable test kits and IoT sensors, have revolutionized real-time water quality assessments, enabling rapid responses to contamination events [57, 107]. Moreover, standardized laboratory procedures are essential for ensuring the reliability and comparability of water samples collected from various sources [12, 19].

In Nigeria, regulatory bodies like NESREA, NAFDAC, and the Federal Ministry of Water Resources are pivotal in enforcing and developing water quality standards that align with national and international benchmarks [57, 102]. Similarly, the United States EPA rigorously implements the Safe Drinking Water Act, ensuring compliance with established national water quality criteria [30, 65]. Regionally, organizations such as the African Ministers' Council on Water (AMCOW) and the European Environment Agency (EEA) play integral roles in establishing frameworks for collaborative water governance [104].

Data transparency and public reporting mechanisms form the backbone of effective water quality management. The establishment of open-access databases allows stakeholders, including researchers and the public, to access vital water quality information [25]. Community-driven reporting systems and citizen science initiatives enhance public engagement and accountability [106]. Furthermore, incorporating Geographic Information Systems (GIS) can significantly aid in identifying contamination hotspots and guiding regulatory action [106, 115].

Periodic assessments and reports issued by regulatory agencies and NGOs bolster public awareness and trust in water quality management practices [25, 34]. By mandating industries to disclose their effluent discharge compliance, regulatory bodies can enhance oversight and improve water quality outcomes [19, 57]. The collaboration between technological advancement and regulation, facilitated through community involvement, underscores a holistic approach to water quality monitoring and management.

7.5 Challenges to Effective Policy Implementation Water Quality and Safety

Effective water quality and safety policies face significant challenges stemming from weak governance, corruption, and limited political will (Table 7.1 and Fig. 7.1). A lack of enforcement of existing regulations often leads to inadequacies in water safety management, necessitating a transformation of governance mechanisms to ensure accountability [24]. Political interference frequently prioritizes short-term financial gains over sustainable environmental practices, compromising long-term water safety [74]. Corruption in procurement processes further exacerbates these issues by misallocating resources intended for water infrastructure projects, leading to ineffective implementation and maintenance [24]. Lastly, poor institutional accountability coupled with inadequate monitoring frameworks undermines the efficacy of regulatory bodies, resulting in unaddressed safety concerns [24].

Funding and infrastructure deficiencies pose further challenges to water quality and safety initiatives. Insufficient budget allocations for water quality monitoring severely restrict the capacity of regulatory agencies to conduct essential tests and enforce compliance with safety standards [74]. Aging infrastructure compounds

Table 7.1 Challenges to effective policy implementation in water quality and safety

Challenge area	Narrative
Weak governance, corruption, and limited political will	– Poor enforcement of water safety policies – Political neglect of water-related issues – Corruption in project execution and procurement – Lack of accountability mechanisms
Inadequate funding and infrastructure	– Insufficient budget for water systems – Degraded or outdated infrastructure – Poor access in underserved communities – Overreliance on unstable donor funding
Limited technical capacity and trained personnel	– Shortage of skilled water professionals – Limited training for field personnel – Poor laboratory and monitoring facilities – Inability to interpret and apply data in policy
Fragmentation across agencies and overlapping mandates	– Lack of coordination among ministries (e.g., health, environment, water) – Policy conflicts and redundancy – Blurred responsibilities across government levels – Weak integration of IWRM approaches

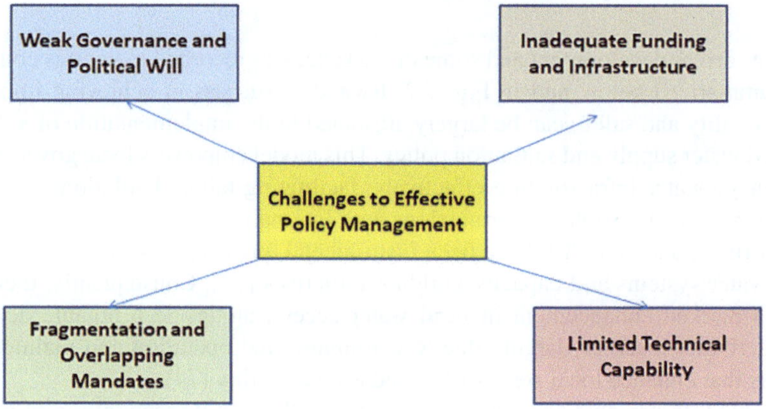

Fig. 7.1 Challenges to effective policy implementation in water quality and safety

these challenges, as outdated or poorly maintained systems can lead to contamination risks, significantly affecting public health [99]. Rural and peri-urban communities, in particular, suffer from limited access to robust distribution systems, further entrenching inequities in water safety [74]. Dependency on inconsistent donor funding exacerbates these challenges by creating vulnerabilities in long-term planning for infrastructure improvements [74].

Limited technical capacity and trained personnel hinder effective water quality management. The shortage of skilled professionals who specialize in water testing and treatment significantly impedes the ability to manage risks related to water safety [74]. This issue is compounded by a lack of training programs for local authorities and community health workers, leaving critical knowledge gaps in water safety protocols [74]. Additionally, inadequate laboratory facilities further restrict the ability to detect contaminants effectively, potentially leading to health hazards [74]. Difficulties in interpreting scientific data for informed policy decisions underscore the need for improved capacity building within water management sectors [24].

Fragmentation across agencies and overlapping mandates contribute to inefficiencies in water governance. Multiple ministries often operate in silos, impeding cohesive water management practices [101]. Conflicting policies at federal, state, and local government levels complicate the execution of essential water safety measures, creating a landscape where responsibilities are neglected due to unclear jurisdictional lines [24]. This fragmentation leads to duplicated efforts and further confusion regarding roles, underscoring the urgent need for integrated water resource management (IWRM) mechanisms to promote collaboration among stakeholders [24].

7.6 Studies on Policy-Driven Successes on Water Safety and Quality

Policies drives development and some of the successes recorded in various countries are summarized below and in Fig. 7.2. Rwanda's success in achieving improved water quality and safety can be largely attributed to the implementation of a decentralized water supply and sanitation policy. This model empowers local governments to manage water infrastructure effectively, facilitating tailored solutions that meet community needs. Strategic partnerships with international donors like USAID, the World Bank, and UNICEF have been instrumental, providing essential funding for rural water systems and capacity-building initiatives [74]. Consequently, there has been a marked enhancement in rural water access, alongside a notable decrease in waterborne diseases, largely due to community-led operation and maintenance models that enhance local stewardship and accountability [74].

Bangladesh presents another compelling case through its recognition of arsenic contamination as a critical public health issue. The government initiated extensive screening of tube wells and launched public health campaigns to address this crisis

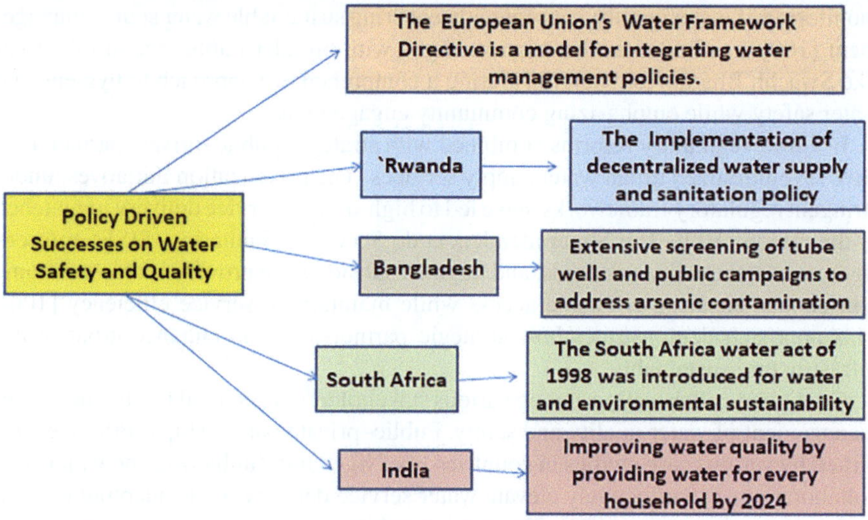

Fig. 7.2 Policy-driven successes on water safety and quality

[100]. The policy shift emphasized alternative water sources such as deep tube wells and rainwater harvesting methods, while collaborations with WHO and international NGOs helped develop the Arsenic Mitigation Policy and Action Plan (AMPAP), bolstering access to safe drinking water [100]. This systemic response has not only addressed immediate health concerns but also transformed water safety strategies nationwide.

On a larger scale, the European Union's Water Framework Directive (WFD) serves as a model for integrated water management policies, mandating all member states achieve "good status" for water bodies through comprehensive river basin management [13]. This directive fosters public participation, enforces pollution controls, and encourages ecosystem-based water management strategies, marking a significant advancement in cooperative water governance and pollution reduction across borders [13]. The long-term monitoring and assessment of chemical, physical, and ecological water parameters demonstrate the effectiveness of integrated approaches to water quality enhancements.

In South Africa, the National Water Act of 1998 has introduced a human-rights-based framework which prioritizes equitable water distribution alongside environmental sustainability [97]. The establishment of catchment management agencies is critical, as it decentralizes water management and ensures local needs are prioritized [97]. Furthermore, the implementation of legal mechanisms for water use licensing has ensured accountability in water-intensive sectors, thereby enhancing overall governance of water resources and safety.

India's ambitious Jal Jeevan Mission aims to deliver piped drinking water to every rural household by 2024, highlighting the government's commitment to improving water quality [100]. This initiative integrates community ownership and village-level monitoring of water quality, subsequently ensuring sustainable water source management [100]. Furthermore, the mission aligns with broader public health initiatives like Swachh Bharat Abhiyan, illustrating a comprehensive approach to hygiene and water safety while emphasizing community engagement.

In Chile, regulatory reforms combined with strategic public–private partnerships have revolutionized urban water supply services. The privatization initiatives, under stringent regulatory frameworks, have led to high-quality service delivery unmatched in the region [100]. The Superintendencia de Servicios Sanitarios (SISS) enforces rigorous water quality standards, and targeted subsidies are provided for low-income households, ensuring equitable access while maintaining service efficiency [100]. This approach demonstrates how strategic partnerships can enhance urban water infrastructure sustainably.

Continued collaboration among various stakeholders is essential for the successful management of water quality and safety. Public–private partnerships (PPPs), exemplified by various case studies in countries like Niger and Zimbabwe, show that such collaborations can effectively elevate water service delivery while adapting to local socio-economic contexts [75]. These partnerships leverage private investment and innovation, thereby improving water access and quality comprehensively.

However, it is important to consider the challenges presented by public–private collaborations. Insights from studies in Bangladesh highlight the necessity of tailoring PPP projects to align with local needs and conditions [11]. Effective communication and clearly defined roles between public entities and private partners are critical to ensuring the sustainability of these investments [11]. Furthermore, continuous monitoring and assessment are essential for adapting these projects to evolving community needs and environmental conditions.

7.7 Community Engagement and Capacity Building

The sustainable management of water quality and safety stands at the intersection of community engagement and institutional frameworks that facilitate collaboration between civil society and local governments (Table 7.2 and Fig. 7.3). These entities play vital roles in both implementing and monitoring policies that govern water resources. The effectiveness of these roles is amplified under participatory governance models, where community input is prioritized, fostering transparency and accountability. It is well documented that the capacity of civil society organizations (CSOs) to engage local populations can lead to improved health outcomes and resource management, providing a platform for inclusive participation in water governance [29, 92].

Local governments are uniquely positioned to implement policies effectively by closely understanding the context in which they operate. Their proximity to the community allows them to tailor interventions based on local needs, ensuring that the governance models employed are both relevant and culturally sensitive. This localized approach can involve mobilizing local resources while fostering accountability

Table 7.2 Community engagement and capacity building for sustainable water quality and safety

Focus area	Strategies and actions
Role of civil society and local governments	– Facilitate community-driven water governance models – Act as intermediaries between national authorities and local populations – Mobilize local resources and advocate for water-related issues – Monitor water quality and infrastructure performance through participatory approaches – Ensure inclusivity by involving marginalized groups in policy decisions
Education, public awareness, and behavioral change	– Promote hygiene and sanitation through outreach programs – Disseminate information via schools, local media, and town halls – Use culturally appropriate messaging to encourage behavior change – Build public trust through transparent communication – Encourage household-level treatment and safe water storage practices
Building technical expertise at grassroots level	– Train local operators and health workers in water quality testing and treatment – Establish and support community-based water committees – Provide vocational education in water and environmental health fields – Promote use of low-cost, locally-adapted purification technologies – Partner with NGOs and academic institutions for ongoing training

mechanisms that empower citizens to actively participate in monitoring water quality standards and delivery systems [29]. Factors such as community-driven initiatives have been shown to evoke stronger public compliance and engagement, enhancing the sustainability of water quality management practices [64].

In acting as intermediaries between national authorities and local populations, civil society facilitates a two-way dialogue that informs policy-making processes while presenting localized perspectives on water governance. This is critical, particularly in diverse socio-political contexts where national policies may not fully contextualize local realities [92]. Engaging communities encourages the development of policies reflective of grassroots realities, which often leads to more equitable treatment of marginalized populations [29].

Furthermore, by focusing on local resource mobilization, civil society can advocate for urgent water-related issues at both grassroots and regional levels. This advocacy can strengthen community networks, equipping them with the tools necessary to address water scarcity and contamination problems collectively [29]. Evidence

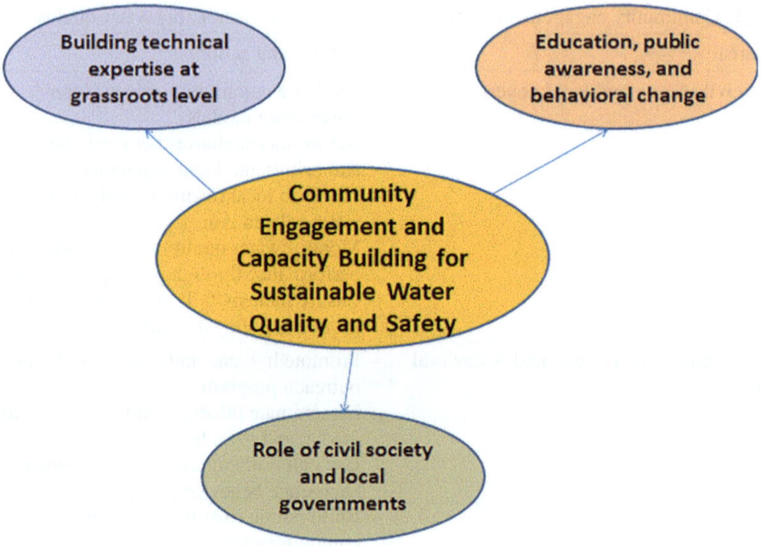

Fig. 7.3 Community engagement and capacity building for sustainable water quality and safety

suggests that when community entities actively engage in resource mobilization, they can significantly improve local capacity for water-related governance, fostering an environment conducive to sustainable management practices.

Citizen science and participatory audits serve as vital tools in monitoring water quality and infrastructure performance. Engaging citizens in data collection, analysis, and interpretation allows communities to become active stakeholders in their water governance, thus enhancing transparency [64]. Such participatory approaches not only empower local residents but also foster trust between communities and authorities, crucial for the effective implementation of water safety policies.

Inclusivity in decision-making processes is essential for achieving equitable water management outcomes, particularly for marginalized and vulnerable groups often overlooked in conventional governance structures. Civil society can play a pivotal role in ensuring that these groups are included, thus enhancing the responsiveness of water policies to diverse community needs [29]. By championing inclusivity, these organizations can drive the agenda for equitable resource distribution, contributing to broader social justice objectives within water governance.

Education and public awareness campaigns are fundamental for fostering community engagement in hygienic practices related to water quality. Effective outreach regarding handwashing and safe water storage practices can significantly mitigate the risks associated with waterborne diseases [64]. This can be further reinforced through consistent messaging distributed via local media and educational institutions that underline the importance of clean water and proper sanitation.

Culturally appropriate strategies should be employed to promote behavior change among communities to achieve meaningful engagement in water safety initiatives.

Tailored messaging that resonates with local customs and practices enhances the likelihood of sustained community buy-in for interventions aimed at improving water quality [64]. Long-term engagement strategies leveraging local knowledge and practices can drive substantial improvements in public health outcomes related to water safety.

Furthermore, consistent and transparent communication is essential for building trust and public support for water safety interventions. Engaging communities with regular updates on water quality and safety measures fosters a shared sense of responsibility that can lead to proactive community engagement [64]. Transparency in decision-making further enhances accountability among stakeholders, encouraging concerted efforts toward improving water quality.

Encouraging household-level water treatment practices is another critical aspect of local-level engagement in promoting water safety. By equipping households with the knowledge and resources necessary to treat their water supplies, civil society and local governments can enhance community resilience against water quality issues [29]. Educational workshops and dissemination of low-cost treatment technologies can empower households while also promoting responsible water usage.

To ensure long-term sustainability in water quality management, it is imperative to build technical expertise at the grassroots level. Training local water operators, health workers, and community volunteers in water quality testing and basic treatment methods ensures that communities have the technical capabilities to sustain their oversight efforts [64]. Establishing community-based water committees can formalize this knowledge-sharing process, creating structures for continued responsibility and hazard reporting within the community.

The promotion of innovation and the adoption of locally-adapted water purification technologies are also critical to enhancing community capacity in water management. Collaborations with NGOs and academic institutions can provide essential resources and knowledge transfer, making sophisticated yet applicable technologies accessible to underserved populations [64]. This approach not only equips communities but also stimulates local economies through the promotion of skills and technologies tailored to their specific contexts.

7.8 Future Directions and Policy Recommendations for Sustainable Water Quality and Safety

Addressing water quality and safety requires comprehensive and sustainable policies that span multiple governance levels (Table 7.3 & Fig. 7.4). Strengthening cross-border cooperation in water management is vital for mitigating transboundary pollution and ensuring equitable access to water resources. Establishing regional governance frameworks can facilitate collaborative management of shared water bodies, as seen in various international treaties that emphasize pollution control and equitable resource sharing [56]. Collaborative efforts can further enhance conflict resolution mechanisms, ensuring that data regarding water quality is transparently shared, which is crucial for informed decision-making [17, 59].

Enhancing climate resilience within water safety policies is equally paramount. Climate change adaptation strategies should be integrated into national water safety frameworks, acknowledging that urban water systems should withstand increasing variability in weather patterns. For instance, the promotion of nature-based solutions, such as wetland restoration, has proven effective in both mitigating climate impacts and enhancing water quality [27, 96]. Furthermore, adaptive infrastructure investments can fortify urban centers against the threats posed by drought and flooding [70]. Including comprehensive climate risk assessments in water management decisions can optimize resource management and disaster preparedness [18, 110].

Accountability in water governance is another critical aspect of ensuring water quality and safety. Establishing measurable water quality targets linked to funding

Table 7.3 Future Directions and Policy Recommendations

Future Direction	Policy Recommendations
Strengthening cross-border cooperation and water diplomacy	- Foster regional governance frameworks for shared water bodies - Promote equitable treaties and pollution-control agreements - Encourage data-sharing and conflict-resolution mechanisms - Establish joint monitoring systems and early-warning networks
Enhancing climate resilience in water safety policies	- Embed climate-adaptation strategies into national water plans - Invest in nature-based solutions (wetland restoration, green infrastructure) - Build adaptive infrastructure to withstand droughts, floods, variability - Mandate climate risk assessments in decision-making
Increasing accountability and performance-based funding	- Set measurable water quality targets tied to budget allocations - Create transparent tracking systems for funds and outcomes - Conduct independent audits with public reporting - Offer incentives or increased funding to entities that meet/exceed benchmarks
Leveraging technology for smarter water quality management	- Deploy IoT sensors for real-time monitoring - Apply AI/ML to predict contamination and optimize treatment - Develop mobile apps for community reporting - Use blockchain to ensure supply-chain traceability - Utilize GIS and satellite imagery for planning

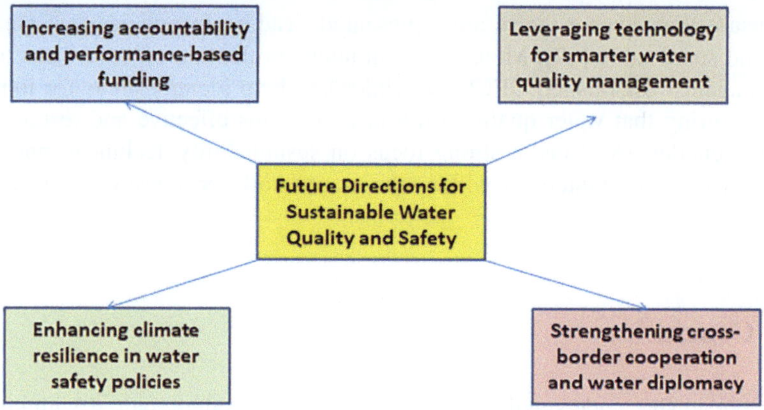

Fig. 7.4 Future directions and policy recommendations for sustainable water quality and safety

can incentivize communities and institutions to meet stringent safety benchmarks [17, 111]. Transparent tracking systems for financial disbursements and independent audits of water quality interventions will bolster public trust and improve governance outcomes [76]. Additionally, rewarding entities that demonstrate exceeded water safety benchmarks can foster a culture of accountability and performance excellence in water management.

Leveraging technology presents a transformative opportunity for improving water quality management practices. The deployment of Internet of Things (IoT) sensors enables real-time monitoring of water quality, enhancing response times to contamination events [51]. Utilizing AI and machine learning can facilitate predictive analytics for contamination risks, optimizing treatment processes and resource allocation. Community engagement through mobile applications for reporting water issues can further democratize water governance by ensuring citizen feedback is integral to water management strategies [5, 17].

Climate resilience, accountability, and technological advancements should converge to form the backbone of any future water quality and safety policies. Both national and localized strategies should recognize the interconnectedness of water governance and climate adaptation [16, 93]. By fostering an inclusive approach that considers diverse stakeholder perspectives (including Indigenous communities' rights and insights) governance frameworks can enhance trust and effectiveness in water management [37, 73]. This holistic strategy not only addresses immediate concerns regarding water safety but also builds a robust foundation for sustainable water resource management in the face of future challenges.

The implementation of these strategies necessitates a commitment to inter-agency cooperation and stakeholder engagement at all levels. Multi-stakeholder dialogues can ensure that diverse views are represented, leading to policies that are equitable and sustainable [17]. Moreover, continuous evaluation of governance mechanisms against established OECD principles can help identify areas for improvement, ensuring that water quality governance remains effective and responsive to evolving challenges. Thus, a strong focus on sustainability, technical innovation, and governance accountability will lay the groundwork for future water quality and safety.

7.9 Conclusion

A comprehensive understanding of water quality and safety, paired with context-specific and strategic policy interventions, is essential for achieving global health, environmental sustainability, and socioeconomic development. This chapter has highlighted how effective national policies, supported by international organizations such as WHO, UNICEF, and the World Bank, can significantly improve access to safe water. However, persistent challenges (such as weak governance structures, limited funding, and inadequate technical capacity) continue to hinder progress. Addressing these issues requires an integrated, multisectoral approach that recognizes the interconnectedness of water with health, education, and environmental systems. Strengthening regulatory frameworks with transparency and accountability, improving infrastructure, and ensuring investment in human resources are all crucial for advancing water quality management. To meet current and future water challenges, sustained international collaboration and locally grounded policies are imperative. Successful case studies illustrate that decentralized governance, cross-sector partnerships, and community engagement are vital to delivering lasting results. International support should go beyond financial assistance to include capacity building, knowledge transfer, and technological innovation. The intersection of civil society participation, local leadership, and technical empowerment presents a viable pathway for achieving inclusive and sustainable water safety outcomes. Thus, addressing the global water crisis demands collective action that embraces sustainability principles and fosters resilience, ensuring that water remains a secure and equitable resource for generations to come.

References

1. Aghoghovwia OA, Ohimain EI, Izah SC (2016) Bioaccumulation of heavy metals in different tissues of some commercially important fish species from Warri river, Niger Delta, Nigeria. Biotechnol Res 2(1):25–32
2. Aghoghovwia OA, Umoru OD, Izah SC (2018a) Physicochemical characteristics of nun river at Gbarantoru and Tombia Axis in Bayelsa State, Nigeria. Biosci Methods 9(1):1–11

3. Aghoghovwia OA, Miri FA, Izah SC (2018b) Impacts of anthropogenic activities on heavy metal levels in surface water of Nun River around Gbarantoru and Tombia Towns, Bayelsa State, Nigeria. Ann Ecol Environ Sci 2(2):1–8
4. Aghoghovwia OA, Izah SC, Miri FA (2018c) Environmental risk assessment of heavy metals in sediment of Nun River around Gbarantoru and Tombia Towns, Bayelsa State, Nigeria. Biol Evid 8(3):21–31
5. Ahmed M, Araral E (2019) Water governance in India: evidence on water law, policy, and administration from eight Indian states. Water 11(10):2071. https://doi.org/10.3390/w11102071
6. Aigberua AO, Izah SC, Richard G (2021) Hazard analysis of trace metals in muscle of *Sarotherodon melanotheron* and *Chrysichthys nigrodigitatus* from Okulu river, rivers state, Nigeria. J Environ Health Sustain Dev 6(3):1340–1356
7. Akinsorotan AM, Iyiola AO, Adejayi OO, Ogunremi IT, Awoniyi OO, Issa AO, Kelau JS (2024) Health risk assessment of heavy metals in smoke-dried African Catfish (*Clarias gariepinus)* in Ikole Ekiti markets, Southwestern Nigeria. Ann Animal Biol Res 4(1):36–47
8. Akinsorotan AM, Iyiola AO, Omotosho FP, Adejayi OO, Adesoye FA (2023) Assessment of heavy metals in water, fish tissues and soil from selected fish farms in Ekiti State, Nigeria. Aceh J Animal Sci 8(1):7–15. https://doi.org/10.13170/ajas.8.1.27520
9. Albrecht TR, Crootof A, Scott CA (2018) The water-energy-food nexus: a systematic review of methods for nexus assessment. Environ Res Lett 13(4):043002. https://doi.org/10.1088/1748-9326/aaa9c6
10. Annisa CN, Purwanti E, Sardjono MA (2023) Water sustainability in the anthropocene: challenges and solutions for a changing world. In: IOP conference series: earth and environmental science, vol 1282(1). IOP Publishing, p 012013. https://doi.org/10.1088/1755-1315/1282/1/012013
11. Arabov N, Voloshyna-Sidei V, Khomitov K, Nasimov D, Ismailov B, Bakayev Z (2023) Assessment of the efficiency of public-private partnership in the service sector. In: E3s web of conferences, vol 452, p 05028. https://doi.org/10.1051/e3sconf/202345205028
12. Barnhart B, Flinders C, Ragsdale R, Johnson G, Wiegand P (2021) Deriving human health and aquatic life water quality criteria in the United States for bioaccumulative substances: a historical review and future perspective. Environ Toxicol Chem 40(9):2394–2405. https://doi.org/10.1002/etc.5130
13. Basuki TM, Nugroho HYSH, Indrajaya Y, Pramono IB, Nugroho NP, Supangat AB, Indrawati DR, Savitri E, Wahyuningrum N, Purwanto et al (2022) Improvement of integrated watershed management in Indonesia for mitigation and adaptation to climate change: a review. Sustainability 14(16):9997. https://doi.org/10.3390/su14169997
14. Ben-Eledo VN, Kigigha LT, Izah SC, Eledo BO (2017a) Bacteriological quality assessment of Epie Creek, Niger Delta region of Nigeria. Int J Ecotoxicol Ecobiol 2(3):102–108
15. Ben-Eledo VN, Kigigha LT, Izah SC, Eledo BO (2017b) Water quality assessment of Epie creek in Yenagoa metropolis, Bayelsa state, Nigeria. Arch Curr Res Int 8(2):1–24
16. Benkov I, Varbanov M, Venelinov T, Tsakovski S (2023) Principal component analysis and the water quality index—a powerful tool for surface water quality assessment: a case study on Struma river catchment, Bulgaria. Water 15(10):1961. https://doi.org/10.3390/w15101961
17. Bertule M, Glennie P, Koefoed Bjørnsen P, James Lloyd G, Kjellen M, Dalton J, Rieu-Clarke A, Romano O, Tropp H, Newton J, Harlin J (2018) Monitoring water resources governance progress globally: experiences from monitoring SDG indicator 6.5. 1 on integrated water resources management implementation. Water 10(12):1744. https://doi.org/10.3390/w10121744
18. Biesbroek R, Dupuis J, Wellstead A (2017) Explaining through causal mechanisms: resilience and governance of social–ecological systems. Curr Opin Environ Sustain 28:64–70. https://doi.org/10.1016/j.cosust.2017.08.007
19. Booij K, Robinson CD, Burgess RM, Mayer P, Roberts CA, Ahrens L, Allan IJ, Brant J, Jones L, Kraus UR et al (2016) Passive sampling in regulatory chemical monitoring of nonpolar organic compounds in the aquatic environment. Environ Sci Technol 50(1):3–17

20. Borowski-Maaser I, Graversgaard M, Foster N, Prutzer M, Roest AH, Boogaard F (2021) WaterCoG: evidence on how the use of tools, knowledge, and process design can improve water co-governance. Water 13(9):1206. https://doi.org/10.3390/w13091206
21. Camkin J, Neto S (2025) Water and climate: building resilient communities. a world water policy journal special issue featuring Cairo water week. World Water Policy 11(2):383–384. https://doi.org/10.1002/wwp2.70004
22. Casallanovo F, Santos GS, Cione AP, Simone DM, de Oliveira Kaminski TS, Chen W (2023) Qualitative analysis of the most locally relevant runoff and erosion parameters for constructing Brazilian scenarios. Integr Environ Assess Manag 19(5):1374–1384. https://doi.org/10.1002/ieam.4741
23. Chen X, Wu F, Wang X (2024) Coupling coordination analysis between reclaimed water utilization capacity and effect in China. Water 16(22):3283. https://doi.org/10.3390/w16223283
24. Chopra A, Ramachandran P (2021) Understanding water institutions and their impact on the performance of the water sector in India. Water Policy 23(2):466–486. https://doi.org/10.2166/wp.2021.207
25. Corona J, Doley T, Griffiths C, Massey M, Moore C, Muela S, Rashleigh B, Wheeler W, Whitlock SD, Hewitt J (2020) An integrated assessment model for valuing water quality changes in the United States. Land Econ 96(4):478–492
26. Cui B, Liang S (2019) Monitoring opportunistic pathogens in domestic wastewater from a pilot-scale anaerobic biofilm reactor to reuse in agricultural irrigation. Water 11(6):1283. https://doi.org/10.3390/w11061283
27. Cunha Libanio PA (2020) Water reforms in Brazil: challenges and opportunities for promoting water security in a continental-sized country. World Water Policy 6(2):230–245. https://doi.org/10.1002/wwp2.12042
28. Dhoba L (2022) Strengthening water, sanitation and hygiene governance: a critical review of Zimbabwe's wash sector institutional arrangements. H2open J 5(2):248–263. https://doi.org/10.2166/h2oj.2022.063
29. Elfithri R, Mokhtar MB, Zakaria S (2019) The need for awareness raising, advocacy, and capacity building in integrated water resources management toward sustainable development: a case study in Malaysia. World Water Policy 5(1):43–54. https://doi.org/10.1002/wwp2.12002
30. Evans SR, Kopp K, Johnson PG, Hopkins BG, Dai X, Schaible C (2022) Comparing smart irrigation controllers for turfgrass landscapes. HortTechnology 32(5):415–424. https://doi.org/10.21273/horttech04985-21
31. Fonseca LM, Domingues JP, Dima AM (2020) Mapping the sustainable development goals relationships. Sustainability 12(8):3359. https://doi.org/10.3390/su12083359
32. Gensemer RW, Gondek JC, Rodriquez PH, Arbildua JJ, Stubblefield WA, Cardwell AS, Santore RC, Ryan AC, Adams WJ, Nordheim E (2018) Evaluating the effects of pH, hardness, and dissolved organic carbon on the toxicity of aluminium to freshwater aquatic organisms under circumneutral conditions. Environ Toxicol Chem 37(1):49–60
33. Gomez-Jaramillo Y, Berrouet L, Villegas-Palacio C, Berrio-Giraldo L (2024) Conceptual framework for analyzing the sustainability of socio-ecological systems with a focus on ecosystem services that support water security. Sustain Dev 32(3):2298–2313. https://doi.org/10.1002/sd.2780
34. Handam NB, Silva ABLGD, Silva RBD, Moura PG, Carvajal E, Sotero-Martins A, Santos JAAD (2022) Sanitary quality of reused water for irrigation in agriculture in Brazil. Revista Ambiente & Água 17(3):e2809. https://doi.org/10.4136/ambi-agua.2809
35. Harjanta SL, Ningrum DP (2023) Inhibiting factors for collaborative water governance: a case study of Mount Merapi ecosystem in Yogyakarta, Indonesia. Channel: Jurnal Komunikasi 11(1):81–86. https://doi.org/10.12928/channel.v11i1.225
36. Houar K, Fetoui M, Bellal SA (2024) Pathways for improving water governance in Western Algeria: a multi-stakeholder analysis. Acta Scientiarum Polonorum Administratio Locorum 23(1):57–70. https://doi.org/10.31648/aspal.9021

37. Hurlbert M, Acharibasam JB, Datta R, Strongarm S, Starblanket E (2024) Decolonizing Indigenous drinking water challenges and implications: focusing on indigenous water governance and sovereignty. Water 16(5):748. https://doi.org/10.3390/w16050748
38. Indriana I (2022) Maintaining legitimacy: the environmental management practices and green innovation on Indonesian manufacturing companies. J Appl Bus Tax Econ Res 2(1):83–99. https://doi.org/10.54408/jabter.v2i1.130
39. Ipinmoroti MO, Iyiola AO, Akanmu OA, Ayinde AT (2022) Assessment of heavy metal contamination in the muscle and gills of *clarias gariepinus* at Osun river in Osogbo, Osun state. J Glob Aff Sustain Dev (JoGASD) 1:107–118
40. Iyiola AO, Akinrinade AJ, Ajayi FO (2022) Effects of water pollution on biodiversity across coastal regions. In: Izah SC (ed) Biodiversity in Africa: potentials, threats and conservation, sustainable development and biodiversity, vol 29. Springer, pp 345–367. https://doi.org/10.1007/978-981-19-3326-4_13
41. Iyiola AO, Setufe SB, Ofori E, Bilikoni J, Ogwu MC (2024a) Biomarkers for the detection of pollutants from the water environment. In: Izah SC, Ogwu MC, Hamidifar H (eds) Biomonitoring of pollutants in the Global South. Springer, Singapore. https://doi.org/10.1007/978-981-97-1658-6_16
42. Iyiola AO, Ipinmoroti MO, Akingba OO, Ewutanure JS, Setufe SB, Bilikoni J, Ofori-Boateng E, Wangboje OM (2024b) Organic chemical pollutants within water systems and sustainable management strategies. In: Izah SC et al (eds) Water crises and sustainable management in the global south, pp 211–251. https://doi.org/10.1007/978-981-97-4966-9_7
43. Iyiola AO, Afolabi OA, Alimi SK, Akingba OO, Izah SC, Ogwu MC (2024c) Climate change and water crisis in the global south. In: Izah SC et al (eds) Water crises and sustainable management in the global south, pp 111–139. https://doi.org/10.1007/978-981-97-4966-9_4
44. Izah SC, Richard G, Stanley HO, Sawyer WE, Ogwu MC, Uwaeme OR (2023) Integrating the one health approach and statistical analysis for sustainable aquatic ecosystem management and trace metal contamination mitigation. ES Food Agrofor 14:1012. https://doi.org/10.30919/esfaf1012
45. Izah SC, Stanley HO, Richard G, Sawyer WE, Uwaeme OR (2024a) Surface water quality: a statistical perspective on the efficacy of environmental and human health assessment tools. Water Air Soil Pollut 235(3). https://doi.org/10.1007/s11270-024-06965-1
46. Izah SC, Stanley HO, Richard G, Sawyer WE, Uwaeme OR (2024b) Environmental health risks of trace elements in sediment using multivariate approaches and contamination indices. Int J Environ Sci Technol. https://doi.org/10.1007/s13762-024-05974-1
47. Izah SC, Jacob DE, Nelson IU, Avez S (2024c) Urban water crisis in the global south. In: Izah SC, Ogwu MC, Loukas A, Hamidifar H (eds) Water crises and sustainable management in the global south. Springer, Singapore, pp 45–83. https://doi.org/10.1007/978-981-97-4966-9_2
48. Izah SC, Etim NG, Alaka OO, Nwachukwu BC, Joshua G (2024d) Public health interventions for cholera control and prevention: water, sanitation, and hygiene (WASH) perspective. Green J Epidemiol Public Health 12(1):1–15
49. Izah SC, Ogwu MC (2025) Modeling solutions for microbial water contamination in the global south for public health protection. Front Microbiol 16:1504829. https://doi.org/10.3389/fmicb.2025.1504829
50. Izah, SC, Richard G, Sawyer WE (2021). Distribution of Fungi density and diversity in a Surface water of Epie Creek in Yenagoa Metropolis, Nigeria. Arch Epidemiol Public Health 3:1–5. https://doi.org/10.15761/AEPH.1000121
51. Jackson M, Stewart RA, Beal CD (2019) Identifying and overcoming barriers to collaborative sustainable water governance in remote Australian indigenous communities. Water 11(11):2410. https://doi.org/10.3390/w11112410
52. Jacob DE, Nelson IU, Izah SC (2024a). Indigenous water management strategies in the global south. In: Izah SC, Ogwu MC, Loukas A, Hamidifar H (eds) Water crises and sustainable management in the global south. Springer, pp 487–525. https://doi.org/10.1007/978-981-97-4966-9_16

53. Jacob DE, Nelson IU, Eniang EA, Izah SC (2024b) Advancing water security and resilience in the global south through recreational development. In: Izah SC, Ogwu MC, Loukas A, Hamidifar H (eds) Water crises and sustainable management in the global south. Springer, pp 527–554. https://doi.org/10.1007/978-981-97-4966-9_17
54. Jacob DE, Nelson IU, Izah SC, Daniel KS (2024c) Rural water crises in the global south: understanding the scope and impact. In: Izah SC, Ogwu MC, Loukas A, Hamidifar H (eds) Water crises and sustainable management in the global south. Springer, Singapore, pp 3–44. https://doi.org/10.1007/978-981-97-4966-9_1
55. Jena M, Mishra S, Moharana H (2023) Water sustainability initiatives to meet the water crisis in India. J Chem Eng Res Updates 10:18–30. https://doi.org/10.15377/2409-983x.2023.10.2
56. Jiménez A, Saikia P, Giné R, Avello P, Leten J, Liss Lymer B, Schneider K, Ward R (2020) Unpacking water governance: a framework for practitioners. Water 12(3):827. https://doi.org/10.3390/w12030827
57. Kalita I, Kamilaris A, Havinga P, Reva I (2024) Assessing the health impact of disinfection byproducts in drinking water. ACS Es&t Water 4(4):1564–1578. https://doi.org/10.1021/acsestwater.3c00664
58. Karamoutsou L, Veliou E, Kvernberg EB, Psilovikos A (2024) Urban stormwater management: a sustainable approach. Water 16(15):2089. https://doi.org/10.3390/w16152089
59. Karimi P, Chapman DV, Valieva S, Li R, Talbi A (2024) Accounting for water quality—a framework for agricultural water use. Front Water 6:1375791. https://doi.org/10.3389/frwa.2024.1375791
60. Kativhu T, Madzivanyika TT, Nunu WN, Macherera M, Chinyama A (2022) Sustainability of water facilities under community based management in Zimbabwe. AQUA—Water Infrastruct Ecosyst Soc 71(1):19–30.https://doi.org/10.2166/aqua.2021.089
61. Korak JA, Brandhuber PJ, Goodwill JE (2024). Lithium in drinking water: review of chemistry, analytical methods, and treatment technologies. AWWA Water Sci 6(6):e70009. https://doi.org/10.1002/aws2.70009
62. Kosoe EA, Ogwu MC, Duniah J, Ndigun FB (2025a) Sustaining indigenous water governance systems in Ayimaye and Lungni, Northern Ghana, through hybrid models. J Sustain 1(2). https://doi.org/10.55845/jos-2025-1235
63. Kosoe EA, Osumanu IK, Ogwu MC (2025b) Reimagining rural and small-town water supply in Ghana through inclusive governance and sustainable models. J Sustain 1(2):1–21. https://doi.org/10.55845/jos-2025-1234
64. Leendertse K, Mitchell S, Harlin J (2008) IWRM and the environment: a view on their interaction and examples where IWRM led to better environmental management in developing countries. Water SA 34(6):691–698. https://doi.org/10.4314/wsa.v34i6.183671
65. Li XF, Mitch WA (2018) Drinking water disinfection byproducts (DBPs) and human health effects: multidisciplinary challenges and opportunities. Environ Sci Technol 52(4):1681–1689. https://doi.org/10.1021/acs.est.7b05440
66. Li X, Dong D, Lin G, Yan R, Li S (2020) Water use for energy production and conversion in Hebei Province, China. Front Energy Res 8:558536. https://doi.org/10.3389/fenrg.2020.558536
67. Lim CH, Wong HL, Elfithri R, Teo FY (2022) A review of stakeholder engagement in integrated river basin management. Water 14(19):2973. https://doi.org/10.3390/w14192973
68. Liu X, Cheng S, Miao Z, Li Z, Li P, Liu T, Zhi H, Zhang S, Wang Y, Zheng X (2023) Analysis of the coupling relationship between water quality and economic development in Hongjiannao Basin, China. Water 15(16):2965. https://doi.org/10.3390/w15162965
69. Liu X, Cheng S, Miao Z, Li Z, Li P, Liu T, Zhi H, Zhang S, Wang Y, Zheng X (2023) Analysis of the coupling relationship between water quality and economic development in Hongjiannao Basin, China. Water 15(16):2965
70. Madonsela B, Koop S, van Leeuwen K, Carden K (2019) Evaluation of water governance processes required to transition towards water sensitive urban design—an indicator assessment approach for the City of Cape Town. Water 11(2):292. https://doi.org/10.3390/w11020292

71. Mariosa DF, Falsarella OM, Siqueira G, da Frigo Silva AC, de Melo Conti D, Álvarez JF (2024) The challenges of universal access to water: some contributions to the sustainable management of water resources in rural areas. Concilium 24(6):140–154. https://doi.org/10.53660/clm-3147-24f22
72. Mateus C, Valencia M, DiFrancesco K, Ochoa-Herrera V, Gartner T, Quiroga D (2020) Governance mechanisms and barriers for achieving water quality improvements in Galapagos. Sustainability 12(21):8851. https://doi.org/10.3390/su12218851
73. McKibbin C (2023) Decolonising Canadian water governance: lessons from indigenous case studies. UCL Open Environ 5:e060. https://doi.org/10.14324/111.444/ucloe.000060
74. Mishra BK, Kumar P, Saraswat C, Chakraborty S, Gautam A (2021) Water security in a changing environment: concept, challenges and solutions. Water 13(4):490. https://doi.org/10.3390/w13040490
75. Mutandwa H, Vyas-Doorgapersad S (2023) The state of water infrastructure development in zimbabwe. Afr J Gov Dev (AJGD) 12(2):40–57. https://doi.org/10.36369/2616-9045/2023/v12i2a4
76. Neto S, Camkin J, Fenemor A, Tan PL, Baptista JM, Ribeiro JM, Schulze R, Stuart-Hill S, Spray C, Elfithri R (2020) OECD principles on water governance in practice: an assessment of existing frameworks in Europe, Asia-Pacific, Africa and South America. In: OECD principles on water governance. Routledge, pp 60–89. https://doi.org/10.1080/02508060.2018.1402650
77. Ngene BU, Nwafor CO, Bamigboye GO, Ogbiye AS, Ogundare JO, Akpan VE (2021) Assessment of water resources development and exploitation in Nigeria: a review of integrated water resources management approach. Heliyon 7(1). https://doi.org/10.1016/j.heliyon.2021.e05955
78. Nwokediegwu ZQS, Ugwuanyi ED, Dada MA, Majemite MT, Obaigbena A (2024) Urban water management: a review of sustainable practices in the USA. Eng Sci Technol J 5(2):517–530. https://doi.org/10.51594/estj.v5i2.829
79. Ogada JO, Krhoda GO, Van Der Veen A, Marani M, van Oel PR (2017) Managing resources through stakeholder networks: collaborative water governance for Lake Naivasha basin, Kenya. Water Int 42(3):271–290. https://doi.org/10.1080/02508060.2017.1292076
80. Ogamba EN, Izah SC, Isimayemiema F (2016a) Bioaccumulation of heavy metals in the gill and liver of a common Niger delta wetland fish, Clarias garepinus. Br J Appl Res 1(1):17–20
81. Ogamba EN, Izah SC, Ofoni-Ofoni AS (2016b) Bioaccumulation of chromium, lead and cadmium in the bones and tissues of Oreochromis niloticus and Clarias camerunensis from Ikoli creek, Niger Delta, Nigeria. Adv Sci J Zool 1(1):13–16
82. Ogamba EN, Izah SC, Omonibo E (2016c) Bioaccumulation of hydrocarbon, heavy metals and minerals in Tympanotonus fuscatus from coastal region of Bayelsa state, Nigeria. Int J Hydrol Res 1:1–7
83. Ogidi OI (2022) Chapter 12: impact of pharmaceutical compounds on the microbial ecology of surface water resources. In: Biodiversity in Africa: potentials, threats and conservation. Springer Nature Singapore Pte Ltd. https://doi.org/10.1007/978-981-19-3326-4_12
84. Ogidi IO (2024) Strategies of sustainable management of water resources in the global south. In: Water crisis and sustainable management in the global south. Springer Nature Singapore Pte Ltd. https://doi.org/10.1007/978-981-97-2428-4_7
85. Ogidi OI, Akpan UM (2022) Chapter 16: aquatic biodiversity loss: impacts of pollution and anthropogenic activities and strategies for conservation. In: Biodiversity in Africa: potentials, threats and conservation. Springer Nature Singapore Pte Ltd. https://doi.org/10.1007/978-981-19-3326-4_16
86. Ogidi OI, Izah SC (2024) Water contamination by municipal solid wastes and sustainable management strategies. In: Water crisis and sustainable management in the global south. Springer Nature Singapore Pte Ltd
87. Ogwu MC (2025) Sustainable development in Africa: challenges, innovations, and policy pathways for the future. Afr Res Rep 1(2):55–69. https://doi.org/10.5897/ARR2025.0005

88. Ogwu MC, Kosoe EA (2024) Place of cultural diversity in sustainable water resource management in Ghana. In: Izah SC, Ogwu MC, Loukas A, Hamidifar H (eds) Water crises and sustainable management in the global south. Springer, Singapore. https://doi.org/10.1007/978-981-97-4966-9_14
89. Ogwu MC, Kosoe EA (2025) Integrating green infrastructure into sustainable agriculture to enhance soil health, biodiversity, and microclimate resilience. Sustainability 17(9):3838. https://doi.org/10.3390/su17093838
90. Ogwu MC, Babafemi OP, Ayilara FD, Ovuru KF, Iyiola AO (2024) Traditional and conventional water treatment methods: a sustainable approach. In: Izah SC et al (eds) Water crises and sustainable management in the global south, pp 461–486. https://doi.org/10.1007/978-981-97-4966-9_15
91. Ogwu MC, Izah SC, Sawyer WE, Amabie T (2025) Environmental risk assessment of trace metal pollution: a statistical perspective. Environ Geochem Health 47:94. https://doi.org/10.1007/s10653-025-02405-z
92. Paudyal HL (2023) Civil society, governance and justice: a critique in global context. J Tilottama 1(1):53–66. https://doi.org/10.3126/jtilottama.v1i1.64554
93. Pipan P, Hribar MŠ, Urbanc M (2023) Motivation, robustness and benefits of water commons: Insights from small drinking water supply systems. Acta Geogr Slov 63(3):85–100. https://doi.org/10.3986/ags.11592
94. Poddar N, Santini JM, Maher MJ (2021. Crystal structure of a complex between the electron-transfer partners arsenite oxidase and cytochrome c552, from the arsenite respiring bacterium Rhizobium sp. NT-26. Acta Cryst 77:C433. https://doi.org/10.1107/s0108767321092539
95. Ricart S, Villar-Navascués R, Squitieri F, Neri M, Hernández-Hernández M, Toth E, Rico-Amorós AM (2025) How to reinforce governance in water–tourism nexus research? Updating the hydrosocial cycle to loop into stakeholders' roles, interactions, and power dynamics. Environ Res Lett 20(3):034021. https://doi.org/10.1088/1748-9326/adb447
96. Rudberg PM, Karpouzoglou T (2022) Using adaptive capacity to shift absorptive capacity: a framework of water reallocation in highly modified rivers. Water 14(2):193.https://doi.org/10.3390/w14020193
97. Ruiters C, Amadi-Echendu J (2022) Public–private partnerships as investment models for water infrastructure in South Africa. Infrastruct Asset Manag 9(4):180–193.https://doi.org/10.1680/jinam.21.00013
98. Sahoo SK, Goswami SS (2024) Theoretical framework for assessing the economic and environmental impact of water pollution: a detailed study on sustainable development of India. TIDEE: TERI Inf Digest Energy Environ 23(1/2):83–83. https://doi.org/10.5267/j.jfs.2024.1.003
99. Sambursky GA, Leont'eva SV, Molchanova YP, Nikitina SV (2022) Some ecological problems of providing the population with high-quality drinking water. In: IOP conference series: earth and environmental science, vol 1076(1). IOP Publishing, p 012077. https://doi.org/10.1088/1755-1315/1076/1/012077
100. Singh A, Naik G (2024) Rural drinking water supply program and societal development: evidence from the early implementation phase of India's Jal Jeevan mission. PLoS ONE 19(11):e0312144. https://doi.org/10.1371/journal.pone.0312144
101. Singh N, Pandey A, Singh V (2024) Impact of institutional overlapping on water governance of Himalayan city, Nainital, India. Water Policy 26(6):601–617. https://doi.org/10.2166/wp.2024.285
102. Tubalawony S, Tuahatu JW, Kalay DE (2023) Water quality of Haruku strait central Maluku Indonesia. In: IOP conference series: earth and environmental science, vol 1207(1), p 012025. IOP Publishing. https://doi.org/10.1088/1755-1315/1207/1/012025
103. Valcourt N, Javernick-Will A, Walters J, Linden K (2020) System approaches to water, sanitation, and hygiene: a systematic literature review. Int J Environ Res Public Health 17(3):702. https://doi.org/10.3390/ijerph17030702
104. Wang CY, Doan ML, Xue L, Barbour AJ (2018) Tidal response of groundwater in a leaky aquifer—application to Oklahoma. Water Resour Res 54(10):8019–8033. https://doi.org/10.1029/2018wr022793

105. Wang Y, Cui X, Zhang X, Wen Q (2022) Exploring the sustainable use strategy of scarce water resources for rural revitalization in Yanchi county from arid region of Northwest China. Int J Environ Res Public Health 19(23):16347. https://doi.org/10.3390/ijerph192316347
106. Warwick CR, Lamm AJ, Warner LAS (2022) Extension and the environment: understanding florida residents' perceptions of environmental water-related topics: WC167/AEC503, rev. 9/ 2022. EDIS (5)
107. Wei J, Wang M, Jiang L, Yu X, Mikelsons K, Shen F (2021) Global estimation of suspended particulate matter from satellite ocean color imagery. J Geophys Res Oceans 126(8):e2021JC017303. https://doi.org/10.1029/2021jc017303
108. Werdiningtyas R, Suminar L, Kusumastuti K (2021) The use of data and information in irrigation water management (case study: Colo irrigation area, Bengawan solo river basin). In: IOP conference series: earth and environmental science, vol 887(1). IOP Publishing, p 012034. https://doi.org/10.1088/1755-1315/887/1/012034
109. Wilbers GJ, de Bruin K, Seifert-Dähnn I, Lekkerkerk W, Li H, Budding-Polo Ballinas M (2022) Investing in urban blue–green infrastructure—assessing the costs and benefits of stormwater management in a Peri-urban catchment in Oslo, Norway. Sustainability 14(3):1934. https://doi.org/10.3390/su14031934
110. Wuijts S, Rijswick HFV, Driessen PP (2021) Achieving European water quality ambitions: governance conditions for more effective approaches at the local-regional scale. Sustainability 13(2):681. https://doi.org/10.3390/su13020681
111. Wuijts S, Friederichs L, Hin JA, Schets FM, Van Rijswick HF, Driessen PP (2022) Governance conditions to overcome the challenges of realizing safe urban bathing water sites. Int J Water Resour Dev 38(4):554–578. https://doi.org/10.1080/07900627.2020.1755617
112. Wutich A, Rosinger AY, Stoler J, Jepson W, Brewis A (2020) Measuring human water needs. Am J Hum Biol 32(1):e23350. https://doi.org/10.1002/ajhb.23350
113. Xie Z, Jianhui J, Wang J, Zhang R, Li S (2023) Application of deep learning techniques in water level measurement: combining improved SegFormer-UNet model with virtual water gauge. Appl Sci 13(9):5614. https://doi.org/10.3390/app13095614
114. Xu N, Zhu Q, Zhu J, Jia J, Wei X, Wang Y (2021) Novel latex microsphere immunochromatographic assay for rapid detection of cadmium ion in asparagus. Foods 11(1):78. https:// doi.org/10.3390/foods11010078
115. Zhang Y, Wang CY, Fu LY, Yang QY (2021) Are deep aquifers really confined? Insights from deep groundwater tidal responses in the North China platform. Water Resour Res 57(11):e2021WR030195. https://doi.org/10.1029/2021wr030195
116. Zheng Y, He J, Zhang W, Lv A (2023) Assessing water security and coupling coordination in the lancang–mekong river basin for sustainable development. Sustainability 15(24):16880

Chapter 8
Education and Behavioral Changes for Safe Water Usage

D. E. Jacob

Abstract Access to safe water is fundamental to public health, yet millions remain exposed to waterborne diseases due to unsafe water usage and sanitation practices. This chapter examines how education and behavioral change can serve as transformative tools in promoting safe water use at individual, household, and community levels. It explores the role of formal curricula, community-based campaigns, digital platforms, and emergency education in reshaping habits and fostering public awareness. The chapter applies behavioral theories such as the Health Belief Model, Theory of Planned Behavior, and Social Cognitive Theory to explain motivations and barriers influencing water practices. It also identifies key risk factors including socio-economic limitations, cultural beliefs, and gender roles. Case studies from global WASH programs highlight successful strategies like Participatory Hygiene and Sanitation Transformation (PHAST) and Community-Led Total Sanitation (CLTS). The chapter further addresses monitoring and evaluation mechanisms, outlining tools and indicators that assess progress. Finally, it presents institutional frameworks and policy recommendations for scaling and sustaining change. Through interdisciplinary integration of education, psychology, and community engagement, the chapter provides a holistic roadmap for fostering water-safe behavior. It concludes that without educational investment and sustained institutional support, infrastructure improvements alone cannot guarantee long-term water safety and public health outcomes.

Keywords Safe water usage · Behavioral change · Health education · Water · Sanitation and hygiene (WASH) · Community engagement · Socio-cultural barriers · Water safety education · Monitoring and evaluation

List of Acronyms

WASH	Water, Sanitation and Hygiene
PHAST	Participatory Hygiene and Sanitation Transformation

D. E. Jacob (✉)
Department of Forestry and Wildlife, University of Uyo, Uyo, Nigeria
e-mail: danieljacob@uniuyo.edu.ng

M. C. Ogwu and S. Chibueze Izah (eds.), *Water Quality and Safety in the Global South*, Springer Water, https://doi.org/10.1007/978-3-032-10602-5_8

CLTS	Community-Led Total Sanitation
HBM	Health Belief Model
TPB	Theory of Planned Behavior
SCT	Social Cognitive Theory
BCC	Behavior Change Communication
SBCC	Social and Behavior Change Communication
FGD	Focus Group Discussion
KII	Key Informant Interview
CBM	Community-Based Monitoring
KAP	Knowledge, Attitude, and Practice
UNICEF	United Nations Children's Fund
WHO	World Health Organization
NGO	Non-Governmental Organization
M&E	Monitoring and Evaluation

8.1 Introduction

Access to safe water remains a fundamental determinant of public health, yet millions of people across the globe continue to face challenges related to water safety and sanitation [44, 45, 48–53]. According to Marinez-Santos (2017), an estimated 1.8 billion individuals worldwide consume drinking water that is contaminated with feces, leading to a wide range of waterborne diseases, including cholera, typhoid, dysentery, and hepatitis A. These risks are further exacerbated in low-income and rural communities, where access to clean water infrastructure is limited and behavioral practices related to water collection, storage, and usage are often shaped by socio-cultural norms, misinformation, or lack of education [8, 51, 53]. Research specifically in Nigeria highlights these dynamics, with studies on the livelihood and resource interaction of households in areas like the Itu Wetland underscoring how local economies and community engagement with their environment influence water safety behaviors [46]. Additionally, the impact of national parks on livelihood and conservation behaviors in Nigerian households further illustrates the complex interplay between human activities and natural resources that affects water usage [50]. Education and behavioral change play a pivotal role in ensuring safe water usage at the household and community levels (Fig. 8.1). While investments in water infrastructure are critical, they are not sufficient on their own to guarantee health outcomes. Behavioral practices such as hand washing, water treatment, hygienic storage, and safe disposal of wastewater are equally essential in preventing water contamination and disease transmission [42, 101]. Without the knowledge and motivation to adopt safe practices, individuals may unknowingly undermine the benefits of improved water infrastructure [56]. Thus, a comprehensive approach that integrates both physical interventions and educational efforts is essential for sustainable water safety.

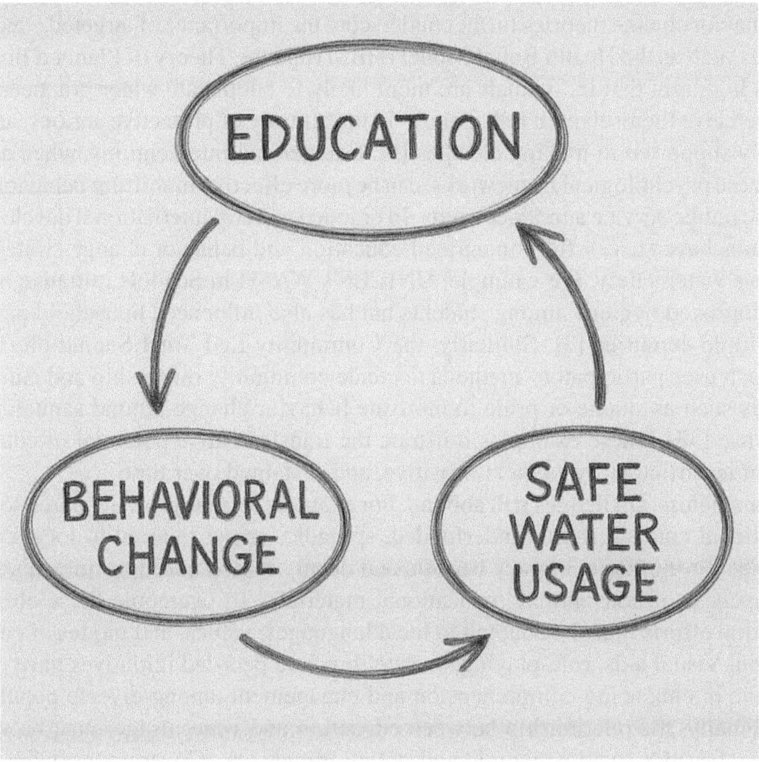

Fig. 8.1 A schematic relationship between education, behavioral change and safe water usage

The connection between behavior and waterborne illness is well-established in public health literature. Unsafe practices such as drawing water from open, unprotected sources, storing water in uncovered or unclean containers, and failing to wash hands with soap are common pathways for microbial contamination [83]. These practices are often rooted in tradition, lack of awareness, or resource limitations. Therefore, behavior change strategies must go beyond simply disseminating information they must be designed to address psychological, social, and environmental factors that influence individual and collective actions [18]. Education, both formal and informal, serves as a critical lever for initiating and sustaining behavior change related to water use. In school settings, water, sanitation, and hygiene (WASH) education can equip children with lifelong habits and also create ripple effects as students transfer knowledge to their families and peers. At the community level, adult education programs, workshops, and awareness campaigns can correct misinformation and challenge harmful practices. For example, community health workers can demonstrate the importance of boiling or treating drinking water and reinforce positive behaviors through regular follow-ups [33].

Behavior change theories further underscore the importance of targeted education. Models such as the Health Belief Model (HBM) and the Theory of Planned Behavior (TPB) highlight that individuals are more likely to adopt safe water practices when they perceive themselves at risk, believe in the efficacy of protective actions, and feel socially supported in making changes [5]. Educational interventions, when aligned with these psychological frameworks, can be more effective in shifting behaviors than generic public service announcements. In practice, several international development programs have successfully integrated education and behavior change strategies to improve water safety. For example, UNICEF's WASH in Schools initiative has not only improved hygiene among students but has also influenced household practices in multiple countries [3]. Similarly, the Community-Led Total Sanitation (CLTS) approach uses participatory methods to create community ownership and emotional triggers such as shame or pride to motivate behavior change around sanitation and water use [38]. These examples illustrate the transformative potential of education when it is participatory, context-sensitive, and sustained over time.

Nonetheless, challenges still abound. For example, in many low-resource settings, educational campaigns are underfunded, sporadic, or not tailored to local cultural contexts. Furthermore, literacy barriers and deeply rooted traditions may hinder the effectiveness of conventional educational materials. To overcome these obstacles, education efforts must be adapted to local languages, values, and modes of communication. Visual aids, role-playing, storytelling, and peer-led initiatives have proven effective in enhancing comprehension and engagement among diverse populations [12]. Finally, the relationship between education and water usage must be viewed within a broader socio-economic and policy framework. Government policies that support curriculum integration, teacher training, and community outreach are crucial in institutionalizing safe water education. Likewise, collaborations between schools, health agencies, and non-governmental organizations (NGOs) can amplify efforts and create sustainable behavior change across generations.

8.2 Education as a Catalyst for Behavioral Change

Education acts as a catalyst for behavioral change by equipping individuals and communities with the knowledge, skills, and motivation to adopt safe water practices. Through formal curricula, community campaigns, digital platforms, and targeted training, education fosters awareness and habit transformation. Behavioral science principles enhance these efforts by addressing underlying motivations and norms. The components and strategies are outlined in Table 8.1, illustrating education's multifaceted role in water safety promotion.

Table 8.1 Multilevel Approaches to Water Safety Education and Behavior Change

Component	Summary	Key strategies/Tools	Example (s)
Formal curricular education	Embedding water safety content in school syllabi to instill early awareness and responsible practices	School subjects (e.g., Science, Geography), practical training, exams, classroom campaigns	"Swim and Survive" program in Australia; WASH in Schools in Sub-Saharan Africa
Community-based campaigns	Utilizing grassroots approaches to raise awareness and build collective responsibility at the community level	Public meetings, posters, drama, local facilitators, traditional leaders	CLTS campaigns in Nigeria; Uganda's Safe Water for All program
Mass media and digital platforms	Spreading water safety messages to wide audiences through modern and traditional communication channels	Radio, TV, social media, SMS alerts, apps, podcasts	Bangladesh arsenic awareness via community radio; CDC "Clean Water" YouTube campaign
Emergency-oriented education	Educating populations quickly during floods, droughts, or disease outbreaks to encourage safe and adaptive behaviors	Flyers, megaphone alerts, mobile warnings, hygiene kits, visual infographics	Haiti cholera outbreak education; Pakistan floods campaign
Stakeholder training programs	Building capacity of teachers, health workers, water agents, and leaders to disseminate water safety education	Workshops, manuals, toolkits, hands-on demonstrations	WSP-trained community health volunteers in Kenya
Behavioral science integration	Using psychology-based methods to shape habits, norms, and emotional responses toward safer water practices	Nudges, emotional appeals, framing techniques, peer influence, social proof	Handwashing with emotional messaging in Ghana; commitment posters in Indonesia

8.2.1 Water Safety Education in Formal Curricula

Integrating water safety education into formal school curricula is pivotal for establishing foundational knowledge about water-related hazards and prevention strategies. Schools provide a structured environment where children can be systematically taught about safe water practices, such as avoiding contaminated sources, understanding flood risks, and recognizing signs of waterborne diseases [7]. For instance, Australia has implemented comprehensive school-based programs like "Swim and Survive," which educate students on swimming proficiency and emergency responses to drowning threats [19]. In Sub-Saharan Africa, the WHO-supported WASH in Schools initiative promotes hygiene and safe water use through school programs, often supported with infrastructure like handwashing stations and clean latrines. When integrated into subjects like biology, geography, or civic education, water safety becomes more than an abstract concern—it is embedded into the lived experience of learners. This education not only increases knowledge but also encourages students to act as safety ambassadors in their homes and communities [7]. Research shows that children who receive water safety education influence household behaviors, such as purifying drinking water or maintaining proper sanitation [55]. Therefore, including water safety in school curricula is both a preventive and promotive health measure, equipping the next generation with the tools to build resilient water-safe societies.

8.2.2 Community-Based Educational Campaigns

Community-based educational campaigns serve as powerful tools for instigating behavioral change, particularly in regions where formal education may not be accessible to all. These campaigns utilize grassroots methods such as town hall meetings, street theatre, radio jingles, and peer educators to disseminate water safety messages. One example is the "I See You" project implemented in peri-urban Zambia, which used visual aids and local storytelling traditions to convey the dangers of open defecation and benefits of water purification [88]. The success of such programs lies in their cultural appropriateness and participatory nature, allowing community members to co-create solutions and feel ownership over behavioral shifts. In Kenya, community-led total sanitation (CLTS) initiatives have been instrumental in changing attitudes toward hygiene and water use, resulting in measurable declines in waterborne diseases [100]. These campaigns are often more effective when they involve respected community leaders, religious heads, and traditional rulers who lend credibility to the messages. Beyond information dissemination, community education fosters social accountability and collective action, as residents begin to self-monitor behaviors like illegal water tapping or pollution. In essence, well-designed campaigns go beyond awareness to stimulate practical, lasting change in communal water practices.

8.2.3 Role of Mass Media and Digital Platforms

Mass media and digital platforms have emerged as indispensable channels for educating vast audiences on water safety. Radio, television, social media, and mobile phone applications can rapidly spread life-saving information, especially during emergencies like floods or waterborne disease outbreaks [63]. For instance, in regions with limited internet access, radio remains a critical medium. In Bangladesh, community radio has played a key role in educating rural populations about arsenic contamination and how to access safe water sources [57]. In sub-Saharan Africa, mobile phone-based platforms such as mWater and U-Report have enabled two-way communication between water safety experts and local users, allowing for real-time reporting of water issues and quick dissemination of precautionary messages [96]. The interactive nature of digital platforms also allows users to participate in quizzes, challenges, or share their water conservation tips, fostering a sense of involvement. Moreover, social media influencers and health advocates can shape public discourse, normalize safety practices, and debunk harmful myths related to water use. Thus, mass media and digital tools have become not just supplementary but central pillars in water safety education strategies.

8.2.4 Education for Behavioral Change in Emergency Contexts

In emergency settings such as floods, droughts, or disease outbreaks, educational initiatives are crucial for rapidly instilling protective behaviors. Crises often disrupt water supply chains, increase the risk of contamination, and create urgency for safe practices. For example, after the 2010 Haiti earthquake, cholera spread rapidly due to poor sanitation. Immediate public education through flyers, mobile messages, and relief workers was vital in promoting handwashing, boiling water, and proper waste disposal 16. Similarly, during the 2010 Pakistan floods, the government and NGOs collaborated to roll out emergency awareness campaigns through megaphones and mobile alerts, advising communities to avoid stagnant water and use chlorine tablets [69]. Education in such contexts focuses on actionable, simple steps that individuals can adopt instantly. The effectiveness of these interventions often hinges on the clarity and urgency of the messaging. Psychological research also indicates that during crises, people are more receptive to behavior change, particularly when messages are consistent and repeated across multiple channels [82]. Therefore, education becomes both a lifeline and a stabilizing force during emergencies, ensuring that misinformation is countered and that communities can respond with knowledge rather than fear.

8.2.5 Training Programs for Local Stakeholders

Capacity-building programs targeting local stakeholders such as teachers, health workers, community leaders, and water utility staff are essential for sustaining water safety education. These individuals serve as multipliers, extending the reach of educational messages and ensuring their local relevance. For instance, in Kenya, the Water and Sanitation Program (WSP) partnered with the county's government to train community health volunteers in water hygiene promotion techniques, leading to improved practices in many households [35]. Such training often includes both theoretical knowledge (e.g., disease transmission mechanisms) and practical skills (e.g., testing water quality, using purification kits). These stakeholders can tailor educational content to specific audiences for example, using agricultural analogies with farmers or religious texts with faith-based communities, thus increasing resonance and impact. Furthermore, training empowers communities to become self-reliant in managing water risks [30, 59, 60]. When local actors are equipped with the right knowledge and resources, they can take proactive roles in inspecting water points, resolving disputes over access, and advocating for policy improvements. This decentralized model of education ensures that water safety becomes embedded in local governance and daily life, rather than being perceived as an external imposition [74, 72].

8.2.6 Behavioral Science Approaches to Water Safety Education

Incorporating behavioral science into water safety education makes interventions more effective by aligning them with how people actually think and behave. Traditional education assumes that people will change behavior when presented with facts. However, behavioral science reveals that habits, social norms, heuristics, and emotional triggers often play a larger role. For example, studies show that people are more likely to wash their hands if signs use emotional appeals rather than purely factual statements [24]. Nudging techniques like placing soap in conspicuous spots or marking safe water containers with bright colors have been successful in low-resource settings. In Indonesia, water users were more likely to purify their water when commitment posters with their names were displayed in homes [43, 92] These insights allow for smarter education design, such as using peer influence, feedback loops, and incentives. Additionally, framing messages to align with cultural beliefs or religious teachings can significantly enhance uptake. For example, aligning water cleanliness with Islamic teachings on purity has increased safe water practices in Muslim communities [86]. Thus, integrating behavioral science doesn't just spread knowledge—it strategically reshapes habits and social dynamics to favor safer water practices.

8.3 Understanding Behavior: Theories and Risk Factors

Understanding human behavior is essential for designing effective water safety education and intervention strategies. Several theoretical models help explain why individuals adopt—or fail to adopt—safe water and hygiene practices. Among the most influential are the Health Belief Model (HBM), Theory of Planned Behavior (TPB), and Social Cognitive Theory (SCT) as indicated in Fig. 8.2.

8.3.1 Theories Related to Behavioral Change

The models as indicated in Table 8.2 are not mutually exclusive; successful water safety programs often draw from multiple theories to tailor interventions that address knowledge gaps, attitudes, and structural constraints simultaneously. For instance, a rural hygiene program may use HBM to raise risk awareness, TPB to engage social networks, and SCT to promote observable household-level changes.

The Health Belief Model posits that individuals will engage in health-promoting behavior if they believe they are susceptible to a disease, that the disease has serious consequences, that taking preventive action would be beneficial, and that barriers to action are outweighed by benefits [37]. For example, a family is more likely to treat their drinking water if they perceive that unsafe water can cause diarrheal diseases, believe that boiling water will prevent illness, and see the process as manageable. Health messages that emphasize personal risk and demonstrate simple, cost-effective

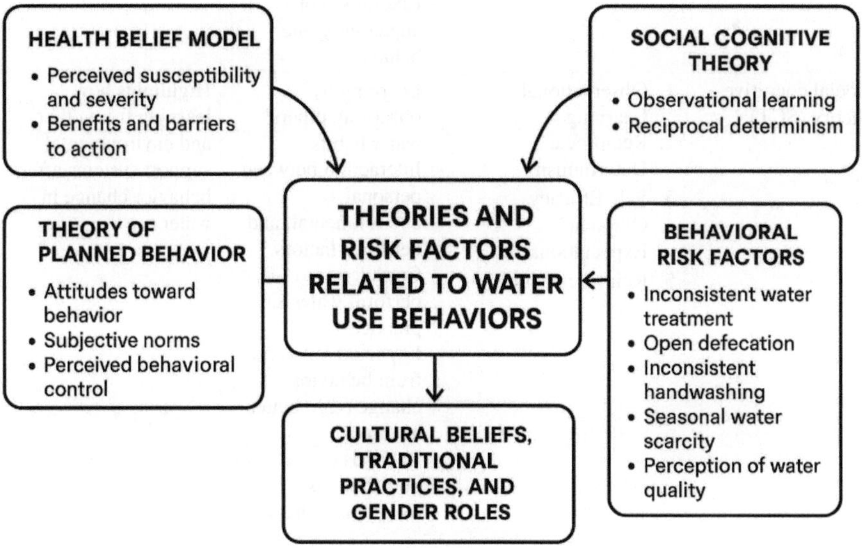

Fig. 8.2 Theories and risk factors related to water use behavior

Table 8.2 Behavioral change theories and their core components

Theory	Core components	Description of components	Relevance to water use behavior
Health belief model (HBM)	1. Perceived Susceptibility 2. Perceived Severity 3. Perceived Benefits 4. Perceived Barriers 5. Cues to Action 6. Self-Efficacy	– Belief about risk of contracting waterborne disease – Belief about seriousness of disease impact – Belief in value of behavior (e.g., boiling water) – Perception of obstacles (e.g., cost, time) – Triggers like health campaigns or community illness – Confidence in ability to take preventive action	Explains how personal risk assessment and belief in the effectiveness of practices influence adoption of safe water habits
Theory of planned behavior (TPB)	1. Attitude toward the Behavior 2. Subjective Norms 3. Perceived Behavioral Control 4. Behavioral Intention	– Personal evaluation of water–safe behavior – Social pressure or approval from peers or family – Individual's perceived control over the behavior – Likelihood of engaging in the behavior	Emphasizes the role of social influence and perceived ease or difficulty of adopting water safety practices
Social cognitive theory (SCT)	1. Observational Learning 2. Reciprocal Determinism 3. Self-Efficacy 4. Outcome Expectations 5. Reinforcement	– Learning by observing others' water habits – Interaction between personal, environmental, and behavior factors – Confidence to perform water-safe practices – Expected outcomes from behavior change (e.g., better health) – Rewards or punishments following behavior	Highlights how learning from others and environmental support systems affect behavior change in water use

solutions have been shown to increase the perceived threat and benefit dimensions of the HBM, thus driving behavioral uptake [20].

The Theory of Planned Behavior (TPB) extends this by asserting that behavioral intention is shaped by attitudes, subjective norms, and perceived behavioral control. In the context of water safety, this means that individuals are more likely to install a latrine or adopt safe water storage if they believe the action is valuable, socially supported, and within their capacity to execute. TPB-based interventions often include community role models, endorsements from local leaders, and skills training to enhance self-efficacy [32].

Social Cognitive Theory (SCT), developed by Bandura, introduces the concept of observational learning, where individuals acquire behaviors by watching others [61]. SCT highlights reciprocal determinism—the dynamic interaction between personal factors, environmental influences, and behavior. In practical terms, seeing neighbors practice handwashing or hearing community champions talk about waterborne disease prevention can lead to adoption of similar practices. SCT interventions often employ drama, storytelling, peer education, and visual demonstrations to create relatable role models and reinforce social norms [13].

8.3.2 Behavioral Risk Factors Contributing to Unsafe Water Use

Despite widespread knowledge of the risks associated with contaminated water, unsafe water use persists due to a constellation of behavioral risk factors. These behaviors are often embedded in daily routines, shaped by generational habits, convenience, social norms, and misinformation. For instance, research into the determinant of households' dependence on community forests in Nigeria provides context on how communities interact with their local environment, which can indirectly influence water collection and usage habits [48].

One common risk factor is inconsistent treatment or storage of drinking water. Even when households have access to clean water, improper handling such as storing water in uncovered or unclean containers can lead to recontamination. This behavior often stems from underestimating the risk or lacking awareness of invisible contaminants like bacteria and viruses [71].

Reluctance to use latrines is another widespread issue, particularly in rural and peri-urban settings. Open defecation persists not only due to infrastructure gaps but also because of cultural preferences, beliefs about cleanliness, or the perception that toilets are for the sick or elderly. This significantly increases the risk of fecal contamination in water sources, especially during rainy seasons [28, 75].

Handwashing practices are similarly inconsistent. While knowledge of the benefits of handwashing with soap may be high, actual behavior is often driven by situational cues rather than health motivations. People may wash hands only when visibly dirty, and not necessarily before eating or after using the toilet. The absence of soap or clean water at the point of use further discourages consistent handwashing behavior [31].

Children's behaviors also play a role in spreading waterborne diseases. Young children are often unaware of hygiene rules and may engage in high-risk activities like playing in contaminated streams or drinking untreated water. Without adult supervision or reinforcement, safe habits are difficult to establish [40].

In many communities, the seasonality of water availability also contributes to risky behavior. During dry seasons, people may resort to unsafe sources such as shallow wells or surface water, even when safer alternatives are available elsewhere. This behavior is often rationalized as temporary necessity and is compounded by long walking distances or cost constraints [99]. Furthermore, the impact of land use and cover changes, such as those observed in Stubbs Creek Forest, Akwa Ibom State, and deforestation trends in the region, can alter water availability and quality, directly affecting community water practices [47, 67].

Finally, the false perception of water clarity as a proxy for safety remains a pervasive issue. Many individuals believe that clear water is safe to drink, regardless of microbial contamination. This misunderstanding often leads to the rejection of chlorinated or treated water, which may have a taste or odor but is microbiologically safer. Behavioral risk factors thus emerge from a mix of misinformation, environmental constraints, and deeply ingrained routines. Without targeted education and sustained engagement, such behaviors continue to undermine water safety efforts.

8.3.3 Cultural Beliefs, Traditional Practices, and Gender Roles Affecting Behavior

Cultural beliefs and traditions profoundly shape water use behaviors. In many societies, water has symbolic meanings associated with purity, ritual, and healing, which may conflict with scientific messages about contamination and treatment [90]. For instance, in some communities, spring water is considered sacred, believed to be blessed or medicinal, and therefore immune to pollution. This reverence can discourage any suggestion of treating the water, as it may be viewed as disrespectful or spiritually unnecessary. In other settings, traditional healers may promote specific water collection rituals or purification practices that lack empirical support but hold strong cultural authority [102]. In South-Southern Nigeria, studies on the socio-cultural importance of sacred forests and the composition of tree species within them highlight how deep-rooted cultural values influence community interaction with natural resources, including water sources, and can present unique challenges to modern water safety interventions [22, 23].

Witchcraft beliefs may also influence water behavior. In parts of Sub-Saharan Africa, diarrhea in children may be attributed to curses or evil spirits rather than poor hygiene, leading families to seek spiritual interventions instead of medical or behavioral solutions [68]. This diversion delays effective treatment and prevention, perpetuating cycles of illness.

Ritual purity practices also play a role. In South Asia, for example, women may avoid using latrines during menstruation due to beliefs about impurity, thereby increasing open defecation rates. Similarly, taboos around discussing menstruation hinder the inclusion of menstrual hygiene education in schools, despite its importance to water safety and female dignity [76].

Gender roles significantly affect who controls water collection, storage, and hygiene behaviors. Women and girls are typically the custodians of water at the household level, responsible for fetching, storing, treating, and using water for cooking, cleaning, and bathing. Yet, they are often excluded from community-level decisions about water infrastructure or maintenance. This marginalization results in infrastructure that may be poorly located, unsafe to access, or misaligned with actual needs [2]. Moreover, women's time poverty due to a disproportionate burden of domestic and caregiving tasks—limits their ability to attend educational meetings or maintain household hygiene. Engaging men in WASH education is thus critical, not only to redistribute tasks but to shift societal norms that constrain behavior change.

Caste, ethnicity, and social hierarchy further influence water access and behavior. In some Indian villages, lower-caste individuals may be barred from accessing communal water points or be forced to use inferior sources, regardless of health risks [84]. This systemic exclusion reinforces unsafe practices while undermining collective water safety goals.

Therefore, addressing cultural and gender dimensions requires culturally sensitive interventions that work with, rather than against, local belief systems. Collaborating with religious leaders, using culturally relevant metaphors, and ensuring equitable participation can help reframe water safety in terms that resonate with local values and social dynamics.

8.4 Strategies for Behavior Change and Community Engagement

Behavior change and community engagement are critical components of successful Water, Sanitation and Hygiene (WASH) programs. Simply building infrastructure (toilets, handwashing stations, etc.) does not guarantee health improvements; people must use and maintain facilities correctly. Effective strategies therefore empower communities to take ownership of sanitation and hygiene, and to change habits such as open defecation or handwashing. Approaches blend participatory methods and behavior change communication (BCC) to shift social norms and daily routines. Research and practice show that combining community-led processes with tailored communication campaigns yields the best results [11, 14]. Some of the key community-based approaches (like PHAST and CLTS) and BCC tools (media campaigns, drama, peer education, etc.).

8.4.1 Community-Based Participatory Approaches

Community-led, participatory approaches engage local people directly in analyzing problems and designing solutions. These methods build **local ownership** and use people's own knowledge and creativity. Two well-known WASH approaches are PHAST and CLTS, both relying on participatory facilitation rather than top-down instruction.

8.4.1.1 Participatory Hygiene and Sanitation Transformation (PHAST)

PHAST is a facilitator-driven, participatory process originally developed by the WHO and partners. Its goal is to help communities "improve hygiene behaviours, prevent diarrhoeal diseases, and encourage community management of water and sanitation facilities [89]. In practice, PHAST uses a series of visual exercises (picture cards, diagrams, mapping) that lead community members through discovery of health links (for example, connecting unsanitary practices to disease). In a typical PHAST workshop, villagers draw a map of their area, plot defecation sites, draw the "F-diagram" of faecal–oral contamination pathways, and calculate how many sick days open defecation causes. The community then plans actions (e.g. building latrines, improving water sources) and agrees on collective management. For example, in Vietnam and other countries, Red Cross agencies adapted PHAST to local cultures and languages; communities became motivated to build latrines and handwashing points after seeing how quickly diarrhoea spreads from unsafe sanitation [78, 81]. Therefore, PHAST increases people's awareness of sanitation issues and boosts self-confidence–communities "share experience, ideas and confidence" and plan their own improvements.

PHAST is based on community participation and empowerment. Facilitators don't lecture but guide locals to analyze their hygiene and sanitation issues such as interpreting pictures of dirty conditions and decide on improvements that fit their realities [89]. This builds ownership and long-term commitment. PHAST emphasizes the link between sanitation and health, boosts self-esteem, and empowers communities to plan environmental changes. It uses visual, interactive tools like the F-diagram (showing feces–water–food transmission), illness charts, and ranking activities to make abstract concepts concrete. These tools engage all members, including non-literate participants, by encouraging active learning and discussion.

PHAST has been applied in Asia, Africa, and Latin America. In Vietnam, the Red Cross found higher latrine adoption and hygiene gains using PHAST over traditional methods [78]. In Africa, WHO/World Bank-led trials in Cameroon, Mauritania, Mozambique, and Eritrea showed improved understanding of sanitation's health impact. In Zambia and Nepal, PHAST-style activities combined with material support have been integrated into WASH programs, resulting in stronger, longer-lasting behavior change [1].

8.4.1.2 Community-Led Total Sanitation (CLTS)

Community-Led Total Sanitation (CLTS) is another widely-used participatory approach, with a different emphasis. Originally devised in Bangladesh and India in the early 2000s, CLTS focuses specifically on ending open defecation (OD) by triggering collective behavioral change. The core idea is to awaken in the community a shared desire to stop OD and achieve "open defecation free" (ODF) status, without subsidies for latrines. CLTS does this through a "triggering" process:

i. **Community appraisal**: A facilitator leads villagers in mapping their settlement and marking defecation areas, and drawing the path of a traveler or stream. As the exercise unfolds, participants realize that fecal matter from a few people ends up on everyone's fields and food, and eventually back into their mouth—often summarized bluntly as *"we are eating each other's feces."* This emotional realization (sometimes involving humor or shock) motivates people to change. [94] explains: *"communities are triggered... through demonstrations that highlight fecal–oral pathways and people coming to the realizations they are eating each other's excreta"*. In practice, facilitators might wash an egg or salt in a small stream downstream from an open defecation site to demonstrate contamination.
ii. **Collective planning**: After triggering, the community works together to stop OD. They set a local date to become ODF (often with a public event or ceremony), and use local materials (wood, plastic sheets, clay) to build simple latrines immediately. The focus is on *behavioral change* rather than hardware. As a WASH program summary notes, CLTS shifts the focus *"from toilet construction for individual households to the creation of open defecation–free villages"* [94]. This communal commitment creates peer pressure: even households without private latrines will help neighbors build theirs, ensuring everyone participates.
iii. **No subsidies**: Unlike older approaches that gave government-funded latrines, CLTS avoids hardware subsidies that can breed dependence. Instead, it fosters a sense of pride and ownership. When all neighbors commit to ending OD, it often leads to 100% participation. CLTS facilitators then support follow-up (encouraging latrine upgrades, checking maintenance) but crucially the impetus remains local. Globally, CLTS has spread to over 50 countries. It has been integrated into national programs like Indonesia's Swachh Desa (Clean Village) campaign [79]. However, it also has limitations (for example, maintaining facilities over time), so practitioners often pair it with later support on latrine upgrades and hygiene promotion.

In addition to PHAST and CLTS, many WASH programs include community-driven elements. Community Health Clubs, often formed by women in rural villages, serve as dynamic platforms where members come together regularly to learn and collectively practice hygiene behaviors. Through participatory activities such as group dramas, songs, and practical home assignments, members engage deeply with themes like safe water use, sanitation, and nutrition [58]. These clubs not only foster knowledge but translate it into action. This include organizing community clean-up days or managing shared water points. A remarkable case in Zimbabwe showed how club members, who contributed small fees to chlorinate borehole water, were far less affected during a cholera outbreak than neighboring communities, illustrating the power of collective behavior change [21].

Similar grassroots structures include WASH committees or local development groups that manage facilities, collect maintenance fees, and enforce hygiene norms. Embedding such oversight into village governance or linking it with health workers reinforces behavioral shifts. Trained community health volunteers, often peers from the same locality, conduct household visits to promote practices like handwashing and latrine use, leveraging trust and familiarity to inspire change [10]. Larger integrated WASH programs combine sanitation with education or health services. School-based interventions often form hygiene clubs or run competitions, encouraging children to model behaviors at home. In Bangladesh, a school program with students led to improved home latrine cleanliness, though challenges like caregivers' engagement and children's feces disposal remained [91]. These outcomes emphasize the importance of adapting even well-designed programs to local contexts.

8.5 Monitoring, Evaluation, and Institutional Support

8.5.1 Tools and Indicators for Tracking Behavioral Change

Monitoring behavioral change in safe water usage requires a blend of quantitative and qualitative tools designed to capture both measurable outcomes and context-specific nuances. Indicators may include the frequency of handwashing at key moments, usage rates of water purification methods, reduction in open defecation, and community attendance at hygiene education sessions. Tools like structured observation checklists, GPS-tagged water point usage records, and self-reported behavior logs are useful. Behavioral indicators must be adapted to local realities factoring in seasonal water availability, household responsibilities, and cultural norms [34]. When used regularly, these tools help track progress, identify bottlenecks, and refine intervention strategies over time.

Moreover, technology-enhanced tools such as mobile phone surveys, remote sensing for water point tracking, and digital dashboards for real-time data visualization have increased the accuracy and responsiveness of monitoring systems [93]. Integrating community-based participatory monitoring—where households themselves contribute to reporting builds ownership and helps triangulate official data. When complemented by baseline and endline assessments, these tools not only measure change but also shed light on what drives or hinders it. Continuous capacity-building for field staff and stakeholders on data collection and analysis further strengthens these systems, ensuring that behavior change is not only documented but understood within its full social and environmental context [26].

8.5.2 Evaluation Methods: Surveys, Observations, Focus Groups

Robust evaluation frameworks help gauge the effectiveness and sustainability of water safety behavior change programs. Surveys provide quantitative insights on knowledge, attitudes, and practices (KAP), revealing trends across demographic segments [66]. Structured observations allow evaluators to verify whether behaviors like water boiling or safe storage are actually being practiced, as opposed to just claimed. Meanwhile, focus group discussions (FGDs) facilitate deeper understanding of beliefs, social pressures, and perceived barriers. These methods, when triangulated, create a well-rounded assessment where numbers reflect the scope and scale of change, and narratives capture context, motivation, and resistance, helping tailor future education and outreach efforts more effectively [62].

Additionally, randomized control trials (RCTs) and quasi-experimental designs can be used to establish causal relationships between interventions and observed behavior changes [15]. Incorporating longitudinal tracking enables evaluation teams to assess the persistence of new behaviors over time. Photovoice, storytelling, and participatory video are emerging qualitative methods that empower communities to express their lived experiences and perspectives. These methods also reveal unanticipated outcomes (positive and negative) that surveys might overlook. A key strength of mixed-methods evaluations is their ability to inform adaptive management, allowing implementers to adjust messaging, delivery platforms, and incentives based on real-world feedback [80]. Ultimately, evaluations must balance scientific rigor with cultural sensitivity, ensuring that insights translate into meaningful and actionable improvements in water safety programs.

8.5.3 Role of Government, NGOs, and Policy Frameworks

Government institutions, NGOs, and international organizations each play critical roles in promoting and sustaining behavior change for water safety. Governments provide legitimacy, scale, and enforcement through public health laws, national WASH policies, and inter-ministerial coordination. NGOs contribute grassroots knowledge, flexible program delivery, and innovation, often working in areas underserved by public systems. Policy frameworks like national sanitation strategies and SDG-aligned water safety plans create a shared roadmap for action. Multi-sector collaboration ensures efforts are harmonized and resources efficiently used. In successful cases, this collaboration leads to community-driven interventions backed by long-term policy and funding commitments, creating an enabling environment for sustained behavior change [98].

Further, regulatory institutions such as public health inspectorates and water resource commissions enforce standards and monitor compliance, bridging policy with community-level outcomes. NGOs often act as intermediaries, translating top-down directives into culturally appropriate community actions, while also feeding local insights back into policy formulation [39]. International agencies such as UNICEF, WHO, and WaterAid provide technical assistance, financial resources, and frameworks for aligning local efforts with global best practices. Multi-stakeholder forums and taskforces allow coordination across education, health, environment, and infrastructure sectors, minimizing duplication and leveraging comparative advantages. Embedding accountability mechanisms, such as citizen report cards and social audits, further enhances transparency and trust. Together, these actors ensure that behavior change is not an isolated goal but a shared, long-term commitment integrated into national development agendas.

8.5.4 Integration of Behavior Change in National Water and Sanitation Strategies

Mainstreaming behavior change into national water and sanitation strategies is essential for lasting impact. Often, infrastructural investments like wells or latrines fail when users revert to unsafe practices due to lack of awareness or motivation. Integrating behavioral science into national plans ensures that interventions are not just physical but psychosocial and educational [36]. This includes allocating dedicated budgets for behavior change communication (BCC), embedding hygiene curricula in schools, and training health workers in interpersonal communication techniques. Additionally, monitoring frameworks must include behavior-specific indicators, and governments should periodically review and revise policies to reflect new findings from behavioral research. Such integration fosters holistic, systems-based improvements in public health.

Equally important is the institutionalization of behavior change units within ministries of health or water, charged with developing messaging strategies, evaluating campaign effectiveness, and maintaining consistency across programs [41]. National development plans should align with behavior-centered design principles, ensuring that interventions address not just knowledge gaps but emotional, cultural, and environmental determinants of behavior. Integrating traditional leaders and religious influencers into national campaigns amplifies reach and cultural resonance. Moreover, donor funding mechanisms should require behavior change components in WASH-related grants, reinforcing their importance. National strategies that embed adaptive learning cycles where behavioral data continuously informs policy revisions build resilience and responsiveness. This ensures that behavior change is not a peripheral activity but a central pillar of water and sanitation policy, reinforcing public health at its foundation.

8.6 Challenges and Barriers to Water Behavioral Change

8.6.1 Barriers to Behavioral Change

Water-related behavioral change is essential for health and sustainability, yet multiple obstacles persist. This section outlines these challenges ranging from socio-economic limitations to cultural resistance and infrastructural gaps. Table 8.3 summarizes the core barriers hindering the adoption of safe water practices across diverse communities and institutional settings.

Table 8.3 Key challenges to sustainable water behavior change

Challenge category	Focus	Implication for water behavior
Socio-economic constraints	Limited household income and financial insecurity restrict ability to afford water treatment, sanitation facilities, and hygiene materials	Intermittent or no adoption of safe water practices due to cost barriers, especially in low-income and rural communities
Cultural and traditional norms	Deep-rooted beliefs in sacred water, traditional purification methods, and taboos around sanitation or illness causation	Resistance to modern water treatment methods; reliance on traditional healers over medical advice; persistence of unsafe habits
Infrastructure gaps	Poor access to clean water sources, lack of latrines, and unreliable water supply systems	Communities revert to contaminated sources; reduced motivation to maintain hygiene when facilities are absent or broken
Institutional weakness	Inadequate coordination between government agencies, poor regulatory enforcement, and underfunded water and sanitation services	Fragmented implementation of behavior change programs; lack of sustainability and trust in public service delivery
Seasonal and environmental variability	Water scarcity during dry seasons, flooding during rains, and climate-induced disruption of supply infrastructure	Irregular water behavior; use of unsafe sources during seasonal shifts; increased contamination risks
Gender and social inequality	Women and girls bear the primary burden of water collection and face social constraints in adopting certain practices (e.g., latrine use, menstruation)	Gendered barriers limit uptake of hygiene behaviors; increased vulnerability of women to disease and physical danger
Programmatic limitations	Short-term, donor-driven interventions without follow-up or community ownership	Initial behavior change may not be sustained post-program; relapse into traditional practices once external support ends

8.6.1.1 Socio-Economic Constraints

Lower income and limited resources constrain the ability of households to adopt safe water behaviors. Fees for water treatment technologies, latrine construction, soap, and fuel for boiling water can be prohibitive, especially for daily wage earners and subsistence-level families [9]. Research on poverty incidence among rural households in Ikot Ondo Community, Nigeria, further illustrates how financial insecurity directly hinders the adoption of safe water practices, even when awareness is raised [49]. Other studies reveal that even when educational programs increase awareness, practical adoption is hindered by cost barriers. For example, in low-income urban settlements, the initial investment for water purifiers such as filters or chlorination packs is often unaffordable, leading to intermittent or incomplete use. Additionally, regular expenditures on fuelwood or charcoal for boiling water may deprioritize water treatment when household budgets are tight and competing needs arise. Also, seasonal income fluctuations due to agricultural cycles or informal labor create irregular access to funds, making consistent adoption of safe practices challenging [27]. Therefore, financial insecurity compounds these issues, especially when coupled with informal water vendors or high-priced private water sources, and behavioral adoption becomes unreliable without financial stability, leading to incomplete prevention of waterborne illnesses.

8.6.1.2 Traditional Norms and Cultural Practices

Deep-rooted cultural beliefs like the perception of sacred water, traditional purification rites, or misconceptions about illness causality greatly influence water-related behaviors [6]. Many communities hold that natural springs or pond water have innate healing properties, undermining acceptance of treatment methods such as chlorination or boiling. Taboos around sanitation, menstruation, and spiritual purity can also limit the use of latrines or consistent practice of hygiene [4, 17]. For example, in many societies, norms restrict latrine use for women during menstruation, reinforcing open defecation and contamination risk. Also, healing rituals that attribute illnesses like diarrhea to spiritual forces drive reliance on traditional healers instead of safe water storage or treatment measures, thereby delaying medical intervention. Cultural deference to tradition can also stigmatize novel behavior. For instance, women, who are often the primary water custodians, may be socially discouraged from adopting practices outside communal norms. Also, peer pressure and fear of social rejection, especially where untreated modes are the norm, can override health education, with households reverting to older customs after external programs end.

8.6.1.3 Infrastructure Gaps and Systemic Barriers

Infrastructure deficits such as distant water sources, inconsistent supply, and broken water points limit both access and motivation to practice safe behaviors. This could cause households to resort to unsafe sources during dry seasons or pump failures, nullifying earlier water treatment efforts [54]. Also, lack of reliable sanitation systems can inhibit community-level behavior change. Where piped water and sewage systems are unavailable, constructing and maintaining latrines becomes costly and impractical. Studies in developing regions consistently show partial adoption of recommended behaviors when infrastructure does not meet reliability and accessibility benchmarks [29]. Furthermore, institutional inadequacies like disconnected agencies, absent enforcement of regulations, and poor public service delivery also pose major obstacles. Without coordinated oversight or structural investment, households lack confidence in sustained behavior change, leading to program fatigue and abandonment of safe water habits.

8.6.2 Best Practices for Sustainable Water Behavior Change

To overcome these barriers, evidence demonstrates that successful behavior-change programs blend contextual analysis, social norms, environmental modification, and continuous support. Key best practices include.

8.6.2.1 Subsidies and Incentives for Low-Resource Communities

Affordable or free distribution of water treatment and hygiene products alongside microfinance options can provide equitable access to behaviors that improve water safety [97]. For instance, pregnancy and drug treatment linked provision of chlorine tablets saw dramatically higher uptake, as removing cost barriers enabled sustained use [70]. Also, incentivizing communities through conditional cash transfers, rewards for open defecation free certification, or communal contests (e.g., best-kept latrine) can foster collective pride and responsibility. Furthermore, recognition, such as public acknowledgment or symbolic incentives, drives competition and cohesion among participants.

8.6.2.2 Participatory and Culturally Sensitive Program Design

Programs grounded in local realities including cost limitations, cultural understanding, and gender dynamics show significantly better outcomes. For example, the use of participatory design methods like CLTS leverage community-led reflection and peer pressure can be used to achieve open-defecation-free status without material subsidies [65]. In addition, environmental modeling such as setting feedback loops and visible prompts in the form of painted handwashing stations or painted water jars reinforces the link between education and behavior [29]. Also, interactive communication strategies such as community theater, role plays and local-language messaging can promote community agency and translate concepts into local idioms.

8.6.2.3 Social Norms and Nudges

Behavioral science offers powerful social tools comparative feedback, peer comparison, and small environmental design changes to shift behavior. For example the Cape Town's "Day Zero" crisis demonstrated how water usage dropped significantly when residents could trace their usage against neighborhood averages and received public recognition for conservation efforts [87]. Moreover, everyday nudges like default provision of small water containers or visible trackers can help to reinforce small habits (e.g., regular water storage treatment). Implementing physical cues such as distinct color-coded chlorine caps or labeled water stations can also embed supportive routines and encourage accountability.

8.6.2.4 Multi-Channel and Long-Term Engagement

Single, isolated interventions rarely create enduring change, as such there is need for effective behavior change which requires sustained engagement through multiple channels such as schools, community health workers, media, religious networks, and service providers [95]. Furthermore, embedding structured behavior-change communication (BCC/SBCC) into existing national education and health programs especially maternal and child health visits and school curricula would help in sustaining dialogue and motivation. Also, reinforcement through periodic campaigns, seasonal reminders, radio dramas, SMS messages, or community celebrations would help to prevent reversion to unsafe practices.

8.6.2.5 Institutional Integration and Capacity Building

Embedding behavior change efforts in governance systems will ensure scale and sustainability. For instance, governments can update WASH codes to mandate water treatment in schools or workplaces and provide dedicated funding through water tariff models or public health budgets [64]. Also, building capacity among local government and utilities for behavior monitoring and social mobilization is equally important. Training local leaders, teachers, and water committees to interpret data on handwashing rates or facility usage empowers communities in maintaining healthy behaviors independently. Furthermore, strengthening multi-stakeholder forums including local councils, NGOs, and technical experts will facilitate resource-sharing and coherence across sectors, thereby reducing duplication and ensuring consistent messaging and capacity.

8.7 Conclusion

The enduring challenge of unsafe water use stems not only from inadequate infrastructure but also from persistent behavioral, socio-cultural, and systemic barriers. As demonstrated throughout this chapter, behavior change is not simply a matter of awareness, it is a dynamic, context-dependent process shaped by perceptions, norms, gender roles, and accessibility. Formal education, mass media, community-led initiatives, and emergency-oriented outreach have each shown measurable success in improving water-related behaviors when aligned with local realities and supported by participatory engagement. Importantly, the application of behavioral science, particularly models such as the Health Belief Model, Theory of Planned Behavior, and Social Cognitive Theory provides a robust theoretical framework for designing and evaluating interventions. Strategies like PHAST and CLTS showcase how communities can lead transformation when empowered through culturally resonant tools. Similarly, integrating safe water messaging into school curricula and health services fosters habit formation across generations. However, without institutional support, monitoring, and long-term investment, such gains may be temporary. Governments, NGOs, and international partners must embed behavior change into national WASH strategies, backed by adequate funding and policy enforcement. This includes capacity building for local actors, regular evaluations, and alignment across sectors. Safe water usage cannot be sustained by infrastructure alone, it requires education-driven behavior change institutionalized through policy and practice. Ultimately, a people-centered, system-oriented approach is essential for realizing the full health and development benefits of safe water access.

References

1. Abadzi H (1994) What we know about acquisition of adult literacy: is there hope? World Bank Publications, vol 245
2. Adeyeye K, Gibberd J, Chakwizira J (2020) Water marginality in rural and peri-urban communities. J Clean Prod 273:122594
3. Ahmed J, Wong LP, Chua YP, Hydrie MZI, Channa N (2022) Drinking water, sanitation, and hygiene (WASH) situation in primary schools of Pakistan: the impact of WASH-related interventions and policy on children school performance. Environ Sci Pollut Res 29:1259–1277
4. Akpabio EM, Takara K (2014) Understanding and confronting cultural complexities characterizing water, sanitation and hygiene in Sub-Saharan Africa. Water International 39(7):921–932
5. Aliabadi V, Gholamrezai S, Ataei P (2020) Rural people's intention to adopt sustainable water management by rainwater harvesting practices: application of TPB and HBM models. Water Supply 20(5):1847–1861
6. Anisfeld SC (2024) Water management: prioritizing justice and sustainability. Island Press
7. Anthonj C, Githinji S, Höser C, Stein A, Blanford J, Grossi V (2021) Kenyan school book knowledge for water, sanitation, hygiene and health education interventions: disconnect, integration or opportunities? Int J Hyg Environ Health 235:113756

8. Anthonj C, Setty KE, Ferrero G, Yaya AMA, Poague KIHM, Marsh AJ, Augustijn EW (2022) Do health risk perceptions motivate water-and health-related behaviour? a systematic literature review. Sci Total Environ 819:152902
9. Asare OMS (2011) How poverty is a fortress without drawbridges and why we must build bridges. Georgetown University
10. Aseyo RE, Mumma J, Scott K, Nelima D, Davis E, Baker KK, Cumming O, Dreibelbis R (2018) Realities and experiences of community health volunteers as agents for behaviour change: evidence from an informal urban settlement in Kisumu, Kenya. Hum Resour Health 16:1–12
11. Awio G, Lawrence S, Northcott D (2007) Community-led initiatives: reforms for better accountability? J Account Organ Chang 3(3):209–226
12. Bakhov I, Terebushko Y, Osaulchyk O, Ryhina O, Vedenieiev V (2024) Cultural challenges in education: strategies for consideration of various intercultural aspects in the educational process. Multidiscip Sci J 6
13. Bandura A (1986) Social foundations of thought and action. Englewood Cliffs, NJ 1986(23–28):2
14. Baptiste S, Manouan A, Garcia P, Etya'ale H, Swan T, Jallow W (2020) Community-led monitoring: when community data drives implementation strategies. Current HIV/AIDS Reports 17:415–421
15. Barnes BR (2019) Quasi-experimental designs in applied behavioural health research. Res Methods Soc Sci 84
16. Beydoun M (2018) Cholera in post-earthquake Haiti: how an outbreak became an epidemic. Doctoral dissertation
17. Bista CB (2023) Socio-cultural practice: a case study on personal hygienic behavior in the Hindu society. Patan Pragya 12(01):1–12
18. Blackstock KL, Ingram J, Burton R, Brown KM, Slee B (2010) Understanding and influencing behaviour change by farmers to improve water quality. Sci Total Environ 408(23):5631–5638
19. Borgonovi F, Seitz H, Vogel I (2022) Swimming skills around the world: Evidence on inequalities in life skills across and within countries. OECD social, employment, and migration working papers, (281):0_1–58
20. Carey RN, Sarma KM (2016) Threat appeals in health communication: messages that elicit fear and enhance perceived efficacy positively impact on young male drivers. BMC Public Health 16:1–16
21. Chiniko BB (2019) Water quality monitoring and vulnerability of the community to Cholera: a case study of Chitungwiza, Zimbabwe Urban Area (Master's thesis, PAUWES)
22. Daniel KS, Jacob DE, Udeagha AU (2015) Tree species composition in selected sacred forests in Nigeria. Intl J Mol Ecol and Conserv 5(7):1–10
23. Daniel KSY, Udeagha AG, Umazi Jacob DE (2016) Socio-cultural importance of sacred forests conservation in South Southern Nigeria. Afr J Sustain Dev 6(2):151–162
24. Davies JF, Miles RE, Haddrell AE, Reid JP (2013) Influence of organic films on the evaporation and condensation of water in aerosol. Proc Nat Acad Sci 110(22):8807–8812
25. Davis OL, Fante RM, Jacobi LL (2013) The effectiveness of sign prompts to increase hand washing behaviors in restrooms. N Amn J Psychol 15(3)
26. DeCorby-Watson K, Mensah G, Bergeron K, Abdi S, Rempel B, Manson H (2018) Effectiveness of capacity building interventions relevant to public health practice: a systematic review. BMC Public Health 18:1–15
27. Devereux S, Sabates-Wheeler R, Longhurst R (eds) (2012) Seasonality, rural livelihoods and development. New York, NY, USA:: Earthscan
28. Dongzagla A, Jewitt S, O'Hara S (2021) Seasonality in faecal contamination of drinking water sources in the Jirapa and Kassena-Nankana municipalities of Ghana. Sci Total Environ 752:141846
29. Dreibelbis R, Winch PJ, Leontsini E, Hulland KR, Ram PK, Unicomb L, Luby SP (2013) The integrated behavioural model for water, sanitation, and hygiene: a systematic review of behavioural models and a framework for designing and evaluating behaviour change interventions in infrastructure-restricted settings. BMC Public Health 13:1–13

30. Durley JL, de Loë RC (2005) Empowering communities to carry out drought contingency planning. Water Policy 7(6):551–567
31. Ezezika O, Heng J, Fatima K, Mohamed A, Barrett K (2023) What are the barriers and facilitators to community handwashing with water and soap? a systematic review. PLOS Global Public Health 3(4):e0001720
32. Fahes F, Dib I, El Haidari R, Nourdine F, Baydoun K, Mansour S, Hoballah A, Fakih M (2025) Chemical, physical and microbiological analyses of different drinking water sources among diverse governorates in Lebanon. Sci Rep 15(1):10539. https://doi.org/10.1038/s41598-025-89048-3. PMID: 40148378; PMCID: PMC11950160
33. Figueroa ME, Kincaid DL (2010) Social, cultural and behavioral correlates of household water treatment and storage. Center Publication HCI, 1
34. Froehlich JE (2011) Sensing and feedback of everyday activities to promote environmental behaviors. University of Washington
35. Githu IW (2022) Where community management works: the evolution and professional management of piped water supplies in rural Kenya (Doctoral dissertation, Johns Hopkins University)
36. Glanz K, Bishop DB (2010) The role of behavioral science theory in development and implementation of public health interventions. Annu Rev Public Health 31(1):399–418
37. Glanz K, Rimer BK, Viswanath K (eds) (2015) Health behavior: theory, research, and practice. John Wiley & Sons
38. Harter M (2018) Understanding mechanisms and effectiveness of community-led total sanitation (CLTS) in promoting the use of safe sanitation services (Doctoral dissertation, University of Zurich)
39. Hulme D (2013) Making a Difference: NGO's and Development in a Changing World. Routledge
40. Huynh HT, Demeter NE, Burke RV, Upperman JS (2017) The role of adult perceptions and supervision behavior in preventing child injury. J Community Health 42:649–655
41. Imas LGM, Rist RC (2009) The road to results. The World Bank 59–74
42. Islam SDU, Mondal PK, Ojong N, Bodrud-Doza M, Siddique MAB, Hossain M, Mamun MA (2021) Water, sanitation, hygiene and waste disposal practices as COVID-19 response strategy: insights from Bangladesh. Environ Dev Sustain 23(8):11953–11974
43. Izah SC, Ogwu MC (2025) Modeling solutions for microbial water contamination in the global south for public health protection. Front Microbiol 16:1504829. https://doi.org/10.3389/fmicb.2025.1504829
44. Izah SC, Jacob DE, Nelson IU, Avez S (2024a) Urban water crisis in the global South. In: Izah SC, Ogwu MC, Loukas A, Hamidifar H (eds) Water crises and sustainable management in the global South. Springer, Singapore. https://doi.org/10.1007/978-981-97-4966-9_2
45. Izah SC, Jacob DE, Nelson IU, Daniel KS, Ogwu MC (2024b) Exploring the influence of protected areas on water crises in the global South: a balancing act. In: Izah SC, Ogwu MC, Loukas A, Hamidifar H (eds) Water crises and sustainable management in the global South. Springer, Singapore. https://doi.org/10.1007/978-981-97-4966-9_20
46. Jacob DE, Nelson IU, Dickson JO (2022) Livelihood and resource interaction of households in Itu Wetland. Nigeria Am J Environ Sci 18(5):105–115
47. Jacob DE, Nelson IU, Okoh KS, Tom AA (2015) Trend in land use/cover changes in Stubbs Creek Forest, Akwa Ibom State, Nigeria. Int J Res Bus, Manag Account 1(1):123–130
48. Jacob DE, Ottong JL, Nelson IU (2018) Determinant of households' dependence on Ikot Ondo community forest. Nigeria J Biol Ser 1(3):062–071
49. Jacob DE, Udeagha AU, Nelson IU (2016) Poverty Incidence among Rural Households in Ikot Ondo Community, Nigeria. J Stud Manag Plan 2(5):80–88
50. Jacob DE, Ityavy J, Nelson IU (2020) Impact national parks on livelihood and conservation behaviours of households in Nigeria. J For, Environ Sustain Dev 6(1):72–85
51. Jacob DE, Nelson IU, Eniang EA, Izah SC (2024c) Advancing water security and resilience in the global south through recreational development. In: Izah SC, Ogwu MC, Loukas A, Hamidifar H (eds) Water crises and sustainable management in the global South. Springer, Singapore. https://doi.org/10.1007/978-981-97-4966-9_17

52. Jacob DE, Nelson IU, Izah SC (2024b) Indigenous water management strategies in the global South. In: Izah SC, Ogwu MC, Loukas A, Hamidifar H (eds) Water crises and sustainable management in the global South. Springer, Singapore. https://doi.org/10.1007/978-981-97-4966-9_16
53. Jacob DE, Nelson IU, Izah SC, Daniel KS (2024a) Rural water crises in the global south: understanding the scope and impact. In: Izah SC, Ogwu MC, Loukas A, Hamidifar H (eds) Water crises and sustainable management in the global South. Springer, Singapore. https://doi.org/10.1007/978-981-97-4966-9_1
54. Jideonwo JA (2014) Ensuring sustainable water supply in Lagos, Nigeria (Doctoral dissertation, University of Pennsylvania)
55. Joshi A, Amadi C (2013) Impact of water, sanitation, and hygiene interventions on improving health outcomes among school children. J Environ Public Health 2013(1):984626
56. Karg H, Drechsel P (2011) Motivating behaviour change to reduce pathogenic risk where unsafe water is used for irrigation. Water International 36(4):476–490
57. Khan MAS, Waliullah M, Hossain M, Mandal BH, Habiba U, Siddique MNA (2016) Safe water access, motivation and community participation at arsenic affected rural areas of Jessore, Bangladesh: insights and realism. J Sci, Technol Environ Inform 4(01):260–269
58. Khatri BK (2025) Impactful activities of user committee (A case study of Dharachaur drinking water and sanitation user committee). Cognition 7(1):14–22
59. Kosoe EA, Osumanu IK, Ogwu MC (2025b) Reimagining rural and small-town water supply in Ghana through inclusive governance and sustainable models. J Sustain 1(2):1–21. https://doi.org/10.55845/jos-2025-1234
60. Kosoe EA, Ogwu MC, Duniah J, Ndigun FB (2025a) Sustaining indigenous water governance systems in Ayimaye and Lungni, Northern Ghana, through hybrid models. J Sustain 1(2). https://doi.org/10.55845/jos-2025-1235
61. Krcmar M (2019) Social cognitive theory. In: Media effects, pp 100–114. Routledge
62. Lowe T (ed) (2023) Advancing student engagement in higher education: reflection, critique and challenge. Taylor & Francis
63. Marsh I, Conley E, Coleman A, Brataas K, Stoneking D, Mocanu R (2024) The Media and Early Warning Systems. In: Oxford research encyclopedia of natural hazard science. Accessed 11 Jun 2025. https://oxfordre.com/naturalhazardscience/view/https://doi.org/10.1093/acrefore/9780199389407.001.0001/acrefore-9780199389407-e-490
64. McGinnis SM, McKeon T, Desai R, Ejelonu A, Laskowski S, Murphy HM (2017) A systematic review: costing and financing of water, sanitation, and hygiene (WASH) in schools. Int J Environ Res Public Health 14(4):442
65. Muchire PW (2014) The role of participatory communication in water, sanitation and hygiene (Wash): a case study of community led total sanitation (Clts) approach in Kaptembwo and Kwa Rhoda settlements in Nakuru (Doctoral dissertation)
66. Muleme J, Kankya C, Ssempebwa JC, Mazeri S, Muwonge A (2017) A framework for integrating qualitative and quantitative data in knowledge, attitude, and practice studies: a case study of pesticide usage in eastern Uganda. Front Public Health 5:318
67. Nelson IU, Eniang EA, Jacob DE (2019) Trend and determinants of deforestation in Akwa Ibom State. J Agric, Environ Resour Manag 4(1):341–347
68. Ngere SH, Akelo V, Ondeng'e K, Ridzon R, Otieno P, Nyanjom M, Omore R, Barr BAT (2022) Traditional medicine beliefs and practices among caregivers of children under five years—the child health and mortality prevention surveillance (CHAMPS), Western Kenya: a qualitative study. PLoS One 17(11):e0276735
69. Niazi JIK (2011) Comparative analysis of emergency response operations Haiti earthquake in January 2010 and Pakistan's flood in 2010 (Doctoral dissertation, Monterey, California. Naval Postgraduate School)
70. Nielsen AM, Garcia LAT, Silva KJS, Sabogal-Paz LP, Hincapié MM, Montoya LJ, Galeano L, Galdos-Balzategui A, Reygadas F, Herrera C, Golden S (2022) Chlorination for low-cost household water disinfection–a critical review and status in three Latin American countries. Int J Hyg Environ Health 244:114004

71. Nwadike BI, Falodun OI, Ogunjobi AA (2024) Bacterial and viral contaminants in drinking water: why do they really matter to us. In: Environmental pollution and public health. Elsevier, pp 3–28
72. Ogwu MC, Kosoe EA (2025) Integrating green infrastructure into sustainable agriculture to enhance soil health, biodiversity, and microclimate resilience. Sustainability 17(9):3838. https://doi.org/10.3390/su17093838
73. Ogwu MC (2025) Sustainable development in Africa: challenges, innovations, and policy pathways for the future. African Research Reports 1(2):55–69. https://doi.org/10.5897/ARR2025.0005
74. Ogwu MC, Kosoe EA (2024) Place of cultural diversity in sustainable water resource management in Ghana. In: Izah SC, Ogwu MC, Loukas A, Hamidifar H (eds) Water crises and sustainable management in the global South. Springer, Singapore. https://doi.org/10.1007/978-981-97-4966-9_14
75. Ogwu MC, Izah SC, Sawyer WE, Amabie T (2025) Environmental risk assessment of trace metal pollution: a statistical perspective. Environ Geochem Health 47:94. https://doi.org/10.1007/s10653-025-02405-z
76. Olsen B, Rodríguez M, Elliott M (2022) Deepening education impact: emerging lessons from 14 teams scaling innovations in low-and middle-income countries. Ongoing Analysis from the ROSIE Project. Center for Universal Education at The Brookings Institution
77. Olson MM, Alhelou N, Kavattur PS, Rountree L, Winkler IT (2022) The persistent power of stigma: a critical review of policy initiatives to break the menstrual silence and advance menstrual literacy. PLOS Global Public Health 2(7):e0000070
78. Ormel I (2021) A partnership approach to improving humanitarian assistance: patient and community narratives of cholera in Haiti. McGill University (Canada)
79. Pakhtigian EL (2020) Air pollution, water, and sanitation: household response to environmental risk (Doctoral dissertation, Duke University)
80. Palinkas LA, Mendon SJ, Hamilton AB (2019) Innovations in mixed methods evaluations. Annu Rev Public Health 40(1):423–442
81. Parkinson J (2009) A review of the evidence base for WASH interventions in emergency responses. Discussion document
82. Porat T, Nyrup R, Calvo RA, Paudyal P, Ford E (2020) Public health and risk communication during COVID-19—enhancing psychological needs to promote sustainable behavior change. Front Public Health 8:573397
83. Potgieter N (2008) Water storage in rural households: intervention strategies prevent waterborne diseases (Doctoral dissertation, University of Pretoria)
84. Ramakrishnan A (2021) Caste realities and the struggles of India's Adivasi and Dalit population in accessing groundwater. Groundwater law and management in India: from an elitist to an Egalitarian Paradigm, pp 31–48
85. Ranaei A, Tavakoly Sany SB, Vahedian Shahroodi M, Sabahi A, Tehrani H, Khajavi A (2025) School-based peer education intervention on physical activity in Iranian adolescent girls: an application of the theory of planned behavior. Front Public Health 13:1558210
86. Raziq MA, Khan N, Maroof MM (2024) Teachings and instructions of Islam about cleanliness and hygiene in the light of Quran and Hadiths. Ann Hum Soc Sci 5(3):515–522
87. Robins S (2019) 'Day Zero', hydraulic citizenship and the defence of the commons in Cape town: a case study of the politics of water and its infrastructures (2017–2018). J South Afr Stud 45(1):5–29
88. Sikopo N, Kataoka Y, Samuel H, Yamauchi T (2024) I See You: tackling Peri-urban water, sanitation, and hygiene by employing visualisation strategies with children and youth in Lusaka, Zambia. Trajectoria 5:1–23
89. Simpson-Hebert M (2011) CRS manager's guide to phast methodology: helping CRS country offices to use the participatory hygiene and sanitation transformation (PHAST) Methodology. Catholic Relief Services
90. Strang V (2015) Water: nature and culture. Reaktion Books

91. Sultana F, Khan SM, Rahman M, Winch PJ, Luby SP, Unicomb L (2023) Sustaining an elementary school-based hygiene intervention in Bangladesh by forming hygiene committees: a pilot study. J Water, SanitIon Hyg Dev 13(10):749–763
92. Sühring N (2021) Cultural and environmental aspects of drinking water supply and treatment: a case study using participatory action research in rural eastern Indonesia (Doctoral dissertation)
93. Tabuenca B, Moreno-Sancho JL, Arquero-Gallego J, Greller W, Hernández-Leo D (2023) Generating an environmental awareness system for learning using IoT technology. Internet of Things 22:100756
94. The Sanitation Learning Hub (2025) The Community-Led Total Sanitation Approach. Institute of Developmental Stidies, Sida. https://sanitationlearninghub.org/practical-support/the-community-led-total-sanitation-approach/
95. Uwizeye G, Mwali AK, Radeny S (2022) Engaging communities and promoting health-seeking behavior through strategic social and behavior change. IntraHealth
96. Vaitla B (2014) The landscape of big data for development: key actors and major research themes. Data2X. https://www.data2x.org/wp-content/uploads/2017/11/LandscapeOfBigDataForDevelopment_10_28.pdf
97. Wachana JK (2017) Impact of microfinance on increasing access to water and improved sanitation of households of Slum Dwellers in Kenya: a case study of Kibera Slum-nairobi County (Doctoral dissertation, University of Nairobi)
98. Wallerstein N, Oetzel JG, Sanchez-Youngman S, Boursaw B, Dickson E, Kastelic S, Koegel P, Lucero JE, Magarati M, Ortiz K, Parker M (2020) Engage for equity: a long-term study of community-based participatory research and community-engaged research practices and outcomes. Health Educ Behav 47(3):380–390
99. Wambua FM (2023) Anthropogenic activities and their influence on water safety in Tseikuru Ward, Kitui County (Doctoral dissertation, University of Nairobi)
100. Wamera EK, Okemwa P, Lodiaga M (2018) Assessing how policy and key implementers influenced gender outcomes of the community led total sanitation approach in selected counties in western Kenya. Int J Dev Sci 7(12):2893–2908
101. World Health Organization (2020) Technical brief on water, sanitation, hygiene and wastewater management to prevent infections and reduce the spread of antimicrobial resistance. World Health Organization
102. Wutich A, Budds J, Jepson W, Harris LM, Adams E, Brewis A, Cronk L, DeMyers C, Maes K, Marley T, Miller J (2018) Household water sharing: a review of water gifts, exchanges, and transfers across cultures. Wiley Interdiscip Rev Water 5(6):e1309

Chapter 9
Water Safety in Sub-Saharan Africa

Sylvester Chibueze Izah, Daniel Etim Jacob, Esther Ugo Alum, and Moses Fayiah

Abstract Water safety is a critical public health and development challenge in Sub-Saharan Africa, where millions lack reliable access to safe drinking water. This chapter examines the status, determinants, and impacts of water safety in the region, focusing on the quality and availability of water sources, contamination risks, and the associated burden of waterborne diseases. Drawing on empirical studies and regional data, it highlights key threats including microbial pathogens, chemical pollutants, inadequate infrastructure, climate variability, urbanization, and displacement. Case studies from countries such as Rwanda, Ghana, Nigeria, and the Horn of Africa illustrate both progress and persistent barriers to water safety. The chapter reviews relevant policy frameworks, such as the WHO Water Safety Plans and regional commitments to Sustainable Development Goal 6, and evaluates community-based and technological interventions aimed at improving water quality. Institutional capacity gaps, fragmented governance, and limited data systems are identified as core obstacles to sustained improvement. The chapter concludes with actionable recommendations focused on strengthening water governance, scaling up investment in infrastructure and hygiene education, and promoting multisectoral collaboration for sustainable, equitable water access.

S. C. Izah
Department of Community Medicine, Faculty of Clinical Sciences, Bayelsa Medical University, Bayelsa State, Yenagoa, Nigeria
e-mail: sylvester.izah@bmu.edu.ng

Department of Microbiology, Faculty of Science, Bayelsa Medical University, Bayelsa State, Yenagoa, Nigeria

D. E. Jacob
Department of Forestry and Wildlife, University of Uyo, Uyo, Nigeria
e-mail: danieljacob@uniuyo.edu.ng

E. U. Alum (✉)
Department of Research and Publications, Kampala International University, Kampala, Uganda
e-mail: esther.alum@kiu.ac.ug

M. Fayiah
Department of Forestry, School of Natural Resources Management, Njala University, Njala, Sierra Leone
e-mail: mfayiah@njala.edu.sl

M. C. Ogwu and S. Chibueze Izah (eds.), *Water Quality and Safety in the Global South*, Springer Water, https://doi.org/10.1007/978-3-032-10602-5_9

Keywords Water safety · Sub-Saharan Africa · Public health · Sanitation · Waterborne diseases · Water policy · Climate change · Rural water supply · Water governance · Environmental health

9.1 Introduction

Water safety in Sub-Saharan Africa is a critical issue involving the adequacy of water supply, its quality, and the accessibility of sanitation services. As defined by the World Health Organization (WHO), water safety refers to the provision of safe water that meets health, environmental, and developmental standards essential for human use and consumption. This definition includes a comprehensive approach recognizing the importance of sustainable management of water resources as well as sanitation and hygiene practices integral for ensuring public health and safety [1]. This holistic framework is particularly paramount in Sub-Saharan Africa, where challenges surrounding water access and quality are pronounced.

The importance of water safety extends beyond mere access to drinking supplies; it significantly influences health outcomes, development, and livelihoods. Unsafe drinking water is a leading contributor to waterborne diseases [2, 3, 4, 5], particularly impacting vulnerable populations such as children. It has been estimated that poor water and sanitation conditions lead to approximately 2.2 million deaths annually due to diarrheal diseases [6]. Moreover, inadequate water supply and sanitation are associated with malnutrition, contributing to significant health disparities in the region. According to the WHO, water-related issues impede progress towards multiple sustainable development goals (SDGs), particularly those focused on health and well-being, economic growth, and reducing inequalities [7]. Access to safe water enhances educational outcomes and economic productivity, thereby improving overall community well-being [8].

A regional overview reveals an urgent need to address pressing water safety concerns in Sub-Saharan Africa. The prevalence of water supply shortages and increased demand due to rapid urbanization further exacerbates the problem. As urban populations surge, the capability of water supply systems to meet demands remains strained, leading to lower coverage rates in informal settlements and rural areas [1]. Current projections suggest that Sub-Saharan Africa will continue to lag behind other regions in expanding access to improved water sources, implicating future public health and economic stability [6]. Thus, effective management strategies must be adopted to ensure equitable distribution of water resources.

Access to clean water is highly unequal across different demographics and geographical areas within the region. Urban areas tend to have greater access to improved water sources, while rural counterparts often fall behind [9]. Recent studies indicate that approximately 42% of the population in Sub-Saharan Africa lacks access to basic water supplies, leading to stark inequalities that undermine health outcomes [10]. Water quality monitoring is further impeded by inadequate infrastructures and insufficient resources, resulting in a reliance on urban piped networks. This often leaves rural areas without proper quality assurance, contributing to widespread contamination and associated health risks [11–13].

Infrastructural development for water safety is also a critical issue. Although several countries have made some progress in water provision and sanitation access, inadequacies remain pervasive, particularly in impoverished neighborhoods [1]. Investment in infrastructure is essential not only for achieving safe water access but also for advancing healthcare and education systems. For example, the lack of proper sanitation in health facilities poses direct risks to public health [14], increasing susceptibility to infectious diseases [7], especially during pandemics like COVID-19. Therefore, increased investments from national governments in water facilities and management practices are crucial for overcoming these barriers [8].

Climate change and environmental factors significantly impact water safety. Droughts, prevalent in several regions, threaten the availability of water resources, adversely affecting agricultural outputs and food security [15]. The need for sustainable practices includes the embrace of groundwater resources, which can provide reliable supplies during water shortages. Groundwater has proven effective in many Sub-Saharan countries, yet reliance on it must be managed wisely to prevent pollution and depletion [16]. Furthermore, public education regarding the benefits of water conservation and prudent usage practices is pivotal for addressing overarching water safety issues.

Furthermore, improving water safety in Sub-Saharan Africa necessitates a multifaceted approach. Engagement with local communities through education and advocacy can lead to lasting behavioral changes concerning hygiene and sanitation practices. Enhancing regulatory frameworks governing water supply and sanitation services can facilitate better monitoring of water quality, thereby instilling trust among consumers [17]. Collaborative efforts between governmental bodies, NGOs, and community organizations can foster resilience and adaptive management models, paving the way for improved water safety in the region.

This chapter focuses on examining the current state of **water safety in Sub-Saharan Africa**, highlighting the challenges associated with water quality, accessibility, and infrastructure. It explores the **public health, environmental, and socioeconomic impacts** of unsafe water, supported by real-world case studies and data. The chapter also outlines existing policies and interventions and provides **strategic recommendations** to improve water safety across the region.

9.2 Sources of Water and Safety Concerns in Sub-Saharan Africa

Water safety and its sources in Sub-Saharan Africa remain critical issues due to ongoing challenges related to access, quality, and health impacts linked to water use. This region predominantly relies on surface water, groundwater, and rainwater harvesting methods, each of which has its advantages and vulnerabilities; often, their safety is compromised by various environmental and infrastructural factors. Table 9.1 presents the overview of the sources of water in Sub-Sahara Africa.

Surface Water

Surface water sources, including rivers, lakes, and ponds, are commonly utilized in rural [19] and peri-urban areas due to their accessibility [18]. Unfortunately, they are also highly susceptible to contamination from human and animal waste, agricultural runoff, and inadequate waste management practices [4, 20–22]. As a result, the public health risks associated with unfiltered surface water are significant, contributing to outbreaks of diseases such as cholera and typhoid [18, 23]. The lack of treatment and regular quality monitoring could exacerbates these health risks, highlighting a critical need for improved sanitation and water safety measures in these communities [18, 22].

Moreover, the reliance on untreated surface water sources can lead to an increased burden on healthcare systems, particularly in low-income settings where access to medical resources is limited [18]. Some studies have emphasized the correlation between the use of unsafe drinking water and high incidences of diarrheal diseases, particularly in children under five years old, stressing the dire implications for child health in Sub-Saharan Africa [11, 18, 24]. The World Health Organization (WHO) has encouraged the implementation of Water Safety Plans (WSPs) specifically tailored

Table 9.1 Overview of sources of water in sub-Saharan Africa

Water source	Narrative description	Safety concerns	Typical users/regions
Surface Water (rivers, lakes, ponds)	Readily available in rural and peri-urban areas	Highly vulnerable to fecal, industrial, and agricultural contamination	Rural and peri-urban communities
Groundwater (boreholes, wells)	Accessed via wells and boreholes; more protected than surface water	Shallow wells prone to contamination; boreholes safer if well maintained	Rural and urban populations
Rainwater Harvesting	Collected from rooftops and stored in tanks	Microbial contamination if poorly stored or harvested	Schools, clinics, households
Municipal/Piped Water	Treated and distributed by government or utilities	Aging pipes, intermittent supply, informal connections introduce contamination risks	Urban and peri-urban settlements
Informal/Unimproved Sources	Includes unprotected springs, scooped holes, vendor water	High contamination risks; rarely monitored	Underserved and informal communities

Source Adapted from [16, 18]

to address these challenges [25]. However, effective application of such frameworks remains limited, especially in peri-urban contexts where rapid urbanization compromises water quality [26].

Groundwater

In contrast, groundwater, accessed through boreholes and wells, is often regarded as a more reliable and safer alternative, particularly during dry seasons. Properly constructed and maintained boreholes can yield clean water, significantly reducing health risks. However, shallow wells pose safety concerns as they are often vulnerable to contamination from nearby latrines and surface runoff [16]. Groundwater is sometimes mistakenly perceived as infallibly safe, however, microbiological contamination remains a pressing issue across many urban centers in Sub-Saharan Africa. This contamination is predominantly driven by inadequate sanitation structures and poorly managed waste disposal, which can leach into aquifers and groundwater sources [26, 27].

The increasing concerns surrounding over-extraction and depletion of groundwater resources are also noteworthy. Excessive reliance on this crucial resource is magnified by rapid urbanization and agricultural demands, especially as many regions face the dual pressures of climate change and population growth [28–30]. A systematic review highlights the urgent need to monitor and manage these groundwater resources more sustainably to ensure long-term safety and availability. Managed groundwater development is essential not only for water supply security but also for enhancing agricultural productivity, which is critical in a region where food insecurity is prevalent [29].

Rainwater Harvesting

Rainwater harvesting serves as an essential supplementary water source, especially in areas characterized by intermittent rainfall patterns [22]. This adaptive strategy is increasingly employed in both rural households and institutional settings, such as schools and clinics, to mitigate temporary shortages. However, the safety of harvested rainwater heavily depends on the cleanliness of storage practices and the maintenance of collection surfaces, such as rooftops [30]. Any lapses in hygiene can lead to microbial contamination, posing significant health risks to users. Thus, implementing guidelines for effective rainwater harvesting and management is critical to promote its safe use within communities [18].

Municipal and Piped Water

In urban settings, piped municipal water is more common but faces its own set of challenges [23, 31]. Although municipal and piped water systems are generally treated and presumed safe, evidence reveals significant inconsistencies in supply, pipe degradation, and intermittent flow that foster microbial regrowth and secondary contamination. A critical gap remains in the effective maintenance and expansion of such systems, especially in peri-urban settlements where informal connections bypass regulatory oversight. Comparative analyses from Lagos and Ghana show that residents often perceive sachet or vendor-supplied water as safer alternatives—despite the absence of proper quality assurance protocols—highlighting a disconnect between public trust and system performance [32, 33].

The relationship between urban growth and water access shows that as population densities increase, the pressure on urban water supply systems intensifies, often leaving marginalized communities underserved [26]. Efforts to enhance resilience in urban water systems through strategic investments and management reforms are paramount in addressing these inequities [29].

Use of Informal and Unimproved Sources

In many instances, marginalized populations resort to informal or unimproved water sources, including unprotected springs and vendor-supplied water [18]. These sources are not subject to regular quality monitoring and thus pose significant health risks. This reliance on unsafe sources reflects broader systemic failures in water provision policies and infrastructure [18]. There is an urgent need to enhance governance of water resources in Sub-Saharan Africa to ensure that even the most vulnerable populations have access to safe drinking water [34].

9.3 Water Safety Challenges in Sub-Saharan Africa

The challenges surrounding water safety in Sub-Saharan Africa are multifaceted, involving microbial and parasitic contamination, chemical pollution, insufficient infrastructure, the impacts of climate change, urbanization pressures, and the disruption caused by conflict and displacement (Table 9.2; [35, 36]). All these factors contribute to significant health risks and alter the region's water quality and availability.

Microbial and Parasitic Contamination

Waterborne diseases remain a leading cause of morbidity and mortality in Sub-Saharan Africa, primarily due to the widespread presence of pathogens such as *Escherichia coli*, *Vibrio cholerae*, and *Giardia* species resulting from poor water treatment practices. The World Health Organization highlights that unsafe drinking water contributes to 10% of global mortality among children under five years old, signaling a grave public health concern for the region [10, 18]. The high incidence of waterborne illnesses fosters a cycle of disease and poverty, as families bear the dual burden of healthcare costs and lost productivity due to illness [18]. The primary sources of microbial contamination include fecal matter from inadequate sanitation practices, exemplified by high rates of open defecation and insufficient waste disposal practices [10]. Significant reports show that approximately 2.2 million deaths in 2001 were linked to diarrheal diseases, which are exacerbated by the reliance on contaminated water sources [18].

Table 9.2 Water safety challenges in sub-saharan Africa

Challenge category	Narrative
Microbial and parasitic contamination	Pathogens like E. coli, Giardia, Vibrio cholerae common in untreated sources. Linked to high child mortality
Chemical pollution	Arsenic, fluoride, and heavy metals from geogenic sources, mining, and agriculture. Chronic exposure causes health effects
Poor infrastructure and sanitation	Limited piped systems, failing sewage, and open defecation drive contamination and reduce supply reliability
Climate change, droughts, and floods	Erratic rainfall and extreme events disrupt availability and increase contamination risks
Urbanization and population growth	Informal settlements outpace infrastructure development; water systems become overstretched and unsafe
Conflict and displacement	War and displacement damage water infrastructure; IDPs face acute shortages and disease risk

Chemical Pollution

Chemical pollutants—including arsenic, fluoride, and heavy metals such as lead and cadmium—pose underreported and poorly monitored threats across groundwater systems. While microbial contamination dominates public discourse and donor-driven interventions, chronic exposure to chemical toxins remains largely unregulated. Notably, the asymptomatic nature of chemical toxicity delays diagnosis and obscures its developmental impact on children and pregnant women. The limited number of community-level toxicological surveillance systems further undermines early detection, exposing a research and policy blind spot. For example, while arsenic contamination in Bangladesh has led to global policy reforms, similar patterns in West African aquifers receive disproportionately less scientific and regulatory attention [37]. Additionally, industrial discharges, mining, and agricultural runoff have considerably contributed to pollution, introducing hazardous compounds like lead and mercury into the water supply. The need for systematic research and monitoring of water quality is underscored by the limited scope of current testing frameworks, which often neglect non-piped water sources where contamination rates may be higher [13].

Poor Infrastructure and Sanitation Systems

Infrastructure deficits play a critical role in perpetuating water safety challenges. With limited access to piped water systems and functional sewage networks, many communities rely on unprotected wells, rivers, and lakes for their water needs. The lack of adequate maintenance of existing infrastructure frequently results in the breakdown of water supply systems, compounding the risks of contamination [25]. A systematic review underscored the alarming statistic that a mere 18% of households in Sub-Saharan Africa demonstrate evidence of effective water treatment practices, illustrating challenges in ensuring safe drinking water for wider populations [38]. This situation is exacerbated by weak governance frameworks and insufficient investment in water infrastructure, which have led to inadequate responses to escalating needs [18].

Impact of Climate Change, Droughts, and Floods

Changing climatic conditions, notably increased frequency and severity of droughts and flooding events, have pronounced impacts on water availability and quality [25]. Erratic rainfall patterns cause water scarcity in many areas, forcing communities to rely on unsafe sources, while floods often contaminate existing water supplies with sewage and chemical runoffs. The added strain of climate change signifies an urgent need to adapt water management strategies to mitigate the impacts of both floods and droughts [25, 39]. These environmental factors contribute to complex vulnerabilities, particularly for vulnerable groups.

Urbanization Pressures and Population Growth

Rapid urbanization has led to the proliferation of informal settlements where piped water infrastructure is either absent or grossly inadequate. In cities like Lagos, Kinshasa, and Nairobi, many residents depend on unsafe surface water or unregulated water vendors [33]. These communities face intersecting vulnerabilities: overcrowding, absence of legal land tenure, poor drainage, and lack of waste disposal. The result is high contamination risk from leaking latrines, open drains, and septic overflow. Studies reveal that these marginalized urban populations often pay more per liter for unsafe water than wealthier, serviced areas, amplifying water injustice [2]. Without targeted investments, slum-dwellers will remain locked out of equitable water access and suffer disproportionately during outbreaks such as cholera or hepatitis E.

Water Access During Conflict and Displacement

Moreover, armed conflicts and political instability significantly interfere with water safety and access. In areas experiencing war or insurgency, water infrastructure is often targeted, leading to diminished access for local populations. Displaced communities, such as internally displaced persons (IDPs) and refugees, frequently face acute challenges regarding water access in overcrowded camps, where the risk of disease transmission, particularly waterborne diseases, is substantially elevated [25]. Reports indicate that the integration of water safety plans in humanitarian responses is inadequately applied in these settings, worsening the public health landscape for displaced populations [25].

9.4 Public Health Impacts

Water safety is a critical public health concern in Sub-Saharan Africa, where inadequate water supply, sanitation, and hygiene (WASH) contribute to severe health complications, including waterborne diseases (Table 9.3). The subregion harbors various pathogenic organisms that thrive in contaminated water [40], leading to substantial morbidity and mortality. This section of the chapter highlights the public health impacts of water safety issues in Sub-Saharan Africa, specifically focusing on waterborne diseases caused by bacteria, viruses, and parasites, and their implications for child health, malnutrition, and health systems.

9.4.1 Public Health Impacts of Water Safety Issues

A range of waterborne diseases disproportionately affect Sub-Saharan Africa, where cholera, typhoid, and other diseases remain prevalent. For instance, Olanrewaju et al.

Table 9.3 Public health impacts of water safety issues

Category	Details
Waterborne diseases	
• *Bacterial Infections*	Cholera (*Vibrio cholerae*), Typhoid/Paratyphoid (*Salmonella spp.*), Shigellosis (*Shigella spp.*), E. coli infections, Leptospirosis (*Leptospira spp.*), Campylobacteriosis (*Campylobacter jejuni*), Legionellosis (*Legionella pneumophila*)
• *Viral Infections*	Hepatitis A & E, Norovirus, Rotavirus, Enteroviruses (e.g., poliovirus, coxsackievirus)
• *Parasitic Infections*	Giardiasis (*Giardia lamblia*), Amoebiasis (*Entamoeba histolytica*), Cryptosporidiosis (*Cryptosporidium spp.*), Schistosomiasis (*Schistosoma spp.*), Dracunculiasis (*Dracunculus medinensis*), Toxoplasmosis (*Toxoplasma gondii*), Balantidiasis (*Balantidium coli*), Isosporiasis (*Cystoisospora belli*), Fascioliasis (*Fasciola hepatica*)
Impact on child health & mortality	- High under-five mortality due to diarrheal diseases - Severe dehydration and nutrient loss - Long-term growth and cognitive deficits
Link to malnutrition & development	- Recurrent infections impair nutrient absorption - Reduced food intake due to illness - Stunting and underweight outcomes
Burden on public health systems	- Increased hospital admissions during outbreaks - Strain on water treatment and health resources - Economic burden on households and governments

(2019) highlights the historical context of flood-related health implications, identifying cholera and typhoid as significant illnesses in Nigeria caused by water contamination. Studies reveal that limited access to treated water and poor sanitation practices contribute to the continued prevalence of waterborne diseases, with Nigeria reporting a high incidence of typhoid fever [41]. Moreover, [42] highlight that despite governmental efforts, outbreaks of these diseases continue in several African countries, exacerbated by reliance on untreated surface water sources.

9.4.2 Waterborne Diseases Due to Contaminated Water Sources

Bacterial Infections

Cholera, caused by the bacterium *Vibrio cholerae*, is endemic in various parts of Sub-Saharan Africa [43, 44]. Poor sanitation, inadequate water quality, and flooding have enhanced the spread of cholera outbreaks, particularly in Nigeria and Ghana. Mina et al. reported an increase in cholera cases linked to contaminated water supplies, further complicating public health initiatives [41]. Typhoid and paratyphoid fevers, attributable to strains of *Salmonella* (e.g., *Salmonella Typhi*), pose similar threats. The presence of *Salmonella* in drinking water resources amplifies human infection rates, underscoring the critical need for safe water hygiene and management practices [41].

Additional bacterial pathogens of concern are *Shigella spp.*, which cause shigellosis, a highly transmissible infection typically linked to fecal contamination in water sources. Shigellosis can lead to severe diarrhea, emphasizing the significance of safe water and sanitation practices [45]. *E. coli* infections, particularly strains such as O157:H7, and other gastrointestinal diseases often arise from foods and water contaminated with fecal matter, again indicating the importance of water safety measures in the prevention of these infections [42]. Moreover, Leptospirosis, caused by *Leptospira interrogans*, can result from contact with water contaminated by animal urine, complicating the public health landscape in regions of Sub-Saharan Africa.

Viral Infections

Viruses also play a pivotal role in waterborne disease transmission in Sub-Saharan Africa. Hepatitis A and E virus infections are commonly associated with contaminated water, leading to liver inflammation. Noroviruses and rotaviruses, especially severe in infants and young children, often lead to outbreaks linked to contaminated recreational and drinking water. Enteroviruses, such as poliovirus, are similarly tied to unsafe water sources, illustrating the intersection of water quality and viral disease risks.

Parasitic Infections

Parasitic infections present additional public health burdens related to water safety. Schistosomiasis, transmitted through freshwater snails, serves as a key example. Over 200 million people in Sub-Saharan Africa are infected with schistosome parasites as a result of contact with contaminated water [46]. The relationship between water bodies inhabited by snails and the prevalence of schistosomiasis highlights fundamental challenges in water management and public health efforts [47].

Moreover, giardiasis, amoebiasis, and cryptosporidiosis are waterborne parasitic infections that exacerbate health conditions, especially in immunocompromised individuals, pointing toward inadequate water treatment systems [42]. The complexity surrounding dracunculiasis or Guinea worm disease, which results from consuming water contaminated with infected copepods, underscores the substantial health risks posed by water safety deficiencies [48].

9.4.3 Impact on Child Health and Mortality

The impact of waterborne diseases extends significantly to child health in Sub-Saharan Africa. Diarrheal diseases attributable to pathogens in contaminated water lead to critical dehydration and electrolyte imbalances in affected children [45]. Waterborne diseases remain one of the leading causes of under-five mortality, highlighting the urgency of addressing water safety issues. Additionally, frequent infections contribute to persistent health problems affecting physical and cognitive development in children, thus creating a cycle of malnutrition and recurrent illness [49].

9.4.4 Link to Malnutrition and Developmental Issues

The inadequacy of clean water leads to malabsorption issues due to intestinal inflammation caused by frequent infections, significantly reducing nutrient intake and retention [49]. Links between persistent diarrheal episodes and the exacerbation of malnutrition reveal the intricate relationship between water quality, food security, and health outcomes, particularly in vulnerable populations such as young children in Sub-Saharan Africa [46, 49].

9.4.5 Burden on Public Health Systems

The burden of waterborne diseases extends into public health systems, where outbreaks lead to overcrowded health facilities, increased demand for sanitation and clean water supplies, and broader economic impacts through lost productivity in the workforce (Olanrewaju et al. 2019). The financial strain placed on already limited health budgets highlights the imperative for sustainable management and infrastructure improvements targeting clean water access [46].

9.5 Socioeconomic and Environmental Impacts of Water Safety in Sub Sahara Africa

The socioeconomic and environmental dimensions of water safety in Sub-Saharan Africa are diverse and complex, significantly influencing public health outcomes and economic productivity (Table 9.4). One of the most pressing issues related to water safety is the prevalence of waterborne diseases, such as cholera, typhoid, and diarrhea, which are significant concerns in region. Poor water quality leads to frequent illnesses that drastically reduce workforce efficiency. For instance, [50] note that inadequate water supply and sanitation can result in high rates of morbidity, which increases absenteeism from work across Sub-Saharan Africa [50]. The economic implications are profound, as families suffer from lost income and increased healthcare costs, with the World Bank estimating that poor water and sanitation conditions contribute significantly to economic losses in these regions [6]. Moreover, children are particularly affected, illnesses caused by unsafe water often lead to school absenteeism, adversely impacting educational outcomes. This educational disruption can hinder future economic opportunities for these children, entrenching them in a cycle of poverty [18].

The gendered burdens of water collection further complicate the socioeconomic dynamics of water safety in Sub-Saharan Africa. Women and girls are often primarily responsible for collecting water, which frequently involves long distances to the nearest source. This labor not only consumes valuable time that could otherwise be allocated for education or income-generating activities but also subjects women and girls to physical strain and safety risks associated with the journey [51]. Menstrual hygiene insecurity, linked to inadequate water and sanitation in schools and public spaces, further compounds educational disruption and emotional distress. The inequities faced by women in the context of water access highlight entrenched gender disparities in societal roles and responsibilities. This is particularly evident in studies from regions such as Zimbabwe, where women reported facing dangerous conditions while gathering water, amplifying their vulnerabilities to violence and societal marginalization [52]. Despite these challenges, WASH policies and infrastructure often neglect gender-specific needs, perpetuating systemic inequities. Therefore, water interventions must mainstream gender-sensitive designs, promote safe access pathways, and empower women in community WASH governance to disrupt entrenched exclusionary practices.

Table 9.4 Socioeconomic and environmental dimensions of water safety in sub-saharan Africa

Dimension	Narrative
Loss of productivity due to water-related illness	- Prevalence of waterborne diseases (e.g., cholera, diarrhea) reduces labor force efficiency - Children miss school, affecting long-term educational and economic outcomes - Sick days and caregiving reduce household income
Gendered burdens of water collection	- Women and girls spend hours daily fetching water, limiting time for education or economic activities - Exposure to physical strain and safety risks, including violence - Reinforces existing gender inequalities
Economic costs to families and communities	- Healthcare expenses for waterborne diseases place financial strain on households - Time lost for water collection reduces productivity - Cost of purchasing or treating unsafe water adds to economic burdens
Ecosystem degradation due to unsafe water practices	- Poor sanitation contaminates rivers and streams, harming aquatic life - Overuse and pollution degrade freshwater ecosystems - Deforestation near water sources disrupts water cycles and reduces natural filtration

"Beyond household-level costs, the macroeconomic impact of unsafe water remains grossly underestimated in regional policy frameworks. Despite well-established links between water insecurity and national GDP losses, many Sub-Saharan African countries fail to integrate water safety investments into long-term economic planning. Current interventions seldom address these overlapping vulnerabilities, reflecting a need for inclusive WASH designs that consider safety, privacy, and dignity. The high healthcare costs incurred from treating waterborne diseases place significant financial strain on families, with many resorting to purchasing unsafe water due to lack of access to safe sources [53]. As time is lost to both water collection and tending to sick family members, community productivity diminishes. This vicious cycle reinforces systemic poverty, making investment in water infrastructure an urgent priority across Sub-Saharan Africa [6, 54]. Furthermore, ecosystem degradation exacerbates this issue, improper waste management and pollution degrade freshwater sources, leading to reduced water availability for agricultural and domestic use, thus threatening food security and livelihoods [26].

Furthermore, the environmental impact of unsafe water practices cannot be overlooked. Pollution of freshwater ecosystems threatens biodiversity and leads to the overexploitation of water resources. The degradation of natural ecosystems, driven by human activities such as deforestation and poor land management, further disrupts the natural water filtration processes that safeguard water quality [8]. As highlighted by studies in Ethiopia and Cameroon, sustainable management of water resources is crucial to mitigate these environmental risks (Tariq and Zhang 2021; [55]). Addressing the intricate web of socioeconomic factors, gender dynamics, and environmental challenges is essential for improving water safety in Sub-Saharan Africa, suggesting that coordinated efforts among policymakers, communities, and international stakeholders are vital for sustainable development in these regions [40].

9.6 Policy and Regulatory Frameworks for Water Safety

In Sub-Saharan Africa, the policy and regulatory frameworks surrounding water safety are pivotal for promoting public health. These frameworks must ensure access to safe water, addressing not only the availability of water but also its quality, infrastructure, and governance systems, which are often challenged by economic constraints and inadequate institutional capacity.

National water safety policies and legal instruments play a crucial role in safeguarding public health and ensuring the availability of safe drinking water. In Nigeria, for instance, the regulatory framework surrounding water supplies is enforced by various agencies, including the National Agency for Food and Drug Administration and Control (NAFDAC) and the Nigerian Standards Organization (NSO). These bodies, however, face challenges related to enforcement and compliance, as highlighted by findings indicating that many households rely on sachet water, often misconceived as safer than piped water, despite substantial bacteriological risks associated with it [56]. Additionally, the regulatory approach in South Africa demonstrates complexities in policy execution; the implementation of the National Water Act faces significant challenges due to misunderstandings among citizens regarding free basic service provisions, which impacts the effectiveness of water resource management.

The adoption of WHO Water Safety Plans (WSPs) offers a structured methodology to assess and mitigate risks in the provision of safe drinking water. In several Sub-Saharan African countries, implementing WSPs is critical for establishing frameworks that effectively monitor water quality and engage communities in the management processes [57]. Regional frameworks such as the African Union and ECOWAS also emphasize the need for collaborative approaches in water resource management, promoting shared water governance across borders. Countries like Ghana and Uganda have begun to align their water safety policies with these regional strategies, working towards fulfilling Sustainable Development Goal (SDG) 6, which stresses the necessity of ensuring availability and sustainable management of water and sanitation for all [58].

Achieving alignment with SDG 6 encompasses a multifaceted approach to water safety, integrating health risks, environmental sustainability, and socio-economic development [59]. Regional efforts to scale up water and sanitation programs are increasingly vital in addressing infrastructure issues in urban settings like Cotonou and Lomé, where public health vulnerabilities are pronounced. Furthermore, the implications of climate change on water resources accentuate the urgency for robust policy adaptations to safeguard water provisions and related agricultural activities, as demonstrated in Southern Africa, where changing climate patterns threaten both water and food security [60].

Climate change intensifies both water scarcity and contamination risks through erratic rainfall, rising temperatures, and frequent extreme events like floods and droughts [40]. Despite increasing evidence of climate–WASH linkages, most national frameworks in Sub-Saharan Africa have yet to mainstream climate resilience in water safety planning. Resilience-building measures include promoting nature-based solutions (e.g., wetland restoration), scaling solar-powered boreholes in arid zones, and enhancing early-warning systems through real-time water quality monitoring [61]. Countries like Ethiopia and Rwanda have begun piloting community drought dashboards and integrating climate forecasts into groundwater management. Embedding such adaptive strategies into policy can reduce disruptions, minimize displacement-related vulnerabilities, and enhance long-term water security.

9.7 Integrated Water Safety Interventions and Country-Based Case Studies in Sub-Saharan Africa

Water safety in Sub-Saharan Africa is defined by a mosaic of socio-environmental challenges and responses tailored to the unique contexts of each country. This section provides a detailed, country-based analysis of interventions, supported by empirical case studies and technological innovations that contribute to improved water quality, community health, and governance (Table 9.5).

Table 9.5 Summary of water safety interventions strategies in selected sub-saharan Africa

Country	Key intervention	Outcome/gaps	References
Rwanda	National WSP rollout in rural areas	Improved monitoring; limited financial support	[62]
Ghana	Decentralized water boards (CWSA)	High local ownership; sustainability concerns	[63]
Nigeria	Lagos WSP pilot; sachet water reliance	Urban leakage control improving; informal sachets remain unregulated	[64]
Uganda	NWSC-led risk-based rainwater harvesting	Adaptive to climate; informal settlement water quality varies	[65]
Sierra Leone	Post-Ebola community WSPs	Increased hygiene awareness; limited long-term capacity building	[66]

Nigeria

Nigeria presents both urban and rural water safety challenges. In rural settings, community-led initiatives that integrate indigenous knowledge with technical solutions have significantly improved access to potable water. These programs foster ownership and sustainability through grassroots participation, especially in underserved regions [21, 67].

In urban Nigeria, especially in Lagos, water quality issues are worsened by rapid urbanization and the expansion of informal settlements. In response, the Lagos Water Corporation has developed phased Water Safety Plans (WSPs), targeting urban treatment facilities, leakage monitoring, and public awareness campaigns. These initiatives aim to mitigate contamination and establish long-term water surveillance frameworks [62, 64]. Despite infrastructure limitations, these urban-focused programs are critical for maintaining water quality in one of Africa's largest cities.

Ghana

Ghana's flagship water management approach is its decentralized model implemented through the Community Water and Sanitation Agency (CWSA). This model empowers local water boards to oversee the functionality of boreholes and small-town supply systems, improving service delivery in rural communities. Hygiene education is emphasized as part of the model to reinforce water safety practices [63–68].

Community-led water safety initiatives in Ghana have also led to behavioral shifts, including improved handwashing practices and safe storage of drinking water, reducing disease outbreaks such as cholera and dysentery [67]. However, a comparative policy review reveals that while decentralization enhances local ownership, gaps remain in areas such as training, infrastructure maintenance, and financial sustainability. Political interference and inconsistent funding also hinder the full potential of this model [66].

Uganda

In Uganda, water safety is approached through a combination of national and community-based interventions. The National Water and Sewerage Corporation (NWSC) plays a vital role in risk assessments and mitigation, especially in informal settlements. The promotion of rainwater harvesting as a safe and alternative water source shows adaptation to climate variability and local environmental conditions [62, 65].

Community engagement is also evident through household-level chlorination and the protection of natural springs, particularly in rural districts. These interventions empower communities to participate actively in safeguarding their water resources and have been linked to improved health outcomes [25]. Uganda's multi-tiered approach blends institutional oversight with grassroots implementation, ensuring broader coverage and accountability.

Rwanda

In Rwanda, the implementation of Water Safety Plans (WSPs) in remote areas has significantly improved monitoring and water quality management. These WSPs employ structured frameworks that integrate community health workers into the surveillance and mitigation processes, ensuring early identification of risks and timely response [25, 62].

Programs focused on household chlorination and spring protection have empowered individuals and families, promoting self-governance and active community participation in maintaining water safety. The inclusion of health education in these interventions has also contributed to sustained behavioral changes at the household level.

South Africa

South Africa offers one of the most technologically advanced frameworks for water quality assurance in Sub-Saharan Africa. The Blue Drop Certification Program serves as a national audit mechanism that evaluates and scores municipal water supplies based on risk management and transparency [69].

Technological interventions, including mobile water quality monitoring tools and Geographic Information Systems (GIS), have been deployed to track water conditions in real-time. These tools are particularly useful in smaller municipalities where capacity constraints exist [70]. The integration of GIS has also enhanced the spatial analysis of water contamination, enabling targeted interventions and effective planning [71, 72].

Cameroon

Cameroon has made significant strides in integrating mobile technology and locally sourced water filtration materials into community water safety programs. These low-cost, scalable interventions have reduced bacterial contamination in drinking water sources, especially in peri-urban and rural communities (Clément et al. 2023; [73]).

Cameroon also benefits from robust public–private partnerships that mobilize funding and technical expertise to address water infrastructure deficits. NGOs play a central role in capacity-building, ensuring that local stakeholders can maintain and replicate water safety systems over time [74]. Additionally, hygiene promotion campaigns informed by cultural norms have helped drive behavior change among school-aged children and families [67].

Zimbabwe and Ethiopia

Both Zimbabwe and Ethiopia have embraced mobile applications for water quality monitoring, providing real-time updates that facilitate quick corrective action by water authorities. These innovations enable the detection of contamination risks early and improve transparency in public reporting [75, 76].

Such mobile-based systems are especially effective in remote and underserved communities, where traditional monitoring mechanisms are difficult to sustain. The use of mobile platforms combined with Geographic Information Systems enhances the spatial mapping of water quality trends and supports data-driven decision-making [71].

Zambia

Zambia's water safety strategy emphasizes community engagement through partnerships with educational institutions. These collaborations include local universities and schools conducting water quality testing and hygiene education campaigns [77]. By building scientific literacy and awareness around waterborne diseases, these initiatives empower communities to recognize and respond to threats independently. Long-term sustainability is supported through community ownership, ongoing public education, and locally driven water governance models.

Sierra Leone

In Sierra Leone, the post-Ebola recovery period prioritized WSP implementation in rural communities. These plans emphasized local council involvement and community participation to restore confidence in public water systems [78]. The interventions also recognized gender roles in water governance, with women often positioned as primary water collectors and hygiene educators. The program integrated gender-sensitive training and roles into governance frameworks, promoting inclusive and resilient water safety systems [66].

Horn of Africa (Somalia, Kenya, Ethiopia)

In the drought-prone regions of the Horn of Africa, water insecurity is exacerbated by climate variability. In response, NGOs have deployed mobile water testing kits and launched rapid-response chlorination programs, especially in refugee camps and informal settlements [79]. These efforts focus on emergency preparedness and response, with the goal of preventing waterborne disease outbreaks among vulnerable populations. The interventions are integrated with broader WASH (Water, Sanitation, and Hygiene) programs, providing both immediate relief and pathways for longer-term infrastructure development.

9.8 Implementation Barriers to Water Safety Efforts

Water safety remains a critical concern across Sub-Saharan Africa, where various implementation challenges undermine efforts to guarantee access to safe and clean drinking water. These barriers manifest in multiple forms, including limited financial and human resources, fragmented governance, inadequate technical training, a lack of reliable data, and cultural barriers that hinder effective community engagement.

Limited Financial and Human Resources

The scarcity of financial and human resources presents a significant obstacle to effective water safety management (Jacob et al., 2024b,c). In many Sub-Saharan African countries, investments in water infrastructure and maintenance are insufficient to meet the needs of rapidly growing populations. For instance, research indicates that inadequate funding leads to poorly maintained facilities, undermining efforts to deliver consistent water quality monitoring and treatment [13]. This is particularly evident in Nigeria, where the reliance on sachet water (often perceived as 'pure water') places many at risk due to the absence of stringent regulatory oversight and resource allocation necessary for consistently ensuring safety [56].

Fragmented Governance and Institutional Inefficiencies

Governance structures in Sub-Saharan African countries often exhibit fragmentation, complicating the implementation of water safety plans (WSPs) and other regulatory frameworks. In Uganda, for example, despite being one of the pioneering nations in WSP development, the execution of these plans is stymied by limited inter-agency coordination and stakeholder involvement, creating gaps in accountability and resource distribution [62]. Such institutional inefficiencies can lead to delays in actions necessary for water safety, thereby exacerbating public health risks associated with waterborne diseases [80].

Inadequate Technical Training and Capacity

The capacity for effective training in water safety management is another pressing challenge [81]. Many regions lack the trained personnel required to conduct thorough water quality assessments [13]. For instance, the training of sanitary inspectors in Uganda has shown that increased instruction can improve agreement on water safety assessments, yet the overall capacity remains limited due to insufficient training programs and resources dedicated to this sector [80]. The result is a reliance on outdated or incorrect safety protocols, which jeopardizes public health efforts.

Lack of Reliable Data and Consistent Monitoring

A critical component of effective water safety is reliable data and monitoring systems. Across Sub-Saharan Africa, water quality testing occurs infrequently, leading to a lack of data that complicates the mapping of water safety challenges and consumer health risks [13]. In Nigeria, the quality of community water supply is typically poorly monitored, with private sources lacking any formal oversight [82]. This data vacuum not only hampers the development of targeted interventions but also prevents the identification of emerging risks that could escalate into public health crises [83].

Cultural Barriers and Community Engagement Gaps

Cultural attitudes and community engagement play crucial roles in the successful implementation of water safety initiatives. In regions like Mbarara, Uganda, local perceptions of water quality are influenced by traditional beliefs and limited awareness of hygiene practices. Consequently, the perceived safety of certain water sources may lead communities to disregard necessary water treatment protocols. Furthermore, the lack of community involvement in decision-making processes compromises the effectiveness of water safety plans, often resulting in mistrust and reluctance to adhere to guidelines [25]. This underscores the need for strategies that integrate local knowledge and improve public outreach regarding water safety.

9.9 Recommendations for Enhancing Water Safety in Sub Sahara Africa

Enhancing water safety in Sub-Saharan Africa requires a multi-faceted approach that aligns with the unique socio-economic and environmental contexts of the region. The challenges stem from inadequate infrastructure, governance issues, climate change impacts, and resultant public health crises, notably cholera outbreaks. The overarching strategy can be broken down into several key recommendations supported by various studies from some countries in Sub-Saharan Africa.

First, strengthening institutional frameworks and governance is essential for improving water safety. [84] report that vulnerabilities related to cholera in Cameroon can be linked to failures in infrastructure and governance, with persistent outbreaks occurring in regions lacking adequate water sanitation systems. To counteract this, a renewed focus on regulatory frameworks governing water safety standards and sanitation services is vital. [85] emphasize the need for multisectoral interventions to control cholera, particularly in urban areas where rapid population growth leads to substandard living conditions, including inadequate access to clean water. By fostering better governance structures, countries can effectively allocate resources and enhance accountability in water management [86].

Increasing investment in water safety infrastructure is equally crucial. The World Health Organization (WHO) suggests that the availability of quality water and sanitation can significantly reduce the burden of diseases such as cholera, which are often exacerbated in areas where water sources are contaminated due to poor infrastructure. For instance, research indicates that contaminated drinking water is a primary cause of cholera epidemics, particularly in Cameroon, where cholera outbreaks have been linked to a lack of access to potable water [84]. Investment in both physical infrastructure (such as pipes, treatment plants, and latrines) and institutional capacity is needed to achieve sustainable water safety.

Furthermore, promoting education and WASH (Water, Sanitation, and Hygiene) programs is paramount in enhancing community understanding of hygiene practices. Despite the recognized benefits of handwashing, studies have shown that schools in various Sub-Saharan African countries report low availability of handwashing facilities and soap [87]. This lack of resources directly correlates with poor hygiene practices, underscoring the need for educational initiatives that inform health practices and foster active community participation. In Ethiopia, increasing access to hygiene education has shown positive trends in community health outcomes, illustrating how awareness programs can directly impact sanitation infrastructure usage [88].

A significant gap in the water safety landscape of Sub-Saharan Africa lies in the absence of integrated, cross-sectoral data systems. Current monitoring is fragmented across ministries, often paper-based, and limited to urban formal water points. This impedes real-time risk identification and resource targeting. Moreover, few longitudinal studies assess the effectiveness and sustainability of implemented WASH interventions. Multidisciplinary research—incorporating environmental microbiology, social science, and climate modeling—is urgently needed to inform equitable, resilient, and evidence-based policy decisions. Expanding regional WASH observatories and data-sharing platforms is critical for tracking progress toward SDG 6.

Fostering innovation and context-specific solutions also plays a significant role in enhancing water safety. Traditional methods may not suffice in addressing the unique challenges posed by various local environments. For instance, community-focused interventions developed with input from local populations in Ethiopia have led to more effective sanitation measures benefiting public health outcomes [89]. Additionally, leveraging technology for real-time monitoring and data collection systems could significantly enhance water safety assessments and responses to emerging threats.

Improving surveillance, data collection, and reporting systems must underpin all strategies aimed at enhancing water safety. Effective surveillance systems are critical in detecting outbreaks early and facilitating timely interventions. In regions like Northern Cameroon, where cholera outbreaks are frequent, enhanced data collection could lead to the identification of high-risk areas and vulnerable populations, allowing for targeted sanitation and hygiene interventions [84]. Moreover, systematic data can help in predicting future outbreaks and effectively allocating resources [85].

9.10 Conclusion

Water safety remains a serious challenge in Sub-Saharan Africa with major public health and environmental consequences. The region faces issues like water contamination, weak infrastructure, and limited access to clean drinking water. These problems increase the burden of waterborne diseases, especially among children and vulnerable groups. Sustainable water management strategies must be adopted across all levels. Investments in infrastructure, sanitation, and hygiene education are essential to improve water safety. Regulatory frameworks must align with international standards like the WHO Water Safety Plans. Effective governance and monitoring will help ensure long-term water security and resilience. Coordinated, multi-sectoral action is crucial to address these complex water safety challenges. Governments, NGOs, and communities must collaborate closely on solutions. Successful examples from some countries in Sub Sahara Africa show that local engagement can drive progress. Technology, risk assessment, and education are key to improving water quality and public health outcomes. A focus on climate adaptation will also help build water resilience. Community-based water safety plans should be developed and scaled up. Cross-border cooperation can strengthen regional efforts and resource sharing. Thus, access to safe water is central to health, development, and sustainability in Sub-Saharan Africa.

References

1. Macharia P, Kitaka N, Yillia P, Kreuzinger N (2021) Assessing future water demand and associated energy input with plausible scenarios for Water Service Providers (WSPs) in Sub-Saharan Africa. Energies 14(8):2169. https://doi.org/10.3390/en14082169
2. Balazs CL, Ray I (2014) The drinking water disparities framework: on the origins and persistence of inequities in exposure. Am J Public Health 104(4):603–611. https://doi.org/10.2105/AJPH.2013.301664
3. Ben-Eledo VN, Kigigha LT, Izah SC, Eledo BO (2017) Water quality assessment of Epie creek in Yenagoa metropolis, Bayelsa state. Nigeria. Arch Curr Res Int 8(2):1–24
4. Seiyaboh EI, Izah SC (2017) Bacteriological assessment of a tidal creek receiving slaughterhouse wastes in Bayelsa state, Nigeria. J Adv Biol Biotechnol 14(1):1–7
5. Seiyaboh EI, Izah SC (2017) A Review of impacts of gas flaring on vegetation and water resources in the Niger Delta region of Nigeria. Int J Econ, Energy Environ 2(4):48–55
6. Fuente D, Allaire M, Jeuland M, Whittington D (2020) Forecasts of mortality and economic losses from poor water and sanitation in sub-Saharan Africa. PLoS ONE 15(3):e0227611. https://doi.org/10.1371/journal.pone.0227611
7. Kanyangarara M, Allen S, Jiwani SS, Fuente D (2021) Access to water, sanitation and hygiene services in health facilities in sub-Saharan Africa 2013–2018: results of health facility surveys and implications for COVID-19 transmission. BMC Health Serv Res 21(1):601. https://doi.org/10.1186/s12913-021-06515-z
8. Matchawe C, Bonny P, Yandang G, Mafo HCY, Nsawir BJ (2022) Water shortages: Cause of water safety in sub-Saharan Africa. In Drought-impacts and management. IntechOpen. https://doi.org/10.5772/intechopen.103927
9. Deshpande A, Miller-Petrie MK, Lindstedt PA, Baumann MM, Johnson KB, Blacker BF, Abbastabar H, Abd-Allah F, Abdelalim A, Abdollahpour I, Abegaz KH (2020) Mapping geographical inequalities in access to drinking water and sanitation facilities in low-income and middle-income countries, 2000–17. Lancet Glob Health 8(9):e1162–e1185. https://doi.org/10.1016/s2214-109x(20)30278-3
10. Obong'o BO, Ogutu FO, Hurley SK, Okiko GM, Mahony J (2024) Exploring the microbial ecology of water in sub-saharan africa and the potential of bacteriophages in water quality monitoring and treatment to improve its safety. viruses, 16(12), 1897.https://doi.org/10.3390/v16121897
11. Alum EU, Uti DE, Agah VM, Orji OU, Ezeani NN, Ugwu OP, Bawa I, Omang WA, Itodo MO (2023) Physico-chemical and Bacteriological Analysis of Water used for Drinking and other Domestic Purposes in Amaozara Ozizza, Afikpo North, Ebonyi State, Nigeria. Niger J Biochem Mol Biol 38(1):1–8
12. Nowicki S, Koehler J, Charles KJ (2020) Including water quality monitoring in rural water services: Why safe water requires challenging the quantity versus quality dichotomy. Npj Clean Water 3(1):14. https://doi.org/10.1038/s41545-020-0062-x
13. Peletz R, Kumpel E, Bonham M, Rahman Z, Khush R (2016) To what extent is drinking water tested in sub-Saharan Africa? A comparative analysis of regulated water quality monitoring. Int J Environ Res Public Health 13(3):275. https://doi.org/10.3390/ijerph13030275
14. Izah SC, Richard G, Sawyer WE (2021) Distribution of Fungi density and diversity in a Surface water of Epie Creek in Yenagoa Metropolis, Nigeria. Arch Epidemiol Public Health 3:1–5. https://doi.org/10.15761/AEPH.1000121Volume3:1-5
15. MacAllister D, MacDonald A, Kebede S, Godfrey S, Calow R (2020) Comparative performance of rural water supplies during drought. Nat Commun, 11(1). https://doi.org/10.1038/s41467-020-14839-3
16. Lapworth DJ, Nkhuwa DCW, Okotto-Okotto J, Pedley S, Stuart ME, Tijani MN, Wright JJHJ (2017) Urban groundwater quality in sub-Saharan Africa: current status and implications for water security and public health. Hydrogeol J 25(4):1093–1116. https://doi.org/10.1007/s10040-016-1516-6

17. Nowicki S, Bukachi SA, Hoque SF, Katuva J, Musyoka MM, Sammy MM, Mwaniki M, Omia DO, Wambua F, Charles KJ (2022) Fear, efficacy, and environmental health risk reporting: complex responses to water quality test results in low-income communities. Int J Environ Res Public Health 19(1):597. https://doi.org/10.3390/ijerph19010597
18. Ntouda J, Sikodf F, Ibrahim M, Abba I (2013) Access to drinking water and health of populations in Sub-Saharan Africa. CR Biol 336(5–6):305–309. https://doi.org/10.1016/j.crvi.2013.06.001
19. Jacob DE, Nelson IU, Izah SC, Daniel KS (2024a) Rural water crises in the global south: understanding the scope and impact. In: Izah SC, Ogwu MC, Loukas A, Hamidifar H (eds) Water Cris Sustain Manag Glob South. (pp 3–44) Springer, Singapore. https://doi.org/10.1007/978-981-97-4966-9_1
20. Ogamba EN, Charles EE, Izah SC (2021) Distributions, pollution evaluation and health risk of selected heavy metal in surface water of Taylor creek, Bayelsa State. Nigeria. Toxicol Environ Health Sci 13(2):109–121. https://doi.org/10.1007/s13530-020-00076-0
21. Rahman Z, Crocker J, Chang K, Khush R, Bartram J (2011) A comparative assessment of institutional frameworks for managing drinking water quality. J Water, SanitIon Hyg Dev 1(4):242–258. https://doi.org/10.2166/washdev.2011.002
22. De Bont C, Börjeson L (2024) Policy over practice. Int J Commons 18(1):82–93. https://doi.org/10.5334/ijc.1313
23. Izah SC, Etim NG, Alaka OO, Nwachukwu BC, Joshua G, (2024a) Public health interventions for cholera control and prevention: water, sanitation, and hygiene (WASH) Perspective. Greener J Epidemiol Public Health. 12(1), pp. 1–15. https://doi.org/10.15580/gjeph.2024.1.102024145
24. Alum EU, Obeagu EI, Ugwu OPC (2024) Enhancing quality water, good sanitation, and proper hygiene is the panacea to diarrhea control and the attainment of some related sustainable development goals: A review. Medicine 103(38): p e39578. https://doi.org/10.1097/MD.0000000000039578
25. Twinomucunguzi FR, Nyenje PM, Kulabako RN, Semiyaga S, Foppen JW, Kansiime F (2020) Reducing groundwater contamination from on-site sanitation in peri-urban Sub-Saharan Africa: Reviewing transition management attributes towards implementation of water safety plans. Sustainability 12(10):4210. https://doi.org/10.3390/su12104210
26. Dos Santos S, Adams EA, Neville G, Wada Y, De Sherbinin A, Bernhardt EM, Adamo SB (2017) Urban growth and water access in sub-Saharan Africa: Progress, challenges, and emerging research directions. Sci Total Environ 607:497–508. https://doi.org/10.1016/j.scitotenv.2017.06.157
27. Numbere AO, Kika EP, Izah SC (2024) Challenges and mitigation of water resource management in rural areas in Nigeria. In: Izah SC, Ogwu MC, Loukas A, Hamidifar H (eds) Water Cris Sustain Manag Glob South. (pp. 373–390) Springer, Singapore. https://doi.org/10.1007/978-981-97-4966-9_12
28. Iyiola AO, Afolabi OA, Alimi SK, Akingba OO, Izah SC, Ogwu MC (2024) Climate change and water crisis in the global south. In Water Crises and Sustainable Management in the Global South (pp. 111–140). Singapore: Springer Nature Singapore. Springer, Singapore. https://doi.org/10.1007/978-981-97-4966-9_4
29. Foster S, Tuinhof A, Van Steenbergen F (2012) Managed groundwater development for water-supply security in Sub-Saharan Africa: investment priorities. Water SA 38(3):359–366. https://doi.org/10.4314/wsa.v38i3.1
30. Turyasingura B, Chavula P, Hirwa H, Mohammed F, Ayiga N, Bojago E, Ngabirano H (2022) A systematic review of climate change and water resources in sub-saharan africa. Int Res J Multidiscip Technovation, 5(2) (2023) 10–18 | 13. https://doi.org/10.21203/rs.3.rs-2281917/v1
31. Ogidi OI, Izah SC (2024) Water contamination by municipal solid wastes and sustainable management strategies. In Water Crises and Sustainable Management in the Global South (pp. 313–339). Singapore: Springer Nature Singapore. https://link.springer.com/chapter/https://doi.org/10.1007/978-981-97-4966-9_10
32. Ahiabor WK, Donkor ES (2025) Microbial safety of sachet water in Ghana: a systematic review. Environ Health Insights 19:11786302241307830. https://doi.org/10.1177/11786302241307830

33. Danert K, Healy A (2021) Monitoring groundwater use as a domestic water source by urban households: analysis of data from Lagos State, Nigeria and Sub-Saharan Africa with implications for policy and practice. Water 13(4):568. https://doi.org/10.3390/w13040568
34. Díaz-Alcaide S, Martínez-Santos P, Villarroya F (2017) A commune-level groundwater potential map for the republic of Mali. Water, 9(11), 839.https://doi.org/10.3390/w9110839
35. Kosoe EA, Osumanu IK, Ogwu MC (2025b) Reimagining rural and small-town water supply in Ghana through inclusive governance and sustainable models. J Sustain, 1(2), 1–21. https://doi.org/10.55845/jos-2025-1234
36. Kosoe EA, Ogwu MC, Duniah J, Ndigun FB (2025a) Sustaining indigenous water governance systems in ayimaye and lungni, northern ghana, through hybrid models. J Sustain, 1(2). https://doi.org/10.55845/jos-2025-1235
37. Alam MG, Allinson G, Stagnitti F, Tanaka A, Westbrooke M (2002) Arsenic contamination in Bangladesh groundwater: a major environmental and social disaster. Int J Environ Health Res 12(3):235–253. https://doi.org/10.1080/0960312021000000998. PMID: 12396524
38. Atalay YA, Gebeyehu NA, Gelaw KA (2024) Household water treatment practice and associated factors in sub-saharan africa: a systematic review and meta-analysis. Environ Health Insights 18:11786302241281652. https://doi.org/10.1177/11786302241281651
39. Omuna D, Obaroh IO, Alum EU, Akiyode OO, Eniru EI, Tiyo CE, Omoding J (2024) Impacts of climate change on water security in Uganda: A review. Int J Adv Multidiscip Res 11(9): 47–60. https://doi.org/10.22192/ijamr.2024.11.09.005
40. Ogwu MC (2025) Sustainable development in africa: challenges, innovations, and policy pathways for the future. (2025). African Research Reports, 1(2), 55–69. https://doi.org/10.5897/ARR2025.0005
41. Mina SA, Debnath P, Hossain AZ, Hasan MZ, Chowdhury AMA (2024) Screening and identification of multiple antibiotic-resistant genes containing Salmonella Typhi from drinking water: A severe public health concern in Bangladesh. Heliyon, 10(22). https://doi.org/10.1016/j.heliyon.2024.e40523
42. Forstinus NO, Ikechukwu NE, Emenike MP, Christiana AO (2016) Water and waterborne diseases: A review. Int J Trop Dis Health 12(4):1–14. https://doi.org/10.9734/ijtdh/2016/21895
43. Alaka OO, Etim NG, Joshua G, Udensi CI, Izah SC, Madu PC (2024) Microbial pathogenesis of vibrio cholerae and antimicrobial resistance trends. ES General. 6, 1358. https://doi.org/10.30919/esg1358
44. Etim NG, Joshua G, Izah SC, Alaka OO, Udensi CI, Etim EN (2024) Cholera vaccine development: progress, efficacy, and public health strategies. Greener J Biomed Health Sci, 7(1), 47–60, https://doi.org/10.15580/gjbhs.2024.1.102024146
45. Louw E, Olanrewaju CC, Olanrewaju OA, Chitakira M (2019). Impacts of flood disasters in Nigeria: A critical evaluation of health implications and management. Jàmbá: J Disaster Risk Stud, 11(1), 1–9. https://doi.org/10.4102/jamba.v11i1.557
46. Asare KK, Mohammed MDW, Aboagye YO, Arndts K, Ritter M (2025) Impact of climate change on schistosomiasis transmission and distribution—scoping review. Int J Environ Res Public Health 22(5):812. https://doi.org/10.3390/ijerph22050812
47. Manyangadze T, Chimbari MJ, Gebreslasie M, Ceccato P, Mukaratirwa S (2016) Modelling the spatial and seasonal distribution of suitable habitats of schistosomiasis intermediate host snails using Maxent in Ndumo area, KwaZulu-Natal Province. S Afr 9(1):572. https://doi.org/10.1186/s13071-016-1834-5
48. Stoler J, Tutu RA, Ahmed H, Frimpong LA, Bello M (2014) Sachet water quality and brand reputation in two low-income urban communities in Greater Accra, Ghana. Am J Trop Med Hyg 90(2):272. https://doi.org/10.4269/ajtmh.13-0461
49. Allan F, Ame SM, Tian-Bi YNT, Hofkin BV, Webster BL, Diakité NR, N'Goran EK, Kabole F, Khamis IS, Gouvras AN, Emery AM (2020) Snail-related contributions from the Schistosomiasis Consortium for Operational Research and Evaluation program including xenomonitoring, focal mollusciciding, biological control, and modeling. Am J Trop Med Hyg 103(1 Suppl):66. https://doi.org/10.4269/ajtmh.19-0831

50. Hutton G, Chase C (2016) The knowledge base for achieving the sustainable development goal targets on water supply, sanitation and hygiene. Int J Environ Res Public Health 13(6):536. https://doi.org/10.3390/ijerph13060536
51. Caruso BA, Mink T, Patrick M, Ogutu E, Dawkins C, Bendit O, Fatima M, Lustig I, Macler A, White J, Zamora A (2025) Women's experiences collecting and accessing water in Guatemala, Honduras, Kenya, and Zimbabwe: A mixed-methods investigation. medRxiv, pp 2025–02. https://doi.org/10.1101/2025.02.12.25322162
52. Akpabio EM, Takara K (2014) Understanding and confronting cultural complexities characterizing water, sanitation and hygiene in Sub-Saharan Africa. Water International 39(7):921–932. https://doi.org/10.1080/02508060.2015.981782
53. Silvestri G, Wittmayer JM, Schipper K, Kulabako R, Oduro-Kwarteng S, Nyenje P, Komakech H, Van Raak R (2018) Transition management for improving the sustainability of WASH services in informal settlements in Sub-Saharan Africa—An exploration. Sustainability, 10(11), p 4052. https://doi.org/10.20944/preprints201808.0546.v1
54. Shandra CL, Shandra JM, London B (2011) World bank structural adjustment, water, and sanitation: A cross-national analysis of child mortality in Sub-Saharan Africa. Organ Environ 24(2):107–129. https://doi.org/10.1177/1086026611413931
55. Esther AQ, Yahaya H, Esther AO, Aboagye-Mensah P, Blessing AJ, Ikome TE (2024) A scoping review on influence of socioeconomic status on antenatal care utilization and pregnancy outcomes in sub-Saharan Africa. MedRxiv, 2024-01. https://doi.org/10.1101/2024.01.11.24301063
56. Edema MO, Atayese AO, Bankole MO (2011) Pure water syndrome: bacteriological quality of sachet-packed drinking water sold in Nigeria. Afr J Food, Agric, Nutr Dev, 11(1). https://doi.org/10.4314/ajfand.v11i1.65885
57. Zindi B, Shava E (2022) COVID-19 and the attainment of Sustainable Development Goal 6 (clean water and sanitation) in South Africa. J Local GovMent Res Innov 3:58. https://doi.org/10.4102/jolgri.v3i0.58
58. Sharma R, Katoch OR (2022) Analysis of the targets and progress toward meeting the 2030 sustainable development goal-6 on clean water and sanitation: Evidence from India. South Asian J Soc Stud Econ 15(3):16–26. https://doi.org/10.9734/sajsse/2022/v15i330407
59. Küfeoğlu S (2022) SDG-6 clean water and sanitation. In emerging technologies: value creation for sustainable development (pp 289–304). Cham: Springer International Publishing. https://doi.org/10.1007/978-3-031-07127-0_8
60. Nhemachena C, Nhamo L, Matchaya G, Nhemachena CR, Muchara B, Karuaihe ST, Mpandeli S (2020) Climate change impacts on water and agriculture sectors in Southern Africa: Threats and opportunities for sustainable development. Water 12(10):2673. https://doi.org/10.3390/w12102673
61. Charnley GE, Kelman I, Murray KA (2022) Drought-related cholera outbreaks in Africa and the implications for climate change: a narrative review. Pathogens and global health 116(1):3–12. https://doi.org/10.1080/20477724.2021.1981716
62. Kanyesigye C, Marks SJ, Nakanjako J, Kansiime F, Ferrero G (2019) Status of water safety plan development and implementation in Uganda. Int J Environ Res Public Health 16(21):4096. https://doi.org/10.3390/ijerph16214096
63. Aigbavboa C, Addo LY, Ebekozien A, Thwala WD, Arthur-Aidoo BM (2023) Developing a framework for effective institutional management of Ghana's urban water supply. Int J Build Pathol Adapt 41(6):290–302. https://doi.org/10.1108/ijbpa-05-2023-0058
64. Tenebe IT, Babatunde EO, Eddy-Ugorji CC, Etu EEE, Ogarekpe NM, Ekeanyanwu CV, Diekola OA, Oladele OS, Ezeudu OB (2023) Bacterial contamination levels and brand perception of sachet water: a case study in some Nigerian urban neighborhoods. Water 15(9):1762. https://doi.org/10.3390/w15091762
65. Habanyama M (2024) Effects of climate change-induced flooding on onsite sanitation services: a case study of Kanyama compound in Lusaka, Zambia (Doctoral dissertation, The University of Zambia)

66. Tutu RA, Gupta S, Busingye JD (2019) Examining health literacy on cholera in an endemic community in Accra, Ghana: a cross-sectional study. Tropical medicine and health 47(1):31. https://doi.org/10.1186/s41182-019-0157-6
67. Yongsi HBN (2010) Suffering for water, suffering from water: access to drinking-water and associated health risks in Cameroon. J Health Popul Nutr 28(5):424. https://doi.org/10.3329/jhpn.v28i5.6150
68. Ogwu MC, Kosoe EA (2024) Place of cultural diversity in sustainable water resource management in ghana. In: Izah SC, Ogwu MC, Loukas A, Hamidifar H (eds) Water Cris Sustain Manag Glob South. Springer, Singapore. https://doi.org/10.1007/978-981-97-4966-9_14
69. Fisher MB, Williams AR, Jalloh MF, Saquee G, Bain RE, Bartram JK (2015) Microbiological and chemical quality of packaged sachet water and household stored drinking water in Freetown. Sierra Leone. PloS one 10(7):e0131772. https://doi.org/10.1371/journal.pone.0131772
70. Olatinwo SO, Joubert TH (2024) Water quality assessment tool for on-site water quality monitoring. IEEE Sens J 24(10):16450–16466. https://doi.org/10.1109/jsen.2024.3383887
71. An MX, Xu JP, Wang XY, Yu JB, Zhang HY, Shao F, Yang B (2014) Remote monitoring system of the rivers and lakes' water quality based on GPRS and 3G technology. Adv Mater Res 898:618–623. https://doi.org/10.4028/www.scientific.net/amr.898.618
72. Ogwu MC, Izah SC, Sawyer WE, Amabie T (2025) Environmental risk assessment of trace metal pollution: a statistical perspective. Environ Geochem Health 47, 94 (2025). https://doi.org/10.1007/s10653-025-02405-z
73. Mabvouna Biguioh R, Adogaye SBB, Nkamedjie Pete PM, Sanou Sobze M, Kemogne JB, Colizzi V (2020) Microbiological quality of water sources in the West region of Cameroon: quantitative detection of total coliforms using Micro Biological Survey method. BMC Public Health 20(1):346. https://doi.org/10.1186/s12889-020-8443-0
74. Mvongo VD, Mabou PB, Defo C, Nguema PF, Mishra AK, Ombolo A (2022) Perspective chapter: Access to rural water and sanitation services in Cameroon within the context of sustainable development goals (SDGs). In Hygiene and Health in Developing Countries-Recent Advances. IntechOpen. https://doi.org/10.5772/intechopen.108113
75. Demetillo AT, Taboada EB (2019) Real-time water quality monitoring for small aquatic area using unmanned surface vehicle. Eng, Technol & Appl Sci Res, 9(2). https://doi.org/10.48084/etasr.2661
76. Kumpel E, Peletz R, Bonham M, Fay A, Cock-Esteb A, Khush R (2015) When are mobile phones useful for water quality data collection? An analysis of data flows and ICT applications among regulated monitoring institutions in Sub-Saharan Africa. Int J Environ Res Public Health 12(9):10846–10860. https://doi.org/10.3390/ijerph120910846
77. Chakanyuka T, Siamabele B, Makungu M, Nsama P, Kalonje V (2025) Interventions by zambian government to prevent the resurgence of cholera in unplanned settlements: lessons from misisi compound in lusaka. IJAMRS, 5(2), 379–388. https://doi.org/10.62225/2583049x.2025.5.2.3851
78. Essilfie GL, Lamptey S, Baddoo RNN, Amenorpe G, Hevi W, Owusu M, Atupra F (2023) Food safety in the horticultural sector in Ghana: challenges, risk factors and interventions. Front Sustain Food Syst 7:1173677. https://doi.org/10.3389/fsufs.2023.1173677
79. Peletz R, Kisiangani J, Bonham M, Ronoh P, Delaire C, Kumpel E, Marks S, Khush R (2018) Why do water quality monitoring programs succeed or fail? A qualitative comparative analysis of regulated testing systems in sub-Saharan Africa. Int J Hyg Environ Health 221(6):907–920. https://doi.org/10.1016/j.ijheh.2018.05.010
80. King R, Okurut K, Herschan J, Lapworth DJ, Malcolm R, McKeown RM, Pond K (2020) Does training improve sanitary inspection answer agreement between inspectors? Quantitative evidence from the Mukono District. Uganda. Resources 9(10):120. https://doi.org/10.3390/resources9100120
81. Izah SC, Jacob DE, Nelson IU, Avez S (2024b) Urban water crisis in the global South. In: Izah SC, Ogwu MC, Loukas A, Hamidifar H (eds) Water crises and sustainable management in the global south. (pp. 45–83). Springer, Singapore. https://doi.org/10.1007/978-981-97-4966-9_2

82. Fatokun EN, Olalekan SM, Fayemi SO (2024) Assessment of antibiotic resistance in groundwater bacterial strains: a community-based study. South Asian J Res Microbiol 18(11):32–45. https://doi.org/10.9734/sajrm/2024/v18i11398
83. Kasozi KI, Namubiru S, Kamugisha R, Eze ED, Tayebwa DS, Ssempijja F, Okpanachi AO, Kinyi HW, Atusiimirwe JK, Suubo J, Fernandez EM (2019) Safety of drinking water from primary water sources and implications for the general public in Uganda. J Environ Public Health 2019(1):7813962. https://doi.org/10.1155/2019/7813962
84. Akoachere JFTK, Masalla TN, Njom HA (2013) Multi-drug resistant toxigenic Vibrio cholerae O1 is persistent in water sources in New Bell-Douala. Cameroon BMC Infect Dis 13(1):366. https://doi.org/10.1186/1471-2334-13-366
85. Ngomba A, Esso L, Fouda Mbarga N, Kenko I, Defo E, Mandeng N, Tonye TA, Mendjime P, Bilounga C, Choupo L, Douba E (2025) Prioritizing areas for multisectoral interventions (PAMIs) for cholera control in Cameroon. medRxiv, pp.2025–01. https://doi.org/10.1101/2025.01.03.25319978
86. Ako AA, Shimada J, Eyong GET, Fantong WY (2010) Access to potable water and sanitation in Cameroon within the context of Millennium Development Goals (MDGS). Water Sci Technol 61(5):1317–1339. https://doi.org/10.2166/wst.2010.836
87. Morgan C, Bowling M, Bartram J, Kayser GL (2017) Water, sanitation, and hygiene in schools: Status and implications of low coverage in Ethiopia, Kenya, Mozambique, Rwanda, Uganda, and Zambia. Int J Hyg Environ Health 220(6):950–959. https://doi.org/10.1016/j.ijheh.2017.03.015
88. Nsagha DS, Atashili J, Fon PN, Tanue EA, Ayima CW, Kibu OD (2015) Assessing the risk factors of cholera epidemic in the Buea Health District of Cameroon. BMC Public Health 15(1):1128. https://doi.org/10.1186/s12889-015-2485-8
89. Deresse, T. (2023). The vulnerability of communities to climate change-induced water scarcity in ethiopia and kenya: a systematic review.Preprint. https://doi.org/10.21203/rs.3.rs-3812532/v1
90. Ogwu MC, Kosoe EA (2025) Integrating green infrastructure into sustainable agriculture to enhance soil health, biodiversity, and microclimate resilience. Sustainability 17(9):3838. https://doi.org/10.3390/su17093838

Chapter 10
Latin America's Water Crisis: Pollution, Governance, and Community Responses

Matthew Chidozie Ogwu, Humberto Merritt, and Maribel Ovando Martínez

Abstract Latin America's water crisis is a multidimensional challenge shaped by environmental degradation, socio-political inequality, and institutional fragmentation. This chapter offers a comprehensive analysis of the causes, impacts, and responses to water pollution across the region. Rapid urbanization, industrial expansion, unsustainable agricultural practices, and inadequate wastewater treatment have severely compromised the quality of rivers, lakes, and aquifers. Climate change further exacerbates these challenges through intensified droughts, floods, and altered hydrological cycles, putting additional strain on already fragile water systems. The chapter highlights how weak regulatory enforcement, corruption, and uneven decentralization have created governance vacuums in which polluters often operate with impunity. Marginalized populations, particularly Indigenous and rural communities, bear the brunt of this crisis, facing disproportionate exposure to contaminated water sources and limited access to basic sanitation and healthcare. Drawing from case studies across Bolivia, Colombia, Mexico, and Brazil, the chapter explores the dynamic role of community-based organizations and grassroots movements in resisting water injustice and advancing alternative water governance models grounded in traditional ecological knowledge. Technological and policy innovations, including decentralized water treatment, nature-based solutions, transboundary cooperation, and participatory monitoring, are analyzed as potential strategies to build resilience and equity in water management. The chapter also identifies critical

M. C. Ogwu (✉)
Goodnight Family Department of Sustainable Development, Appalachian State University, Boone, NC, USA
e-mail: ogwumc@appstate.edu

H. Merritt
Centro de Investigaciones Económicas, Administrativas y Sociales (CIECAS), Instituto Politecnico National Mexico City, Mexico City, Mexico
e-mail: hmerritt@ipn.mx

M. O. Martínez
Departamento de Investigaciones Científicas y Tecnológicas, Facultad Interdisciplinaria de Ciencias Biológicas y de Salud, Universidad de Sonora, Hermosilo, Mexico
e-mail: Mexicomaribel.ovando@unison.mx

M. C. Ogwu and S. Chibueze Izah (eds.), *Water Quality and Safety in the Global South*, Springer Water, https://doi.org/10.1007/978-3-032-10602-5_10

gaps in data infrastructure, indicator harmonization, and citizen science integration that hinder effective decision-making. By situating Latin America's water crisis within a broader Global South perspective, this work contributes to interdisciplinary discourse on environmental justice, sustainable development, and the democratization of water governance, offering pathways toward more inclusive and accountable water futures.

Keywords Water governance · Environmental justice · Water pollution · Indigenous communities · Climate change · Grassroots movements · Latin America

10.1 Introduction

Water pollution remains a significant challenge across Latin America, where rapid urbanization, industrial growth, and inadequate infrastructure intensify environmental and public health risks. Many of the region's rivers, lakes, and aquifers are increasingly contaminated due to unchecked industrial discharges, agricultural runoff, and insufficient wastewater treatment systems. The World Health Organization (WHO) underscores the importance of clean water as a fundamental determinant of health. Poorly managed water sources contribute to outbreaks of waterborne diseases such as cholera, disproportionately affecting children and low-income communities across the region [49]. Access to clean water remains uneven, with marginalized populations, particularly Indigenous and rural communities, facing the greatest burden, thereby perpetuating cycles of poverty and environmental injustice [48].

Beyond health, water quality is deeply intertwined with ecological and economic systems. Freshwater ecosystems sustain biodiversity and agricultural productivity, but when polluted, these systems degrade rapidly. Deteriorating water quality diminishes crop yields, affects livestock, and weakens the resilience of rural economies that depend on clean water for survival [22]. At the heart of this water crisis are failures in governance. Many Latin American countries struggle to develop and enforce effective regulatory frameworks, leaving water resources vulnerable to degradation and exploitation [8]. The consequences are far-reaching, impacting health, livelihoods, ecosystems, and long-term development goals. Toxic pollutants in water bodies are widespread across the region. Industrial effluents and agricultural chemicals introduce heavy metals, pesticides, and other hazardous substances into the environment, posing severe health and ecological risks. Weak enforcement of environmental regulations often leads to a culture of impunity, where polluters operate without consequence. In many contexts, pollution control frameworks exist only nominally and are rarely implemented effectively (OECD, 2020). This institutional failure breeds growing conflict, as affected communities mobilize against powerful industries that prioritize economic gain over public welfare [29].

Climate change compounds the crisis. Increasingly erratic weather patterns, such as intensified droughts and floods, further stress fragile water systems, threatening availability and quality. These events strain existing infrastructure and complicate water management strategies, especially in low-capacity regions. According to the Pan American Health Organization (PAHO), thousands of families continue to suffer from water-related diseases caused by poor infrastructure and governance [49]. The convergence of pollution and climate variability creates a complex and urgent water security dilemma that demands integrated, multisectoral solutions. Governance of water resources in Latin America is fragmented, with significant disparities across countries and regions. In some cases, decentralization has resulted in inconsistent policy implementation and varying capacities at the local level [75]. Many municipalities lack the technical resources or legal authority to enforce national environmental laws effectively [1]. Corruption remains a critical barrier, weakening institutional accountability and undermining public confidence. The absence of transparent decision-making and enforcement mechanisms inhibits sustainable water management and fuels public dissatisfaction [5].

As a response to institutional inaction, grassroots and civil society movements have gained momentum across the region. These movements often emerge in reaction to pollution incidents and seek to hold corporations and governments accountable. Community-based organizations are increasingly involved in water monitoring, public education, and policy advocacy, highlighting the power of local knowledge and participation. For example, water user associations in rural areas have demonstrated success in managing local water supplies collaboratively, improving both environmental outcomes and social cohesion [66]. Such models foster accountability and enable communities to participate directly in decision-making. Nevertheless, systemic obstacles remain. Community initiatives frequently encounter resistance from entrenched political and economic interests, and limited funding hampers their scalability. Institutional biases may also marginalize grassroots actors and prevent them from gaining meaningful influence over water governance. To overcome these challenges, stronger partnerships between government agencies, NGOs, and communities are essential. These alliances can promote knowledge-sharing, increase resource mobilization, and reinforce participatory governance frameworks. Equitable water governance depends not only on infrastructure investment but also on inclusive decision-making processes that empower affected communities [65, 74].

Technology offers promising tools for enhancing water quality monitoring and community engagement. Innovations such as remote sensing, GIS mapping, and mobile applications facilitate real-time data collection, enabling both government and community actors to track pollution sources and respond swiftly. These tools also improve transparency and compliance with environmental regulations. In addition, digital platforms provide new avenues for civic engagement. Through social media and online networks, communities can rapidly mobilize in response to environmental crises, hold polluters accountable, and pressure public institutions to act [75]. However, technological interventions are not without limitations. Many of these tools require financial investment and technical expertise, which may be lacking in under-resourced areas. The digital divide, particularly in rural and low-income regions, poses a serious risk of deepening existing inequalities [43, 44]. For technology to be truly transformative, stakeholders must prioritize inclusive access and capacity-building to ensure that all communities can benefit from innovation.

The aim of this chapter is to examine Latin America's ongoing struggle with water pollution through the lens of governance failures, socio-environmental inequities, and community resilience. Despite the region's rich freshwater endowment, unchecked industrial discharge, agricultural runoff, and poor wastewater management have severely compromised water quality across both rural and urban landscapes. This crisis is compounded by weak regulatory enforcement and exclusionary governance structures that often neglect the needs and rights of vulnerable populations. The chapter argues that while state-led efforts have largely fallen short, community-led initiatives—particularly among Indigenous and rural groups—are emerging as powerful forces of resistance and innovation. These localized efforts not only highlight alternative models of water stewardship but also reveal the importance of inclusive governance and culturally rooted approaches in tackling pollution. By situating these issues within a broader Global South framework, the chapter contributes to interdisciplinary discourse on water justice, sustainable resource management, and the democratization of environmental decision-making.

10.2 Sources and Drivers of Water Pollution in Latin America

Water pollution in Latin America is a multifaceted challenge driven by both anthropogenic activities and environmental factors. The region's diverse geography, rapid urbanization, industrial development, and agricultural intensification have collectively exerted mounting pressure on freshwater ecosystems. While national policies and international agreements exist to regulate pollutants, enforcement remains uneven, particularly in areas with limited governance capacity or where marginalized populations reside. In many cases, Indigenous and rural communities bear the disproportionate burden of degraded water quality, often relying on contaminated rivers and lakes for drinking, fishing, and agricultural needs. Table 10.1 presents a comprehensive overview of the major sources and underlying drivers of water pollution across Latin America. It includes examples of significantly impacted water bodies and concise notes on the environmental and social implications of each category.

10.2.1 Industrial Discharge and Aquatic Pollution

Industrial activities, particularly in sectors like mining and manufacturing, are pivotal contributors to water pollution in Latin America. Mining operations are associated with severe ecological disruptions, including the release of heavy metals and

Table 10.1 Cause of water pollution in Latin America and their mechanism

Source of pollution	Primary drivers	Examples of affected water bodies	Notes
Agricultural runoff	Excessive use of fertilizers and pesticides; lack of riparian buffer zones	Lake Atitlán (Guatemala), Laguna de Bay (Mexico), Itajaí River (Brazil)	Nutrient over-enrichment leads to eutrophication, algal blooms, and oxygen depletion
Industrial discharges	Ineffective enforcement of environmental regulations; legacy contamination from factories and mines	Rímac River (Peru), Bogotá River (Colombia), Coatzacoalcos River (Mexico)	Contamination includes heavy metals, solvents, and persistent organic pollutants
Urban wastewater	Rapid population growth and limited sewage treatment infrastructure	Tietê River (Brazil), Mapocho River (Chile), La Paz River (Bolivia)	High microbial loads increase risk of waterborne diseases; chronic sanitation issues
Solid waste and plastics	Poor solid waste management; waste dumping near or in waterways	Guaire River (Venezuela), Lago de Maracaibo (Venezuela), Río Ozama (Dominican Republic)	Plastic debris clogs drainage systems, disrupts aquatic habitats, and releases toxins
Oil spills and hydrocarbon leaks	Offshore exploration and pipeline leaks; accidents involving fuel transport	Napo River (Ecuador), Marañón River (Peru), Orinoco River Delta (Venezuela)	Oil pollution affects aquatic biodiversity and threatens Indigenous water sources
Deforestation and land degradation	Large-scale logging, cattle ranching, and infrastructure expansion	Madeira River (Brazil), San Juan River (Nicaragua), Usumacinta River (Mexico/Guatemala)	Soil erosion and sedimentation reduce water quality and alter river morphology
Climate change	Intensified droughts and floods; glacier retreat disrupting hydrological cycles	Santa River (Peru), Paraná River Basin (Brazil, Paraguay, Argentina), Patagonian Rivers (Chile/Argentina)	Changes in precipitation patterns increase water scarcity and degrade water quality
Illegal mining (e.g., gold)	Informal, small-scale mining operations in remote regions lack oversight	Madre de Dios River (Peru), Caquetá River (Colombia), Guainía River (Colombia/Venezuela)	Mercury and cyanide used in gold extraction contaminate rivers and bioaccumulate in fish

toxic waste into freshwater bodies, significantly impairing aquatic ecosystems. For instance, contaminants from mining processes have been found in rivers and lakes, leading to concerns about ecosystem health and human exposure [33, 44]. Additionally, the lack of stringent regulatory frameworks in various countries has exacerbated the situation, allowing unchecked industrial discharge. Studies highlight how rapid industrial development, especially in the last few decades, has outpaced environmental regulations, thus compromising water quality [12].

10.2.2 Agricultural Runoff and Pesticide Contamination

Agricultural activities further compound water pollution issues through runoff that carries fertilizers and pesticides into water bodies. In regions characterized by extensive agricultural land, such as parts of Brazil and Argentina, pesticides have been identified as critical contaminants affecting surface and groundwater sources. The introduction of chemicals like malathion and carbendazim poses significant risks to aquatic biodiversity, indicating that pesticide utilization in agriculture must be carefully assessed to mitigate environmental harm [62]. Furthermore, the dependence on monoculture farming and chemical inputs has led to significant ecological imbalances that affect water quality and threaten food security in vulnerable communities [12].

10.2.3 Urbanization and Untreated Sewage

Urbanization serves as a key driver of water pollution, primarily through the inadequate treatment of sewage. As cities expand rapidly, especially in Latin America where urban populations have surged, the infrastructure necessary for proper waste management often lags [58]. Urban areas face extraordinary challenges regarding untreated sewage discharge, which contaminates drinking water supplies and fosters the proliferation of waterborne diseases. For example, the capital cities of several Latin American nations exhibit alarming rates of untreated wastewater being released directly into rivers, heightening public health risks [44]. The scale of urban pollution reflects both infrastructural inadequacies and a prioritization of economic growth over environmental sustainability [40, 53, 54].

10.2.4 Climate Change and Hydrological Stress

Climate change acts as a significant underlying driver of water pollution by altering hydrological patterns across Latin America. Increasing temperatures and erratic precipitation lead to severe droughts and flooding events, exacerbating the effects of pollutants in the water from various sources, including industrial and agricultural runoff [10]. Furthermore, climate-induced variations in water availability exacerbate competition for water resources, intensifying risks related to water quality deterioration. For instance, as droughts become more prevalent, the concentration of pollutants in limited water supplies increases, leading to contamination events [21, 37]. In addition to direct effects, climate change also influences the distribution of water resources, challenging effective management of both quantity and quality. The Intergovernmental Panel on Climate Change has highlighted the necessity of adapting water management strategies to account for these changes, advocating for a multifaceted approach that encompasses industrial, agricultural, and urban water management considerations [21, 51].

10.2.5 Impacts on Human Health and Biodiversity

The ramifications of water pollution in Latin America extend beyond environmental concerns to include serious human health risks. Various studies have established links between exposure to polluted water and chronic health conditions, particularly among vulnerable populations such as children and the elderly [40]. The increase in vector-borne diseases, exacerbated by water quality deterioration, further underscores the interplay between pollution and public health. The World Health Organization has identified various waterborne illness outbreaks attributed to poor water quality in urban areas [44]. Biodiversity also suffers due to high pollution levels, as aquatic organisms struggle to thrive amid toxic environments, leading to disrupted ecosystems that can affect local populations reliant on these resources for food and livelihoods [12].

10.2.6 Interconnectedness of Drivers

These sources of water pollution do not operate in isolation; they are interlinked. For instance, industrial discharges can exacerbate agricultural runoff effects by increasing pollutant loads in water bodies, leading to eutrophication—a phenomenon where excess nutrients cause oxygen depletion and harm aquatic life [12, 44]. Urbanization intensifies agricultural effects by increasing impervious surfaces in cities, leading to more intense runoff during rain events. This interconnectedness highlights the need for synergistic policy approaches that address each sector's contributions to ensure comprehensive solutions to water pollution challenges in Latin America.

10.2.7 Policy Implications and Future Directions

To effectively mitigate water pollution, Latin American nations must prioritize the development of robust environmental policies that consider the multifaceted nature of these pollution sources. Research indicates that coordinated efforts between industries, agricultural sectors, and urban planners, integrating sustainable practices with regulatory measures, are essential [52, 58]. Moreover, investment in wastewater treatment infrastructure is critical in managing urban pollution, alongside promoting sustainable agricultural practices that minimize pesticide use and enhance soil health [12]. Adaptation strategies must also consider climate change impacts, integrating hydrological management with broader environmental goals to safeguard water resources for future generations [51].

The intricate dynamics of water pollution in Latin America underline the pressing need for a comprehensive understanding of its sources and drivers, facilitating targeted interventions that safeguard both public health and ecological integrity. Without concerted efforts to address these issues, the region risks escalating challenges to its water resources, health outcomes, and overall sustainability.

10.3 Socioeconomic and Health Impacts of Water Crisis in Latin America

Access to clean and safe water remains a critical challenge in many parts of Latin America, disproportionately affecting vulnerable populations. The impacts of water scarcity and contamination extend far beyond environmental degradation, influencing public health, economic stability, gender equity, and social cohesion. Table 10.2 outlines key categories of impact, with regional examples and brief explanatory notes to illustrate the breadth of the crisis. Indigenous communities, in particular, have faced systemic inequalities, resulting in heightened vulnerabilities during water crises, as their traditional livelihoods depend on the natural environment, which is increasingly threatened by pollution and climate change [27, 31]. Regarding the socioeconomic ramifications, the lack of reliable water sources has been directly linked to food insecurity and declines in fishery resources, critical to the diets and economies of many Latin American communities. Strategies for food production heavily rely on access to clean water for irrigation and sustenance, particularly in Indigenous populations where agriculture forms the backbone of cultural and economic practices. Studies show that inadequate water quality exacerbates malnutrition and health disparities in these communities as essential nutrient sources become compromised due to contamination [2, 71].

Water pollution is a primary public health concern, with significant associations between contaminated water sources and disease outbreaks such as cholera, diarrhea, and skin diseases. The re-emergence of cholera in the 1990s serves as a poignant example; it highlighted how lapses in water quality and sanitation can catalyze

Table 10.2 Socioeconomic and health impacts of the water crisis in Latin America

Impact category	Summary	Examples from the region	Notes
Public health	Increased prevalence of waterborne diseases such as diarrhea, cholera, leptospirosis, and dengue	High infection rates in rural Bolivia, Haiti, and Amazonian Peru	Poor sanitation and lack of access to clean water contribute to preventable illness and mortality
Child mortality and malnutrition	Undernutrition due to recurrent diarrhea and unsafe drinking water	Rural Guatemala, Andean Ecuador, and northeastern Brazil	WASH (Water, Sanitation, and Hygiene) deficits exacerbate child development challenges
Gender inequality	Women and girls are disproportionately affected due to water collection burdens and caregiving roles	Indigenous communities in Mexico and Guatemala	Time spent fetching water limits girls' education and women's economic participation
Economic productivity	Lost workdays due to illness or water-fetching duties; reduced agricultural output	Urban slums in Lima, Peru and Manaus, Brazil; farming communities in Honduras	Water scarcity and contamination directly lower labor efficiency and economic growth
Migration and displacement	Water scarcity and pollution drive rural-to-urban migration and environmental displacement	Semi-arid northeast Brazil (Sertão), Andean highlands of Bolivia, Venezuela's Orinoco Delta	"Water refugees" increase pressure on urban infrastructure and housing
Education disruption	School attendance drops due to illness or domestic water duties	Rural schools in Nicaragua and Colombia	Girls often miss school due to lack of WASH facilities and menstrual hygiene infrastructure
Social unrest and conflict	Water privatization and scarcity spark protests and local disputes	Cochabamba Water War (Bolivia); ongoing tensions in Chile's Atacama region	Water is increasingly politicized as a resource, triggering civil resistance
Food insecurity	Contaminated irrigation and droughts reduce agricultural yields and access to clean food	Dry corridors in Central America (El Salvador, Honduras, Guatemala)	Food prices rise as safe water availability declines, affecting poor households first

widespread health crises. Many Latin American nations still grapple with inferior infrastructure, leading to increased susceptibility to cholera and other waterborne diseases [15, 16, 60]. The interplay of poor sanitation practices and environmental factors further underscores the complexity of disease transmission in this region, as conditions conducive to cholera's survival often prevail during certain climatic events, such as El Niño [36, 81].

Furthermore, studies indicate a direct relationship between ecological conditions and cholera occurrences, with significant evidence suggesting that the Vibrio cholerae bacterium thrives in coastal regions where water pollution is rampant. Historical data demonstrate that pollution from urban runoff and industrial waste has directly contributed to a rise in cholera outbreaks across various countries in Latin America, revealing a precarious health landscape shaped by environmental neglect and infrastructural deficits [13, 17]. Indigenous populations not only face a higher likelihood of encountering contaminated water but also have limited access to healthcare resources necessary for addressing ensuing health crises. Research has shown that marginalized communities often experience barriers in healthcare access, which can lead to inadequate responses to cholera outbreaks and exacerbate health inequalities [7, 83]. The intersectionality of race, class, and geographical location further complicates these disparities, disadvantaged groups inhabit regions most vulnerable to natural disasters and environmental pollutants and are often the last to receive governmental support during health crises [27].

Food security in relation to the water crisis also poses significant challenges. Overfishing, driven by both collapsing fish populations and contaminated water bodies, sharply impacts food availability, particularly for those reliant on fish as a primary protein source. Studies illustrate that disadvantaged communities, including Indigenous groups, are more impacted by declines in fishery resources, leading to compounded issues of malnutrition and economic instability [27, 35]. Consequently, food scarcity will continue to rise if water quality deteriorates, establishing a cyclical pattern wherein impoverished communities bear the brunt of the crisis.

Further compounding these issues is the lack of effective policy frameworks aimed at preserving Indigenous rights and promoting sustainable water management practices. While environmental policies exist, their efficiency is often undermined by systemic neglect and the prioritization of industrial interests over community needs. Efforts to amplify Indigenous participation in decision-making related to water governance have been noted as crucial, as they can foster more equitable solutions that align with the ecological knowledge inherent in these communities [27]. The documentation of this acute water crisis illustrates a larger narrative—one that connects the dots between water as a human right and the precarities faced by marginalized and Indigenous groups within Latin America. Impediments to access clean water and sanitation, coupled with the health risks posed by pollution, create a multivocal crisis demanding urgent collective action. Inhabiting regions marked by historical sociopolitical marginalization, these communities exemplify the broader consequence of anthropogenic actions on water quality and public health, leading to dire health outcomes and undermined societal structures [2, 35, 71].

On a broader scale, policy recommendations call for strengthened frameworks that prioritize environmental justice and recognize the unique position of Indigenous communities in the water crisis narrative. The emphasis must shift towards sustainable water management strategies that integrate Indigenous knowledge systems, promoting resilience as well as empowering these communities to address their health and environmental challenges [31, 81]. Aligning public health initiatives with environmental sustainability will not only mitigate the socioeconomic impacts of the water crisis but will also champion the rights of marginalized groups, ensuring their voices and needs are central to the discourse [13, 60]. The water crisis in Latin America reflects a profound public health challenge that embodies socio-political and environmental dimensions requiring comprehensive and inclusive approaches. The marginalization of Indigenous communities amplifies health disparities and food insecurity while underlining the urgency for systemic change within water governance frameworks. By weaving together these disparate elements—from pollution and disease to food security and socio-political dynamics—a more holistic understanding of the crisis emerges, illuminating pathways toward sustainable solutions that uplift all communities impacted by the water crisis in Latin America.

10.4 Governance and Policy Frameworks for Managing Water Crisis in Latin America

Latin America faces a significant water crisis characterized by inadequate management, infrastructural deficits, and governance challenges in its water resources. The governance and policy frameworks for managing this crisis differ significantly across the region, reflecting a complex interplay of national policies, regional dynamics, and varying levels of institutional effectiveness.

10.4.1 Regional Overview of Water Management Institutions and Regulations

Water governance in Latin America is increasingly recognized for its complexity, necessitating a multidimensional approach that incorporates local, national, and regional considerations. A systematic review indicates that participatory water governance is becoming prevalent in the region, emphasizing community engagement and the importance of stakeholder inclusion in decision-making processes related to water management [73]. Numerous countries have established institutional frameworks aimed at coordinating water management efforts, often involving decentralized governance structures that leverage local knowledge and stakeholder involvement [11, 73]. Despite the establishment of various governance frameworks, Latin American nations still grapple with challenges rooted in historical inequalities and socio-political dynamics. The lack of accountability, compounded by insufficient rule of law, has hindered effective water governance and policy enforcement across many jurisdictions [24]. Countries such as Brazil and Mexico exhibit varying levels of governance success, with persistent challenges in effective policy implementation reflecting broader systemic issues [11, 72]. The reliance on centralized control and top-down approaches has often led to ineffective management practices, diminishing local autonomy in water resource decisions [26, 72].

10.4.2 Successes and Failures of National Water Policies

Brazil, with its extensive water resources, provides a compelling case study of both successes and persistent failures in water policy. Successful initiatives, such as the National Water Agency (ANA), illustrate the potential for integrated water resources management. However, systemic issues such as corruption, bureaucratic inefficiencies, and lack of comprehensive infrastructure have undermined these successes, leading to recurring water crises in various regions [11, 72]. The Brazilian model emphasizes basin committees and participatory governance, however, many of these measures face bureaucratic inertia and lack sufficient authority to enforce compliance during crises [72]. In Mexico, the situation varies as governance around wastewater management highlights both innovative solutions and significant deficits [24]. The political landscape has seen a concentration of powers at the national level, resulting in challenges for subnational entities in implementing effective water policies. The political misuse of infrastructure projects has also impeded accountability, as investments often reflect political agendas rather than genuine infrastructural needs [24]. Despite the challenges, Mexico's National Water Law advocates for participative governance, which has led to localized initiatives that demonstrate potential for improved outcomes [11]. Peru presents another dimension to the water governance narrative in Latin America. The country has implemented policies aimed at enhancing water access in rural areas, resulting in improvements in potable water services. Nonetheless, these successes have been undermined by structural inequalities and insufficient enforcement of water regulations, particularly in informal settlements [61]. Peru's governance framework, while progressive, remains stymied by inadequate investment, bureaucratic red tape, and corruption, which further exacerbate water resource management challenges [11, 72].

10.4.3 Corruption, Weak Enforcement, and Political Barriers

Corruption and weak enforcement mechanisms pose significant roadblocks to effective water governance across Latin America. Studies reveal that political corruption pervades many national and local administrations, severely impacting water management initiatives [24, 67]. This corruption often manifests in the misallocation of resources and the perpetuation of inequality, where marginalized communities lack access to clean water and sanitation [24, 73]. The lack of robust legal enforcement further complicates these dynamics, as many water management policies exist in theory but fail in practice due to the absence of political will and proper funding for enforcement agencies [11, 72]. Regions characterized by weak governance structures often exhibit a cyclical pattern where political barriers obstruct legislative initiatives aimed at fostering accountability and transparency in water management practices [11, 61].

Furthermore, political barriers, including bureaucratic inertia and a lack of public engagement, hinder collaborative governance approaches that are essential for effective water resource management [26, 73]. The interplay between local and national entities is often strained by conflicts of interest and competition for resources, leading to fragmented policies that do not effectively address community needs [11, 72]. While Latin America has made strides toward establishing participatory and integrated governance frameworks for water management, significant barriers remain. The successes observed in countries like Brazil, Mexico, and Peru showcase the potential of effective governance but also underscore the necessity of addressing systems of corruption, enforcement weaknesses, and political obstacles to achieve sustainable water management in the region. Governance and policy frameworks for managing the water crisis in Latin America remain intricate and fraught with challenges. While the push toward participatory governance shows promise, the underlying issues of corruption, weak policy enforcement, and political barriers necessitate urgent attention. Future efforts must focus not only on enhancing institutional framework efficiencies but also on fostering a culture of accountability and transparency to ensure equitable access to water resources across the diverse and complex landscapes of Latin America.

10.5 Data, Indicators, and Monitoring Gaps in Latin America's Water Systems

Fragmented data systems, inconsistent indicators, and significant gaps in long-term monitoring hinder effective water governance in Latin America. Many countries lack robust, centralized databases, while others rely on outdated or inaccessible records. The absence of standardized metrics for water quality, availability, and infrastructure coverage complicates regional comparisons and limits timely policy interventions. Figure 10.1 visualizes the key dimensions of the region's data and monitoring challenges, highlighting institutional fragmentation, underfunded monitoring networks, and gaps in transparency and interoperability. It also emphasizes the critical need for integrated data systems to inform sustainable water management and ensure accountability.

The existing systems are often incomplete and inconsistent, leading to a lack of reliable data necessary for assessing the actual state of water resources. For example,

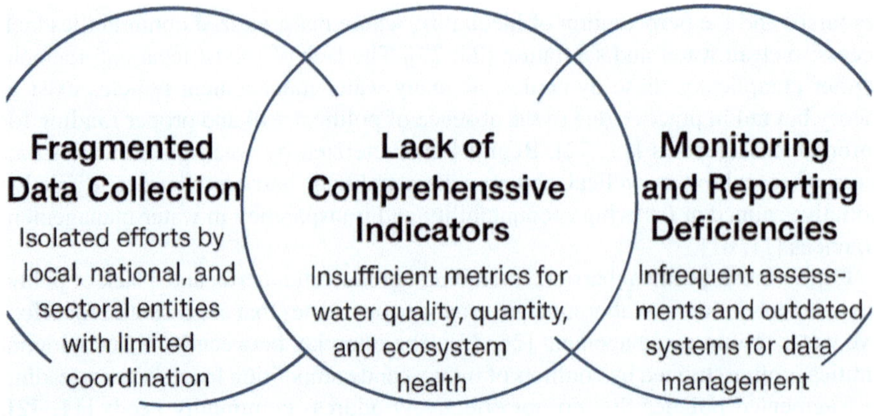

Fig. 10.1 Data, indicators, and monitoring gaps in Latin America's water systems

in many areas, particularly rural regions, the absence of systematic monitoring has resulted in an inability to understand the full scope of water pollution and its health impacts on communities. Consequently, children and vulnerable populations remain at risk due to undetected contaminants in the water supply [49]. Furthermore, urban areas also face challenges, such as underfunded and outdated monitoring infrastructure, which restricts real-time data availability and obscures the efficacy of water quality interventions [49].

Central to this issue is the absence of harmonized national and regional indicators for water quality. The critical global frameworks, such as the Sustainable Development Goals (SDGs), particularly Goal 6 regarding clean water and sanitation, rely on standardized indicators for effective tracking and assessment. Unfortunately, Latin America has struggled with the alignment and implementation of these indicators, complicating regional assessments and policymaking efforts. This misalignment limits the ability of stakeholders to prioritize actions or allocate resources effectively based on health and environmental needs [22, 49]. In addition, without consistent data reported through these standardized indicators, policymakers struggle to develop strategies that meet urgent population needs regarding water access, quality, and sustainability.

Moreover, barriers to data collection persist across technical, financial, and institutional dimensions. Technically, there often lacks appropriate technologies for accurate monitoring, leading to questions about the reliability of available data. Financial constraints prevent municipalities from investing in necessary infrastructure or personnel training [49]. Institutionally, fragmented governance structures and insufficient political will often impede unified approaches to data collection and sharing. Many Latin American countries have historically lacked comprehensive water governance frameworks that integrate various sectors and stakeholders necessary for effective data management and environmental protection [22]. These multifaceted barriers result in a system where disjointed efforts lead to sporadic and unreliable data collection.

In recent years, the role of citizen science has emerged as a significant complement to formal data collection efforts. Citizen science initiatives can help address existing gaps by involving local communities in water monitoring efforts, thereby increasing data volume and participation [25]. These projects empower residents to engage with environmental care while contributing valuable data to local and regional databases. In this way, citizen science not only increases data availability but also enhances community awareness of water quality issues and promotes local advocacy for improvements in water management [25, 63]. Additionally, utilizing open data platforms for citizens to upload findings helps create a dynamic and participatory data-rich environment where stakeholders can make informed decisions.

Integrating Indigenous and local knowledge into formal monitoring frameworks is also essential for comprehensive water management. Indigenous communities often possess a profound understanding of local ecosystems, providing invaluable insights into water quality and the impacts of pollution [22]. Incorporating local knowledge systems alongside scientific approaches can bolster the effectiveness of water monitoring programs and foster trust between communities and governmental agencies. This integration enhances data reliability and enriches resource management practices, ensuring that solutions respect local contexts and customs [22].

Strengthening regional water observatories or transboundary monitoring networks presents another opportunity to enhance water data collection in Latin America. Establishing such networks can enable shared resources, improved data frameworks, and intergovernmental collaboration on transboundary water issues, which are becoming increasingly critical due to climate change and population pressures [22]. Collaborative efforts might include joint data collection initiatives, shared technologies, and cross-border policies that facilitate seamless monitoring across regions. By pooling expertise and resources, Latin American countries can better understand regional water dynamics and implement cohesive strategies benefiting all involved nations.

As Latin America grapples with water quality challenges exacerbated by climate change, urbanization, and economic pressures, addressing data infrastructure gaps becomes imperative. A robust monitoring framework built on reliable, harmonized indicators and enhanced through citizen engagement will play a crucial role in informing policy decisions, aligning with global sustainability goals, and ultimately safeguarding public health. Continuous investment and innovation in water data systems can foster resilience against water crises and promote sustainable resource management across the region in the long term.

10.6 Community Responses and Grassroots Movements in Latin America to Address Waste Crisis

Across Latin America, communities have mobilized in response to growing water scarcity, privatization, pollution, and inequitable access to clean water. Grassroots movements have played a pivotal role in resisting harmful policies, defending water as a common good, and promoting sustainable local stewardship. Table 10.3 highlights some notable community-led initiatives and movements that have shaped water justice efforts across the region. Communities have risen in response to water shortages, pollution, and privatization of water supplies, showcasing diverse, often indigenous-led, initiatives.

In Bolivia, the Water Wars of the early 2000s stand out as a seminal case of grassroots resistance against water privatization. The public uproar began after the Bolivian government privatized water services in Cochabamba, leading to a dramatic

Table 10.3 Examples of community responses and grassroots movements addressing water crises in Latin America

Country	Community/Movement	Response strategy	Impact
Bolivia	Coordinadora del Agua (Water Coordinating Committee)	Anti-privatization protests, public water advocacy	Reversed water privatization in Cochabamba (2000 Water War)
Mexico	Movimiento Mexicano de Afectados por las Presas y en Defensa de los Ríos (MAPDER)	Advocacy against dams, defense of rivers, and local water rights	Halted harmful dam projects and promoted water justice
Brazil	Movement of People Affected by Dams (MAB)	Legal advocacy, community organizing, water rights campaigns	Raised national awareness on water injustice and displacement
Colombia	Ríos Vivos Colombia	Defense of river ecosystems, women-led water activism	Strengthened community resistance to extractivist projects
Chile	MODATIMA (Movement for the Defense of Access to Water, Land, and Environmental Protection)	Campaigns for constitutional water reform and water as a public good	Brought water rights into the national policy debate
Peru	Rondas Campesinas	Rural community patrols monitoring water sources	Reduced water theft and improved local stewardship
Argentina	Asamblea Ambiental de Gualeguaychú	Mobilization against industrial water pollution	Pressured the government to act on cross-border water disputes

increase in water prices and prompting widespread protests. The mobilization demonstrated that grassroots movements could effectively challenge neoliberal policies that prioritize profit over human rights and social equity. Activists and ordinary citizens collectively rejected the privatization model, bringing international attention to the broader crisis of access to clean water in Latin America [70].

Colombia provides another poignant example, particularly through the efforts of river defenders who have emerged in response to environmental degradation caused by extraction industries. These individuals and community organizations rally to protect their waterways from pollution and over-exploitation, highlighting the intimate relationships between indigenous communities and their surrounding ecosystems. The activists employ a range of strategies, from legal actions to direct action protests, demonstrating the power of grassroots movements in environmental advocacy [38]. Such efforts effectively bridge the gap between local ecological knowledge and formal environmental governance, instigating changes at multiple levels of policy discourse.

Central to these grassroots movements is the integration of indigenous knowledge and traditional water stewardship practices. Indigenous communities possess invaluable knowledge of local ecosystems accrued over generations, enabling them to manage water resources sustainably. Research demonstrates that indigenous land management practices can lead to more efficient water conservation and biodiversity preservation [50]. For instance, collaborative management frameworks in indigenous territories have been shown to enhance ecological and community resilience, ultimately assisting in the recovery of water systems impacted by climate change [78].

Moreover, environmental justice movements in Latin America advocate for the right to clean water as a fundamental human right. The prevailing environmental injustices disproportionately affect marginalized communities, particularly indigenous peoples, who often lack access to safe drinking water and sanitation. Activists have argued that the state's failure to provide access to clean water violates the rights of these communities and infringes upon their ability to thrive. Evidence suggests that movements advocating for clean water often intersect with broader human rights campaigns, aiming not only for environmental justice but also for socioeconomic reforms aimed at alleviating poverty [68]. These frameworks align closely with the principles of climate justice, emphasizing that all people, regardless of their socioeconomic status, should have an equal stake in environmental governance. The new wave of regionalism across Latin America has significantly influenced grassroots movements advocating for environmental justice and better water governance. Politically progressive governments in some countries have embraced more inclusive policies, allowing grassroots activism to gain traction. This context has fostered a resurgence of collective activism and community engagement—essential components in the fight for equitable access to water resources [56]. In some areas, citizens have leveraged these political changes to push for legislative reforms that recognize indigenous rights and environmental justice principles.

Nevertheless, challenges remain as neoliberal policies persist in many parts of Latin America. Grassroots movements often confront not only systemic disadvantages in accessing resources but also state repression of their activism. Organizations involved in defending water rights frequently face legal challenges, violence, or outright criminalization by authorities aligned with large corporations [76]. These dynamics underscore the importance of international solidarity and engagement in supporting local struggles for water rights and environmental justice.

Additionally, educational initiatives within these grassroots frameworks are vital for empowering local communities. Popular education movements have flourished outside the formal education system, fostering awareness regarding the significance of water protection and sustainable resource management [38]. This grassroots educational outreach has become a pivotal tool for building awareness and legitimacy around the struggles faced by these communities, enhancing collective agency and encouraging advocacy for their rights. The water crisis in Latin America elicits a multi-dimensional response from grassroots movements, rooted in local knowledge and framed by a commitment to social and environmental justice. In contexts like Bolivia's Water Wars and Colombia's river defenders, local communities are leveraging traditional practices and collective action to resist external pressures that threaten their access to clean water. Distinctly, the integration of indigenous knowledge into contemporary environmental movements underscores the importance of sustainable practices for managing water resources. This synthesis of traditional stewardship and grassroots advocacy is essential in confronting ongoing challenges posed by neoliberal policies and systemic inequities in access to clean water.

10.7 Innovations and Pathways Forward for Latin America's Water Crisis

The water crisis in Latin America is multifaceted, requiring comprehensive strategies that incorporate innovations and cross-sectoral approaches. Key among these innovations are nature-based solutions, decentralized water treatment technologies, transboundary cooperation, and the vital participation of diverse actors such as NGOs and academic institutions. Each of these components plays a significant role in fostering resilience and sustainability in water resource management across the region.

To begin with, nature-based solutions (NbS) such as wetlands restoration and the implementation of green infrastructure can play a crucial role in managing water resources sustainably. By leveraging natural ecosystem processes, these solutions enhance the ability of the landscape to retain and filter water, thereby improving water quality and availability [73]. Evidence from various studies highlights how restoring wetlands can mitigate flooding, enhance biodiversity, and sequester carbon, thus addressing both water management issues and climate change [30]. Furthermore, the integration of NbS into urban planning not only addresses water scarcity but also promotes social cohesion by involving local communities in the management and stewardship of their water resources [73].

Research also emphasizes the potential of decentralized and low-cost water treatment technologies as critical innovations in addressing water access and quality issues in Latin America. These technologies are particularly vital in rural and marginalized urban areas where conventional infrastructure may be lacking or economically unfeasible [9]. For instance, local water purification systems and greywater recycling solutions empower communities to manage their water resources more effectively, reducing reliance on centralized systems that often fail to meet local needs. This approach enhances social equity by providing safe water access to underserved populations and promotes community engagement in water management [57].

In parallel, transboundary cooperation is essential for managing shared water resources across countries in Latin America. Many river basins in the region transcend national boundaries, resulting in complex governance challenges [19]. Frameworks such as the Escazú Agreement highlight the importance of integrating environmental governance with human rights and transboundary cooperation to address shared waters sustainably [28]. The utilization of cooperative agreements can facilitate equitable resource allocation, enhance conflict resolution mechanisms, and foster collaboration among neighboring states. Studies on transboundary cooperation indicate that successful water-sharing agreements can lead to significant improvements in water security and environmental health across borders [47].

Moreover, the roles of NGOs, academia, and international agencies cannot be overlooked in the efforts to mitigate the water crisis. NGOs often act as intermediaries, bridging gaps between communities and government authorities, while ensuring that local voices are heard in decision-making processes [73]. Academic institutions contribute by providing research and technical expertise, facilitating knowledge transfer, and fostering innovation in water management practices [48]. International agencies support this effort by offering funding, capacity-building programs, and strategic guidance for integrated water resource management [70]. Together, these actors create a robust collaborative network essential for advancing sustainable water management initiatives in Latin America.

Furthermore, the fragmented policy landscape in many Latin American countries necessitates a shift towards integrated watershed management as a way to enhance governance structures. Effective integrated water resource management (IWRM) requires not only a multi-stakeholder approach but also the use of composite indices to assess the health of freshwater ecosystems [9]. By assessing various indicators such as water quality, availability, and service equity across different regions, stakeholders can develop better-informed, data-driven management practices that respond to the unique challenges faced by diverse locales [80].

The evolving nature of water governance also underscores the importance of addressing political, social, and economic dimensions within integrated water management frameworks. The historical context of privatization in many Latin American countries has exacerbated inequalities, underscoring the need for policies that prioritize social justice and equitable resource distribution [57]. Addressing the discrepancies in water access among different socio-economic groups is fundamental for achieving sustainable development goals, particularly in the context of climate change and increased competition for water resources [70].

Climate change adds another layer of complexity to the water crisis in these nations, highlighting the urgent need for adaptive management strategies that anticipate and respond to climate variability. For example, participatory approaches in watershed management can empower local communities to engage in climate resilience efforts, enhancing their capacity to respond to water scarcity and protect their ecosystems [30]. Implementing localized adaptation measures, such as rainwater harvesting and soil conservation techniques, can fortify communities against the impacts of climate-induced stressors on water supply [64].

Within the academic sphere, ongoing research into socio-hydrological dynamics continues to inform our understanding of water management in a transboundary context. A closer examination of human-water interactions reveals that equitable transboundary governance must account for the diverse social, cultural, and economic contexts affecting different riparian states [32]. This approach not only enriches the discourse around water security but also encourages policymakers to consider the intricate relationships between human activities and hydrological systems in their governance frameworks [20].

Furthermore, the need for comprehensive data and information sharing across borders is essential for fostering effective transboundary cooperation. A unified socio-environmental geodatabase can facilitate the integration of diverse datasets that address cross-border water issues, thus supporting more informed and collaborative decision-making [59]. By overcoming data disparities and inconsistencies, countries can enhance their collective ability to manage water resources collaboratively, reducing potential conflicts over shared waters [59].

Tackling the water crisis in Latin America requires an innovative approach that embraces nature-based solutions, decentralized technologies, cooperative management, and active participation from a broad range of stakeholders. The multifaceted nature of this challenge reveals how intertwined governance, climate resilience, and social equity are in achieving sustainable water resource management. Consequently, it is crucial that policymakers adopt holistic strategies that reflect these complexities, aiming for a sustainable and equitable future for water management across the region.

10.8 Methodological Considerations for Water Crisis Research and Policy in Latin America

In addressing the water crisis in Latin America, methodological considerations are crucial, particularly the deployment of mixed-methods approaches that combine advanced technologies with community-based insights (Fig. 10.2). Utilizing a suite of methodologies, including remote sensing, policy analysis, case studies, and community surveys, provides a comprehensive framework to understand and address the complexities surrounding water resource management in this region.

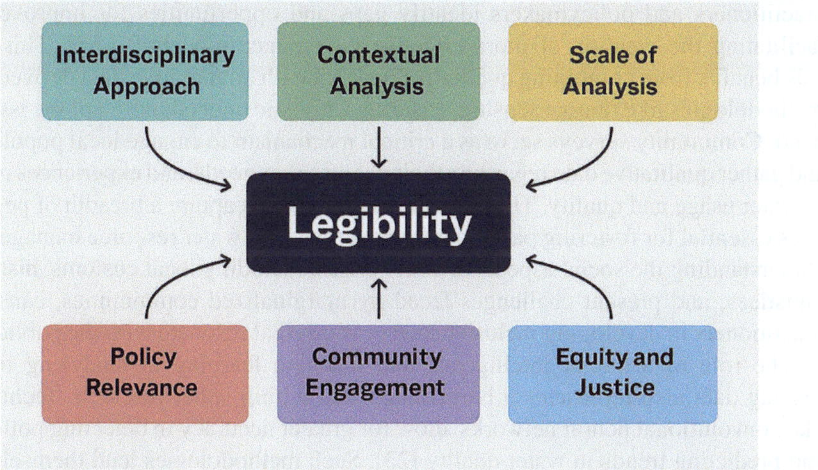

Fig. 10.2 Methodological considerations for water crisis research and policy in Latin America

Remote sensing has evolved as a valuable tool for environmental monitoring, particularly in hydrology and water quality assessment. Innovations in remote sensing technologies, including the use of drones and UAVs, allow for higher spatial and temporal resolution in collecting water quality data. For example, autonomous UAV-assisted systems have advanced the collection of water quality indicators, enabling researchers to analyze water bodies with increased precision and reduced human error [41, 79]. This shift is critical given the challenges associated with traditional in-situ measurements, which are often labor-intensive and less comprehensive [39]. High-resolution remote sensing data, as demonstrated in studies by [6], shows that such approaches can lead to better accuracy in capturing water quality parameters like turbidity, a vital factor in assessing water cleanliness and ecological health. Moreover, these technologies facilitate the monitoring of small water bodies, which are often overlooked yet play crucial roles in local ecosystems [6]. As geographic and environmental challenges intensify, utilizing techniques such as the Modified Normalized Difference Water Index (MNDWI) can improve water body delineation and enhance water resource management strategies [18].

In addition to remote sensing, case study methodologies are vital for understanding specific local contexts and governance structures surrounding water resources. Case studies enable a nuanced examination of policy frameworks and community engagement practices, essential for developing tailored strategies that address the unique configurations of water access and quality in various Latin American regions. Longitudinal studies that capture interactions between regulatory environments and community behavior can reveal insights crucial for policy formulation [4].

Policy analysis remains an indispensable component of research methodologies as it aids in dissecting existing regulations and their efficacy concerning water quality and access. Evaluating the effectiveness of current water management policies helps practitioners and policymakers identify gaps and opportunities for improvement, facilitating the creation of more effective governance models [3, 55]. This analysis benefits from combining qualitative insights with quantitative data derived from methodologies like remote sensing, ensuring a holistic understanding of the issues at hand. Community surveys serve as a critical mechanism to engage local populations and gather qualitative data regarding their perceptions, needs, and experiences related to water usage and quality. They enable researchers to capture a breadth of perspectives essential for fostering participatory governance in water resource management. Understanding the social aspects of water crises, including local customs, historical injustices, and present challenges faced by marginalized communities, can guide practitioners in developing inclusive interventions that resonate with the public [14].

The role of artificial intelligence and machine learning in analyzing remote sensing data also represents a burgeoning field within water research. Techniques like convolutional neural networks allow for greater accuracy in detecting pollutants and predicting trends in water quality [23]. Such methodologies lend themselves to forecasting potential crises before they escalate, which is vital in regions where water scarcity and quality issues are anticipated to worsen due to climate change and other anthropogenic influences [34]. This predictive capacity can greatly enhance both the resilience of water systems and community preparedness.

Furthermore, integrating cross-disciplinary approaches enhances the richness of water crisis research. For example, combining ecological studies with remote sensing can yield significant insights into the health of aquatic ecosystems, informing better water management practices [82]. Understanding how urbanization and land use changes impact water quality and availability enables timely adaptations to plan and policies affecting resource management. In Latin America, characterized by diverse geographical challenges, applying multifaceted methodologies is essential for effective policy implementation. The exploration of water scarcity, pollution, and social equity in access to clean water requires a sophisticated analytical framework that bridges technological approaches with grounded community experiences. By employing a mixed-methods strategy that encompasses remote sensing, qualitative community engagement, and empirical policy analysis, researchers can significantly enhance the rigor and relevance of their findings [4, 77].

Methodological diversity is essential in conducting rigorous water crisis research in Latin America. The integration of advanced technologies like remote sensing, comprehensive case studies, policy evaluations, and community surveys creates a robust platform for understanding and effectively addressing the myriad challenges associated with water resources. Enhanced methodological frameworks will not only inform sound policy decisions but also elevate community voices, aligning research outputs with the urgent realities of water management in this critical region.

10.9 Conclusion

The water crisis in Latin America presents an urgent humanitarian, ecological, and governance challenge that requires coordinated responses at multiple levels. The intricate dynamics of pollution underscore the necessity for comprehensive water management strategies that consider the unique socio-political landscapes of individual countries. Investing in community-driven initiatives and advancing technological solutions can bolster efforts to mitigate water pollution and improve governance quality. Ultimately, sustainable progress hinges upon an equitable collaboration between civil society, governmental entities, and private sectors to ensure access to clean water, safeguarding health, livelihoods, and the environment for future generations. This analysis highlights that while the region possesses substantial freshwater resources, widespread contamination from industrial effluents, agricultural runoff, and untreated urban wastewater continues to undermine water quality and human well-being. These challenges are compounded by institutional weaknesses, fragmented governance frameworks, and deeply rooted socio-environmental inequalities that limit effective responses and policy enforcement.

A key finding of this review is the emergence of community-led initiatives and Indigenous environmental stewardship as critical drivers of change. These localized efforts challenge dominant, top-down governance paradigms and offer innovative, context-specific solutions grounded in traditional knowledge and participatory monitoring. In parallel, policy innovations in select countries demonstrate the potential for integrated water resource management (IWRM), green infrastructure, and decentralized treatment systems to enhance water quality outcomes. To address the persistent water pollution crisis in Latin America, this review underscores the need for inclusive governance structures that prioritize equity and participation, long-term investment in sustainable water infrastructure, and enhanced regional cooperation. Moreover, cross-sectoral approaches that bridge scientific knowledge with local experience are essential to achieving resilient and just water systems. The lessons drawn from Latin America's water crisis have broader relevance for the Global South, where similar environmental and institutional conditions prevail. By advancing frameworks that combine community empowerment, regulatory reform, and ecological sustainability, Latin America can inform future research and policymaking aimed at securing safe water access and environmental justice on a global scale.

References

1. Aguerre C, Bonina C (2023) Open government, civic tech, and digital platforms in Latin America: a governance study of Montevideo's urban app 'por mi barrio.' Inf Syst J 34(4):1037–1067. https://doi.org/10.1111/isj.12468
2. Alam M, Nusrin S, Islam A, Bhuiyan N, Rahim M, Delgado G, Morales R, Mendez JL, Navarro A, Gil AI, Watanabe H, Cravioto A (2010) Cholera between 1991 and 1997 in Mexico was

associated with infection by classical, El Tor, and El Tor variants of Vibrio cholerae. J Clin Microbiol 48(10):3666–3674. https://doi.org/10.1128/jcm.00866-10
3. Altenburger R, Aït-Aïssa S, Antczak P, Backhaus T, Barceló, Seiler T, Brion F, Busch W, Chipmen K, de Alda ML, de Aragao Umbuzeiro, Brack W (2015) Future water quality monitoring—adapting tools to deal with mixtures of pollutants in water resource management. Sci Total Environ 512–513, 540–551. https://doi.org/10.1016/j.scitotenv.2014.12.057
4. Andrés L, Boateng K, Borja-Vega C, Thomas E (2018) A review of in-situ and remote sensing technologies to monitor water and sanitation interventions. Water 10(6):756. https://doi.org/10.3390/w10060756
5. Appe S, Layton M (2016) Government and the nonprofit sector in Latin America. Nonprofit Policy Forum 7(2):117–135. https://doi.org/10.1515/npf-2014-0028
6. Avdan Z, Kaplan G, Göncü S, Avdan U (2019) Monitoring the water quality of small water bodies using high-resolution remote sensing data. ISPRS Int J Geo Inf 8(12):553. https://doi.org/10.3390/ijgi8120553
7. Balasubramanian D, Murcia S, Ogbunugafor C, Gavilán R, Almagro-Moreno S (2021) Cholera dynamics: lessons from an epidemic. J Med Microbiol 70(2). https://doi.org/10.1099/jmm.0.001298
8. Balderacchi C, Cassani A, Tomini L (2024) Shut up! Governments' popular support and journalist harassment: evidence from Latin America. Lat Am Polit Soc 66(4):23–39. https://doi.org/10.1017/lap.2024.3
9. Bezerra M, Vollmer D, Acero N, Marques M, Restrepo D, Mendoza E, Coutinho B, Encomenderos I, Zuluaga L, Rodriguez O, Shaad K, Serrano L (2021) Operationalizing integrated water resource management in Latin America: insights from application of the freshwater health index. Environ Manag 69(4):815–834. https://doi.org/10.1007/s00267-021-01446-1
10. Blanco-Gómez P, Jimeno-Sáez P, Senent-Aparicio J, Pérez-Sánchez J (2019) Impact of climate change on water balance components and droughts in the Guajoyo River Basin (El Salvador). Water 11(11):2360. https://doi.org/10.3390/w11112360
11. Cárdenas G, García S, Salas A (2018) Institutional framework and governance in Latin America. Int J Emerg Mark 13(5):1088–1107. https://doi.org/10.1108/ijoem-09-2017-0371
12. Carriquiriborde P, Mirabella P, Waichman A, Solomon K, Brink P, Maund S (2014) Aquatic risk assessment of pesticides in Latin America. Integr Environ Assess Manag 10(4):539–542. https://doi.org/10.1002/ieam.1561
13. Chin C, Sorenson J, Harris J, Robins W, Charles R, Jean-Charles, Bullard J, Webster DR, Kasarskis A, Peluso P, Paxinos EE, Waldor M (2011) The origin of the Haitian cholera outbreak strain. N Engl J Med 364(1):33–42. https://doi.org/10.1056/nejmoa1012928
14. Cornejo M, Bustamante J, Río M, Toro X, Latorre M (2023) Researching with qualitative methodologies in the time of coronavirus: clues and challenges. Int J Qual Methods 22. https://doi.org/10.1177/16094069221150110
15. Delgado G, Morales-Espinosa R, Méndez J, Cravioto A (2010) The re-emergence of cholera in the Americas, pp 79–95. https://doi.org/10.1007/978-1-60327-265-0_5
16. Domman D, Quilici M, Dorman M, Njamkepo E, Mutreja A, Mather A, Delgado G, Morales-Espinosa R, Grimont PA, Lizarraga-Partida ML, Bouchier C, Thomson N (2017) Integrated view of vibrio cholera in the Americas. Science 358(6364):789–793. https://doi.org/10.1126/science.aao2136
17. Dorman M, Domman D, Poklépovich T, Tolley C, Zolezzi G, Kane L, Vinas MR, Panagopulo M, Moroni M, Binsztein N, Caffer MI, Thomson N (2020) Genomics of the Argentinean cholera epidemic elucidate the contrasting dynamics of epidemic and endemic Vibrio cholerae. Nat Commun 11(1). https://doi.org/10.1038/s41467-020-18647-7
18. Du Y, Zhang Y, Ling F, Wang Q, Li W, Li X (2016) Water bodies' mapping from Sentinel-2 imagery with modified normalized difference water index at 10-m spatial resolution produced by sharpening the SWIR band. Remote Sens 8(4):354. https://doi.org/10.3390/rs8040354
19. Empinotti V, Budds J, Aversa M (2019) Governance and water security: the role of the water institutional framework in the 2013–15 water crisis in são paulo, Brazil. Geoforum 98:46–54. https://doi.org/10.1016/j.geoforum.2018.09.022

20. Epperly J, Witt A, Haight J, Washko S, Atwood T, Brahney J, Brothers S, Hammill E (2018) Relationships between borders, management agencies, and the likelihood of watershed impairment. Plos One 13(9):e0204149. https://doi.org/10.1371/journal.pone.0204149
21. Feld B, Galiani S (2015) Climate change in Latin America and the Caribbean: policy options and research priorities. Lat Am Econ Rev 24(1). https://doi.org/10.1007/s40503-015-0028-4
22. Félix-López M, Ormaza-Murillo M, Santana C, Banchón C (2023) Exploring the impact of reclaimed water on Latin America's development. J Ecol Eng 24(10):157–173. https://doi.org/10.12911/22998993/169962
23. Feng L, Yu Y, Zhang D, Li F, Guo J, Zhang Y, Fang F (2020) Water remote sensing eutrophication inversion algorithm based on multilayer convolutional neural network. J Intell Fuzzy Syst 39(4):5319–5327. https://doi.org/10.3233/jifs-189017
24. Flores C (2022) Toward a contextualized research agenda: governance challenges of the wastewater treatment policy in Mexico and the role of subnational governments. Wiley Interdiscip Rev Water 10(1). https://doi.org/10.1002/wat2.1617
25. Fraisl D, Campbell J, See L, Wehn U, Wardlaw J, Gold M, Moorthy I, Arias R, Peira J, Oliver JL, Maso J, Fritz S (2020) Mapping citizen science contributions to the unsustainable development goals. Sustain Sci 15(6):1735–1751. https://doi.org/10.1007/s11625-020-00833-7
26. Geddes M (2014) Neoliberalism and local governance: radical developments in Latin America. Urban Stud 51(15):3147–3163. https://doi.org/10.1177/0042098013516811
27. Gómez-Betancur L, Vilardy S, D Torres R (2021) Ecosystem services as a promising paradigm to protect environmental rights of indigenous peoples in Latin America: the constitutional court landmark decision to protect Arroyo Bruno in Colombia. Environ Manag 69(4):768–780. https://doi.org/10.1007/s00267-021-01483-w
28. Gorgoglione A, Crisci M, Kayser R, Chreties C, Collischonn W (2019) A new scenario-based framework for conflict resolution in water allocation in transboundary watersheds. Water 11(6):1174. https://doi.org/10.3390/w11061174
29. Gurgatz B, Carvalho-Oliveira R, Antoniaconi G, Saldiva P, Huergo L, Reis R (2018) Spatial distribution of atmospheric pollutants through biomonitoring in tree bark using X-ray fluorescence. Ecletica Quim 43(2):59. https://doi.org/10.26850/1678-4618eqj.v43.2.2018.p59-64
30. Harsha J (2025) Unlocking the blue frontier: saline water footprint as a path to sustainability. World Water Policy 11(1):165–171. https://doi.org/10.1002/wwp2.12257
31. Hesketh C (2024) Indigenous environmentalism: for a political economy where people (and planet) matter. Eur J Soc Theory 27(3):398–417. https://doi.org/10.1177/13684310241244524
32. Hussein H, Menga F, Greco F (2018) Monitoring transboundary water cooperation in SDG 6.5.2: how a critical hydropolitics approach can spot inequitable outcomes. Sustainability 10(10):3640. https://doi.org/10.3390/su10103640
33. Izah SC, Ogwu MC (2025) Modeling solutions for microbial water contamination in the global south for public health protection. Front Microbiol 16:1504829. https://doi.org/10.3389/fmicb.2025.1504829
34. Jiao Z (2024) The application of remote sensing techniques in ecological environment monitoring. Highlights Sci Eng Technol 81:449–455. https://doi.org/10.54097/7dqegz64
35. Jiménez C, Tercan E, Horstick O, Igboegwu E, Dambach P, Louis V, Winkler V, Deckart A (2022) Prevalence of anemia among indigenous children in Latin America: a systematic review. Revista De Saúde Pública 56:99. https://doi.org/10.11606/s1518-8787.2022056004360
36. Jutla A, Whitcombe E, Hasan N, Haley B, Akanda A, Huq A, Alam M, Sack RB, Colwell R (2013) Environmental factors influencing epidemic cholera. Am J Trop Med Hyg 89(3):597–607. https://doi.org/10.4269/ajtmh.12-0721
37. Kaenzig R, Piguet É (2013) Migration and climate change in Latin America and the Caribbean, pp 155–176. https://doi.org/10.1007/978-94-007-6985-4_7
38. Kane L (2007) Conflict and co-operation between 'popular' and 'state' education in Latin America. J Adult Contin Educ 13(1):53–67. https://doi.org/10.7227/jace.13.1.5
39. Kaplan G, Avdan Z, Göncü S, Avdan U (2019) Evaluation of RapidEye-3 satellite data for assessing water turbidity of Lake Borabey 14. https://doi.org/10.3390/ecws-4-06424

40. Kephart J, Ávila-Palència I, Bilal U, Gouveia N, Caiaffa W, Roux A (2021) COVID-19, ambient air pollution, and environmental health inequities in Latin American cities. J Urban Health 98(3):428–432. https://doi.org/10.1007/s11524-020-00509-8
41. Koparan C, Koc A, Privette C, Sawyer C (2019) Autonomous in situ measurements of noncontaminant water quality indicators and sample collection with a UAV. Water 11(3):604. https://doi.org/10.3390/w11030604
42. Kosoe EA, Osumanu IK, Ogwu MC (2025) Reimagining rural and small-town water supply in Ghana through inclusive governance and sustainable models. J Sustain 1(2):1–21. https://doi.org/10.55845/jos-2025-1234
43. Kosoe EA, Ogwu MC, Duniah J, Ndigun FB (2025) Sustaining indigenous water governance systems in ayimaye and Lungni, Northern Ghana, through hybrid models. J Sustain 1(2). https://doi.org/10.55845/jos-2025-1235
44. Laborde A, Tomasina F, Bianchi F, Bruné M, Buka I, Comba P, Corra L, Cori L, Duffert CM, Harari R, Iavarone I, Landrigan P (2015) Children's health in Latin America: the influence of environmental exposures. Environ Health Perspect 123(3):201–209. https://doi.org/10.1289/ehp.1408292
45. Lome-Hurtado A, Touza J, White P (2019) Environmental injustice in Mexico City: a spatial quantile approach. Expo Health 12(2):265–279. https://doi.org/10.1007/s12403-019-00310-2
46. López-Cubillos S, Muñoz-Ávila L, Roberson L, Suárez-Castro A, Ochoa-Quintero J, Crouzeilles R, Gallo-Cajiao E, Rhodes J, Dresseler W, Martinez-Harms M (2021). El histórico acuerdo de escazú: una oportunidad para integrar democracia, derechos humanos y conservación transfronteriza. https://doi.org/10.32942/osf.io/m8fd2
47. López-Cubillos S, Muñoz-Ávila L, Roberson L, Suárez-Castro A, Ochoa-Quintero J, Crouzeilles R., Gallo-Cajiao E, Rhodes J, Dresseler W, Martinez-Harms M, Runting R (2021) The landmark escazú agreement: an opportunity to integrate democracy, human rights, and transboundary conservation. Conserv Lett 15(1). https://doi.org/10.1111/conl.12838
48. Mantilla J, Feldmann A (2021) Criminal governance in Latin. America. https://doi.org/10.1093/acrefore/9780190264079.013.697
49. Miguel J (2018) Water management in Europe and Latin America. Manag Environ Qual Int J 29(2):348–367. https://doi.org/10.1108/meq-05-2017-0044
50. Mistry J, Bilbao B, Berardi A (2016) Community-owned solutions for fire management in tropical ecosystems: case studies from indigenous communities of South America. Philos Trans R Soc B Biol Sci 371(1696):20150174. https://doi.org/10.1098/rstb.2015.0174
51. Naspoli G, Libra J (2024) Fs 3.5: water footprint estimation in Latin America. https://doi.org/10.18235/0005516
52. Ogwu MC (2025) Sustainable development in Africa: challenges, innovations, and policy pathways for the future. (2025). Afr Res Rep 1(2):55–69. https://doi.org/10.5897/ARR2025.0005
53. Ogwu MC, Kosoe EA (2024) Place of cultural diversity in sustainable water resource management in Ghana. In: Izah SC, Ogwu MC, Loukas A, Hamidifar H (eds) Water crises and sustainable management in the global south. Springer, Singapore. https://doi.org/10.1007/978-981-97-4966-9_14
54. Ogwu MC, Kosoe EA (2025) Integrating green infrastructure into sustainable agriculture to enhance soil health, biodiversity, and microclimate resilience. Sustainability 17(9):3838. https://doi.org/10.3390/su17093838
55. Ogwu MC, Izah SC, Sawyer WE, Amabie T (2025) Environmental risk assessment of trace metal pollution: a statistical perspective. Environ Geochem Health 47:94. https://doi.org/10.1007/s10653-025-02405-z
56. Orjuela D, Chenou J (2018) Regionalism and presidential ideology in the current wave of Latin American integration. Int Area Stud Rev 22(1):41–63. https://doi.org/10.1177/2233865918815008
57. Páez A, Piedrahíta C (2021) Channeling water conflicts through the legislative branch in Colombia. Water 13(9):1214. https://doi.org/10.3390/w13091214

58. Pino P, Iglesias V, Garreaud R, Cortés S, Canals M, Folch W, Burgos S, Levy K, Naeher LP, Steenland K (2015) Chile confronts its environmental health future after 25 years of accelerated growth. Ann Glob Health 81(3):354. https://doi.org/10.1016/j.aogh.2015.06.008
59. Plassin S, Koch J, Paladino S, Friedman J, Spencer K, Vaché K (2020) A socio-environmental geodatabase for integrative research in the transboundary Rio Grande/río bravo basin. Sci Data 7(1). https://doi.org/10.1038/s41597-020-0410-1
60. Poirier M, Izurieta R, Malavade S, McDonald M (2012) Re-emergence of cholera in the Americas: risks, susceptibility, and ecology. J Glob Infect Dis 4(3):162. https://doi.org/10.4103/0974-777x.100576
61. Quenan C (2013) América latina frente a la crisis económica internacional: Buena resistencia global y diversidad de situaciones nacionales. Ideas (4). https://doi.org/10.4000/ideas.780
62. Rico A, Waichman A, Geber-Corrêa R, Brink P (2011) Effects of malathion and carbendazim on Amazonian freshwater organisms: comparison of tropical and temperate species sensitivity distributions. Ecotoxicology 20(4):625–634. https://doi.org/10.1007/s10646-011-0601-9
63. Rigler G, Dokou Z, Khadim F, Sinshaw B, Eshete D, Aseres M, Amera W, Zhou W, Wang X, Moges M, Azage M, Anagnostou E (2022) Citizen science and the sustainable development goals: building social and technical capacity through data collection in the Upper Blue Nile Basin, Ethiopia. Sustainability 14(6):3647. https://doi.org/10.3390/su14063647
64. Riveros P, Meriño J, Crespo F, Baptista B (2022) Situated transdisciplinarity in university policy: lessons for its institutionalization. High Educ 84(5):1003–1025. https://doi.org/10.1007/s10734-021-00812-6
65. Rodríguez-Villamizar L, Belalcázar L, Fernández-Niño J, Marín D, Rojas-Sánchez O, Acuña-Merchán L, Ramirez-Garcia N, Mangones-Matos SC, Vargas-Gonzalez JM, Herrera V (2021) Air pollution, sociodemographic and health conditions effects on COVID-19 mortality in Colombia: an ecological study. Sci Total Environ 756:144020. https://doi.org/10.1016/j.scitotenv.2020.144020
66. Sainz-Santamaría J, Martínez-Cruz A (2021) Governance of urban green spaces across Latin America—insights from semi-structured interviews to managers amid COVID-19. SSRN Electron J. https://doi.org/10.2139/ssrn.3782285
67. Sanders M, Miller L, Bhagwat S, Grient J, Rogers A (2019) Practitioner insights as a means of setting a context for conservation. Conserv Biol 34(1):113–124. https://doi.org/10.1111/cobi.13394
68. Silva E (2009) Challenging neoliberalism in Latin. America. https://doi.org/10.1017/cbo9780511803222
69. Stahler-Sholk R (2007) Resisting neoliberal homogenization. Lat Am Perspect 34(2):48–63. https://doi.org/10.1177/0094582x06298747
70. Taraky Y, Liu Y, McBean E, Daggupati P, Gharabaghi B (2021) Flood risk management with transboundary conflict and cooperation dynamics in the Kabul River basin. Water 13(11):1513. https://doi.org/10.3390/w13111513
71. Taylor D, Kahawita T, Cairncross S, Ensink J (2015) The impact of water, sanitation, and hygiene interventions to control Cholera: a systematic review. PLoS ONE 10(8):e0135676. https://doi.org/10.1371/journal.pone.0135676
72. Trimble M, Olivier T, Anjos L, Tadeu N, Giordano G, Donnell L, Laura R, Salvadores FJ, Santana Chavas IM, Torres PH, Pascual MA, Pahl-Wostl C (2022) How do basin committees deal with water crises? Reflections for adaptive water governance from South America. Ecol Soc 27(2). https://doi.org/10.5751/es-13356-270242
73. Valer J (2023) Participation in water resources management in Latin America 2017–2022: a systematic review. https://doi.org/10.56238/devopinterscie-144
74. Vázquez P, Carrera A, Cornejo M (2020) Corporate governance in the largest family firms in Latin America. Cross-Cult Strat Manag 27(2):137–163. https://doi.org/10.1108/ccsm-11-2018-0194
75. Verbyla M, Iriarte M, Mercado A, Coronado O, Almanza M, Mihelcic J (2016) Pathogens and fecal indicators in waste stabilization pond systems with direct reuse for irrigation: fate and transport in water, soil, and crops. Sci Total Environ 551–552:429–437. https://doi.org/10.1016/j.scitotenv.2016.01.159

76. Weesley J, Allen A. Zárate L, Emanuelli M (2021) Generative pedagogies from and for the social production of habitat. Planext-Next Gener Plan 11:26–43. https://doi.org/10.24306/plnxt/72
77. Wendt L, Hilberg S, Robl J, Braun A, Rogenhofer E, Dirnberger D, Strasser T, Fureder P, Lang S (2015) Using remote sensing and GIS to support drinking water supply in refugee/IDP camps. Gi_forum 1:449–458. https://doi.org/10.1553/giscience2015s449
78. Winter K, Vaughan M, Kurashima N, Wann L, Cadiz E, Kawelo A, Cypher M, Kaluhiwa L, Springer H (2023) Indigenous stewardship through novel approaches to collaborative management in Hawai'i. Ecol Soc 28(1). https://doi.org/10.5751/es-13662-280126
79. Xiao Y, Guo Y, Yin G, Zhang X, Shi Y, Hao F, Fu Y (2022) UAV multispectral image-based urban river water quality monitoring using stacked ensemble machine learning algorithms—a case study of the Zhanghe River, China. Remote Sens 14(14):3272. https://doi.org/10.3390/rs14143272
80. Yalew S, Kwakkel J, Doorn N (2021) Distributive justice and sustainability goals in transboundary rivers: case of the Nile basin. Front Environ Sci 8. https://doi.org/10.3389/fenvs.2020.590954
81. Yeates E, Cotterman K, Rhodes A (2020) Hydrologic impacts on human health: El Niño Southern Oscillation and cholera. https://doi.org/10.21079/11681/39483
82. Zhao Z (2019) Analysis of remote sensing technology applied on hydrology and water resources–taking Weihe's ecology as an example. Appl Ecol Environ Res 17(5). https://doi.org/10.15666/aeer/1705_1188711899
83. Zuluaga L, Montoya-Díaz J, Scott C, Bautista-Gómez M (2025) Equitable community participation in health initiatives among indigenous women in Latin America: a systematic review. BMJ Innov 11(2):98–106. https://doi.org/10.1136/bmjinnov-2024-001282

Chapter 11
South and Southeast Asia's Flooding and Water Contamination Crisis

Matthew Chidozie Ogwu and Attachai Jintrawet

Abstract South and Southeast Asia is facing a compounded environmental crisis at the intersection of climate-induced flooding and widespread water contamination, with profound implications for human health, ecological stability, and regional development. This chapter critically examines the systemic drivers and cascading consequences of recurrent flooding across diverse typologies, including flash floods, riverine inundation, urban waterlogging, and coastal surges, and their role in mobilizing biological, chemical, and emerging contaminants. Drawing from interdisciplinary evidence and region-specific case studies from Bangladesh, Indonesia, the Philippines, Thailand, and Vietnam, the chapter demonstrates how infrastructural fragilities, governance failures, and socio-spatial inequalities exacerbate exposure to waterborne pathogens, heavy metals, and persistent pollutants such as arsenic and microplastics. The analysis foregrounds vulnerable populations, particularly those in informal settlements, who bear the brunt of both infrastructural neglect and institutional exclusion. By bridging insights from environmental science, urban planning, and public health, the chapter advances a holistic framework for resilience grounded in adaptive infrastructure, inclusive policy design, and regionally coordinated responses. It also underscores the strategic relevance of community-based knowledge systems, decentralized wastewater solutions, and digital innovations (e.g., geospatial flood mapping) in strengthening disaster preparedness and water security. In reframing Southeast Asia's water crisis as a multidimensional governance and equity challenge, the chapter contributes novel pathways for research and policy integration in climate-vulnerable regions. The chapter repositions flooding and water contamination as interconnected governance and justice challenges—rather than isolated environmental events—providing a platform for rethinking integrated water

M. C. Ogwu (✉)
Goodnight Family Department of Sustainable Development, Appalachian State University, Boone, NC, USA
e-mail: ogwumc@appstate.edu

A. Jintrawet
Faculty of Agriculture (Retired), Chiang Mai University, Muang Chiang Mai, Thailand
e-mail: attachai.j@cmu.ac.th

M. C. Ogwu and S. Chibueze Izah (eds.), *Water Quality and Safety in the Global South*, Springer Water, https://doi.org/10.1007/978-3-032-10602-5_11

management under climate uncertainty. It calls for an urgent, interdisciplinary reorientation of water and disaster management systems to avert a deepening cycle of socio-environmental risk.

Keywords **South and** Southeast Asia · Flooding · Water contamination · Environmental justice · Climate resilience · Public health · Integrated governance

11.1 Introduction

South and Southeast Asia, distinguished by its intricate ecological systems and rapid urbanization, faces a growing crisis rooted in flooding and water contamination. The region's unique geographical and climatic conditions, alongside its socio-economic dynamics, make it particularly vulnerable to these challenges. This chapter explores the multifaceted nature of flooding and water contamination in South and Southeast Asia, examining the significance of these issues and their connections to global trade, health, and climate systems.

Flooding in South and Southeast Asia is exacerbated by climate change, urbanization, and deforestation, resulting in increased frequency and intensity of flood events. Major urban centers in countries such as Indonesia, Thailand, and Vietnam are situated in low-lying areas that are vulnerable to inundation, particularly during monsoonal rains. The World Bank has identified several South and Southeast Asian cities, including Jakarta, Bangkok, and Ho Chi Minh City, as among the most vulnerable globally to future flood events due to their geographical positioning and socio-economic conditions [1]. This increased flood risk presents significant implications for local populations, infrastructure, and economies, necessitating the development of effective flood resilience strategies [2].

Historically, flooding has significantly altered the landscapes and livelihoods within South and Southeast Asia. Annual flooding events disrupt agricultural production, threaten food security, and disseminate waterborne pathogens, adversely affecting public health. Diseases such as cholera have been linked to contaminated water sources exacerbated by flood conditions, highlighting the urgent public health implications tied to flooding [3]. The relationship between water quality and health is compounded by cultural practices, such as the reliance on street food vendors who often use contaminated water for food preparation, increasing the risk of gastrointestinal diseases [4]. The interconnectedness of these challenges points to a broader crisis where localized flooding and water contamination have far-reaching consequences, impacting regional stability, trade networks, and health systems. Rapid urbanization and population growth across South and Southeast Asia is accompanied by inadequate sanitation infrastructure and wastewater management, escalating the challenges posed by flooding and water quality degradation [5]. Without adequate intervention, these interlinked crises threaten to destabilize economic and social structures, necessitating comprehensive strategies that address both flooding risk management and water quality improvement.

South and Southeast Asia's geographical and economic importance is underscored by its role in global trade, biodiversity, and climate systems. Critical waterways, such as the Mekong River, serve as vital trade routes while supporting millions who depend on them for sustenance and economic activity. The degradation of these waterways due to flooding and contamination has implications that extend beyond local communities, affecting international trade and regional economies [2].

Moreover, the looming threat of climate change demands immediate action, as changing weather patterns contribute to increased flooding risks and water scarcity [6]. These changes are reflected in agricultural impacts, particularly in the context of rice production, which is the staple food for millions in the region. Reports indicate significant shortfalls in rice yield due to flooding-induced waterlogging and salinity issues, threatening food security and economic stability across South and Southeast Asia [7]. Emerging contaminants, including heavy metals like arsenic, further complicate the water pollution crisis in the region. Contaminated groundwater sources, primarily due to industrial activities and inadequate waste management, endanger millions at risk of chronic health issues associated with long-term exposure [8]. The intertwining challenges of flooding, water contamination, and public health create a nexus requiring interdisciplinary responses that involve environmental science, public health, and urban planning initiatives [9].

This chapter aims to provide a comprehensive examination of the complex and interlinked challenges of flooding and water contamination in South and Southeast Asia. By analyzing the environmental, infrastructural, and socio-political drivers of recurrent flooding events and their consequences on water quality, the chapter sheds light on a pressing regional crisis with global health and ecological implications. It contributes to existing knowledge by integrating case studies from multiple South and Southeast Asian countries, offering insights into both common patterns and localized experiences. Moreover, the chapter highlights gaps in policy, infrastructure, and community resilience, while emphasizing the role of climate change and urbanization in exacerbating risks. Through this multidisciplinary analysis, the chapter advances scholarly and policy-oriented understanding of how flood-related water contamination impacts public health, ecosystems, and sustainable development in one of the world's most climate-vulnerable regions.

11.2 Water Crisis in South and Southeast Asia: Geographic and Climatic Context

The water crisis affecting South and Southeast Asia, particularly in the Mekong Delta, is deeply intertwined with the region's complex physical geography, climate variability, and escalating human activity. This dynamic interplay shapes the hydrological and ecological profiles of the area, intensifying vulnerabilities to flooding while complicating strategies for effective flood management and conservation.

The Mekong River represents one of the principal hydrological arteries flowing through Southeast Asia, providing essential resources to millions, either directly through water supply or indirectly via ecosystem services. Its delta, encompassing an extensive floodplain in Vietnam, plays a crucial role in the region's economy by supporting agriculture, particularly rice cultivation, and sustaining fisheries, which are among the most productive in the world, driven by the annual flood pulse that inundates the delta during the wet season [10, 11],. However, recent manipulations in the region's hydrology, particularly through dam constructions aimed at hydropower generation, threaten to disrupt this balance. Existing dams on the Mekong tributaries have already exhibited significant impacts, altering hydrological patterns crucial for the ecosystems reliant on floodplain dynamics [12, 13]. This alteration has rendered the Mekong Delta increasingly susceptible to sediment starvation, leading to enhanced erosion and ecological degradation as sediment deposition fails to keep pace with rising sea levels [11].

Urbanization within flood-prone areas has further exacerbated the situation. Rapid urban growth, particularly in cities such as Phnom Penh and Ho Chi Minh City, has led to heightened population density in regions that are traditionally vulnerable to flooding. This demographic pressure facilitates the over-extraction of water resources and increased waste discharge, compounding the stress on local ecosystems [14]. Consequently, the intricate relationship between urban development and flood risk is of paramount concern, with informal settlements in particular representing high-risk areas during flooding events. A considerable section of Metro Manila's population, consisting of informal settlers, has been experiencing severe flooding consequences due to typhoons, which have increased in frequency and intensity since the 1970s, illustrating the dire implications of unmanaged urbanization in the face of climate variability [15].

The effects of monsoon patterns and climatic variations, notably those influenced by the El Niño Southern Oscillation (ENSO), further complicate water resource management across Southeast Asia. Seasonal flood regimes, driven primarily by monsoonal rains, are crucial for the maintenance of freshwater systems in the Mekong Delta and related floodplains like the Tonle Sap. Studies indicate a potential shift in the timing and intensity of these rainfalls due to climate change, leading to irregular flooding patterns [16, 17]. This fluctuation threatens agricultural yields and increases the risk of waterborne diseases, as prolonged flooding may lead to discharges of pollutants into freshwater supplies, adversely affecting community health [18].

Climate change also poses another significant threat through rising sea levels, which are projected to inundate coastal regions within the Mekong Delta, diminishing the area suitable for agriculture while simultaneously increasing salinity levels detrimental to freshwater ecosystems [19]. The interdependence of flooding patterns and climatic conditions underscores the complexity with which ecological resilience must be assessed, as conditions shift, livelihoods dependent on both predictable flooding dynamics and ecological health face increasing uncertainty [20, 21].

Land use practices have also been implicated in worsening flooding conditions. Notably, deforestation contributes to adverse hydrological outcomes, such as increased runoff and reduced soil retention capabilities, resulting in elevated flood risks. The destruction of mangroves and other natural buffers has substantially diminished the landscape's capacity to absorb storm surges and high-water levels, leading to greater inundation during extreme weather events [22]. Furthermore, the loss of these crucial ecosystems directly impacts local fisheries and broader biological diversity as habitat zones contract due to anthropogenic pressures [23].

Moreover, the water management frameworks in place to combat flooding must adapt to accommodate both natural and socio-economic fluctuations. Aggressive infrastructure development, often executed with insufficient regard for ecological consequences, has altered the natural hydrological configurations of rivers and lakes in the region [13, 24, 25]. The hydrologic models employed so far have proved inadequate in anticipating the nuances of climate-induced hydrological challenges, indicating a need for a reevaluation of flood forecasting methodologies to incorporate resilience-based approaches. This paradigm shift can help prepare communities and infrastructures for what is increasingly likely to be a more variable and chaotic hydrological future [26]. Hence, as South and Southeast Asia grapples with the dual pressures of climate change and urbanization, comprehensive strategies that merge environmental conservation with socio-economic development are crucial in mitigating risks associated with floods. Collaborative governance at regional levels is essential, bringing together stakeholders that are directly affected by or can influence water management practices. Such collaboration is fundamental to creating and implementing policies that safeguard both human livelihoods and natural ecosystems in this flood-prone area [10, 27].

The interplay of these multifaceted influences—geographical, climatic, and anthropogenic—underscores the urgent need for an integrated approach to managing the water crisis in South and Southeast Asia. Strategies must emphasize sustainable land use, ecosystem restoration, and enhanced flood management infrastructure tailored to the unique hydrological features of the Mekong Delta and its surrounding areas. This requires foresight and collaboration among governments, communities, and researchers to envision a resilient future that can withstand the inevitable pressures of climate volatility and human development.

11.3 Typology and Drivers of Flooding in Southeast Asia

Southeast Asia is confronted by a multitude of flooding events that can be classified into distinct types: flash floods, riverine floods, urban floods, and coastal surges (Table 11.1). Various natural and anthropogenic factors drive each of these flood types. Table 11.1 presents a classification of major flood types affecting Southeast Asia, along with their underlying drivers and geographic patterns: trends, land-use dynamics, and socio-ecological vulnerabilities.

Flash floods: Flash floods, characterized by their rapid onset and short duration, are often triggered by intense rainfall events that exceed the capacity of drainage systems. These floods can result from sudden, localized storms or the rapid melting of snow in adjacent regions. Studies indicate that these phenomena are exacerbated

Table 11.1 Flood types and drivers of flooding in Southeast Asia

Flood type	Explanation	Key drivers	Typical locations
Fluvial floods	River-based flooding caused by rising water levels in rivers and tributaries	Monsoon rains, upstream deforestation, poor watershed management, and dam discharge	Mekong delta (Vietnam), Chao Phraya basin (Thailand)
Riverine floods	Flooding due to overflow from large river systems and floodplains	Prolonged rainfall, snowmelt (in upland areas), catchment degradation, sedimentation	Irrawaddy Basin (Myanmar), Red River (Vietnam)
Pluvial floods	Surface flooding due to high-intensity rainfall overwhelming drainage systems	Urbanization, inadequate stormwater infrastructure, and convective storms	Jakarta (Indonesia), Metro Manila (Philippines)
Coastal Floods	Flooding caused by storm surges, tidal inundation, or seawater intrusion	Sea-level rise, tropical cyclones, coastal erosion, and land subsidence	Coastal Bangladesh, Mekong Delta (Vietnam)
Flash floods	Rapid-onset floods with high flow velocities in steep or hilly terrain	Intense local rainfall, deforestation, and land instability	Northern Luzon (Philippines), Central Sulawesi (Indonesia)
Urban floods	Flooding in built-up areas is often linked to infrastructure failure	Impervious surfaces, drainage blockage, and poor urban planning	Bangkok (Thailand), Ho Chi Minh City (Vietnam)

by climate change, as increasing atmospheric temperatures lead to higher moisture levels, which can intensify rainfall [28, 29]. Southeast Asia, particularly during the monsoon season, frequently experiences such extreme weather, resulting in flash flooding that poses significant danger to human life and property. For instance, regions in the Philippines are particularly susceptible due to geographic and climatic conditions [30, 31].

Riverine floods: Riverine floods occur gradually, following the overflow of rivers due to prolonged rainfall or snowmelt. In Southeast Asia, many of the major rivers, such as the Mekong and the Chao Phraya, are prone to flooding during the wet season [1, 32]. Riverine flooding is critically influenced by changes in land use, including deforestation and urbanization, which reduce the land's ability to absorb water [33]. Furthermore, infrastructure limitations in flood management exacerbate these issues. For instance, inadequate levees and flood control systems can fail to contain rising water levels during prolonged rainy periods, leading to catastrophic events [34, 35].

Urban floods: Urban floods have emerged as a significant concern in rapidly growing cities such as Jakarta, Bangkok, and Ho Chi Minh City. Rapid urbanization has outpaced the capacity of urban drainage systems to manage excessive runoff, leading to widespread flooding [1, 32]. The combination of concrete surfaces and inadequate drainage exacerbates the flood risk in metropolitan areas. In urban settings, the risk of flooding is further heightened by climate change, which can increase the severity and frequency of extreme rainfall events [36]. Moreover, urban areas often lack sufficient green spaces that could mitigate runoff, thus highlighting the need for integrated urban planning and better stormwater management systems [32].

Coastal surges: Coastal surges are another significant flooding concern, particularly for countries like Vietnam, Thailand, and the Philippines, which have extensive coastlines vulnerable to typhoons and rising sea levels [29]. Cyclones can lead to storm surges that inundate coastal areas, and this risk is amplified by climate change, as rising sea levels increase the baseline for storm surges [37]. The interplay between storm intensity and frequency due to climatic shifts presents a growing challenge for coastal communities, necessitating proactive adaptation strategies to mitigate impact [30, 38].

Natural drivers of flooding in Southeast Asia are multifaceted. Intense rainfall, cyclones, and climate change are significant contributors to flooding events in the region. The occurrence of cyclones is notably pronounced in Southeast Asia's geography, where several cyclone-prone areas converge [29]. Recent trends show that not only the frequency but also the intensity of cyclones is increasing, leading to more severe flooding incidents and posing threats to infrastructure and livelihoods [38, 39].

Anthropogenic drivers also play a critical role in exacerbating flood conditions in Southeast Asia. Deforestation is a significant issue throughout the region [40], where forest ecosystems are converted into agricultural land or urban areas [33]. The removal of trees reduces the land's natural ability to absorb rainfall, increasing runoff and contributing to flood risks. Additionally, poor urban drainage infrastructure is prevalent, particularly in informal settlements where rapid population growth outstrips the development of adequate water management systems [32, 34].

Furthermore, water mismanagement is a critical factor influencing flood dynamics [41]. The construction of dams, while aimed at managing water resources, can inadvertently lead to altered water flow patterns and increased flood risks downstream when not managed properly [42]. Inadequate zoning and land-use planning further complicate the issue, as areas prone to flooding may be developed without sufficient constraints or flood control measures in place [1]. Community vulnerability in flood-prone areas is influenced not only by the physical environment but also by social and economic factors. Engagement in flood risk management policies and local participation can significantly affect a community's resilience to flooding [34]. Addressing social inequalities and enhancing community involvement in flood planning and response mechanisms is crucial to improving overall flood resilience [43].

11.4 Flooding-Induced Water Contamination in Southeast Asia: Mechanisms and Pathways

Flooding-induced water contamination poses a significant environmental and public health risk in Southeast Asia, a region increasingly affected by extreme weather events due to climate change. The mechanisms contributing to this contamination are multifaceted, stemming from municipal sewage overflow, agricultural practices, industrial activities, and infrastructural failures. This table outlines the key mechanisms through which flooding events contribute to water contamination across Southeast Asia (Table 11.2). Municipal sewage systems in urban areas of Southeast Asia are often overwhelmed during heavy rainfall and flooding events, leading to the overflow of untreated sewage into surrounding waterways. This phenomenon is particularly pronounced in densely populated regions where infrastructure may be outdated or inadequately maintained. Flooding exacerbates the influx of pathogens and pollutants into the water supply, significantly affecting public health and safety. A study in Semarang, Indonesia, found that flooding led to the contamination of drinking water sources, with over 55% of water samples containing Escherichia coli due to runoff from sewage during such events [44].

Agricultural runoff is another critical mechanism contributing to water contamination in flood-ravaged regions. In Southeast Asia, heavy rainfall can lead to the leaching of fertilizers and pesticides from agricultural fields directly into waterways. The chemical loading resulting from these agricultural practices can create hazardous conditions in surface waters, as highlighted by recent reviews discussing the threat posed by emerging contaminants, including pesticides and pharmaceuticals, to both ecosystems and human health [45]. Furthermore, the intensive practices associated with rice cultivation—especially in paddy-field-reliant countries—increase the likelihood of nutrient runoff during floods [42]. These agricultural practices, combined with flooding, underscore the need for a comprehensive approach to water management that considers the chemical dynamics during extreme weather events.

Industrial runoff further complicates the landscape of water contamination during floods. Industries located near water bodies can inadvertently release toxic substances into the environment when flooding occurs [46]. Evidence suggests that floodwaters can transport these toxic substances across considerable distances, affecting water quality downstream and posing long-term ecological risks. The release of pollutants from industrial sites not only contaminates drinking water but also significantly impacts aquatic ecosystems, potentially leading to decreased biodiversity and the disruption of fish populations, which are critical for local economies and food security.

A primary concern also arises from the breakdown of water treatment facilities during flooding. Many regions in Southeast Asia suffer from inadequate water treatment infrastructure, which is often compromised during flood events. This vulnerability can result in the introduction of untreated water into drinking supplies, thereby exacerbating health risks associated with waterborne diseases [44]. The reliance on

Table 11.2 Mechanisms and pathways of flooding-induced water contamination in Southeast Asia

Contamination source	Mechanism during flooding	Primary pollutants	Public health and environmental impacts
Sewage overflow	Inundation of septic systems and pit latrines leading to discharge into open water	Fecal coliforms, E. coli, viruses, helminths	Diarrheal diseases, cholera, typhoid, and ecosystem degradation
Industrial spillage	Floodwaters breach containment of industrial sites and storage facilities	Heavy metals (Hg, Pb, and Cd), solvents, and hydrocarbons	Toxic exposure, long-term soil and water pollution
Agricultural runoff	Mobilization of fertilizers, pesticides, and livestock waste into waterways	Nitrates, phosphates, pesticides, pathogens	Eutrophication, harmful algal blooms, and contamination of drinking water
Solid waste leachate	Flooding of open dumpsites and informal landfills, releasing leachates	Organic waste, microplastics, endocrine disruptors	Surface and groundwater contamination, vector-borne disease risks
Oil and fuel leakage	Flood-induced damage to fuel stations and storage tanks	Petroleum hydrocarbons, PAHs	Surface film on water, toxicity to aquatic organisms
Saline intrusion (Coastal)	Coastal flooding introduces seawater into freshwater and agricultural systems	Salts, suspended solids	Soil salinization, reduced freshwater availability

electric power for treatment processes means that power outages, common during extreme flooding, render these systems ineffective. Scholars emphasize the need for improvements in the resilience of water management infrastructure to ensure that treatment systems remain functional, even during flooding [46].

Additionally, the pathways through which floodwater infiltrates drinking water sources are varied and complex. Contaminated floodwaters can easily seep into wells and aquifers, particularly in rural areas where groundwater serves as the primary source of drinking water. The susceptibility of aquifers to surface contaminants is alarming, as demonstrated by findings that highlight the link between flooding and groundwater pollution [44]. Significant urban areas like Bangkok experience increased vulnerability due to their geographic and infrastructural layout, leading to systemic risks during flood events [1].

Addressing flooding-induced water contamination throughout Southeast Asia requires a multifaceted and integrated management approach. Effective policies should encompass improvements in urban planning to include resilient infrastructure capable of withstanding flood events, alongside innovations in agricultural practices to minimize runoff. Recent proposals advocate for the adoption of low-impact development strategies that enhance water quality and control stormwater runoff [47]. By promoting sustainable land-use practices and implementing stronger regulatory frameworks for industrial emissions, the region can mitigate the negative impacts of floods and protect public health.

11.5 Public Health and Environmental Impacts of Water Crisis in Southeast Asia

The water crisis in Southeast Asia presents significant public health and environmental challenges that are critical for understanding the region's socio-economic dynamics. This crisis is marked by rising instances of waterborne and vector-borne diseases, the emergence of environmental contaminants, and pervasive ecological degradation. Each of these issues disproportionately affects vulnerable populations, including children, the elderly, and those living in informal settlements, highlighting the need for comprehensive intervention strategies.

Waterborne diseases, such as cholera, typhoid fever, leptospirosis, and hepatitis A and E, are prevalent due to inadequate water quality and access. The relationship between unsafe water access and these diseases is especially evident in studies such as that by Mogasale et al., which systematically reviews the increased typhoid fever risk associated with insufficient access to clean water in Southeast Asia [48]. The review noted that the region suffering from water scarcity and lacking sanitation facilities are significantly more inclined to experience outbreaks of these diseases. Moreover, water sources contaminated with pathogens due to insufficient sanitation infrastructure contribute meaningfully to these public health issues [49]. In addition to waterborne diseases, vector-borne illnesses are exacerbated by flooding, a common occurrence during Southeast Asia's monsoon season. Dengue fever and malaria outbreaks increase as stagnant water creates optimal breeding conditions for mosquitoes [50]). The cyclical nature of these outbreaks underscores the need for integrative public health strategies that take environmental factors and infrastructure challenges into account.

The emergence of environmental contaminants in Southeast Asia further complicates public health issues. Emerging contaminants, including microplastics and per- and polyfluoroalkyl substances (PFAS), have infiltrated water supplies, posing substantial threats to both ecological and human health [45, 51]. Bhagat et al. outline the pressing concerns regarding these pollutants, as their presence can result in serious health implications, particularly for communities reliant on contaminated water sources [45]. Their examination of the region's environmental challenges reveals a disturbing trend of increasing pollution levels in both terrestrial and aquatic ecosystems. Microplastics, in particular, have been documented widely across Southeast Asia's drinking water systems. New et al. report that microplastics have been detected in various drinking water sources, indicating broader issues concerning plastic waste management in the region [52]. These pollutants not only compromise water quality but also adversely affect aquatic life, leading to ecosystem imbalances [53]. The systemic shortcomings, particularly in wastewater treatment facilities, exacerbate the issue, as many facilities are ill-equipped to effectively remove microplastics [54].

The environmental degradation driven by wastewater discharge and plastic pollution is further exemplified by eutrophication. High nutrient levels in contaminated water bodies lead to algal blooms that deplete oxygen levels and disrupt aquatic ecosystems [55]. The ecological health of rivers such as the Chao Phraya, Citarum, and Saigon is threatened by this degradation, as evidenced by data collected from these vital waterways [56]. The impact of such environmental challenges could reverberate through socioeconomic systems, affecting fisheries, livelihoods, and ultimately food security [53].

Vulnerable populations bear the brunt of these crises, as economic and social disparities greatly influence access to clean water and sanitation. According to research by Babel et al., inadequate water management systems further exacerbate health risks in regions heavily populated by informally settled communities [56]. These populations often experience heightened susceptibility to health risks from both waterborne diseases and environmental contaminants. Children and the elderly are particularly vulnerable, given their greater sensitivity to waterborne pathogens and contaminants [48]. Climate-related factors also play a significant role in exacerbating water crises in Southeast Asia. Rising temperatures, prolonged droughts, and extreme weather events contribute to water scarcity, while flooding events increase the risk of contaminating potable water resources [57]. The interconnectedness of climate change, environmental justice, and public health indicates that socio-political considerations must be integrated into regional strategies aimed at addressing these crises [58]. The interplay between environmental degradation and public health is nuanced and multifaceted across Southeast Asia. Interdisciplinary approaches are essential in crafting sustainable solutions. For example, improving wastewater treatment infrastructure could alleviate public health burdens by reducing waterborne illnesses and environmental impacts by minimizing pollutant discharge [51].

Future studies should further explore perceptions and coping strategies toward water scarcity and quality challenges among Southeast Asian populations. Raising public awareness about clean water and sanitation issues can significantly influence consumer behavior towards water use, thereby enhancing overall community resilience [57]. This behavioral adaptation, combined with infrastructural improvements, could potentially bridge the gap between adequate water availability and health outcomes for affected populations. Given the urgency of the water crisis in South and Southeast Asia, proactive measures encompassing policy, education, and community engagement are essential in mitigating associated public health risks and environmental impacts. Future planning must prioritize sustainable water governance, robust public health outreach, and environmental stewardship to ensure a resilient response to these challenges.

11.6 Case Studies of Flooding and Water Contamination in South and Southeast Asia

Flooding and water contamination pose significant threats to public health across South and Southeast Asia, driven by urbanization, climatic conditions, and socio-economic vulnerabilities. This analysis examines several pivotal case studies from Bangladesh, Indonesia, the Philippines, Thailand and Vietnam, focusing on the multi-faceted impacts of flooding and accompanying water quality deterioration. The case studies collectively highlight the multifaceted challenges South and Southeast Asian countries face regarding flooding and water quality. Each scenario, while unique in local dynamics, illustrates a common theme where inadequate infrastructure, rapid urbanization, and climatic factors converge to present compounded risks for public health. Addressing these challenges requires collaborative approaches involving environmental management, urban planning, and health education. Further research and policy efforts are crucial for developing resilient systems that can withstand flooding impacts and mitigate water pollution in the region.

11.6.1 Bangladesh and the Ganges Delta: Chronic Flooding and Arsenic Contamination

Bangladesh, particularly the Ganges Delta, is notoriously susceptible to chronic flooding, primarily due to its geographic position and climate change effects, including increased rainfall and rising sea levels. These recurrent flooding events significantly enhance the mobilization of environmental toxins, particularly arsenic, which leaches into groundwater supplies from natural geologic formations. Water safety in Bangladesh is further compromised by the scarcity of standard arsenic removal technologies in rural areas, leading to a public health crisis characterized by arsenicosis and other health issues [59]. Research indicates that the failure of drainage systems during floods exacerbates the situation, allowing accumulated contaminants from urban runoff to infiltrate potable water sources. Arsenic concentrations in drinking water can soar beyond recommended limits after heavy flooding events. Chronic exposure to arsenic raises significant health risks, including skin lesions and heightened cancer susceptibility, demonstrating that flooding not only disrupts daily life but also catalyzes long-term health issues among the population.

11.6.2 Vietnam (Mekong Delta): Rising Sea Levels, Salinization, and Fecal Contamination

In Vietnam's Mekong Delta, rising sea levels pose a growing threat to fresh-water resources, leading to the salinization of agricultural land and disturbances in local ecosystems [117]. The saline intrusion challenges traditional farming and affects food security, particularly as rice cultivation—the backbone of the local economy—becomes unsustainable. This condition is exacerbated during the wet season when heavy rains lead to flooding, which further spreads fecal contamination from poorly constructed sanitation systems into waterways [60]. Research highlights that inadequate waste management in informal settlements allows for significant fecal pathogen transmission, especially during and after flooding periods. The resulting fecal contamination in water supplies can precipitate outbreaks of gastroenteritis and other waterborne diseases, reinforcing the urgent need for improved infrastructural responses to manage both salinity and pathogen presence in drinking water sources [60].

11.6.3 Indonesia (Jakarta): Subsidence, Clogged Drainage, and Fecal Pollution

Jakarta faces critical challenges characterized by land subsidence and inadequate drainage systems. Rapid urbanization, marked by construction and population growth, has exacerbated the city's vulnerability to flooding. The combination of subsidence impacted by excessive groundwater extraction and inadequate drainage results in frequent inundations, especially during the rainy season [61]. Clogged drainage channels due to waste accumulation not only impede water flow but also lead to the accumulation of fecal matter and pollutants in floodwaters. A study confirms that Jakarta's floods are intertwined with urban density and insufficient waste management practices, leading to severe public health risks, including waterborne diseases and contamination of drinking water supplies [61]. Without a comprehensive management strategy addressing subsidence, drainage, and sanitation, Jakarta remains susceptible to future water quality crises.

11.6.4 Philippines (Manila): Typhoons, Informal Settlements, and Leptospirosis Outbreaks

The Philippines is highly vulnerable to natural disasters, particularly typhoons that cause heavy flooding. These events severely impact communities, especially in Manila, where a significant portion of the population resides in informal settlements characterized by inadequate infrastructure [62]. Studies have shown a direct correlation between extreme weather events and outbreaks of leptospirosis, a bacterial infection often contracted through contaminated water [63, 64]. Leptospirosis outbreaks frequently surge following heavy rainfall, as observed in the aftermath of major typhoons. Conditions in slum areas often lead to stagnant pools of water, creating an ideal environment for bacteria to thrive [64, 65]. In the context of urban health, the convergence of socioeconomic instability, suboptimal sanitation, and climate-related vulnerabilities necessitates urgent interventions to mitigate leptospirosis risks and improve overall community health resilience [62, 65]. The Philippine experience highlights how environmental health is linked to both natural and anthropogenic factors.

11.6.5 Thailand: The 2011 Flood Event and Industrial Chemical Spills

The catastrophic floods of 2011 in Thailand illustrated the intricate relationship between flooding and industrial pollution. The extensive inundation disrupted industrial operations, resulting in chemical spills that contaminated surrounding water bodies [66]. A study conducted after the floods revealed alarming concentrations of heavy metals in water and sediments near industrial sites, raising concerns about long-term ecological impacts and human health risks associated with exposure [15]. The economic impacts of such flooding are profound, with estimated losses affecting local industries, particularly in manufacturing and agriculture. Researchers developed a modeling framework to analyze both direct and indirect economic losses resulting from flooding, underlining the need for preemptive industrial regulations aimed at minimizing chemical pollution during flood events [67]. The Thai case exemplifies the necessity for integrated flood management strategies that also account for environmental protection and public health considerations.

11.6.6 Infrastructure and Technological Responses to Flooding and Water Contamination in Southeast Asia

Flooding and water contamination present significant challenges in Southeast Asia, exacerbated by climate change and rapid urbanization that often outpaces infrastructural development. Countries in this region experience recurrent flooding during monsoon seasons, primarily due to geographic vulnerability and inadequate water management systems. An integrated response involving early warning systems, flood-resilient infrastructure, urban upgrades, decentralized wastewater treatment technologies, and geospatial mapping tools is crucial for enhancing flood resilience and managing water quality in the region. Effective early warning systems are essential for disaster preparedness and response to flooding events. These systems enable the timely dissemination of information that can significantly reduce casualties and economic losses associated with floods. Research has underscored the importance of using advanced technologies such as remote sensing and geospatial data analysis in developing these early warning systems [68]. Such systems can utilize satellite imagery and local sensor data to predict flood risks and monitor water levels in real-time, empowering communities to respond proactively. Moreover, incorporating community engagement in early warning systems has been shown to enhance local preparedness [1]. Innovatively designed frameworks that integrate governmental, private, and community resources are necessary for maximizing the effectiveness of early warning initiatives [6].

In terms of flood-resilient water and sanitation systems, Southeast Asia is witnessing increasing adoption of solutions like floating toilets and raised tanks. These technologies are particularly relevant in informal settlements where conventional sanitation systems are insufficient and vulnerable to flooding. Utilizing such innovative infrastructure can help mitigate the direct impacts of flooding on sanitation, thereby reducing public health risks related to waterborne diseases [42]. Additionally, studies emphasize the potential benefits of decentralized wastewater treatment systems, which enable local communities to manage their wastewater sustainably, particularly during flood events when centralized systems may fail [69, 70]. These decentralized systems not only contribute to disaster resilience but also facilitate the recovery of nutrients and water, promoting a circular economy within communities [71].

Urban infrastructure upgrades play a crucial role in managing stormwater and minimizing the risk of flooding. Increasingly, governments across Southeast Asia are investing in green infrastructure, such as permeable pavements and green roofs, alongside traditional gray infrastructure solutions like enhanced drainage systems [2]. This dual approach is critical for improving stormwater management since green infrastructure can significantly reduce runoff and enhance water quality during flood conditions. The feasibility and effectiveness of integrating blue-green infrastructure into urban planning have been supported by studies highlighting case examples in cities like Jakarta and Ho Chi Minh City, where urban flooding has necessitated innovative drainage solutions [72]. However, evidence suggests that while green infrastructure can enhance urban resilience, its efficacy is contingent on careful implementation and consideration of local conditions [73].

The adoption of decentralized wastewater treatment technologies serves as a complementary strategy in Southeast Asia, addressing both flood resilience and water contamination issues. Such technologies allow for local processing of wastewater, which is particularly pertinent in communities where centralized facilities are impractical due to logistical or financial constraints [70, 74]. Various types of decentralized systems, including anaerobic–aerobic treatment combinations and constructed wetlands, have been identified as effective solutions for managing industrial and domestic wastewater [75]. These systems alleviate the burden on existing treatment facilities during floods and empower communities to protect their local water resources from pollution [76]. Moreover, the economic feasibility of decentralized solutions has been examined, illustrating their capacity to meet the unique needs of urban, peri-urban, and rural areas within Southeast Asia.

The use of Geographic Information Systems (GIS) and remote sensing plays a pivotal role in flood mapping, risk assessment, and management across Southeast

Asia. These technologies enhance the ability to visualize, analyze, and prioritize flood-prone areas for proactive interventions [6, 77]. By integrating GIS with hydrological models, decision-makers can more accurately forecast flooding scenarios and allocate resources effectively to mitigate potential impacts. Research into operational flood monitoring utilizing deep learning models has also shown promise, demonstrating how geospatial data can improve situational awareness needed to respond to flood crises swiftly [68]. These technological advancements signify a crucial shift towards data-driven, evidence-based planning in addressing the multifaceted challenges posed by flooding and water contamination in the region. The synergistic implementation of early warning systems, flood-resilient water and sanitation solutions, urban infrastructure upgrades, decentralized wastewater treatment technologies, and sophisticated geospatial tools represent an integrated approach to managing the dual threats of flooding and water contamination in Southeast Asia. Not only do these strategies enhance the region's resilience to climate-induced disasters, but they also promote sustainable water management practices that can support ongoing economic development and public health initiatives. Increased investment in infrastructure, capacity building, and community engagement will be essential for fully realizing these opportunities.

11.6.7 Flooding and Water Contamination Policy and Governance Challenges in Southeast Asia

National disaster and water management frameworks in Southeast Asia significantly influence how countries respond to water-related challenges. Development in urban centers is juxtaposed against vulnerabilities linked to climate change and rapid urbanization. As cities expand, integrating water security into urban planning becomes crucial. Lorenzo and Kinzig articulate that Southeast Asian cities face unique water management challenges, necessitating dedicated efforts to improve sanitation and flood resilience, while addressing the socio-economic disparities unique to the region [5]. Policy frameworks must encapsulate these aspects, promoting holistic approaches that emphasize sustainable urban development while addressing inequities in water management.

A critical examination reveals that these frameworks often suffer from significant coordination gaps between environmental, water, and urban planning sectors. Laeni et al. highlight the fragmented institutional landscapes that impede effective flood resilience strategies, as different agencies frequently operate in silos without coherence among their respective mandates [78]. The inability of various sectors to collaborate can lead to inefficient responses during disaster events, worsening the impacts of flooding and increasing the risk of water contamination due to poorly managed urban runoff. Moreover, Laeni et al. underscore the need for integrated flood management frameworks that foster inclusivity and participation from local communities, thereby enhancing resilience by leveraging local knowledge [79].

Inadequate funding remains a pervasive issue impacting institutional capacities across Southeast Asia. Local governments often struggle to allocate the necessary financial resources for robust flood management infrastructure and sustainable practices due to competing development needs exacerbated by rapid urbanization [80]. The lack of financial support translates into weakened institutional mechanisms that could have otherwise facilitated effective disaster response and water management. Sithirith argues that Cambodia's vulnerability to flooding stems not only from environmental factors but also from insufficient governance structures and financial constraints in effective water resource management [81]. These limitations necessitate innovative funding mechanisms to bridge financial gaps and support sustainable water management initiatives.

The challenge of transboundary water governance presents additional complexities to water management in the region, particularly regarding the Mekong River, which serves as a lifeblood for multiple countries, including China, Myanmar, Laos, Thailand, Cambodia, and Vietnam. Transboundary governance arrangements often fall short due to competing national interests and a lack of cooperative frameworks. Yong's exploration of hydropower development highlights how inequities from state-led initiatives can exacerbate environmental injustices, undermining local livelihoods and aggravating water security issues across national borders [82]. The political dynamics surrounding the exploitation of the Mekong's resources, especially concerning large hydropower projects, expose the fragility of regional governance systems, these projects may disproportionately impact downstream states.

Addressing transboundary challenges requires engaging regional organizations such as ASEAN, which plays a pivotal role in fostering cooperative governance across member nations. Initiatives promoting collaborative resource management and shared regulations can bolster regional resilience and enhance collective disaster response strategies. While Armitage et al. emphasizes the importance of cooperative governance and policymaking in transboundary contexts, they advocate for strengthening science-policy interactions to improve decision-making processes [83]. Such frameworks can contribute to effective governance that aligns with local needs, enhancing both ecological and socio-economic resilience in the face of climate-induced challenges.

In this regulatory landscape, the effective implementation of policies remains obstructed by various operational challenges. The need for robust disaster risk reduction mechanisms that prioritize equity and sustainability has been emphasized in multiple studies. Laeni et al. stress that building local institutional capacities is vital for successful water adaptation frameworks and can significantly improve responses to climate-related events [79]. As Southeast Asian cities increasingly confront unpredictable weather patterns, efforts must center not only on structural infrastructural development but also on nurturing the institutional capacity necessary for effective governance.

The stark realities presented by climate change and urbanization underscore the urgency for integrated disaster management, effective coordination, and innovative governance structures that can evolve to meet the increasing complexities of water security. Recommendations from various scholars suggest that Southern Southeast Asia would greatly benefit from mobilizing resources and knowledge-sharing mechanisms to create resilient urban environments.

11.7 Community-Based and Indigenous Responses

11.7.1 Community-Based and Indigenous Responses: The Role of Local Knowledge in Flood Prediction and Mitigation

In recent years, there has been increasing recognition of the role that local knowledge, particularly indigenous knowledge systems, play in flood prediction and mitigation. Local and indigenous communities have accumulated extensive knowledge regarding their environments over generations, often leading to effective management strategies for flood risks that are deeply rooted in their cultural practices and environmental relationships [84, 85]. These systems of knowledge, developed through observation and interaction with local ecosystems, provide critical insights into the timing and impact of floods, as well as suitable responses [84]. One notable aspect of local knowledge is its adaptive quality, allowing communities to respond efficiently to changing environmental conditions and increasing flood risks attributed to climate change [86]. Traditional ecological practices, such as the planting of specific crops that thrive in flood-prone areas and the construction of indigenous water management systems, demonstrate a profound understanding of local hydrological conditions [87]. Combining this traditional knowledge with modern scientific methods can lead to improved flood management practices, fostering resilience within vulnerable communities [86]. Moreover, community-led approaches to monitoring and mapping flood-prone areas leverage local insights and participation, enhancing the accuracy of flood risk assessments [88, 89]. Initiatives that incorporate local stakeholders into flood management frameworks facilitate a more comprehensive understanding of flood dynamics and help prioritize community needs and preferences [90, 91]. This participatory approach not only empowers local residents but also promotes ownership of flood mitigation strategies, aligning them closely with indigenous methodologies that have historically proven effective [92].

11.7.2 Traditional Water Storage and Purification Methods

Indigenous communities have long utilized culturally significant methods for water storage and purification, particularly in flood-prone regions. Such practices often include the use of natural filtration systems, rainwater harvesting techniques, and communal storage facilities that take advantage of local resources [87]. In many cultures, specific vegetation is planted around water bodies to enhance water quality through natural filtration processes [86]. These traditional methods are not only sustainable but also resilient to environmental changes. In the context of climate change, indigenous water management systems may offer considerable advantages over modern, technologically dependent systems, which sometimes struggle in drastically altered climates [93]. For example, studies in the Mekong Delta have highlighted how integrating traditional water management systems with contemporary practices can help adapt to new flood risk patterns [86]. Additionally, traditional water purification techniques often utilize indigenous knowledge of local flora for biofiltration. This has been shown to provide safe drinking water for communities while being environmentally sustainable [87, 88]. Such practices can significantly contribute to community resilience during flooding, which often disrupts regular water supplies and complicates access to clean water [89].

11.7.3 Community-Led Monitoring and Mapping

Community-led initiatives for monitoring flood risks are pivotal in enhancing preparedness and response capabilities. Indigenous knowledge systems serve as a valuable resource for mapping flood-prone regions, as they are based on local experiences and historical flood records [88, 94]. The integration of these indigenous mapping techniques into formal governance systems presents both opportunities and challenges, offering richer, context-specific information that is often lacking in standard scientific models [85, 89]. Participatory mapping workshops, where community members use their knowledge to define flood zones and critical infrastructure, have proven effective in various regions. Such initiatives improve the accuracy of flood risk assessments and build communal ties, promoting collective action in flood preparedness [87, 91]. However, this community engagement must navigate existing bureaucratic structures that can impede the implementation of locally derived solutions, causing tensions between informal practices and formal governance systems [90, 92].

11.7.4 Challenges in Integration with Formal Governance Systems

The integration of community-based and indigenous flood management practices into formal governance structures remains fraught with challenges. Often, conventional flood risk management approaches focus heavily on technological solutions and centralized governance, leaving little room for local knowledge [86, 90]. This top-down approach can marginalize indigenous voices, leading to missed opportunities for improving flood resilience [87, 89]. Moreover, there is frequently a disconnection between the scientific knowledge produced by formal institutions and the experiential knowledge held by local communities [89, 91]. For effective integration, policymakers must recognize the value of participatory flood management frameworks that genuinely incorporate local knowledge. This necessitates establishing channels for dialogue and collaboration between communities and government agencies [95, 96].

In some cases, existing policies may not accommodate traditional practices or may lack the flexibility needed to incorporate community insights into design and implementation strategies [90, 92]. Moreover, institutional inertia and resistance to change often hamper the incorporation of indigenous practices, meaning that communities must continuously advocate for their recognition and integration into flood management policies [97, 98]. Traditional methods of water storage, purification, and community-led mapping provide indispensable tools for effective flood management, demonstrating the value of integrating these practices into formal governance systems. However, transforming these theories into practice requires overcoming the challenges posed by existing governance structures, necessitating a commitment to participatory approaches that recognize and utilize local insights. Fostering an inclusive dialogue between indigenous knowledge holders and formal institutions can lead to more comprehensive, effective, and culturally relevant flood resilience strategies that empower vulnerable communities and enhance the sustainability of their environments.

11.7.5 Climate Change and the Future of Water Security in Southeast Asia

Climate change poses a profound challenge to water security in Southeast Asia, exacerbating existing vulnerabilities through projected increases in extreme rainfall and sea-level rise. Diverse climatic conditions and a high dependence on water resources for agriculture, industry, and domestic use characterize Southeast Asia. The Intergovernmental Panel on Climate Change (IPCC) projects that, by 2050, extreme weather events, including heavy rainfall and flooding, will increase significantly, leading to potential disruptions in water supply and quality in countries across the region, including Indonesia and the Philippines [99, 100]. This trend underscores an urgent need for adaptive strategies in water infrastructure management.

To address the imminent challenges posed by climate change, adaptation strategies must prioritize upgrading existing water infrastructure to withstand extreme weather. For instance, the efficacy of enhanced water treatment facilities has been evaluated in regions like Sabah, Malaysia, where historical data highlights disruption caused by extreme rainfall [101]. Adapting to more variable rainfall and improving water purification systems are key steps for resilience. Moreover, the need for policies mandating climate-resilient infrastructure is becoming a recognized necessity in countries such as Indonesia and Chile, where the formulation of legal frameworks to support adaptation measures varies significantly, with Chile demonstrating more progress [102].

Risk modeling and scenario planning are pivotal in formulating effective adaptive strategies for water management. The dynamic nature of climate variables necessitates a robust risk assessment framework; employing advanced hydrological models can illuminate the potential impact of differing climate scenarios on water supply systems [103]. Such models allow stakeholders to visualize the repercussions of extreme weather patterns on water availability and infrastructure functionality, thereby informing necessary adjustments and investments in climate-resilient infrastructure [104]. Various studies point to a fundamental shift in how water governance is approached, emphasizing the necessity for integrated water resource management (IWRM) under conditions of climate uncertainty [105, 106].

Moreover, IWRM helps harmonize the conflicting interests of various water users while promoting sustainability. This approach is crucial in managing water resources within a changing climate context, particularly where competition for water may intensify due to agriculture, urbanization, and industrial demands. To that end, comprehensive governance structures are necessary to facilitate cooperation among stakeholders and this includes farmers, municipal planners, and policymakers to ensure equitable water distribution and access [100, 103]. The case of Bangladesh illustrates the impacts of climate change on water security, highlighting the urgent need for adaptations in this vulnerable region that is highly susceptible to flooding risks and water scarcity aggravated by socio-economic factors [99]. Adaptation strategies, such as building dikes and reservoirs, have shown potential in mitigating the adverse effects of climate change, yet they require substantial financial commitments and long-term planning [99].

Furthermore, an essential aspect of planning involves the assessment of infrastructure investments through scenario analyses that reflect evolving climate conditions. Evaluating the financial sustainability of water management companies can provide critical insights into the cost-effectiveness of adaptation projects [107]. Decisions made today about water infrastructure investment directly affect future resilience, therefore, careful consideration of climate resilience in the planning process is essential to enhance the adaptive capacity of systems under varying climatic conditions [104, 108].

11.7.6 Recommendations and Pathways Forward to Flooding and Water Contamination in Southeast Asia

To address the persistent and intensifying crisis of flooding and water contamination in Southeast Asia, a multi-pronged and integrated strategy is essential

(Table 11.3). The region's vulnerability stems not only from its unique geographical and climatic conditions but also from institutional, infrastructural, and socio-economic constraints. The following key recommendations provide a roadmap for mitigating risks, protecting public health, and fostering long-term resilience.

11.7.7 Strengthening Regional Collaboration and Data Sharing

Southeast Asia comprises multiple nations sharing transboundary rivers, climatic patterns, and socio-environmental vulnerabilities. Regional cooperation through platforms such as the Association of Southeast Asian Nations (ASEAN), the Mekong River Commission, and other multilateral arrangements is vital for sharing hydrological data, coordinating early warning systems, and aligning disaster response protocols. Integrated flood-risk databases and open-access platforms for water quality monitoring should be developed and maintained to support timely decision-making and cross-border planning [109].

11.7.8 Investment in Resilient Infrastructure and Public Health Systems

Governments and development partners must prioritize funding for flood-resilient infrastructure—such as elevated sanitation facilities, climate-smart drainage systems, flood-retention basins, and robust water treatment plants. Additionally, investment in public health surveillance systems and mobile healthcare units is critical to respond to outbreaks of waterborne diseases following flood events rapidly. Infrastructure planning should embed climate adaptation and equity considerations to serve both urban and marginalized rural populations [25, 110].

11.7.9 Mainstreaming Water Quality Monitoring into Flood Management

Water quality assessment is often overlooked in flood disaster response. There is a need to integrate routine microbial and chemical water testing into early warning systems and post-flood interventions [111, 112]. Deployment of low-cost sensor technologies, mobile testing kits, and geospatial mapping can enhance local capacity to detect contamination in real time. This integration will enable faster remediation actions and minimize long-term health impacts [113].

Table 11.3 Strategic focus areas for addressing flooding and water contamination in Southeast Asia

Strategic focus area	Key actions	Expected outcomes
Regional collaboration and data integration	Enhance cooperative mechanisms among ASEAN member states and transboundary river basin authorities	Strengthened cross-border coordination, improved disaster preparedness, and informed regional decision-making
	Establish open-access platforms for real-time hydrological and water quality data exchange	
Investment in resilient infrastructure and health systems	Prioritize the development of flood-adaptive water, sanitation, and drainage infrastructure	Reduced infrastructure vulnerability, minimized post-flood disease outbreaks, and enhanced health system responsiveness
	Expand public health surveillance, early disease detection, and mobile response units in flood-prone zones	
Mainstreaming water quality monitoring in flood management	Integrate microbial and chemical water quality assessments into early warning and disaster response systems	Timely identification of water contamination, targeted remediation efforts, and enhanced public health protection
	Promote the use of mobile testing kits, GIS mapping tools, and community-based monitoring	
Community empowerment and participatory governance	Involve local communities, particularly marginalized groups, in flood risk assessment, planning, and adaptation strategies	Increased community resilience, equitable risk reduction, and enhanced local capacity to manage water-related hazards
	Implement culturally tailored education programs on hygiene, water safety, and emergency preparedness	
Research advancement and policy integration	Support interdisciplinary research on climate change, urbanization, and their roles in water contamination and flooding	Evidence-informed policy development, coherent governance frameworks, and long-term sustainability of water and flood management initiatives
	Institutionalize multi-sector policy dialogues among environmental, health, water, and disaster management agencies	

11.7.10 Empowering Local Communities and Inclusive Planning

Community-based adaptation strategies, rooted in local knowledge and participatory planning, are essential for sustainable risk reduction. Governments and NGOs should engage regional stakeholders, including women, youth, and indigenous groups, in designing and implementing flood defense systems, sanitation technologies, and public health campaigns. Education programs on hygiene, water safety, and flood preparedness must be localized and continuous to build community resilience [114].

11.7.11 Enhancing Research and Policy Dialogue on Water Contamination Risks

There remains a critical need for interdisciplinary and region-specific research that examines the intersection of flooding, water contamination, climate change, and socio-economic inequality. Academic institutions, think tanks, and policy bodies must collaborate to produce actionable knowledge that informs governance frameworks and legislative reforms [115]. Cross-sectoral policy dialogue, including ministries of environment, health, water, and disaster management, should be institutionalized to ensure coordinated and sustained responses [116].

11.8 Conclusion

South and Southeast Asia's flooding and water contamination crisis represents one of the most urgent and multifaceted environmental challenges in the region, with far-reaching implications for public health, ecological sustainability, and socio-economic resilience. This chapter has explored the geographic, climatic, and anthropogenic drivers that contribute to recurring floods and the subsequent contamination of water sources, highlighting the vulnerability of densely populated urban centers, low-lying delta regions, and underserved communities. The interconnectedness of poor infrastructure, weak governance, rapid urbanization, and inadequate disaster preparedness has exacerbated the impacts of flooding, resulting in frequent outbreaks of waterborne diseases, degradation of ecosystems, and threats to food and water security.

Drawing on detailed case studies from countries such as Bangladesh, Indonesia, the Philippines, Thailand and Vietnam, the chapter has shown how both structural and systemic shortcomings, coupled with climate-induced stressors, contribute to a growing crisis. While technological interventions, such as improved drainage systems and decentralized water treatment technologies, offer promising solutions, their success depends heavily on inclusive governance, coordinated regional policies, and investment in adaptive infrastructure. Ultimately, addressing this crisis requires an integrated and proactive approach—one that combines early warning systems, resilient urban planning, community participation, and climate adaptation strategies. South and Southeast Asia's experience serves as a critical case for global learning, emphasizing the need to treat flooding and water contamination not as isolated phenomena but as deeply interrelated challenges requiring cross-sectoral collaboration and long-term commitment.

References

1. Laeni N, Brink M, Arts J (2019) Is Bangkok becoming more resilient to flooding? A framing analysis of Bangkok's flood resilience policy combining insights from both insiders and outsiders. Cities 90:157–167. https://doi.org/10.1016/j.cities.2019.02.002
2. Hamel P, Tan L (2021) Blue–green infrastructure for flood and water quality management in Southeast Asia: evidence and knowledge gaps. Environ Manage 69(4):699–718. https://doi.org/10.1007/s00267-021-01467-w
3. Mahapatra T, Mahapatra S, Babu G, Tang W, Banerjee B, Mahapatra U, Das A (2014) Cholera outbreaks in South and Southeast Asia: descriptive analysis, 2003 and 2012. Jpn J Infect Dis 67(3):145–156. https://doi.org/10.7883/yoken.67.145
4. Mulyodarsono N, Kristopo H (2024) The water hygiene of street food vendors in Southeast Asia: a review. Iop Conf Ser Earth Environ Sci 1324(1):012104. https://doi.org/10.1088/1755-1315/1324/1/012104
5. Lorenzo T, Kinzig A (2019) Double exposures: future water security across urban Southeast Asia. Water 12(1):116. https://doi.org/10.3390/w12010116
6. Wang X, Yu X, Yu X (2022) Flood disaster risk assessment based on DEA model in southeast Asia along "The Belt and Road." Sustainability 14(20):13145. https://doi.org/10.3390/su142013145
7. Hussain M (2018) Comprehensive analysis of effect of submergence on rice grain quality. Curr Investig Agric Curr Res 5(5). https://doi.org/10.32474/ciacr.2018.05.000223
8. Rahman M, Naidu R, Bhattacharya P (2009) Arsenic contamination in groundwater in the Southeast Asia region. Environ Geochem Health 31(S1):9–21. https://doi.org/10.1007/s10653-008-9233-2
9. Meharg A, Rahman M (2002) Arsenic contamination of Bangladesh paddy field soils: implications for rice contribution to arsenic consumption. Environ Sci Technol 37(2):229–234. https://doi.org/10.1021/es0259842
10. Arias M, Cochrane T, Norton D, Killeen T, Khon P (2013) The flood pulse as the underlying driver of vegetation in the largest wetland and fishery of the Mekong basin. Ambio 42(7):864–876. https://doi.org/10.1007/s13280-013-0424-4
11. Mạnh N, Dũng N, Hung N, Merz B, Apel H (2014) Large-scale suspended sediment transport and sediment deposition in the Mekong Delta. Hydrol Earth Syst Sci 18(8):3033–3053. https://doi.org/10.5194/hess-18-3033-2014
12. Arias M, Piman T, Lauri H, Cochrane T, Kummu M (2014) Dams on Mekong tributaries as significant contributors of hydrological alterations to the Tonle Sap floodplain in Cambodia. Hydrol Earth Syst Sci 18(12):5303–5315. https://doi.org/10.5194/hess-18-5303-2014
13. Cochrane T, Arias M, Piman T (2014) Historical impact of water infrastructure on water levels of the Mekong River and the Tonle Sap system. Hydrol Earth Syst Sci 18(11):4529–4541. https://doi.org/10.5194/hess-18-4529-2014
14. Phy S, Sok T, Try S, Chan R, Uk S, Hen C, Oeurng C (2022) Flood hazard and management in Cambodia: a review of activities, knowledge gaps, and research direction. Climate 10(11):162. https://doi.org/10.3390/cli10110162

15. Quetulio-Navarra M (2023) Philippines' housing commitment as a preventive resettlement strategy for informal settler families. Int Migr 61(4):345–348. https://doi.org/10.1111/imig.13164
16. Hoang L, Lauri H, Kummu M, Koponen J, Vliet M, Supit I, Ludwig F (2016) Mekong river flow and hydrological extremes under climate change. Hydrol Earth Syst Sci 20(7):3027–3041. https://doi.org/10.5194/hess-20-3027-2016
17. Van P, Popescu I, Griensven A, Solomatine D, Nguyen T, Green A (2012) A study of the climate change impacts on fluvial flood propagation in the Vietnamese Mekong Delta. Hydrol Earth Syst Sci 16(12):4637–4649. https://doi.org/10.5194/hess-16-4637-2012
18. Phung D, Huang C, Rutherford S, Chu C, Wang X, Nguyen M (2014) Association between annual river flood pulse and paediatric hospital admissions in the Mekong Delta area. Environ Res 135:212–220. https://doi.org/10.1016/j.envres.2014.08.035
19. Sato T, Yamazaki D, Lim W, Koirala S, Kanae S (2014) Modeling the impact of sea level rise on potential flood damage in the Mekong. J Jpn Soc Civ Eng Ser B1 (Hydraul Eng) 70(4):I_103-I_108. https://doi.org/10.2208/jscejhe.70.i_103
20. Kummu M, Tes S, Yin S, Adamson P, Józsa J, Koponen J, Sarkkula J (2013) Water balance analysis for the Tonle Sap Lake–floodplain system. Hydrol Process 28(4):1722–1733. https://doi.org/10.1002/hyp.9718
21. Västilä K, Kummu M, Sangmanee C, Chinvanno S (2010) Modelling climate change impacts on the flood pulse in the lower Mekong floodplains. J Water Clim Chang 1(1):67–86. https://doi.org/10.2166/wcc.2010.008
22. Xu Z, Li C, Li A, Zhongqiong Y, Yao W, Chen Y, Huang L (2020) Morphological characteristics of Cambodia Mekong Delta and Tonle Sap Lake and its response to river-lake water exchange pattern. J Water Resour Prot 12(04):275–302. https://doi.org/10.4236/jwarp.2020.124017
23. Stas S, Spracklen B, Willetts P, Le T, Tran H, Le T, Spracklen D (2022) Implications of tropical cyclones on damage and potential recovery and restoration of logged forests in Vietnam. Philos Trans R Soc B Biol Sci 378(1867). https://doi.org/10.1098/rstb.2021.0081
24. Ogwu MC, Kosoe EA (2024) Place of cultural diversity in sustainable water resource management in Ghana. In: Izah SC, Ogwu MC, Loukas A, Hamidifar H (eds) Water crises and sustainable management in the global South. Springer, Singapore. https://doi.org/10.1007/978-981-97-4966-9_14
25. Ogwu MC, Kosoe EA (2025) Integrating green infrastructure into sustainable agriculture to enhance soil health, biodiversity, and microclimate resilience. Sustainability 17(9):3838. https://doi.org/10.3390/su17093838
26. Oddo P, Ahamed A, Bolten J (2018) Socioeconomic impact evaluation for near real-time flood detection in the lower Mekong River basin. Hydrology 5(2):23. https://doi.org/10.3390/hydrology5020023
27. Frappart F, Biancamaria S, Normandin C, Blarel F, Bourrel L, Aumont M, Darrozes J (2018) Influence of recent climatic events on the surface water storage of the Tonle Sap Lake. Sci Total Environ 636:1520–1533. https://doi.org/10.1016/j.scitotenv.2018.04.326
28. Olcese G, Bates P, Neal J, Sampson C, Wing O, Quinn N, Probyn I (2024) Developing a fluvial and pluvial stochastic flood model of Southeast Asia. Water Resour Res 60(6). https://doi.org/10.1029/2023wr036580
29. Tran T, Ritchie E, Perkins-Kirkpatrick S (2022) A 50-year tropical cyclone exposure climatology in Southeast Asia. J Geophys Res Atmos 127(4). https://doi.org/10.1029/2021jd036301
30. Chan F, Mitchell G, Adekola O, McDonald A (2012) Flood risk in Asia's urban mega-deltas. Environ Urban ASIA 3(1):41–61. https://doi.org/10.1177/097542531200300103
31. Liu J, Xiong J, Cheng W, Li Y, Cao Y, He Y, Yang G (2021). Assessment of flood susceptibility using support vector machine in the Belt and Road region. https://doi.org/10.5194/nhess-2021-80
32. Salsabila P, Maarif S, Sari D (2023) Strategy for improving green open space (rth) based on community participation in reduce the risk of flood disaster in Jakarta. Iop Conf Ser Earth Environ Sci 1173(1):012061. https://doi.org/10.1088/1755-1315/1173/1/012061

33. Koh L, Sodhi N (2010) Conserving Southeast Asia's imperiled biodiversity: scientific, management, and policy challenges. Biodivers Conserv 19(4):913–917. https://doi.org/10.1007/s10531-010-9818-9
34. Torti J (2012) Floods in Southeast Asia: a health priority. J Glob Health 2(2). https://doi.org/10.7189/jogh.02.020304
35. Winsemius H, Aerts J, Beek R, Bierkens M, Bouwman A, Jongman B, Ward P (2015) Global drivers of future river flood risk. Nat Clim Chang 6(4):381–385. https://doi.org/10.1038/nclimate2893
36. Mahto S, Mishra V (2024) Global evidence of rapid flash drought recovery by extreme precipitation. Environ Res Lett 19(4):044031. https://doi.org/10.1088/1748-9326/ad300c
37. Láng-Ritter J, Berenguer M, Dottori F, Kalaš M, Sempere-Torres D (2021). Compound flood impact forecasting: integrating fluvial and flash flood impact assessments into a unified system. https://doi.org/10.5194/hess-2021-387
38. Singh S, Mackill D, Ismail A (2014) Physiological basis of tolerance to complete submergence in rice involves genetic factors in addition to the sub1 gene. Aob Plants 6. https://doi.org/10.1093/aobpla/plu060
39. Pongpiachan S, Settacharnwit T, Chalangsut P, Hirunyatrakul P, Kittikoon I (2012). Impacts and preventive measures against flooding and coastal erosion in Thailand. https://doi.org/10.2495/friar120131
40. Zeng Z, Estes L, Ziegler AD, Chen A, Searchinger T, Hua F, Guan K, Jintrawet A, Wood EF (2018) Highland cropland expansion and forest loss in Southeast Asia in the twenty-first century. Nat Geosci 11:556–562. https://doi.org/10.1038/s41561-018-0166-9
41. Poapongsakorn N, Meethom P (2013) Impact of the 2011 floods, and flood management in Thailand. ERIA Discussion Paper Series, p 72. https://www.eria.org/ERIA-DP-2013-34.pdf
42. Tolentino P, Williams R, Hurst M (2025) Natural flood risk management in tropical southeast Asia: prospects in the biodiverse archipelagic nation of the Philippines. Wiley Interdiscip Rev Water 12(1). https://doi.org/10.1002/wat2.70000
43. Gain A, Mojtahed V, Biscaro C, Balbi S, Giupponi C (2015) An integrated approach of flood risk assessment in the eastern part of Dhaka city. Nat Hazards 79(3):1499–1530. https://doi.org/10.1007/s11069-015-1911-7
44. Budiyono B, Ginandjar P, Saraswati L, Pangestuti D, Martini M, Jati S, Rahfiludin Z (2015) Risk assessment of drinking water supply system in the tidal inundation area of Semarang, Indonesia. Procedia Environ Sci 23:93–98. https://doi.org/10.1016/j.proenv.2015.01.014
45. Bhagat J, Singh N, Shimada Y (2024). Southeast Asia's environmental challenges: emergence of new contaminants and advancements in testing methods. Front Toxicol 6. https://doi.org/10.3389/ftox.2024.1322386
46. Arrighi C, Tarani F, Vicario E, Castelli F (2017) Flood impacts on a water distribution network. Nat Hazards Earth Syst Sci 17(12):2109–2123. https://doi.org/10.5194/nhess-17-2109-2017
47. Pyke C, Warren M, Johnson T, LaGro J, Scharfenberg J, Groth P, Main E (2011) Assessment of low-impact development for managing stormwater with changing precipitation due to climate change. Landsc Urban Plan 103(2):166–173. https://doi.org/10.1016/j.landurbplan.2011.07.006
48. Mogasale V, Ramani E, Mogasale V, Park J, Wierzba T (2018) Estimating typhoid fever risk associated with lack of access to safe water: a systematic literature review. J Environ Public Health 2018:1–14. https://doi.org/10.1155/2018/9589208
49. Akter M (2016) Conceptualizing environmental governance on the GMB Basin. Bandung 3(1):1–17. https://doi.org/10.1186/s40728-015-0025-7
50. Mliless M (2024) Nexus of climate change and policing. Int J Police Sci 3(1). https://doi.org/10.56331/03132024.31
51. Ali A, Khalid A, Razak N, Maulana N, Roslan N, Razmi R, Anuar S (2024) A review on the presence of microplastics in environmental matrices within Southeast Asia: elucidating risk information through an analysis of microplastic characteristics such as size, shape, and type. Water Emerg Contam Nanoplastics 3(2). https://doi.org/10.20517/wecn.2023.73

52. New W, Kristanti R, Manik H, Wijayanti Y, Adeyemi D (2023) Occurrence of microplastics in drinking water in South Southeast Asia: a short review. Trop Environ Biol Technol 1(1):14–24. https://doi.org/10.53623/tebt.v1i1.221
53. Haberstroh C, Arias M, Yin Z, Sok T, Wang M (2021) Plastic transport in a complex confluence of the Mekong River in Cambodia. Environ Res Lett 16(9):095009. https://doi.org/10.1088/1748-9326/ac2198
54. Reza T, Riza Z, Abdullah S, Hasan H, Ismail N, Othman A (2023) Microplastic removal in wastewater treatment plants (WWTPs) by natural coagulation: a literature review. Toxics 12(1):12. https://doi.org/10.3390/toxics12010012
55. Gandhi S, Jones T (2019) Identifying mangrove deforestation hotspots in South Asia, Southeast Asia, and Asia-Pacific. Remote Sens 11(6):728. https://doi.org/10.3390/rs11060728
56. Babel S, Ta A, Nguyen T, Sembiring E, Setiadi T, Sharp A (2022) Microplastics pollution in selected rivers from Southeast Asia. Apn Sci Bull 12(1):5–17. https://doi.org/10.30852/sb.2022.1741
57. Yu J, Lin H, Lo Y, Tseng K, Hsu C (2021) Measures to cope with the impact of climate change and drought in the island region: a study of the water literacy awareness, attitude, and behavior of the Taiwanese public. Water 13(13):1799. https://doi.org/10.3390/w13131799
58. Hassaniyan A (2024) Iran's water policy: environmental injustice and peripheral marginalisation. Prog Phys Geogr Earth Environ 48(3):420–437. https://doi.org/10.1177/03091333241252523
59. Mingkhwan R, Worakhunpiset S (2018) Heavy metal contamination near industrial estate areas in Phra Nakhon Si Ayutthaya Province, Thailand, and human health risk assessment. Int J Environ Res Public Health 15(9):1890. https://doi.org/10.3390/ijerph15091890
60. Arbiol J, Borja M, Yabe M, Nomura H, Gloriani N, Yoshida S (2013) Valuing human leptospirosis prevention using the opportunity cost of labor. Int J Environ Res Public Health 10(5):1845–1860. https://doi.org/10.3390/ijerph10051845
61. Radi M, Hashim J, Jaáfar M, Hod R, Ahmad N, Nawi A, Ayub N (2018) Leptospirosis outbreak after the 2014 major flooding event in Kelantan, Malaysia: a spatial-temporal analysis. Am J Trop Med Hyg 98(5):1281–1295. https://doi.org/10.4269/ajtmh.16-0922
62. Jalilov S, Kefi M, Kumar P, Masago Y, Mishra B (2018) Sustainable urban water management: application for integrated assessment in Southeast Asia. Sustainability 10(1):122. https://doi.org/10.3390/su10010122
63. Tanoue M, Taguchi R, Nakata S, Watanabe S, Fujimori S, Hirabayashi Y (2020) Estimation of direct and indirect economic losses caused by a flood with long-lasting inundation: application to the 2011 Thailand flood. Water Resour Res 56(5). https://doi.org/10.1029/2019wr026092
64. Zhang H, Zhang C, Zhu Y, Mehmood K, Liu J, McDonough S, Chang Y (2019) Leptospirosis trends in China, 2007–2018: a retrospective observational study. Transbound Emerg Dis 67(3):1119–1128. https://doi.org/10.1111/tbed.13437
65. Nazir A, Oduoye M, Nazir A, Uzoamaka C, Muzammil M, Sakr S, Suheb M (2023) Unraveling the leptospirosis epidemic: tales from the Philippine outbreak—a short communication. Ann Med Surg 86(2):1238–1242. https://doi.org/10.1097/ms9.0000000000001535
66. Mateo C, Hanasaki N, Komori D, Tanaka K, Kiguchi M, Champathong A, Oki T (2014) Assessing the impacts of reservoir operation to floodplain inundation by combining hydrological, reservoir management, and hydrodynamic models. Water Resour Res 50(9):7245–7266. https://doi.org/10.1002/2013wr014845
67. Berlaer G, Jong F, Das T, Gundran C, Samyn M, Gijs G, Hubloue I (2018) Clinical characteristics of the 2013 Haiyan typhoon victims presenting to the Belgian first aid and support team. Disaster Med Public Health Prep 13(02):265–278. https://doi.org/10.1017/dmp.2018.54
68. Kim J, Kim H, Jeon H, Jeong S, Song J, Vadivel S, Kim D (2021) Synergistic use of geospatial data for water body extraction from Sentinel-1 images for operational flood monitoring across Southeast Asia using deep neural networks. Remote Sensing 13(23):4759. https://doi.org/10.3390/rs13234759

69. Igoud S, Zeriri D, Aoudjit L, Boutra B, Sebti S, Khene F, Mameche A (2020) Climate change adaptation by solar wastewater treatment (SOWAT) for reuse in agriculture and industry*. Irrig Drain 70(2):243–253. https://doi.org/10.1002/ird.2540
70. Milićević D, Milićević M, Trajković R (2024) Decentralized wastewater treatment - a sustainable solution for protecting water resources from pollution. ZRGAF 39(1):35–46. https://doi.org/10.62683/zrgaf39.35-46
71. Gómez-Román C, Cameselle J, Álzate M, Medina B (2021) Environmental concern priming and social acceptance of sustainable technologies: the case of decentralized wastewater treatment systems. Front Psychol 12. https://doi.org/10.3389/fpsyg.2021.647406
72. Lâm S, Nguyen-Viet H, Yên N, Huong N, Harper S (2015) Evidence for public health risks of wastewater and excreta management practices in Southeast Asia: a scoping review. Int J Environ Res Public Health 12(10):12863–12885. https://doi.org/10.3390/ijerph121012863
73. Razman K, Hanafiah, Ramli A, Harun S (2023) Industrial wastewater treatment methods are employed in Southeast Asian countries. Iop Conf Ser Earth Environ Sci 1167(1):012020. https://doi.org/10.1088/1755-1315/1167/1/012020
74. Xue X (2022) Practical research on the modular equipment for the treatment of rural decentralized domestic sewage. Iop Conf Ser Earth Environ Sci 1011(1):012034. https://doi.org/10.1088/1755-1315/1011/1/012034
75. Sotelo T, Satoh H, Mino T (2019) Assessing wastewater management in the developing countries of Southeast Asia: underlining flexibility in appropriateness. J Water Environ Technol 17(5):287–301. https://doi.org/10.2965/jwet.19-006
76. Waqas S, Harun N, Sambudi N, Bilad M, Abioye K, Ali A, Abdulrahman A (2023) A review of rotating biological contactors for wastewater treatment. Water 15(10):1913. https://doi.org/10.3390/w15101913
77. An Y, Tan X, Gu B, Zhu K (2020) Flood risk assessment using the CV-TOPSIS method for the Belt and Road Initiative: an empirical study of Southeast Asia. Ecosyst Health Sustain 6(1). https://doi.org/10.1080/20964129.2020.1765703
78. Laeni N, Brink M, Arts J (2021) Institutional conditions for inclusive, flood resilient urban deltas: a comparative institutional analysis of two international resilience programs in Southeast Asia. Water 13(18):2478. https://doi.org/10.3390/w13182478
79. Laeni N, Brink M, Busscher T, Ovink H, Arts J (2020) Building local institutional capacities for urban flood adaptation: lessons from the water as leverage program in Semarang Indonesia. Sustainability 12(23):10104. https://doi.org/10.3390/su122310104
80. Coker R, Hunter B, Rudge J, Liverani M, Hanvoravongchai P (2011) Emerging infectious diseases in Southeast Asia: regional challenges to control. The Lancet 377(9765):599–609. https://doi.org/10.1016/s0140-6736(10)62004-1
81. Sithirith M (2021) Downstream state and water security in the Mekong region: a case of Cambodia between too much and too little water. Water 13(6):802. https://doi.org/10.3390/w13060802
82. Yong M (2023) Hybrid governance, environmental justice, and hydropower development in the Mekong transboundary commons. Wiley Interdiscip Rev Water, 10(5). https://doi.org/10.1002/wat2.1665
83. Armitage D, Loë R, Morris M, Edwards T, Gerlak A, Hall R, Wolfe B (2015) Science–policy processes for transboundary water governance. Ambio 44(5):353–366. https://doi.org/10.1007/s13280-015-0644-x
84. Bwambale B, Muhumuza M, Nyeko M (2018) Traditional ecological knowledge and flood risk management: a preliminary case study of the Rwenzori. Jàmbá J Disaster Risk Stud 10(1). https://doi.org/10.4102/jamba.v10i1.536
85. Hazarika N, Tayeng T, Das A (2016) Living in troubled waters: stakeholders' perception, susceptibility and adaptations to flooding in the upper Brahmaputra plain. Nat Hazards 83(2):1157–1176. https://doi.org/10.1007/s11069-016-2366-1
86. Hoang L, Biesbroek R, Trí V, Kummu M, Vliet M, Leemans R, Ludwig F (2018) Managing flood risks in the Mekong Delta: how to address emerging challenges under climate change and socioeconomic developments. Ambio 47(6):635–649. https://doi.org/10.1007/s13280-017-1009-4

87. Dewan T (2015) Societal impacts and vulnerability to floods in Bangladesh and Nepal. Weather Clim Extrem 7:36–42. https://doi.org/10.1016/j.wace.2014.11.001
88. Ahmad M (2019) Role of indigenous knowledge in managing floods projects. Adv Soc Sci Res J 6(9):87–96. https://doi.org/10.14738/assrj.69.7074
89. Ridhoi R, Hadi A, Hudiyanto R (2022) Traditional flood mitigation by the indigenous people of Sampang Madura. Kne Soc Sci. https://doi.org/10.18502/kss.v7i16.12162
90. Meem, R., Ahmed, A., and Hossain, M. (2021). A review on adaptation practices of affected people with riverine floods in Bangladesh. Int J Environ Clim Chang 251–260. https://doi.org/10.9734/ijecc/2021/v11i1230575
91. Solín Ľ, Madajová M (2023) Spatial planning as a tool of flood risk management in rural landscapes? position, limitations, and other findings: the case of Myjava region (Slovakia) Moravian Geographical Reports 31(2):106–117. https://doi.org/10.2478/mgr-2023-0010
92. Snel K, Hegger D, Mees H, Craig R, Kammerbauer M, Doorn N, Wamsler C (2022) Unpacking notions of residents' responsibility in flood risk governance. Environ Policy Gov 32(3):217–231. https://doi.org/10.1002/eet.1985
93. Mabahwi N, Nakamura H (2020) The issues and challenges of flood-related agencies in Malaysia. Environ-Behav Proc J 5(13):285–290. https://doi.org/10.21834/e-bpj.v5i13.2069
94. Munawar H, Khan S, Anum N, Qadir Z, Kouzani A, Mahmud M (2021) Post-flood risk management and resilience building practices: a case study. Appl Sci 11(11):4823. https://doi.org/10.3390/app11114823
95. Hedelin B (2016) The EU floods directive tricking down: tracing the ideas of integrated and participatory flood risk management in Sweden. Water Policy 19(2):286–303. https://doi.org/10.2166/wp.2016.092
96. Nakamura S, Nakai F, Ito Y, Okada G, Oki T (2023) Levee system transformation in coevolution between human and water systems along the Kiso River. Japan. https://doi.org/10.5194/egusphere-2023-2866
97. Edelenbos J, Buuren A, Roth D, Winnubst M (2016) Stakeholder initiatives in flood risk management: exploring the role and impact of bottom-up initiatives in three 'room for the river' projects in the Netherlands. J Environ Planning Manage 60(1):47–66. https://doi.org/10.1080/09640568.2016.1140025
98. Liefferink D, Wiering M, Crabbé A, Hegger D (2017) Explaining stability and change. Comparing flood risk governance in Belgium, France, the Netherlands, and poland. J Flood Risk Manag 11(3):281–290. https://doi.org/10.1111/jfr3.12325
99. Haque S, Nahar N (2023) Bangladesh: climate change issues, mitigation, and adaptation in the water sector. Acs Esandt Water 3(6):1484–1501. https://doi.org/10.1021/acsestwater.2c00450
100. Lindner A, Stamm J (2025) Integrating climate change adaptation and water resource management: a critical overview. Standards 5(1):4. https://doi.org/10.3390/standards5010004
101. Leong M, Ngui M, Ayog J (2024) Assessment on future rainfall variability for adaptive water resource management in Sabah. Bio Web of Conferences 131:05012. https://doi.org/10.1051/bioconf/202413105012
102. Maskun M, Mukarramah N, Angi J (2023) Policy strategies for climate-resilient infrastructure in climate change adaptation: jurisdictional comparison between Chile and Indonesia. E3s Web Conf 467:05001. https://doi.org/10.1051/e3sconf/202346705001
103. Momblanch A, Beevers L, Srinivasalu P, Kulkarni A, Holman I (2020) Enhancing production and flow of freshwater ecosystem services in a managed Himalayan river system under uncertain future climate. Clim Change 162(2):343–361. https://doi.org/10.1007/s10584-020-02795-2
104. Pot W (2023) Deciding for resilience: utilizing water infrastructure investments to prepare for the future. Wiley Interdiscip Rev Water 10(5). https://doi.org/10.1002/wat2.1661
105. Etukudoh E, Ilojianya V, Ayorinde O, Daudu C, Adefemi A, Hamdan A (2024) Review of climate change impact on water availability in the USA and Africa. Int J Sci Res Arch 11(1):942–951. https://doi.org/10.30574/ijsra.2024.11.1.0169

106. Li J, Zhang C, Huang Q, Ding M, He Y, Liu M, Yang C (2024) Water infrastructure impacts of agricultural industry in China under extreme weather: a system dynamics model of a multi-level, climate resilience perspective. Systems 12(12):562. https://doi.org/10.3390/systems12120562
107. López-Moreno S (2024) Assessment of the financial performance and sustainability of water management companies in Spain from a quantitative and technological perspective (2018–2022). Eur Public Soc Innov Rev 9:1–22. https://doi.org/10.31637/epsir-2024-338
108. Rodríguez M, Delpla I (2017) Climate changes and drinking water in sustainable cities: impacts and adaptation 1:573–582. https://doi.org/10.2495/sc170501
109. Delgado JM, Merz B, Apel H (2021) Collaborative data sharing for transboundary flood risk governance in Southeast Asia. Hydrol Earth Syst Sci 25(4):1947–1963
110. Asian Development Bank (2020) Climate risk country profile: Southeast Asia. ADB, Manila
111. Izah SC, Ogwu MC (2025) Modeling solutions for microbial water contamination in the global south for public health protection. Front Microbiol 16:1504829. https://doi.org/10.3389/fmicb.2025.1504829
112. Ogwu MC, Izah SC, Sawyer WE, Amabie T (2025). Environmental risk assessment of trace metal pollution: a statistical perspective. Environ Geochem Health 47:94. https://doi.org/10.1007/s10653-025-02405-z
113. World Health Organization (WHO) (2019) Water quality and health review of evidence 2019: an updated synthesis of the health effects of water contamination. WHO, Geneva
114. United Nations Office for Disaster Risk Reduction (UNDRR) (2022) Localizing disaster risk reduction: community engagement toolkit for Southeast Asia. UNDRR, Geneva
115. Ogwu MC (2025) Sustainable development in africa: challenges, innovations, and policy pathways for the future. Afr Res Rep 1(2):55–69. https://doi.org/10.5897/ARR2025.0005
116. Chaudhary M, Haque M, Nugroho A (2023) Integrated flood risk and water quality management in Southeast Asia: Challenges and policy responses. Environ Manag Policy 67(2):145–163
117. Park E, Tran DD, Minderhoud PSJ, Clarke R, Chan FKS (2025) Practical paths to halt elevation loss in Vietnamese Mekong Delta. Geogr Sustain 6(5):100335. https://doi.org/10.1016/j.geosus.2025.100335

Chapter 12
South Asia's Solutions to Water Scarcity

Prasann Kumar

Abstract A major socio-economic and environmental challenge, reflecting population pressure, climate change, unsustainable agriculture, and poor governance, is the water shortage in South Asia. This chapter examines both conventional and redesigned approaches to addressing the water shortages in South Asia. The chapter also discusses the importance of transboundary cooperation, particularly regarding standard river basins such as the Indus and the Ganges–Brahmaputra-Meghna systems. It also emphasizes the need for inclusive and gender-sensitive water management policies, as well as the importance of incorporating local knowledge into contemporary scientific research. Through an analysis of several strategies, the chapter aims to provide insight into how South Asian nations are addressing a major contemporary environmental and developmental issue.

Keywords Agriculture · Population · Environment · Climate change · South Asia · Water · Gender

12.1 Introduction

Despite having historically rich monsoon-fed river systems and diverse hydrological resources, South Asia, home to nearly a quarter of the world's population, increasingly faces severe water shortages. This region comprises eight nations: India, Pakistan, Bangladesh, Nepal, Bhutan, Sri Lanka, the Maldives, and Afghanistan, each with distinct geographical, climatic, and socioeconomic contexts that influence their water availability and usage patterns, revealing significant disparities [12, 13, 16]. While some areas are susceptible to seasonal flooding, others endure chronic droughts, highlighting the coexistence of two extremes of water abundance and scarcity within the same region. Rapid urbanization, industrial growth, unsustainable farming practices, and rising pollution exacerbate this paradox. In particular, the

P. Kumar (✉)
Department of Agronomy, School of Agriculture, Lovely Professional University, Phagwara, Punjab, India
e-mail: prasann0659@gmail.com

M. C. Ogwu and S. Chibueze Izah (eds.), *Water Quality and Safety in the Global South*, Springer Water, https://doi.org/10.1007/978-3-032-10602-5_12

Indo-Gangetic Plain, which supports some of the world's most intensive agricultural activities, suffers from the over-extraction of groundwater for irrigation, a leading cause of water scarcity in South Asia. More than 60% of India's irrigated agriculture relies on groundwater, resulting in considerable depletion rates in states such as Punjab, Haryana, and Rajasthan. Similarly, in Pakistan, ineffective canal systems and unregulated extraction through uncontrolled tubewells have led to declining water tables. Climate change has altered precipitation patterns, reduced glacial melt contributions to essential rivers such as the Indus and Ganges, and increased the frequency of extreme weather events, all of which jeopardize long-term water security. The rising population intensifies these stresses [3, 25].

With an estimated 1.9 billion people and growing numbers, South Asia's per capita water availability continues to decline. The Falkenmark Index, which measures water stress, indicates that many areas in the region are nearing or have already exceeded the threshold of water scarcity, defined as less than 1,000 cubic meters per person per year. Urban areas, including Delhi, Dhaka, Kathmandu, and Colombo, are increasingly struggling to provide their growing populations with reliable and safe water, resulting in unequal access and a decline in service quality. Moreover, inadequate cross-sectoral coordination, fragmented institutional frameworks, and poor governance hinder efficient water management across the region. Many national water policies are outdated or poorly enforced, and transboundary conflicts over shared river basins complicate the development of cooperative solutions [2, 6]. Environmental degradation, including deforestation, wetland loss, and contamination of surface and groundwater sources by industrial effluents and agrochemical runoff, further deteriorates the quality and utility of available water resources. Thus, water scarcity in South Asia must be viewed not only as a physical issue but also as a social and economic problem.

Numerous rural and peri-urban areas face economic water scarcity, where inadequate infrastructure and investment hinder communities from accessing sufficient water, even when it is physically present. Particularly concerning gender and caste, social inequalities often dictate who has access to clean water and who bears the burdens of water-related hazards and labor [18, 20, 27, 28] Recognizing water scarcity in South Asia requires a nuanced regional perspective that considers the diverse ecosystems, cultures, governments, and development trajectories across the subcontinent. Unlike homogeneous policy approaches that might work in more uniform areas, South Asia's complexity demands localized strategies that acknowledge the interactions among geography, politics, economy, and society. Each nation in the region faces different yet interconnected water challenges, influenced by shifting socio-environmental conditions, geopolitics, and historical legacies. For instance, Pakistan faces declining flows in the Indus River system due to upstream diversions and climate variability. At the same time, India grapples with groundwater overuse and disputes over the use of interstate rivers. In contrast, Bangladesh, particularly in coastal areas affected by sea-level rise, suffers from both salinity intrusion and waterlogging [11, 24]. Meanwhile, island countries like the Maldives face existential threats from declining freshwater resources due to rising sea levels. At the same time, Nepal and Bhutan depend primarily on mountainous watersheds that are vulnerable to glacial retreat and land-use changes. Analyzing transboundary water issues, crucial for water security in South Asia, also requires a regional perspective. Shared river basins, including the Indus and the Ganges–Brahmaputra-Meghna (GBM), involve multiple riparian states with conflicting interests and insufficient means of coordinated management. Though cooperative efforts could yield more sustainable outcomes, political tensions, particularly between India and Pakistan, often impede these initiatives. Through a regional approach, stakeholders can identify commonalities, foster trust, and develop basin-wide plans that balance equity, efficiency, and ecological integrity. Additionally, regional studies promote knowledge and best practices through interactions and collaboration. From stepwells (baolis) in India to tank cascade systems in Sri Lanka and community-managed irrigation projects in Nepal, traditional water collection systems provide insightful analyses of locally adapted, low-cost, and robust water management models [5, 15]. By recording and supporting these indigenous innovations alongside modern technologies, South Asian nations can create integrated solutions that respect local contexts and address broader developmental goals. Furthermore, addressing the impacts of climate change, which transcend political borders, necessitates regional cooperation. Coordinated monitoring, forecasting, and adaptive planning are essential for tackling climate-induced changes in rainfall patterns, glacier melting, and sea-level rise. Regional platforms, including the International Centre for Integrated Mountain Development (ICIMOD) and the South Asian Association for Regional Cooperation (SAARC), play crucial roles in promoting dialogue and data sharing among member states. However, to adequately address the transnational nature of water-related hazards, these institutions require stronger mandates, funding, and political support [8, 23].

This chapter provides a comprehensive analysis of water shortages in South Asia, examining their causes, impacts, and potential solutions. By using evidence-based insights from case studies, policy analyses, and multidisciplinary research, it aims to bridge the gap between academic research and pragmatic policymaking. To enhance water resilience across the subcontinent, the chapter underscores the importance of integrating traditional knowledge with scientific innovation, promoting inclusive governance, and fostering regional collaboration. Specifically, it aims to: contextualize water scarcity in South Asia by analyzing the biophysical, socio-economic, and institutional factors that contribute to water stress; assess traditional and modern water management strategies by evaluating both indigenous practices and contemporary technological interventions; explore policy and governance frameworks by critically examining national and regional water policies, legal instruments, and institutional arrangements that affect water allocation and use; analyze transboundary water dynamics by investigating the complexities of shared river basins, existing treaties, and opportunities for cooperative management in the region; highlight gender and social inclusion in water governance by underscoring the role of marginalized groups, especially women and lower-caste communities, in shaping equitable water access and decision-making processes; present country-specific case studies by illustrating successful initiatives and lessons learned through detailed examples from India, Pakistan, Bangladesh, Nepal, and Sri Lanka; and finally, propose future directions

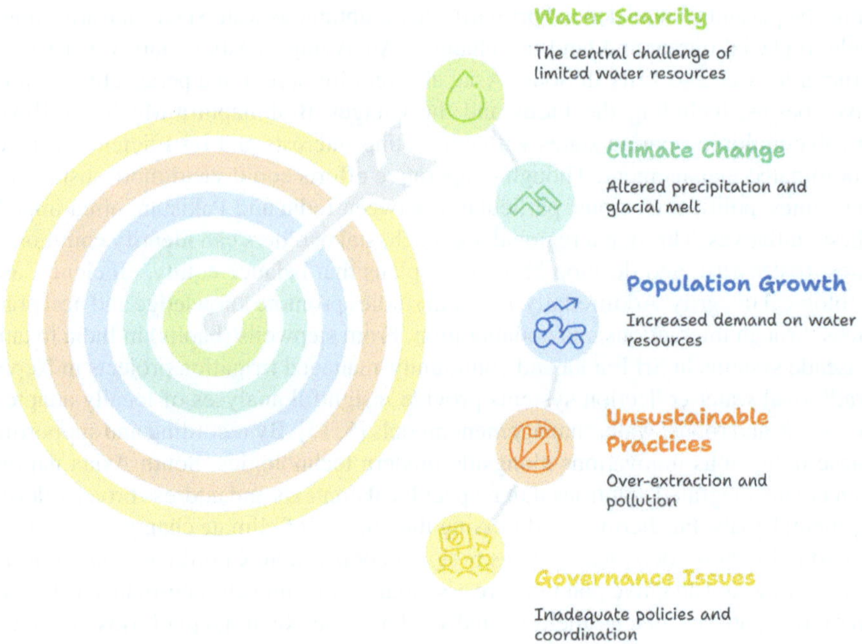

Fig. 12.1 Water scarcity and its possible cause

for sustainable water management by offering policy recommendations and strategic pathways for improving water security in South Asia [14, 21, 22]. Beginning with an overview of water shortages in the area, the chapter's structure logically progresses from traditional and modern solutions to policy responses, transboundary issues, and socio-cultural dimensions. Each section builds upon the previous one to produce actionable suggestions tailored to the South Asian context. Through a multidisciplinary and multi-scalar approach, the chapter contributes to the broader conversation on water sustainability, providing a valuable resource for academics, practitioners, and policymakers engaged in the field (Fig. 12.1).

12.2 Understanding Water Scarcity in South Asia

Physical and economic water scarcity, which has complex underlying environmental, socio-economic, and institutional determinants, appears in two interrelated forms in South Asia. Physical water scarcity occurs when there is insufficient water to meet all demands, including those required for ecosystems to function effectively. This is particularly critical in some of the northwestern states of India, including Punjab, Haryana, and Rajasthan, as well as in Pakistan's Punjab and Sindh provinces, where the rate of groundwater depletion is among the highest in the world [1, 4, 17, 19].

The unsustainability of extraction: The root of the over-extraction problem lies in the dependence on groundwater for irrigation, especially under the Green Revolution model, which promotes water-intensive crops such as rice and wheat. In these regions, receding water tables have rendered traditional shallow wells unviable, placing a significant burden on farmers, many of whom lack steady cash flow or access to electricity, which forces them to incur the costs of lifting water. Adding to physical scarcity downstream, the retreat of glaciers in the Hindu Kush-Himalayan region jeopardizes long-term river flows in strategic basins such as the Indus, Ganges, and Brahmaputra.

An economic water shortage occurs when a region has a sufficient water supply to meet its needs, but the supply is not utilized effectively. Yet, the available water is inaccessible due to various direct or indirect factors, including inadequate infrastructure, misgovernance, insufficient funding, or ineffective policy-making regarding the development and distribution of water [29, 32]. This kind of deficit is common in overcrowded urban slums and rural areas of South Asia, especially in Bangladesh, Nepal, and certain regions of eastern India. Despite having ample monsoon and surface water resources, millions of people lack access to clean drinking water and adequate sanitation. For instance, despite its proximity to several perennial rivers, the Kathmandu Valley has experienced erratic water supplies and seasonal shortages due to poor governance, outdated infrastructure, and rapid, unplanned urbanization. Similarly, in Dhaka and Kolkata, "peri-urban" areas face severe water stress caused by the siltation and encroachment of rivers and wetlands, coupled with inadequate delivery of urban services. Economic water scarcity disproportionately impacts impoverished communities, as well as women, children, and lower-caste groups, who are more likely to bear the responsibility of collecting water from distant sources, thus hindering their education, health, and livelihood opportunities [7–10, 23].

In the vast matrix of environmental and developmental factors, the underlying causes of water scarcity in South Asia include climate change, population density, agricultural expansion, and contamination. Climate change has profoundly impacted hydrological cycles across the region, leading to frequent and severe droughts, floods, and erratic monsoons. Increasing temperatures have caused the Himalayan glaciers, which are the source of water for many of the region's primary rivers, to retreat at some of the highest rates globally, posing a long-term threat to water supplies. Additionally, higher temperatures increase evapotranspiration, thereby reducing soil moisture and raising agricultural water demand. Disrupted rainfall patterns have also disrupted traditional farming cycles, prompting farmers to rely more heavily on groundwater resources to sustain crop yields, thereby exacerbating the problem of depletion [8, 17, 22, 24].

Another key reason for water shortages is population growth. Countries such as India, Pakistan, and Bangladesh are experiencing rapid population growth, resulting in South Asia accounting for nearly a quarter of the world's population. The region's population is projected to reach 2.2 billion by 2025, putting extreme stress on already limited water supplies. Particularly in urban areas, there is growing concern about the rising demand for industrial and household water. This unprecedented pressure on aging water infrastructure has led cities such as Delhi, Mumbai, Karachi, and

Dhaka to allocate water unfairly and to face service challenges [8, 10, 15, 19]. The issue worsens with migration from rural to urban areas, where unplanned community growth occurs without adequate preparation, leading to a lack of basic services. Inadequate wastewater treatment systems are causing varied types of pollution, making it impossible to use most available surface water for human needs or to maintain aquatic health in both rural and urban settings.

Agriculture is a significant contributor to water scarcity, accounting for 80–90% of total water withdrawals in South Asia. Strong incentives for increased water use in water-stressed regions have led to excessive resource extraction, particularly in heavily subsidized systems that cultivate water-intensive crops, such as paddy rice and sugarcane, in countries like India and Pakistan. These government pricing policies enable irrigation pumps to operate on highly subsidized or even free electricity, resulting in the over-extraction of groundwater. Moreover, the widespread use of pesticides and chemical fertilizers not only contaminates water resources but also degrades soil structure, which reduces natural infiltration and recharge capacity, particularly in Bangladesh and the Indian Sundarbans—coastal regions where intensive agriculture, combined with rising sea levels, has permitted saltwater to infiltrate freshwater aquifers [11, 14, 17, 23]. These interconnected issues highlight the need to adopt sustainable agricultural water-use practices, including precision irrigation, agroforestry, and crop diversification, to mitigate the impact of climate change.

Another key factor undermining water access and quality across South Asia is pollution. Industrial effluents, treated sewage, and agricultural runoff laden with toxins have contaminated rivers, lakes, and groundwater reservoirs. Several rivers, including the Ganges, Yamuna, and Buriganga, are heavily polluted, posing potentially severe public health risks due to high levels of heavy metals, pathogens, and organic pollutants. Arsenic contamination of groundwater, a naturally occurring phenomenon worsened by over-extraction, has affected millions in Bangladesh and led to chronic and premature deaths. Urbanization also significantly contributes to water pollution through the unregulated discharge of industrial and domestic waste. Even in rural areas, open defecation and animal waste entering the environment contaminate local water sources, perpetuating cycles of poverty and disease. Combating water pollution requires more than stricter enforcement of environmental regulations; it also necessitates investments in community-based sanitation programs and sewage treatment facilities [1, 22].

While geography, economic development, political organization, and institutional resources are central to the pattern of regional disparity in water use and governance across South Asia, these countries share common vulnerabilities. The largest country in the region, India, varies greatly within its borders; water-stressed Rajasthan and Gujarat could hardly be further from water-rich Assam and Meghalaya in the north. The provinces in Pakistan also display imbalances in water availability; Balochistan is predominantly arid and lacks adequate infrastructure, while Punjab benefits from extensive canal irrigation systems [5, 9, 19, 24]. Despite having abundant surface water, Bangladesh faces seasonal floods, saline seawater intrusion, and arsenic contamination in its water cycle; therefore, it must manage its limited water resources carefully. Even though they have Himalayan-fed rivers and significant hydropower potential, Nepal and Bhutan also struggle with local water scarcity due to the rugged nature of their terrain, land degradation, and inadequate storage facilities.

Water scarcity poses unique challenges for island nations such as the Maldives and Sri Lanka. Comprising coral atolls, the Maldives depends almost entirely on desalination and rainwater collection, as it lacks permanent rivers or significant groundwater sources. However, the accumulation of saltwater and rising sea levels pose a threat to the sustainability of freshwater lenses. Therefore, immediate mitigation solutions are essential. Although Sri Lanka is generally wet, regional disparities exist; the dry zone often faces water stress, with frequent droughts and reduced reservoir capacity due to sedimentation and the growth of aquatic weeds, such as hyacinth. Decades of conflict and instability have left Afghanistan without robust water governance structures, rendering much of its population vulnerable to both water scarcity and flood risks [12, 14, 15, 18, 23].

These contrasts underscore the importance of designing locally tailored policy interventions. While transboundary cooperation and regional integration can enhance water security, national governments should prioritize context-specific strategies for long-term sustainability alongside immediate priorities. New water management investments that promote the adoption of closed ecological systems, community engagement, and transparent governance will be crucial for addressing gaps and ensuring equitable access to water resources. Furthermore, while international cooperation on shared river basins, including the Indus, the Ganges–Brahmaputra-Meghna (GBM), and the Kaveri, must be reevaluated in light of political solutions and cooperative water-sharing mechanisms to resolve political conflicts, it is also essential to tackle the underlying issues.

Ultimately, understanding water scarcity in South Asia requires an integrated strategy that considers both physical and financial shortages, the cumulative effects of climate change, population dynamics, water accumulation, and pollution, as well as significant regional disparities affecting water access and vulnerability [6, 8, 11, 22]. Integrated approaches involving community participation, policy reform, institutional development, and technological innovation are essential for addressing these challenges. Only a multidimensional paradigm can ensure sustainable development in South Asia for future generations and maintain a reliable water supply in the years ahead (Fig. 12.2).

12.3 Traditional and Indigenous Water Management Systems

Historical water practices and traditional water management systems in South Asia are adaptations that have evolved over hundreds of years, rooted in local contexts and based on flexible water management practices to address regionalized climatic variations, seasonal water availability, and geographical limitations [12, 14, 20, 23].

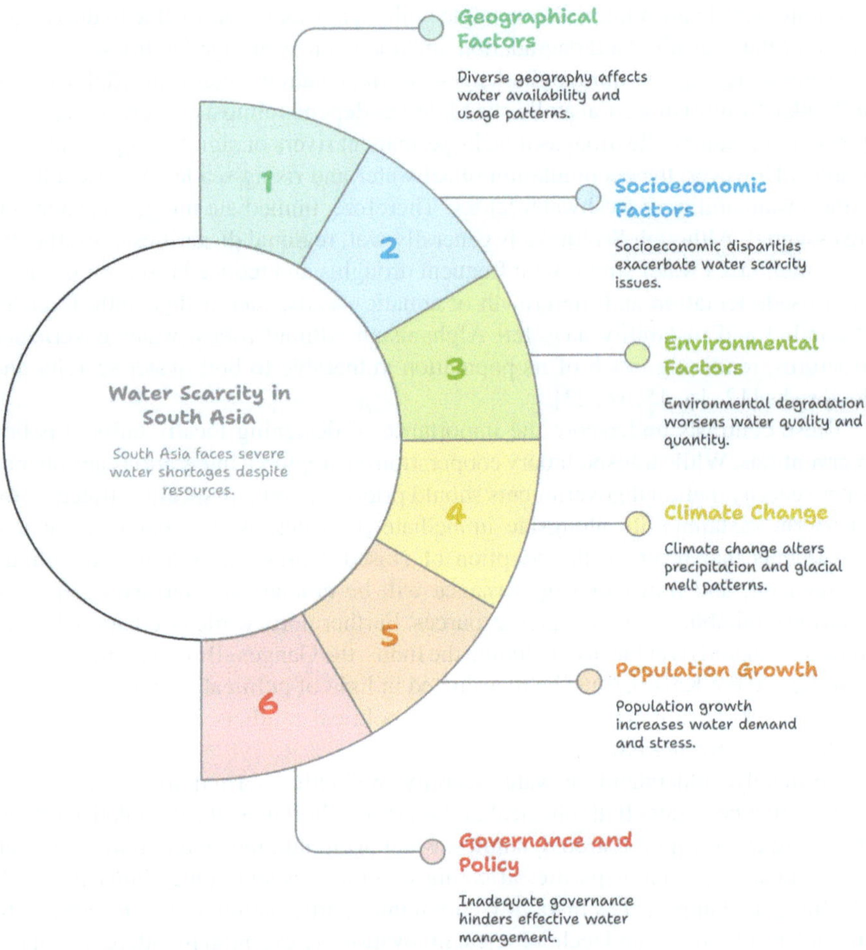

Fig. 12.2 Unveiling South Asia's water scarcity challenges

These systems are rooted in decades and centuries of inherited knowledge and traditional practices, which draw on a profound understanding of hydrological cycles, land use, and community needs. The most evident historical systems were stepwells (baolis or vavs), tanks, and cascade systems, which not only served as stores of water but also functioned as social and religious spaces. A work of art in itself, stepwells like the ones scattered across the arid landscapes of Rajasthan and Gujarat in India were architectural marvels with multiple tiers, built to provide year-round access to what was otherwise an untidy and unreliable source of water. These were works of wonder that demonstrated that architecture could be borne beautifully, she said, frequently embellished with delicate carvings of leaves and local sayings, that represented a repository of local lore and ecological knowledge. Likewise, historical tank systems—such as the tanks of South India and the wewas of Sri Lanka were designed to capture monsoon rainwater and control its release for irrigation, potable water, and household use. These tanks were part of extensive watershed systems, linked through overflow and underflow, and participated in flood moderation and groundwater recharge, thus minimizing the impact of droughts [6, 8, 14, 15, 24].

However, the other side of these massive interventions in the natural landscape consists of community-focused irrigation and conservation systems that have sustained agriculture and rural livelihoods across various landscapes. In the Himalayan regions of Nepal and Bhutan, ancient irrigation technologies, known as kohls and guls, respectively, divert glacial meltwater through gravity-fed canals to terraced fields, ensuring the water is distributed equitably among farmers. These systems are regulated by traditional rules and maintained by village institutions, with maintenance, repair, and distribution carried out collectively [3, 5, 20, 22]. Such decentralized governance systems promoted social inclusion and environmental sustainability thousands of years before the emergence of nation-states and bureaucracies. In the rain-fed agroclimatic regions of eastern India and Bangladesh, indigenous communities practiced flood recession farming, growing crops in the nutrient-rich silt left behind after seasonal floods receded. This practice has also increased soil fertility, reduced reliance on external inputs, and better aligned agricultural production with natural hydrological cycles. Likewise, in the drylands of western India, pastoralist communities have historically utilized johads, small earthen check dams, to store monsoon rainwater and recharge aquifers, which in turn have facilitated the settlement of human populations and the grazing of livestock.

The rehabilitation and reappropriation of traditional techniques have recently garnered increased attention due to their sustainability, cost-effectiveness, and adaptability to local socio-ecological contexts. In India, the Watershed Organisation Trust (WOTR) and Tarun Bharat Sangh are excellent examples of national grassroots movements that have rejuvenated traditional water harvesting structures, such as johads, stepwells, and check dams, in some of the most water-scarce areas of Rajasthan, Maharashtra, and Madhya Pradesh. These programs demonstrate how the restoration of traditional methods, combined with contemporary scientific techniques such as Geographic Information Systems (GIS) and community-based hydrology models, can significantly restore groundwater, enhance crop production, and improve rural incomes [1, 8, 9, 11, 14, 30, 31]. Additionally, current implementations of classic systems have been adapted to incorporate new materials and technologies, thereby improving response times and scalability. Incorporating plastic lines or concrete linings into traditional ponds and tanks has reduced seepage losses and extended the duration for which the structure can store water, making them more practical in semi-arid areas, for example. In cities, rooftop rainwater harvesting a practice dating back to ancient times has gained legal recognition through policies in towns such as Chennai and Bangalore, where it is now mandatory for new buildings to adopt this practice Similarly, managed aquifer recharge (MAR), a low-cost and straightforward technique based on the traditional recharge process, is being implemented in arsenic-affected regions of Bangladesh and West Bengal to enhance water quality and quantity. The existence of these hybrid traditions highlights the need for a convergence of traditional knowledge and modern scientific research to address contemporary challenges, including climate change, human population growth, and environmental degradation [4, 15, 18, 21].

In the past, TWM has encountered obstacles to its widespread adoption and formal recognition, including the lack of policy integration, weak institutional support, and inadequate financial backing. Decentralized and community-based options are often overlooked in Favor of significant, centralized, infrastructure-heavy interventions, such as large dams and canal networks. Moreover, increasing urbanization, land-use changes, and the submergence of traditional water bodies have all contributed to the significant destruction of several heritage water systems. Nevertheless, an expanding body of field and academic evidence, along with policy dialogues, shows that integrating traditional knowledge into domestic water strategies is cost-effective, enhances ecosystem integrity, and promotes equity. Development organizations and international funders are realizing that these indigenous knowledge systems can bolster climate resilience and adaptive capacity, particularly in low-income communities [26].

Established, ecologically friendly, and socially embedded responses to current water scarcity issues are evident in the traditional and indigenous water management systems of South Asia. Support from policy frameworks, technological advancements, and community engagement can help revive and re-engineer these ancient technologies in modern times, contributing to the much-needed water security in the region [9, 19, 23, 24]. Bringing these practices into the public eye and incorporating them into broader approaches to water governance will not only help preserve cultural heritage but also contribute to achieving Sustainable Development Goal 6: ensuring clean water and sanitation for all (Fig. 12.3).

12.4 Modern Technological Innovations

As concerns about water security mount and traditional sources of water distribution become increasingly inefficient, modern technology and innovation in alternatives to freshwater systems are gaining traction in South Asia. Through these methods, including drip and sprinkler irrigation, remote sensing, GIS-based water resource management, and desalination and wastewater recycling technologies, water needs in both agricultural and urban areas were effectively addressed in the region. They are technologies that help ensure the sustainable availability of water, and they are part of the adaptation strategies that will help us achieve climate resilience by reducing reliance on unpredictable rainfall and over-extraction of underground water [9, 18, 19, 23].

Drip and sprinkler irrigation, known as precision irrigation technologies, is among the most revolutionary advances in on-farm water use. In contrast to inefficient flood applications, which typically result in a loss of over 50% of water to evaporation, runoff, and leaching beyond the root zone, irrigating with drip and sprinkler systems provides a controlled supply of water directly to the roots, resulting in significant water savings. Drip irrigation, for instance, moves water over long distances through

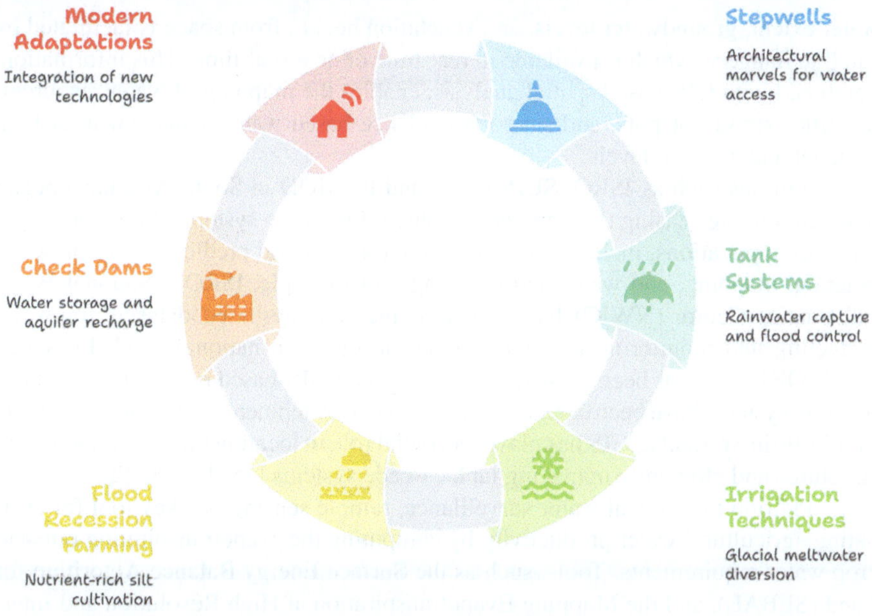

Fig. 12.3 South Asia's water management systems

a series of pipes and emitters that slowly and constantly dribble water to meet the crop's needs, thereby reducing waste. This technique has been particularly successful with high-value crops, such as fruits, vegetables, and sugarcane, where precise water control has improved yield and root development while conserving water.

In India, there has been an impetus for drip irrigation driven by farmer subsidies and mass extension plans, such as the PMKSY (Pradhan Mantri Krishi Sinchayee Yojana), aimed at expanding micro-irrigation coverage to millions of hectares of land. The water savings and productivity of states like Maharashtra, Karnataka, and Tamil Nadu have surged following the introduction of these systems. Likewise, in Pakistan, the Punjab and Sindh provinces have undertaken pilot projects to demonstrate the use of drip irrigation in cotton and citrus; however, scaling is hampered by high upfront costs and a lack of awareness among farmers [8, 10, 11, 19]. In Bangladesh, efforts to promote low-cost drip kits among smallholder farmers have been practical, particularly in drought-prone areas of the northwest. However, a lack of technical knowledge, operational hurdles, and cost-effectiveness continue to constrain the adoption of TFH, especially among small and marginal farmers, who comprise more than two-thirds of the agrarian community in South Asia.

The application of remote sensing (RS) and Geographic Information Systems (GIS) in water resources management has transformed the way governments and researchers monitor, evaluate, and plan for water availability and processes across large and diverse territories. The capacity of remote sensing technology to continuously and remotely monitor hydrological variables, such as soil moisture, surface water extent, groundwater levels, and vegetation health, from space is facilitated by satellite imagery, which is available in real-time or near real-time. This information, combined with GIS-based spatial analysis, enables the mapping of water resources, identifies stress hotspots, and supports evidence-based water management at both regional and national levels.

Institutions such as ISRO, SUPARCO, and ICIMOD in South Asia have begun to utilize remote sensing (RS) and geographic information systems (GIS) for a wide range of applications, including drought monitoring, flood prediction, groundwater potential mapping, and watershed planning. For example, ISRO's National Water Informatics Centre (NWIC) has been utilizing satellite-based datasets for water budgeting and monitoring reservoir storage levels on a national level. In Nepal, the GLOF hazard has been evaluated using GIS and GIS-based modelling, and early warning systems have been designed for Himalayan catchments vulnerable to GLOF. Similarly, in Sri Lanka, GIS has played a crucial role in locating rainwater harvesting structures and efficiently managing tank cascade systems [15, 17, 18, 22].

In addition to physical water surveillance, remote sensing is a key tool for evaluating agricultural water productivity by computing the evapotranspiration rate and crop water requirements. Tools, such as the Surface Energy Balance Algorithm for Land (SEBAL), and the Mapping Evapotranspiration at High Resolution and Internalizing Spatial Variability (METRIC) model provide the capability to estimate water consumption patterns in different land uses, that inform policy makers in the more efficient allocation of water and the promotion of efficient operational practices. However, while the potential of these technologies is vast, their use in South Asia is often hindered by comparatively low institutional capacity, limited access to data, and inadequate training for regional stakeholders. Addressing the digital divide between science and the field is essential to ensure the effective use of RS and GIS tools in support of water governance and climate adaptation.

With depleting freshwater resources and increasing demand, both desalination and wastewater recycling technologies have garnered significant attention in South Asia, particularly in coastal cities and arid regions, which suffer from water scarcity or pollution in their natural water sources. Desalination, the process of removing salt and impurities from seawater or brackish water, has been practiced for years in Middle Eastern countries. Its use in South Asia has slowly gained popularity but remains underutilized due to the high cost of energy and environmental concerns. Reverse Osmosis (RO), the most widely used desalination method, is expensive due to high energy costs, making it ideal for industrial and municipal desalination purposes, but not for rural households [4, 19, 23, 24].

India's use of Chennai and some parts of Mumbai and Tamil Nadu have desalination plants supplying the municipal drinking water source, especially in drought conditions. The Chennai Metropolitan Water Supply and Sewage Board operates two large-scale reverse osmosis (RO)-based desalination plants, which produce a total of 180 million litres per day to meet the city's demand. The Maldives also heavily rely on desalination to supply drinking water to their widely dispersed island populations, due to the lack of perennial rivers and the fact that salt water permeates most freshwater lenses [5]. Yet, the environmental toll produced by desalination, including marine pollution caused by brine discharge and carbon emissions resulting from energy-intensive operations, remains a contentious issue that demands careful management and technological innovation to mitigate ecological impacts.

Concomitant with desalination, the recycling and reuse of wastewater have become vital elements of integrated urban water strategies, with the most significant impact being felt in large, highly populated metropolitan regions that face severe water shortages. Treated wastewater can be used for non-drinking purposes, such as industrial cooling, landscape irrigation, and toilet flushing, thereby reducing pressure on freshwater supplies. Over the past few years, many South Asian cities have constructed costly advanced wastewater treatment (AWT) facilities featuring tertiary treatment systems using membrane bioreactor (MBR), ultrafiltration (UF), and ultraviolet (UV) disinfection for the production of high-grade reclaimed water [14]. For instance, Delhi's NEWater Project, modelled on Singapore's success, is replicating the treatment of sewage for reuse in industry and horticulture. Bangalore has also implemented decentralized wastewater treatment systems in residential layouts and commercial buildings, enabling on-site reuse and reducing reliance on a centralized water supply.

However, the full-scale application of wastewater reuse is hindered by social and cultural barriers, as well as a lack of related regulations, financial problems, and other issues. Concerns regarding the safety and acceptability of recycled water, particularly for drinking, limit public acceptance. Additionally, the absence of a common authority to regulate and monitor the discharge and use of wastewater curtails the scalability of such initiatives [8]. These issues need to be addressed through strong public education, investment in small-scale treatment plants and other facilities, and also through policy instruments that incentivize action from the private sector to treat wastewater.

Cutting-edge technologies, including those for precision irrigation, satellite imagery, and alternative water purification methods, are becoming increasingly important in combating water scarcity across South Asia. Although these solutions present some promising opportunities for sustainable water use, their effective transition requires the tackling of economic, institutional, and social challenges with the goals of multi-stakeholder engagement, policy coherence, and capacity-building. With the region caught between the complex challenges of climate change and rapid development, realizing the potential of these innovations is crucial to creating a secure water future and resilient livelihoods in diverse ecosystems (Fig. 12.4).

12.5 Policy and Institutional Responses

The legal and institutional framework for water resources management in South Asia has evolved over the last few decades, influenced by colonial legacies, post-independence development goals, and an increasing recognition of the need for sustainable and equitable water governance. Most countries in the region have also developed national water policies and legal frameworks to regulate water usage

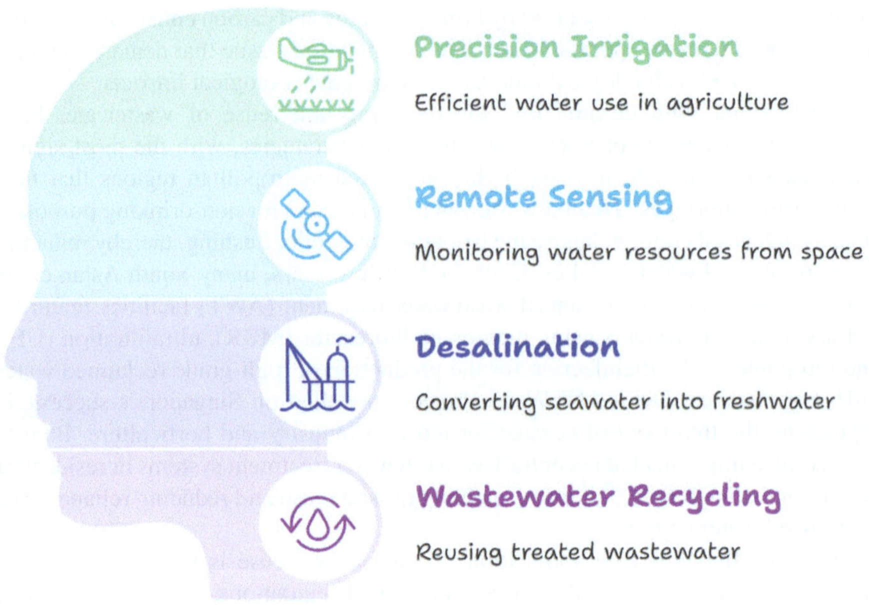

Fig. 12.4 Enhancing water security in South Asia

across various economic sectors, promote environmental sustainability, and address the rising demands from agriculture, industry, and domestic use. For example, India's National Water Policy (2012) emphasizes integrated water resources management, prioritizes the supply of drinking water, and supports decentralized planning [14, 23, 24]. Similarly, Pakistan's National Water Policy (2018) aims to tackle inter-provincial differences, encourage effective irrigation techniques, and enhance water storage capacity. The National Water Policy (1999) in Bangladesh presents a comprehensive plan for managing surface and groundwater, flood management, and pollution control. In Nepal, the National Water Resources Strategy (2058 B.S.) highlights the development of hydropower potential and ensuring equitable access to water. However, despite these policy measures, implementation has been inadequate due to fragmented authority, inefficient monitoring agencies, and insufficient budgetary provisions.

A significant development over the past few years has been the growing emphasis on decentralized governance and participatory water management, which acknowledges the shortcomings of the traditional, top-down bureaucratic approach and underscores the need for local-level involvement in water planning and management. Decentralization entails the distribution of power, resources, and responsibilities from central governments to regional and local institutions, enabling more context-sensitive and responsive water governance [15, 18, 22]. This method aligns with the community-based natural resource management (CBNRM) model. It has been implemented in various projects, such as watershed development initiatives, tank rehabilitation programs, and groundwater user associations. In India, local self-government institutions, including panchayats and municipalities, are empowered to manage water resources at the gram sabha (village council) level. The 73rd and 74th Constitutional Amendments (1992) reinforce this framework. Some states have adopted participatory irrigation management (PIM), where farmers' associations assume responsibility for maintaining canals and administering water resources. This strategy has demonstrated increased efficiency and reduced conflict, as seen in Andhra Pradesh and Maharashtra. Similarly, in Nepal, the Community-based Flood and Glacial Lake Outburst Risk Reduction Project has allowed local communities to establish warning systems and undertake small-scale water infrastructure projects. However, the success of decentralized systems hinges on strengthening regional institutions, ensuring transparency in decision-making, and maintaining ongoing financial support from higher levels of government [14, 17, 23, 24].

Complementing state-led initiatives, NGOs and other civil society organizations have significantly influenced water governance in South Asia by intervening where the state has been reticent in areas such as advocacy, infrastructure delivery, and grassroots mobilization. These organizations operate at various levels, from village-level advocacy to national and transboundary water dialogues, to amplify the voices of those experiencing poverty and support inclusive water policies. Civil society organizations also serve a vital watchdog role, monitoring government performance, lobbying for legal changes, and litigating against projects that harm the environment. In Sri Lanka, environmental NGOs have successfully lobbied for tighter restrictions on the discharge of industrial effluent into rivers. In Pakistan, civil society coalitions have advocated for greater transparency regarding the construction of dams and their social and environmental impacts.

Moreover, NGOs engage in capacity-building, education, and empowerment interventions, particularly in remote and underserved areas where formal water governance structures are weak or nonexistent. By training citizens to use water safely and test its quality, as well as conducting brainstorming and mapping exercises to identify problems and solutions, these organizations can enhance individuals' understanding of water rights and drive collective advocacy for better service delivery [1, 8, 10, 15, 22]. Digital media and mobile technology have further improved the dissemination of civil society interventions, enabling real-time data collection, grievance redressal, and citizen journalism to flourish about water issues. Another example is the South Asia Network on Dams, Rivers and People (SANDRP), which serves as a regional partner hub that unites activists, academics, and policymakers to share knowledge on basin governance, climate adaptation, and indigenous water rights. International NGOs, such as WaterAid, Oxfam, and the World Resources Institute, collaborate with local partners to enhance the implementation of WASH programs, influence policy discussions, and address the role of gender in water planning [5, 11, 12, 18, 25].

However, despite their importance, NGOs and civil society organizations are often hindered by regulatory constraints, financial difficulties, and political opposition. Additionally, repressive laws governing foreign funds and civic engagement in several South Asian countries limit the operational space for independent organizations working on environmental and human rights issues. Furthermore, cooperation between civil society and government agencies is at best ad hoc, with many water-related programs characterized by duplication of effort, a lack of sustainability, and inadequate integration into mainstream planning. To address these challenges, there is an urgent need to institutionalize multi-stakeholder platforms that bring various stakeholders together to deliberate, co-design policies, and ensure accountability in water governance. Substantial collaboration between state and non-state actors, supported by appropriate legal frameworks and financial investments, will be crucial for advancing inclusive and adaptive water governance in the region.

The development of national water matrices and legal frameworks in South Asia reflects an emerging awareness of the complex and dynamic socio-hydrological aspects of the region [1, 3, 8, 11, 20]. Centralized forms of water governance, however, are the dominant approach. Still, they are increasingly being challenged, and alternatives are emerging with strong potential to empower local communities and respond more effectively to diverse water needs. Additionally, the active involvement of NGOs and civil society has been crucial in bridging implementation gaps, fostering innovation, and advocating for equitable access to water. With South Asia confronting increasing impacts from climate change, population growth, and resource degradation, building upon sound legal structures, decentralizing decision-making, and strengthening civil society are essential for advancing water futures that are equitable and resilient for all (Fig. 12.5).

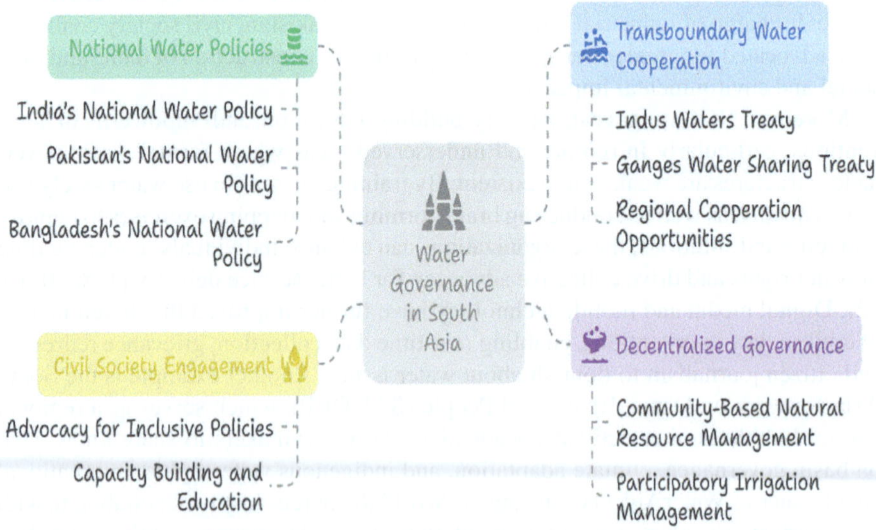

Fig. 12.5 Water governance in south Asia: challenges and opportunities

12.6 Transboundary Water Cooperation

South Asia is home to some of the largest transboundary river networks in the world, including the Indus, Ganges, Brahmaputra, and Meghna (GBM), as well as the Kaveri, among others, which flow through several countries before emptying into the Bay of Bengal and the Arabian Sea. These mighty river basins are the lifeblood of millions, fueling agriculture, hydropower, fisheries, transportation, and domestic water supplies. The Indus Basin, spanning over 1.1 million square kilometers and shared by India, Pakistan, Afghanistan, and China, originates in the Tibetan Plateau and nourishes the fertile agricultural heartlands of Punjab and Sindh. One of the largest and most complex hydrological systems in the world, the Ganges–Brahmaputra-Meghna (GBM) Basin is a transboundary river system that spans across India, Bangladesh, Nepal, Bhutan, and China, covering an area of more than 1.7 million square kilometers. Seasonal monsoon flows and melting glaciers from the Himalayas primarily feed these rivers, making them highly dynamic and sensitive to the impacts of climate change (i.e., altered precipitation patterns, glacier retreat, and the increased frequency of extreme weather events).

Despite their ecological and economic importance, transboundary rivers in South Asia are sites of geopolitical tension due to competing visions of the nation-state, power asymmetries, and historical claims to water sharing [2, 8, 11, 12, 14, 23]. Historically, there is also the Indus Waters Treaty (IWT) of 1960, as well as historical ties that, when not violated, were grounds for regional cooperation. The IWT, brokered by the World Bank and signed by India and Pakistan, divides the Indus water system between the countries to settle disputes arising from partition, via reverse osmosis. The treaty assigns the three eastern rivers, Ravi, Beas, and Sutlej, to India, and the three western rivers, Indus, Jhelum, and Chenab, to Pakistan, with some control vested in India for the use of the western banks of the rivers. The treaty also sets forth distinct dispute resolution procedures, which include the establishment of a Court of Arbitration to resolve disputes that cannot be resolved through the Permanent Indus Commission, a group of officials from India and Pakistan that meet twice a year. While it has been criticized for being rigid, constrained within the straightjacket of its original form and for not addressing the challenges of the contemporary period of climate and environmental variability and flow, the IWT remains an anchor in the area of bilateral water relations and has weathered (and even withstood) the force of multiple conflicts, including wars and moments of deep diplomatic strain [5, 14, 24, 25].

Across the Indus, the hydro-institutional regimes of other transboundary river basins in South Asia are less established. The Ganges–Brahmaputra-Meghna (GBM) Basin, for example, features a fragmented framework of bilateral agreements that are not always conducive to fostering basin-wide cooperation. The Ganges Water Sharing Treaty of 1996 was a breakthrough as it provided a minimal flow sharing mechanism at the Farakka Barrage during the lean season between India and Bangladesh. However, its realization has been uneven, hindered by unilateral dam building, upstream diversions, and inadequate attention to downstream ecological needs. Similarly, the 2011 Framework Agreement on Cooperation for the Sustainable Development of the GBM Basin, signed by India and Bangladesh, outlines general principles for cooperative management but lacks binding legal obligations or mechanisms for dispute resolution with enforcement capabilities [6, 14, 23, 24]. In comparison, Nepal and India have numerous bilateral treaties regarding the rivers they share, including the Kosi, Gandak, and Mahakali, most of which are viewed in Nepal as unfair, as they are accused of sidelining Nepalese sovereignty and environmental safety in their surface and sea-stage operations in Nepal to meet their irrigation and energy goals.

The Cauvery River dispute between the Indian states of Karnataka and Tamil Nadu illustrates the challenges of managing transboundary rivers within national borders, much less between sovereign territories. This case of contested waters, while not an international one, exemplifies the broader governance dilemmas of South Asia, characterized by divided institutional roles, inadequate adaptation, and minimal attention to climate risk in water sharing. Efforts to establish a South Asian river-basin organization, similar to the one in place for the Mekong in Southeast Asia, have remained mainly unattained due to deep-rooted mistrust among the river-sharing countries, a lack of robust regional multilateral architecture, and the shadows of other political, economic, and security issues [2, 8, 12, 24].

Indeed, various obstacles impede regional cooperation concerning the shared use of water. Political antagonisms especially between India and Pakistan have generally hindered collaboration, as water becomes a pawn in larger geopolitical struggles. Nationalism, electoral politics, and bureaucratic inertia at home further obstruct efforts to establish a consensus on collective water governance models. Moreover, the absence of standardized data-sharing mechanisms hampers scientific assessments and early warnings related to floods and droughts. Pollution, sedimentation, and the loss of wetlands and floodplains all exacerbate water stress, diminishing the natural capacity of rivers to adapt to change. Additionally, climate change adds another layer of uncertainty, as shifting rainfall patterns, the gradual decline of snowpack, and rising temperatures are projected to disrupt river flows and heighten competition for limited resources.

However, beneath these challenges lie numerous opportunities for regional cooperation. Strengthening existing bilateral agreements with modernization clauses that include climate adaptation, ecosystem protection, and participatory governance would enhance their relevance and effectiveness. Establishing regional forums for discussion, such as a South Asian Transboundary Water Commission under the framework of the South Asian Association for Regional Cooperation (SAARC) or the Bay of Bengal Initiative for Multi-Sectoral Technical and Economic Cooperation (BIMSTEC), will facilitate the exchange of information, collaborative research, and capacity-building activities. Investment in transboundary watershed management initiatives, supported by regional financial mechanisms and technical assistance from international organizations such as ICIMOD and the United Nations Development Programme (UNDP), can promote sustainable land and water use practices across international borders [5, 18, 21, 24, 25].

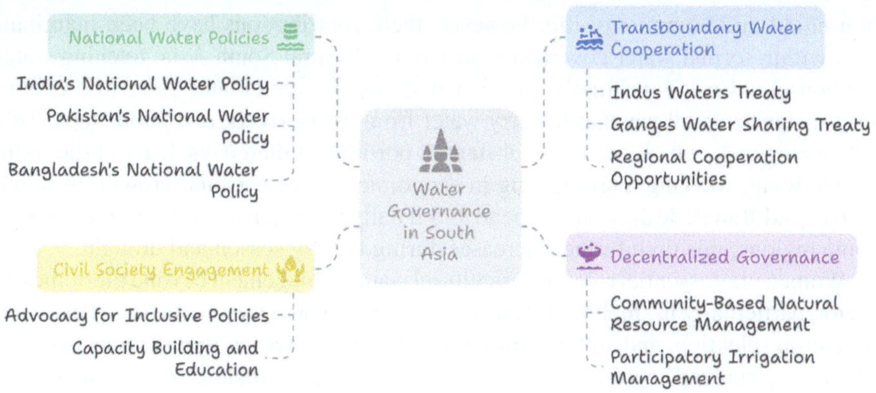

Fig. 12.6 Water governance in South Asia: challenges and opportunities

Additionally, involving non-state actors (civil society organizations, academic institutions, and the private sector) in transboundary water governance can help develop solutions that are more transparent, inclusive, and innovative. People-centered frameworks, which recognize the rights and knowledge of so-called 'poor' groups and communities left behind in border regions, can foster more equitable and sustainable water sharing. Technological innovations, such as satellite-based surveillance, real-time hydrological models, and digital platforms for sharing cross-border data, also provide new avenues for building trust and reducing information asymmetry among riparian states.

The broad evolutionary patterns form the foundation for a dynamic of "competing cooperation", where, based on a combination of riparian position, geopolitical stature, power asymmetries, and discursive resources, the rules of riparian engagement appear more competitive than cooperative. While longstanding treaties and agreements, including the Indus Waters Treaty, provide crucial examples of stability and continuity, they will need to be updated to address the new challenges of climate change, environmental degradation, and increased water demand. There is a necessity to advance with more inclusive, evidence-based, and adaptive forms of governance supported by regional structures, legal instruments, and multi-stakeholder participation, which can transform the source of tension around our water into a source of cooperation for peace and sustainability in South Asia (Fig. 12.6).

12.7 Gender and Social Inclusion in Water Management

The nexus of gender and water in South Asia indicates a complex landscape where 'customary roles,' socio-cultural patterns, and institutional mechanisms play pivotal roles in access, control, and decision-making over water resources. Women's involvement in water collection and management has always been a crucial aspect of both household and community life; however, their contributions have been marginalized within formal water governance structures. In rural South Asia, fetching water for household use is primarily the responsibility of women and girls, who endure long distances in all weather to carry water from wells, rivers, and public taps. This exhausting task, which occupies substantial portions of their days, leaves little room for studying, working, participating in economic life, or personal growth. In India, Nepal, and Bangladesh, women spend 1–3 h daily on water-related activities, and in some regions, this time further increases during the dry season and drought.

Women play essential roles in agricultural water management beyond the domestic sector, particularly in smallholder farming systems, where they are involved in irrigation, crop selection, and soil moisture preservation. However, their perspectives are often marginalized in water allocation decisions, infrastructure development, and policy formulation. Women are frequently overwhelmed in traditional water user associations, village committees, and local self-government groups, which perpetuates male-centered institutions [1, 11, 21, 25]. This exclusion not only reinforces gender disparities but also undermines water management, given the critical role of women who possess specific knowledge of seasonal fluctuations, water quality, and household consumption that inform sustainable resource management.

At stake is not just an equitable distribution of water but its connection to large-scale social hierarchies based on class, caste, ethnicity, and region. Across vast areas of South Asia, among the most dispossessed groups, Dalits, Adivasis, Scheduled Tribes, and landless laborers, systemic obstacles deny them access to clean, reliable water. Discriminatory customs of caste-based exclusion prevent low-caste households from utilizing familiar water sources, such as hand pumps, wells, and community tanks. In rural India, for instance, there are ongoing reports of Dalit women being denied the right to draw water from wells controlled by higher castes, serving as reminders of social hierarchies that continue despite constitutional guarantees and affirmative action. Similarly, in Nepal, as a result of past exclusion and marginalization, marginalized groups and Madhesi communities face more significant barriers in accessing water fairly [11, 12, 20, 21, 23, 25].

Further dimensions of inequality arise in urban settings, as informal settlements and slum residents (often migrant or low-income households) lack legal entitlement to safe water services. These communities rely on unreliable and contaminated water sources, including illegal taps, private vendors, and polluted streams, which heightens the risk of disease and perpetuates the cycle of poverty. Women and children experience the most severe impacts from these shortages and face a higher risk of water-related illnesses due to inadequate hygiene and sanitation facilities. These issues are exacerbated by a lack of gender-inclusive planning in urban areas, with public water points often located far from residential clusters and lacking adequate safety features, such as sufficient lighting and secure access (e.g., at night).

Given these disparities, there has been an increasing emphasis on inclusive policies and grassroots actions that promote gender equality and women's empowerment in water governance. Some national and subnational government policies in South Asia have introduced reforms to mainstream gender issues in water management. For instance, India's National Water Policy (2012) recognizes the role of women in watershed development, particularly in participating in irrigation projects. Similarly, Nepal's National Strategy for Sanitation and Hygiene (2018) provides for the representation of women in local water user groups and sanitation committees. Yet, implementing such policy commitments remains an ongoing challenge that will require continual advocacy, institutional reform, and capacity-building efforts at various levels [2, 5, 13, 14, 24].

On the ground, many citizens and community-based organizations (including women-led cooperatives and other civil society organizations) have proven to be effective players in promoting an inclusive water governance agenda. In Bangladesh, the involvement of women in polder management committees (which are responsible for maintaining coastal embankments and controlling the inflow of tidal water) has resulted in improved flood risk reduction and enhanced livelihood activities, as well as modern adaptation. In Pakistan's Punjab province, the Aga Khan Rural Support Program (AKRSP)-promoted Community-Based Organizations (CBOs) have successfully engaged women in water harvesting and micro-irrigation systems, resulting in improved crop production and enhanced household water security. Similarly, in Sri Lanka, women's participation in the restoration campaign for the tank cascade system has led to improved maintenance of traditional irrigation infrastructure, resulting in increased food production in drought-prone areas.

These bottom-up efforts demonstrate the change that can be achieved by prioritizing the empowerment of women and marginalized communities through collective action and participatory governance systems. By establishing channels for dialogue, negotiation, and co-management, they create opportunities for greater accountability, transparency, and responsiveness in the provision of water services. More importantly, they also enable people experiencing poverty to demand justice, claim a redressal process or mechanism, and advocate for policy making that impacts their lives and livelihoods.

These documents are complemented by international frameworks, including the Sustainable Development Goals (SDGs), notably Goals 6 (Clean Water and Sanitation) and 5 (Gender Equality), which serve as an overarching framework for incorporating gender and social aspects into water management interventions at the global level. In (100 lakh ha), state cooperation mechanisms such as SAARC and BIMSTEC are also beginning. They also incorporate gender-sensitive approaches in transboundary water dialogues, recognizing that ensuring just water governance is crucial for peace, stability, and sustainable development [2, 8, 11, 18, 24, 25].

However, there is still much work to be done to incorporate the values of gender justice and social inclusion into water policies and practices in South Asia. A long road remains ahead, as the inertia of bureaucracy, social norms that lean toward patriarchy, and limited political will must be addressed. Greater legal protection, access to education, and leadership training for women and other marginalized groups, along with multi-stakeholder partnerships, are essential components for equitable water futures. Ultimately, achieving equity in water management will require not only an engineering solution but also a reassessment of power relations, cultural values, and institutional priorities. This necessary shift demands the collaborative action of both state and non-state actors, ensuring that no one is left behind in the pursuit of water security.

12.8 Agricultural Water Use and Efficiency

Agriculture is the largest consumer of freshwater in South Asia, accounting for over 80% of the region's total water withdrawals. The sector's water consumption is driven by large-scale irrigation systems, the cultivation of water-intensive crops, and inefficient water management that exacerbates both physical and economic water stress. Understanding the patterns of crops and water use is key to addressing the problem of agricultural water use in South Asia. Crops such as paddy, wheat, and sugarcane consume an excessively high amount of water, to the detriment of more water-efficient crops like millets, pulses, and oilseeds. For instance, rice cultivation in the Indian Punjab and Haryana regions requires about 4,000–5,000 L of water per kilogram of grain, thereby significantly increasing pressure on groundwater reserves. Similarly, cotton cultivation in the Punjab and Sindh provinces of Pakistan substantially contributes to local water stress due to its deep root system and long growing period.

The location of these crops is often dictated by historical policy decisions, market incentives, and subsidies, rather than by agroecological suitability. Water-intensive crops usually dominate farmholdings due to decades of state procurement prices, primarily for staple grains such as wheat and rice, under the public distribution system [5, 13, 20, 21]. This creates a self-perpetuating cycle, as farmers continue to grow water-dependent crops despite declining water tables and rising input costs. Additionally, millions of smallholder farmers who depend on traditional crops face significant economic risks if changes in market infrastructure, processing facilities, and consumer preferences do not facilitate the adoption of alternative crops.

In light of the unfeasible trend of agricultural water use, sustainable farming techniques and irrigation methods, such as optimizing water use, investments, and overall allocation, have increasingly been prioritized as part of a broader initiative to enhance efficiency and sustainability in future climate change scenarios. Specific approaches, including drip and sprinkler irrigation, alternate wetting and drying (AWD) in rice cultivation, conservation agriculture, and agroforestry, have proven effective in conserving water without significantly affecting yields. Drip irrigation, for instance, can conserve up to 60% of water compared to flood irrigation, particularly for orchard crops and vegetables. AWD has made progress in parts of Bangladesh and eastern India, where substantial water savings and reductions in methane emissions have been achieved through cyclic draining and re-flooding of paddies, rather than maintaining continuous submergence.

Another point is that conservation agriculture, which involves minimal soil disturbance, permanent organic soil cover, and diversified crop rotation, is the most effective way to increase soil moisture retention while reducing loss through evaporation. An example from Nepal and Bhutan is an integrated farming system, where a combination of livestock, aquaculture, and crops has been shown to enhance water use efficiency through nutrient cycling and waste recycling. Furthermore, rainwater harvesting and on-farm storage structures (for example, farm ponds, check dams, and contour trenches) have provided a buffer against erratic monsoon rains, serving as a vital backup in drought-prone areas of semi-arid India and Pakistan. In addition to conserving water, these methods enhance soil health, supporting biodiversity and carbon sequestration, which contribute to broader environmental sustainability goals [2, 11, 15].

Despite this, several institutional, financial, and behavioral challenges hinder the widespread adoption of sustainable farming systems. The majority of farmers in South Asia are smallholder farmers who often lack access to credit, technical skills, and extension services necessary for adopting new technologies. Additionally, the cost of investing in precision irrigation systems or soil conservation measures may pose a barrier to investment for resource-poor farmers. A strong cultural preference for traditional agricultural practices, combined with unclear perceptions of harvest and market prices, also discourages the shift to water-efficient approaches. Hence, it is essential to implement targeted capacity-building programs, such as FFS and demonstration plots, to build trust and promote the adoption of best practices.

Another significant factor influencing agricultural water use in South Asia has been the impact of subsidies on water consumption, particularly for electricity, fertilizers, and canal irrigation. Subsidized power for agricultural pumping has encouraged widespread overdrafting of groundwater, especially in India and Pakistan, where flat-fee power supply removes price signals for efficient water use. For instance, in India, several states have extensive electricity subsidy programs that provide nearly free electricity to farmers, which encourages the overuse of groundwater and the cultivation of water-intensive crops. Together, these trends have led to negative consequences, including declining water tables, increased electricity usage, and growing financial liabilities for state utilities. A similar situation has arisen in Pakistan, where unchecked tubewell extraction, supported by inexpensive diesel and electricity, has caused a significant decline in aquifers in vital agricultural areas.

Fertilizer subsidies, particularly those for nitrogenous fertilizers like urea, have also indirectly impacted water use by promoting intensive cropping systems that require heavier irrigation [15, 19, 20, 22, 23]. Overapplication of fertilizers leads to nutrient runoff and groundwater contamination, which in turn deteriorate water quality and increase costs associated with water treatment. Conversely, balanced nutrient management and organic farming can mitigate these effects by enhancing nutrient uptake efficiency and reducing losses from leaching.

Acknowledging the unwelcome consequences of these subsidies, some governments have begun to seek alternative policy instruments to promote sustainable water use. For example, India's Pradhan Mantri Krishi Sinchayee Yojana (PMKSY) focuses on "per drop more crop" by attempting to align irrigation efficiency with financial incentives for adopting micro-irrigation systems. Both Andhra Pradesh and Gujarat have also undertaken pilot projects for direct benefit transfers (DBT) in electricity consumption, wherein farmers pay costs based on usage rather than receiving a flat-rate subsidy. These reforms aim to increase the costs and energy associated with water use, encouraging people to use it more efficiently.

Still, subsidy reform is a politically sensitive issue because it affects millions of small farmers who depend on this assistance for their lives and livelihoods. The withdrawal or reduction of subsidies should be accompanied by well-designed safety nets, alternative financing options, and effective communication, all of which are essential to ensure social acceptance and mitigate harm to the poorest segments of the population.

In the face of increasing challenges to agricultural water use and efficiency in South Asia, crop diversification, sustainable farming practices, and reform of subsidies must be pursued simultaneously. Although technological solutions offer exciting opportunities for conserving water and enhancing productivity, their realization relies on an appropriate policy environment, institutional capacity, and active participation from farmers. Reconfiguring agricultural incentives away from water-intensive crops towards water-sensitive varieties, adopting conservation-focused farming practices, and aligning subsidies that support agrarian methods with actual environmental costs are vital prerequisites for developing resilient and sustainable food systems in the region. Given the challenges of water in a changing climate, the urgency for comprehensive responses that create coherence between agricultural development and water governance is critical.

12.9 Urban Water Management Challenges and Solutions

Rapid urbanization in South Asia has overwhelmed water infrastructure, exposing already fragile urban water supply, sanitation, and wastewater management systems. The region is one of the fastest urbanizing areas in the world, and many cities are facing the simultaneous challenges of rapidly increasing demand from expanding populations while ensuring that their existing water systems are sustainable and resilient [8, 15, 18, 19, 25]. Currently, more than half a billion people reside in urban areas in South Asia, a figure projected to surpass 1 billion by 2050, according to the United Nations. This demographic change has led to unplanned suburban development, squatter settlements, and strained urban services, particularly in the distribution of water and sewage treatment.

Among the most pressing challenges is the outdated and neglected water infrastructure, which cannot keep up with the growing demand. Non-revenue water (NRW) in supply networks, primarily due to leakage and inefficient billing, typically ranges from 30 to 40% (or higher) in many South Asian cities. Furthermore, in Dhaka, only 60% of the water supplied throughout the town is available for use at end sources, with the remainder lost before reaching households. This unsustainable over-pumping of groundwater to compensate for city shortfalls has worsened the situation, leading to land subsidence, declining water levels, and increased risk of contamination from industrial discharges and untreated sewage. This is especially true in low-income communities and slums, where access to clean and safe water is limited, forcing residents to rely on unsafe sources or purchase water from expensive private vendors [3, 5, 8, 24].

In response to these systemic shortages, the cities of South Asia are increasingly focusing on rainwater harvesting in urban areas as a local and sustainable solution to the water crisis. Rainwater capture and storage in cisterns for the irrigation of kitchen gardens and groundwater recharge are essential methods described. Due to the monsoon-driven nature of the climate, where the majority of rainfall occurs over just a few months, taking advantage of these seasonal fluctuations can help reduce vulnerability to shortages during the dry season and reliance on centralized water supply systems. Chennai, India, has fostered a culture of rooftop rainwater harvesting through a mandate for new buildings, significantly contributing to the high level of groundwater achieved after it was alarmingly low in Chennai and Bangalore, India. In the Kathmandu Valley, community-based activities have similarly restored old ponds and constructed artificial recharge structures to store groundwater and ensure water security during dry periods [2, 15, 21, 23].

In addition to individual building-level initiatives in the private sector, municipal-scale urban planning projects have also incorporated rainwater harvesting into stormwater and green infrastructure programs. In Hyderabad and Pune, urban parks and public spaces are being redesigned with bioswales, permeable pavements, and retention basins that collect runoff and direct it to recharge wells underground. These green infrastructure solutions conserve water, prevent flooding, improve water quality, and enhance urban biodiversity. However, the widespread adoption of urban rainwater harvesting is hindered by the lack of enforcement, insufficient technical expertise among property owners, and inadequate maintenance of existing systems [3, 11, 14, 18, 23]. Strengthening legal frameworks and promoting financial incentives, along with knowledge sharing, should be essential enablers for implementing a rainwater strategy in urban water resilience.

In cities where water scarcity is often most acute, the reuse of treated wastewater is playing an increasingly important role as a partial solution to the growing gap between supply and demand, particularly in densely populated urban areas. As the production of wastewater increases daily, with a significant portion remaining untreated, cities not only face environmental degradation but also the loss of valuable resources. Reclaimed or recycled wastewater (i.e., treated water used for non-potable applications, such as industrial heating and cooling, landscape irrigation, building construction, and toilet flushing) can alleviate some of the pressure on diminishing freshwater reserves. Several South Asian cities have initiated wastewater recycling in recent years, addressing both water scarcity and pollution simultaneously.

In India, the National Green Tribunal (NGT) and several state governments have mandated that private developers establish decentralized sewage treatment plants (STPs) in large residential complexes, commercial complexes, and industrial parks. The 'NEWater' project in Delhi, inspired by Singapore's reclaimed water system, is under consideration. It aims to treat and utilize water for industrial and horticultural purposes, thereby reducing the need to discharge wastewater into the Yamuna River, which flows into the Yamuna. In Ahmedabad and Surat, the textile industry employs a closed-loop water recycling process, allowing for multiple cycles before discharge, thereby reducing the input and pollution of freshwater. In Bangladesh, pilot initiatives in peri-urban settings of Dhaka demonstrate the potential of greywater reuse in urban agriculture, not only sustaining livelihoods but also conserving drinking water.

Despite these positive advances, the reuse of wastewater still encounters significant challenges, including being capital-intensive, technologically complex, and a complicated issue from a social resistance standpoint. Public perception remains a significant barrier, particularly regarding the use of treated sewage for purposes other than direct potable use. Health risks, odour, and appearance typically prevent less developed reuse techniques from being implemented, particularly for residential uses. Moreover, the absence of general guidelines and effective monitoring procedures for wastewater reuse hinders its consideration within broader urban planning efforts. Addressing these challenges with firm policy commitments, investments in innovative treatment technologies, and targeted communication to foster trust and encourage behavioural change is essential [2, 19, 22–25].

Additionally, when combining wastewater reuse with a circular economy approach that emphasizes resource efficiency, waste recovery, and systemic connectivity, urban water management can reveal new synergies. By recovering energy, nutrients, and water from waste, cities have the opportunity to transform liabilities into assets. Metropolitan areas could also adopt more natural water-use systems, such as converting wet waste into biogas through anaerobic digestion, transforming nutrient-rich sludge into agricultural fertilizer, and capturing heat from wastewater flows.

With the increasing urbanization and overwhelmed water infrastructure in South Asia, solutions must be responsible and adaptable, utilizing decentralized systems based on strategies such as rainwater harvesting and wastewater recycling. While these methods offer affordable and environmentally friendly alternatives to traditional water supply models, their success depends on coordinated action across policy, technology, finance, and community participation [6, 8, 15, 16, 22, 24]. Enhancing various legal and institutional frameworks, providing incentives for private sector involvement, investing broadly in research and capacity development, and promoting social marketing will be necessary to mainstream sustainable urban water management practices. As countries in South Asia rapidly urbanize, it is increasingly vital for cities to reassess how they use, manage, and conserve water to achieve equitable access, environmental sustainability, and long-term resilience amid growing climate and development pressures.

12.10 Future Directions and Recommendations

As an increasingly water-stressed, environmentally degraded, and climate-impacted region, South Asia needs reform in water governance and management. The future security of the subcontinent's water supply relies not only on technological innovations and policy reforms but also on merging traditional wisdom with new technologies, adopting a multilateral approach to tackle water issues, continuing investment in human resources, and preparing for climate change adaptation. The integration of traditional and modern methods, enhanced cooperation in the region, investment in research, education, and capacity building, along with the inclusion of climate resilience in water planning, form the basis for a sustainable and equitable water future for South Asia.

The hybrid combination of traditional, scientific, and technological solutions provides a robust mechanism for addressing water-related problems by leveraging proven, time-honoured indigenous knowledge systems for water planning and management, and integrating them with cutting-edge scientific and technological innovations. Their revival, facilitated by modern hydrological modeling systems, GIS-based mapping, and remote sensing, can significantly enhance water resource management, water harvesting, and drought relief efforts. For example, the restoration of ancient johads (ponds) in Rajasthan and Maharashtra using geo-informatic tools has improved aquifer recharge and agricultural yields. Likewise, in Sri Lanka, the connection of community-managed irrigation systems with digital monitoring systems has enhanced transparency and accountability in tank maintenance and water allocation. Future strategies should emphasize hybrid models that are attuned to local ecological realities and can be scaled up with technological, financial, and institutional support.

Also essential is the need to strengthen regional cooperation in managing transboundary waters, which are particularly challenging in shared river basins like the Indus, Ganges–Brahmaputra-Meghna (GBM), and Kaveri. While bilateral agreements, such as the Indus Waters Treaty between India and Pakistan, may impact joint water management, they often do not explicitly address adaptive management, ecosystem conservation, or collaborative decision-making regarding cross-border rivers. Introducing the dimensionality of scale to these mechanisms through multilateral engagement with regional institutions, such as SAARC, BIMSTEC, or the proposed South Asian River Basin Commission, could offer a more inclusive and scientific foundation for negotiations. Regional collaboration should also include data sharing, joint monitoring of water quality and flow, and the collective development of flood and drought forecasting systems. In this regard, trust-building activities, such as academic exchange programs, collaborative research projects, and civil society dialogues, can serve as vital pathways for depoliticizing water issues and fostering cooperative arrangements that prioritize long-term sustainability challenges over competitive considerations.

A third pillar for future actions is investment in research, education, and capacity building at all levels of water governance. However, despite the growing complexity of water issues, a gap remains in the availability of trained professionals, technical expertise, and institutional capacity to implement Integrated Water Resource Management (IWRM) effectively. University and research institution curricula should offer more interdisciplinary programs that integrate hydrology, engineering, environmental economics, and social sciences, thereby cultivating a new generation of well-trained water professionals who can think in both technical and socio-political frameworks. Additionally, these governments and development agencies should support field-based training programs, farmer field schools, and extension services that transform scientific knowledge into practical applications at the community level. To further democratize information, access to digital literacy and open-source resources can enable more communities to engage in water governance and disaster preparedness.

Last but not least, the mainstreaming of water-based climate resilience should be a core goal of national and subnational water policies. Future climate scenarios suggest increased variability in monsoon patterns and temperatures, glacial retreat, and more frequent severe weather events (floods and droughts). These changes will pose significant challenges for socio-economically "underprivileged" groups, such as those who rely on rain-fed agriculture or live in slums in rapidly growing cities. To enhance resilience, water infrastructure should be designed flexibly and redundantly, incorporating natural alternatives and nature-based solutions (e.g., wetland restoration, green roofs, permeable pavements, and floodplain reconnection). Climate risk assessments must be integrated into zoning regulations to prevent new developments from encroaching on ecologically valuable areas. Additionally, early warning systems, community-based disaster response teams, and insurance instruments for water-related risks can improve preparedness and reduce vulnerability.

Financial and institutional arrangements also need to be developed to promote investments in climate-resilient water schemes. This includes integrating climate finance into national budgets, utilizing international climate funds like the Green Climate Fund (GCF), and encouraging private sector involvement in developing resilient infrastructure. Policy coherence across agriculture, energy, environment, and disaster management sectors is essential for water investments to contribute to broader sustainability rather than intensifying existing trade-offs.

To achieve these future directions, political resolve, intersectoral coordination, and public involvement are essential. Civil society, media, academia, and international partners must continue to advocate for transparent, inclusive, and evidence-based water governance. At the same time, revisions of the legal frameworks are necessary to incorporate principles of fairness, sustainability, and intergenerational justice, as well as to recognize water rights for both people and ecosystems.

In short, the path to water sustainability in South Asia relies on various factors that blend historical wisdom with technological innovation, cross-border cooperation, the development of human and institutional capabilities, and the integration of climate resilience into every aspect of water planning. By following these recommendations, South Asian countries can transform water from a source of trouble and turmoil into a foundation of peace, abundance, and ecological balance. The urgency of the moment demands decisive leadership, long-term investment, and comprehensive collaboration to secure a water guarantee for everyone, both today and in the future.

12.11 Conclusion

The examination of water scarcity and its management options in South Asia presents a complex yet instructive landscape shaped by a myriad of ecological, sociopolitical, and economic forces. The key findings summary highlights that water stress in the area is not merely about physical scarcity but is also intertwined with governance failures, unsustainable agriculture, rapid urbanization, pollution, and climate change. Historically, traditional water management systems have served as robust solutions; however, their gradual deterioration has been counterbalanced by the direct extraction of groundwater from local aquifers and the development of civil infrastructure. Newer technologies, including drip irrigation, remote sensing, wastewater reuse, and decentralized rainwater harvesting, demonstrate promise, especially when combined with indigenous knowledge systems and participatory governance structures.

Cross-border river basins remain a source of both tension and opportunity; the existence of treaties, such as the Indus Waters Treaty, shows that it's possible to pursue converging activities despite political disagreements. However, broader regional cooperation is still in its early stages, indicating a need for more integrated and responsive frameworks. Access to water is further complicated by issues of gender and social equity, heightening the importance of including marginalized groups, particularly women and lower-caste communities, in policy decisions. Increasing urbanization exerts greater pressure on cities and aging infrastructure, necessitating new investments in sustainable water supply, sanitation, and global reuse.

These lessons offer value in various areas, particularly those that address sustainability challenges such as water scarcity resulting from population growth and habitat decline. South Asia demonstrates the importance of combining traditional and modern solutions, prioritizing locally led approaches alongside technology. The region also highlights that the key to equitable and effective water use involves policy coherence, institutional cooperation, and multi-stakeholder participation. Moreover, the ongoing challenges of cross-border disputes and climate vulnerability emphasize the need for trust-building through data-sharing mechanisms, early warning systems, and collaborative research platforms—blueprints that can be easily replicated in other politically sensitive river basins globally.

Lastly, there is a call to action for policymakers, practitioners, and researchers to prioritize water sustainability in development. Governments need to implement legal and institutional reforms to promote Integrated Water Resources Management (IWRM), encourage water conservation in agriculture, and assert the human right to water and sanitation. Investment in distributed infrastructure, sanitation, and climate-resilient planning must be increased through public–private partnerships and international climate finance instruments.

Third, stakeholders—such as NGOs, local governments, and civil society organizations—must continue to bridge the gap between policy and implementation by expanding successful local models, building capacity and resources in communities, and articulating a vision of more inclusive governance. The training programs must be scaled up to equip local stakeholders with the technical knowledge necessary for water auditing, hydrological modeling, and disaster risk reduction.

At the same time, researchers play a crucial role in developing evidence-based solutions tailored to local contexts. Such inclusive studies, which combine hydrology, climate science, sociology, and economics, are necessary to support adaptive policy processes. Further interaction among universities, think tanks, and regional institutions serves as a foundation for knowledge sharing and innovation in water governance.

Ensuring water for the future of South Asia demands collective responsibility, sustained efforts, and coordinated action among all levels of society. As the region's population grows and faces escalating climate-related risks, the time to act is now, before water scarcity becomes a barrier to peace, security, and sustainable development.

References

1. Ansari R, Liaqat MU, Grossi G (2024) Improving flood and drought management in transboundary Upper Jhelum Basin-South Asia. Sci Total Environ 945:174044. https://doi.org/10.1016/j.scitotenv.2024.174044
2. Arfanuzzaman M (2025) Transboundary water cooperation and joint river basin management are pivotal for climate resilient development in South Asia. World Dev Perspect 38:100681. https://doi.org/10.1016/j.wdp.2025.100681
3. Asif M, Imran Khan M, Pandey A (2024) Navigating the inclusive and sustainable energy transitions in South Asia: progress, priorities and stakeholder perspectives. Energy Convers Manag 313:118589. https://doi.org/10.1016/j.enconman.2024.118589
4. Breyer C, Oyewo AS, Gulagi A, Keiner D (2023) Renewable energy enabling pathways towards prosperity in Africa and South Asia. Sol Compass 8:100057. https://doi.org/10.1016/j.solcom.2023.100057
5. Chowdhury P, Das P, Yeassin R, Bonah Agyekum E, Irshied Al-Maaitah M, Odoi-Yorke F (2025) Exploring the potential of solar and wind-powered green hydrogen: production, costs and environmental impacts in South Asia. Int J Hydrogen Energy 137:288–302. https://doi.org/10.1016/j.ijhydene.2025.05.100
6. Do SK, Akhtar F, Goffin B, Aryal A, Tran TND, Lipscomb M, Lakshmi V (2024) Assessing terrestrial water storage variations in Afghanistan using GRACE and FLDAS-Central Asia data. J Hydrol Reg Stud 55:101906. https://doi.org/10.1016/j.ejrh.2024.101906
7. Fattah MA, Hasan MM, Dola IA, Morshed SR, Chakraborty T, Kafy A- Al, Alsulamy S, Khedher KM, Shohan AAA (2024) Implications of rainfall variability on groundwater recharge and sustainable management in South Asian capitals: an in-depth analysis using Mann Kendall tests, continuous wavelet coherence, and innovative trend analysis. Groundw Sustain Dev 24:101060. https://doi.org/10.1016/j.gsd.2023.101060
8. Goyal MK, Rakkasagi S, Surampalli RY, Zhang TC, Erumalla S, Gupta A, Dubey S, U-tapao C (2024) Enhancing sustainable development through spatiotemporal analysis of Ramsar wetland sites in South Asia. Technol Soc 79:102723. https://doi.org/10.1016/j.techsoc.2024.102723

9. Gupta V, Singh SP (2025) Exploring the multiple dimensions of solar irrigation in South-Asian countries: insights from a systematic review. Renew Energy Focus 54:100711. https://doi.org/10.1016/j.ref.2025.100711
10. Jamal A, Tarek YA, Siddique MAB, Shaikh MAA, Debnath SC, Uddin MR, Ahmed S, Akbor MA, Al-Mansur MA, Islam ARMT, Khan R, Moniruzzaman M, Sultana S (2023) Development of a fabricated first-flush rainwater harvested technology to meet up the freshwater scarcity in a South Asian megacity, Dhaka, Bangladesh. Heliyon 9(1):e13027. https://doi.org/10.1016/j.heliyon.2023.e13027
11. Maharjan S, Li W, Fazli S, Ansari A, Tiwari S, Thakurathi R, Thomas R, El-Askary H (2025) Unfolding cascading impacts of changing South Asia monsoon on a Hindu Kush Himalayas basin. J Hydrol Reg Stud 57:102155. https://doi.org/10.1016/j.ejrh.2024.102155
12. Peng Y, Zhang H, Zhang Z, Tang B, Shen D, Yin G, Li Y, Chen X, Hu Z, Habib Nazrollozoda S (2024) Future challenges of terrestrial water storage over the arid regions of Central Asia. Int J Appl Earth Obs Geoinf 132:104026. https://doi.org/10.1016/j.jag.2024.104026
13. Poh J, Tjiawi H, Chidire A, Veerasamy BB, Massier T, Romagnoli A, Wu W, Lu D, Lim JWM, Yang L, Chuah KEC (2025) Geothermal development in South, Southeast and East Asia: a review. Renew Sustain Energy Rev 209:115043. https://doi.org/10.1016/j.rser.2024.115043
14. Prateek G (2025) Productivity versus sustainability: paradigms of climate-resilient development in South Asian smallholder agriculture. Curr Opin Environ Sustain 74:101520. https://doi.org/10.1016/j.cosust.2025.101520
15. Prinja S, Purohit N, Kaur N, Rajapaksa L, Sarker M, Zaidi R, Bennett S, Rao KD (2024) The state of primary health care in south Asia. Lancet Glob Heal 12(10):e1693–e1705. https://doi.org/10.1016/S2214-109X(24)00119-0
16. Rahut DB, Mishra R, Bera S (2024) Geospatial and environmental determinants of stunting, wasting, and underweight: empirical evidence from rural South and Southeast Asia. Nutrition 120:112346. https://doi.org/10.1016/j.nut.2023.112346
17. Rasheed K, Saad S, Zaland A, Khan MW, Ali Z, Haris M, Ammad S, Sadiq T (2024) Circular economy and critical barriers: mapping the pathways and success metrics for sustainable circular success in industrialised South Asian developing nations. Results Eng 24:102995. https://doi.org/10.1016/j.rineng.2024.102995
18. Saha SK (2025) Empowering rural South Asia: off-grid solar PV, electricity accessibility, and sustainable agriculture. Appl Energy 377:124639. https://doi.org/10.1016/j.apenergy.2024.124639
19. Shah MH, Hassan ST, Ullah I, Wang Y, Shah AA (2025) Sustainable water management practice in South Asian Countries: the importance of water resources, temperature, and bioenergy on ecological footprint. Gondwana Res 144:109–122. https://doi.org/10.1016/j.gr.2025.04.004
20. Shenoy S, Jena A, Levinson C, Sharma V, Deepak P, Aswani-Omprakash T, Sebastian S, Colombel J-F, Agrawal M (2025) Inflammatory bowel disease in south Asia: a scoping review. Lancet Gastroenterol Hepatol 10(3):259–274. https://doi.org/10.1016/S2468-1253(24)00341-8
21. Thakur S, Jayaram D (2024) Resilience in the anthropocene: discourses of development, climate change, and security in South Asia. Curr Opin Environ Sustain 67:101425. https://doi.org/10.1016/j.cosust.2024.101425
22. Wang F, Rani T, Razzaq A (2024) Resource curse, energy consumption, and moderating role of digital governance: insights from South Asian countries. Resour Policy 98:105329. https://doi.org/10.1016/j.resourpol.2024.105329
23. Xue P, Liu H, Zhao D, Liu J (2024) Mineral resources and equitable economic development: South Asian mineral resources policy perspective with innovation for resources efficiency. Resour Policy 96:105151. https://doi.org/10.1016/j.resourpol.2024.105151
24. Yuan F, Ospina R, Babu A, Noguchi N, He Y, Liu Y (2025) Smart agriculture in Asia. Plant Commun 101377. https://doi.org/10.1016/j.xplc.2025.101377
25. Zhou Y, Zaitchik BF, Kumar S V, Nie W, Loomis BD, Richey McLarty AS, Appana R (2024) Satellite-informed simulation of irrigation in South Asia: opportunities and uncertainties. J Hydrol 641:131758. https://doi.org/10.1016/j.jhydrol.2024.131758

26. Ogwu MC (2025) Sustainable development in Africa: challenges, innovations, and policy pathways for the future. African Research Reports 1(2):55–69. https://doi.org/10.5897/ARR2025.0005
27. Kosoe EA, Ogwu MC, Duniah J, Ndigun FB (2025a) Sustaining indigenous water governance systems in Ayimaye and Lungni, Northern Ghana, through hybrid models. J Sustain 1(2). https://doi.org/10.55845/jos-2025-1235
28. Kosoe EA, Osumanu IK, Ogwu MC (2025b) Reimagining rural and small-town water supply in Ghana through inclusive governance and sustainable models. J Sustain 1(2):1–21. https://doi.org/10.55845/jos-2025-1234
29. Ogwu MC, Kosoe EA (2025) Integrating green infrastructure into sustainable agriculture to enhance soil health, biodiversity, and microclimate resilience. Sustainability 17(9):3838. https://doi.org/10.3390/su17093838
30. Izah SC, Ogwu MC (2025) Modeling solutions for microbial water contamination in the global south for public health protection. Front Microbiol 16:1504829. https://doi.org/10.3389/fmicb.2025.1504829
31. Ogwu MC, Izah SC, Sawyer WE, Amabie T (2025) Environmental risk assessment of trace metal pollution: a statistical perspective. Environ Geochem Health 47:94. https://doi.org/10.1007/s10653-025-02405-z
32. Ogwu MC, Kosoe EA (2024) Place of cultural diversity in sustainable water resource management in Ghana. In: Izah SC, Ogwu MC, Loukas A, Hamidifar H (eds) Water crises and sustainable management in the global South. Springer, Singapore. https://doi.org/10.1007/978-981-97-4966-9_14

Chapter 13
Emerging Technologies for Water Quality Monitoring

Matthew Chidozie Ogwu

Abstract Water quality monitoring is increasingly critical for environmental and public health, yet traditional methods often suffer from limitations such as high costs, labor intensity, and delayed responsiveness. This chapter explores the transformative impact of emerging technologies on water quality assessment, highlighting how innovations like smart sensors, biosensors, Internet of Things (IoT), remote sensing, artificial intelligence (AI), machine learning (ML), microfluidics, and blockchain are revolutionizing the field. These technologies enable real-time, cost-effective, and scalable solutions that overcome the spatial and temporal limitations of conventional monitoring techniques. Smart sensors and biosensors offer continuous monitoring of key water parameters, while IoT and wireless sensor networks (WSNs) allow for remote and automated data collection. AI and ML enhance predictive analytics and early warning systems, and satellite technologies provide spatially extensive insights into aquatic ecosystems. Microfluidic and nanotechnology platforms enable rapid and portable water testing, especially in low-resource settings. Moreover, blockchain ensures data integrity and transparency, supporting trust in multi-stakeholder systems. The chapter also discusses integration frameworks, ethical considerations, and global case studies—from smart urban water systems in Singapore and the Netherlands to community-based initiatives in India and Kenya. It concludes by emphasizing the importance of interdisciplinary collaboration, policy support, and inclusive technological deployment to advance global water security and meet Sustainable Development Goal 6. These innovations represent a paradigm shift in monitoring, offering robust tools for safeguarding water resources amidst growing environmental and societal challenges.

Keywords Smart sensors · Internet of things · Remote sensing · Biosensors · Microfluidics · Blockchain · Real-time monitoring · Machine learning

M. C. Ogwu (✉)
Goodnight Family Department of Sustainable Development, Appalachian State University, Boone, NC, USA
e-mail: ogwumc@appstate.edu

M. C. Ogwu and S. Chibueze Izah (eds.), *Water Quality and Safety in the Global South*, Springer Water, https://doi.org/10.1007/978-3-032-10602-5_13

13.1 Introduction

Water quality has emerged as a critical concern, deeply intertwined with environmental health and public welfare. As water serves as a vital resource for drinking, sanitation, and agriculture, its quality directly affects human health and ecological stability. Inadequate monitoring practices can lead to serious health repercussions, including outbreaks of waterborne diseases. Traditional water quality monitoring approaches, predominantly reliant on manual sampling and laboratory analysis, provide valuable insights yet are inherently limited. Despite their significance, these conventional methodologies often result in delayed responses to water quality deterioration, lacking the immediacy that real-time data can provide [1, 2].

Water quality monitoring plays a pivotal role in managing water resources effectively and ensuring public safety. The traditional approaches primarily encompass field sampling and periodic laboratory testing, which, while effective in some contexts, yield high operational costs and labor-intensive practices, often hampering timely interventions [3]. Most traditional methods only capture snapshots of water quality data from specific locations, which may lead to inadequate assessments of spatial variability or temporal changes in water quality [2]. The integration of real-time monitoring emerges as a necessary evolution, as it allows for continuous data collection, facilitating timely responses to pollution events and other changes in water quality parameters [1, 4].

The challenges associated with conventional methods of water quality monitoring are multifaceted. Firstly, the high costs involved in manual sampling and the subsequent analysis can strain budgets, especially in resource-limited regions [3]. Additionally, the spatial and temporal coverage provided by traditional methods often falls short, sampling locations might not represent the broader environmental conditions accurately. Consequently, high costs and limited coverage necessitate innovative approaches that can bridge these gaps [4, 5]. Labor intensiveness constitutes another significant barrier, as personnel must be dedicated to regular sampling and analysis, limiting the scalability of monitoring efforts. Furthermore, the lack of real-time data hinders predictive capabilities, preventing timely interventions that could mitigate adverse effects on public health [6].

As the threats posed by pollution increase alongside global challenges such as climate change and population growth, it becomes imperative to innovate in the realm of water quality monitoring. Rising pollution levels necessitate a shift towards more efficient and proactive monitoring technologies that can provide immediate feedback on water quality conditions [4, 7]. Digital transformation, characterized by advancements in automation, big data analytics, and the Internet of Things (IoT), presents a promising frontier. These technologies not only enhance the capacity for data collection but also facilitate real-time analytics, decision-making, and automated responses to shifts in water quality [1, 8]. By harnessing these emerging technologies, it becomes feasible to develop a comprehensive water quality monitoring framework that can adapt to the evolving demands of environmental integrity and public health [9, 10].

The implementation of automated and remotely operated systems is gaining traction, leveraging wireless sensor networks (WSNs) to continuously monitor multiple water quality parameters, including pH, turbidity, and contaminant levels. Such systems offer significant advancements over traditional practices, enabling the collection of data in real-time [2, 4]. For instance, the utilization of a remote monitoring system based on Geographic Information Systems (GIS) and cloud computing enhances accessibility and integration of data across various platforms, thus improving oversight and management of water quality [11, 12]. Furthermore, integration of predictive models utilizing artificial intelligence and machine learning can significantly improve forecasting of water quality events, enabling proactive measures to safeguard water resources [1, 7]. The significance of these advancements extends beyond mere convenience, they actively facilitate compliance with environmental regulations through enhanced monitoring capabilities. Regulatory frameworks across various regions are increasingly requiring robust monitoring systems that can provide accurate data to justify decisions and manage water bodies sustainably [11, 13]. The convergence of technological innovations also presents opportunities to engage local communities in participative monitoring efforts, encouraging greater stewardship and involvement in maintaining water resource quality [9, 11].

This chapter aims to explore and critically examine emerging technologies that are transforming water quality monitoring, with a particular focus on innovations that offer real-time, cost-effective, and scalable solutions. By examining advancements such as smart sensors, IoT-enabled systems, remote sensing platforms, AI-based analytics, and nanotechnology, the chapter offers a comprehensive overview of how technological integration can solve long-standing challenges in water monitoring and management. Its contribution to knowledge lies in synthesizing recent interdisciplinary developments while emphasizing practical applications, barriers to implementation, and future research directions across various geographic and socioeconomic contexts.

13.2 Smart Sensors and Biosensors

The advent of advanced sensor technologies, particularly smart sensors and biosensors is transforming the landscape of water quality monitoring. These modern sensors, employing various methodologies, are crucial in maintaining safe water standards in a dynamic environment characterized by increasing pollution due to urbanization and industrial activities. This discussion delves into the types of sensors, biosensor technologies, their applications in water quality monitoring, and illustrative case studies showcasing their practical implementations across diverse contexts.

13.2.1 Types of Sensors

Figure 13.1 provides an extensive overview of sensor classifications, illustrating the diverse types and functions of sensors across multiple domains. According to [14] sensors are categorized based on various criteria: the measurand (what is being measured such as acoustic, biological, chemical, or thermal signals), energy/power type (active vs. passive), signal conversion mechanisms (biological, physical, chemical), output signals (analog or digital), and sensor material used. Additional classifications include physical contact (contact vs. non-contact), application domains (e.g., agriculture, medicine, defense, and manufacturing), and technical specifications such as sensitivity, stability, and accuracy.

Sensors for water quality monitoring can be classified into various categories, including optical, electrochemical, piezoelectric, and thermal sensors. Optical sensors utilize the principle of light interaction with substances, making them suitable for detecting a range of contaminants, including heavy metals and organic pollutants. For instance, electrochemical sensors, such as microbial fuel cells (MFCs), measure the electrical signal resulting from bacterial metabolism when exposed to

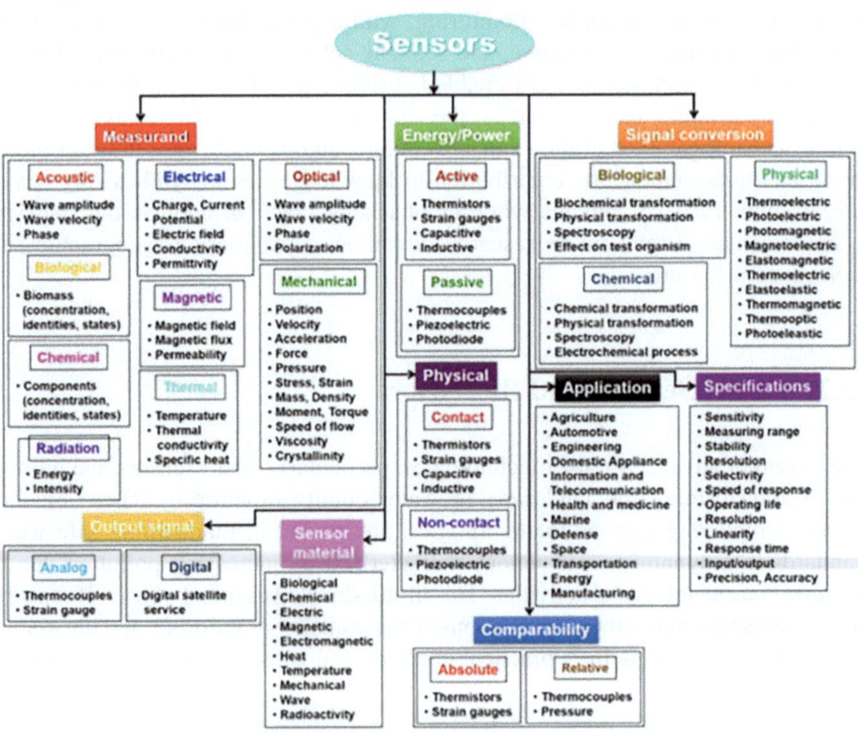

Fig. 13.1 Comprehensive classification of sensors based on measurand, signal conversion, energy source, output type, and application. *Source* [14]

toxic substances or pollutants, providing real-time monitoring capabilities for water quality [15, 16]. Electrochemical sensors are particularly advantageous because they can afford high sensitivity, often outperforming other sensor types in detecting low concentrations of contaminants [15, 17]. Piezoelectric sensors respond to changes in mass or pressure, offering potential applications in measuring chemical concentrations in water, while thermal sensors leverage temperature changes to provide insights into microbial activity and metabolic processes in aquatic environments [18, 19].

13.2.2 Biosensor Technologies

The evolution of biosensor technologies has ushered in various forms of detection mechanisms, most notably microbial biosensors, enzyme-based biosensors, and immunosensors. Microbial biosensors exploit the unique metabolic activities of microorganisms to detect pollutants, particularly useful for monitoring heavy metals and organic compounds in water [20, 21]. In particular, genetically engineered microbial systems, such as *E. coli* modified with luminescent genes, have shown promise in detecting environmental toxins like mercury and thiosulfate, helping to assess microbial responses to pollutants [20, 22]. Enzyme-based biosensors utilize specific biochemical reactions catalyzed by enzymes to quantify contaminants or substrates in water samples. This type of biosensor benefits from specificity and sensitivity due to the targeted nature of enzyme–substrate interactions [23, 24]. Immunosensors, on the other hand, utilize antibodies that bind specifically to target analytes and are particularly effective in detecting pathogens and organic pollutants, rendering them advantageous for health-related water quality assessments [23, 24].

Figure 13.2 illustrates the fundamental components and working flow of a biosensor system. The process begins with the detection of various analytes such as biomolecules, microorganisms, pharmaceuticals, hormones, metabolites, and toxins [25]. These analytes interact with specific bioreceptors, including whole cells, enzymes, antibodies, aptamers, and microbes, which are immobilized on transducers. Depending on the bioreceptor used, the transduction mechanism may be optical, electrochemical, or piezoelectric. The biorecognition event is converted into an electric signal by the transducer, which is subsequently acquired and processed digitally for analytical interpretation and display [25].

13.2.3 Applications in Water Monitoring

With the pressing need for efficient water quality monitoring, the integration of smart sensors stands at the forefront of real-time detection methodologies for contaminants such as heavy metals, pathogens, pesticides, and nutrients (Table 13.1). Smart sensors and biosensors offer advanced, real-time, and cost-effective solutions for detecting a wide range of physical, chemical, and biological contaminants in water systems

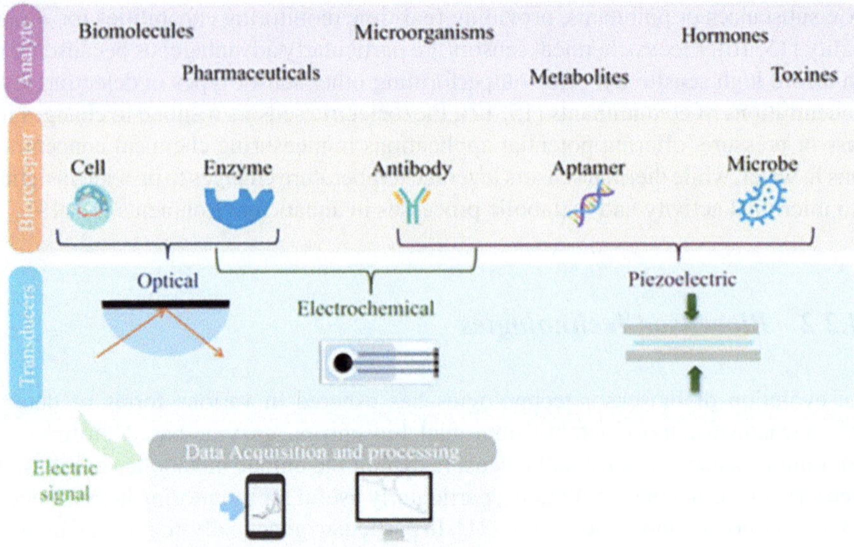

Fig. 13.2 Schematic representation of a biosensor system: from analyte detection to data processing. *Source* [25]

(Table 13.1). These technologies facilitate continuous monitoring and immediate alerts for water quality deviations, a critical capability for mitigating health risks associated with contaminated water supplies [26, 27]. For instance, biosensor applications in urban water systems not only enable the detection of pathogens but also allow for the monitoring of chemical parameters such as dissolved oxygen, biochemical oxygen demand (BOD), and organic pollutants through electrochemical transduction [16, 27, 28]. Signal degradation can often compromise data integrity, however, advancements in microfabrication and automation have enhanced the reliability of sensor readings. Moreover, smart sensors capable of *in-situ* and online monitoring enable the characterization of water quality in dynamic environments, such as during industrial effluent discharges, by providing critical data on microbial dynamics and pollutant interactions [29–31].

Table 13.1 Applications of smart sensors and biosensors in water monitoring and quality assessment

Application area	Type of sensor	Target analyte/parameter	Example use
Water quality assessment	Electrochemical biosensor	Nitrate, Nitrite, Phosphate	Monitoring nutrient levels to prevent eutrophication
Pathogen detection	Optical biosensor	E. coli, Salmonella, Vibrio spp.	Early warning for microbial contamination in drinking and recreational water
Heavy metal monitoring	Piezoelectric biosensor	Lead (Pb), Cadmium (Cd), Mercury (Hg)	Detecting toxic metals from industrial discharge
Pesticide detection	Enzyme-based biosensor	Organophosphates, Carbamates	Tracking agricultural runoff in surface water
pH and conductivity	Smart chemical sensors	pH, Electrical Conductivity (EC)	Real-time assessment of water's chemical balance and salinity
Dissolved oxygen	Optical (luminescent) sensor	Dissolved Oxygen (DO)	Assessing water body health for aquatic organisms
Turbidity monitoring	Optical turbidity sensor	Suspended solids	Evaluating sediment loads and erosion effects in rivers and lakes
Temperature logging	Thermistor or RTD sensor	Water temperature	Assessing thermal pollution or seasonal water quality dynamics
Toxin detection	Cell-based biosensor	Cyanotoxins, Mycotoxins	Monitoring harmful algal blooms (HABs) in reservoirs and lakes
Remote sensing & IoT	Wireless smart sensor nodes	Multiparametric	Continuous, real-time water monitoring using IoT-enabled sensor networks
Chlorine monitoring	Electrochemical sensor	Free and total chlorine	Monitoring disinfection levels in treated drinking water
Oil pollution detection	Fluorescence-based sensor	Hydrocarbons	Detecting oil spills or petrochemical contamination in marine environments

13.3 Internet of Things (IoT) and Wireless Sensor Networks (WSNs)

13.3.1 Architecture of IoT for Water Monitoring

The architecture of the Internet of Things (IoT) for water monitoring is constructed to enhance data acquisition, real-time processing, and efficient resource management. At the foundational level, sensor nodes play a critical role in collecting environmental data, such as water quality parameters (pH, turbidity, temperature, etc.) and flow rates. Figure 13.3 illustrates the layered architecture of a smart water quality monitoring system. The system integrates distributed sensing components (e.g., pH, turbidity, temperature, EC, DO, and TDS sensors) with edge computing devices (Arduino, NB-IoT) for real-time data collection. These sensor nodes are typically deployed in strategic locations to maximize coverage and data accuracy [32]. Each sensor node is designed to communicate wirelessly with data acquisition units, which process the incoming data before transmitting it to cloud servers for further analysis [33].

Data acquisition units serve as intermediaries that manage the data gathered from sensor nodes. The role of these units is significant as they not only gather and preprocess information but also ensure that data integrity is maintained during transmission. This process is critical in remote or rural areas where infrastructure might be limited and consistent data flow is crucial for effective monitoring [35–37]. Moreover, cloud servers are employed to store and analyze large sets of data, enabling advanced data analytics and machine learning applications that inform decision-making in water management [38]. Therefore, the IoT architecture for water monitoring is not only

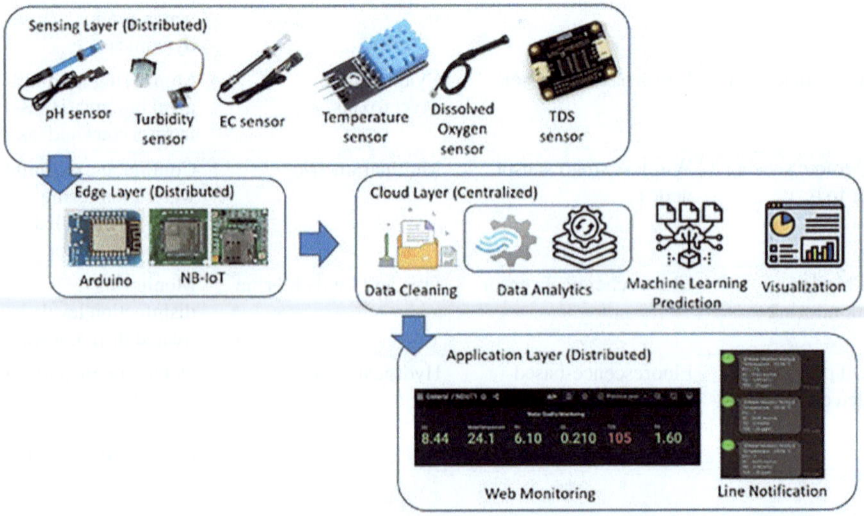

Fig. 13.3 Architecture of a smart water quality monitoring system. *Source* [34]

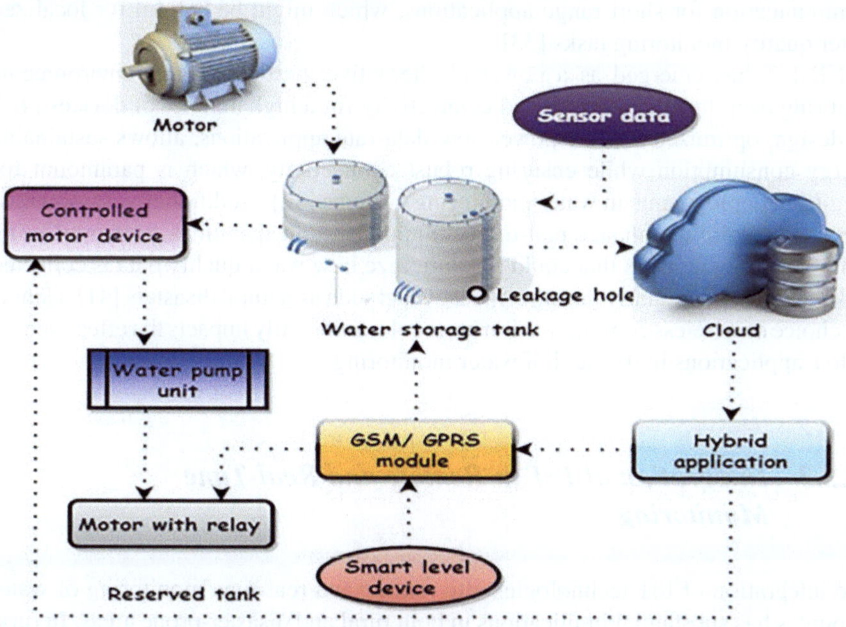

Fig. 13.4 Smart water tank monitoring and control system architecture. *Source* [39]

composed of sensor nodes and data acquisition units but also extends to sophisticated cloud computing capabilities that facilitate data-driven insights.

Figure 13.4 presents a smart water management system for automated monitoring and control of water storage tanks. The system integrates sensor data from smart level devices with a GSM/GPRS communication module, enabling remote tracking via a hybrid cloud-based application [39]. A controlled motor device and pump unit regulate water flow between tanks, responding to real-time data on water levels and potential leakage. This setup enhances water use efficiency, leak detection, and automated pumping through integrated IoT and mobile technologies [39].

13.3.2 Wireless Communication Protocols

Wireless communication protocols form the backbone of IoT frameworks, particularly in applications such as water monitoring, where logistical and operational constraints may limit traditional communication methods. Among the prominent protocols, LoRaWAN (Long Range Wide Area Network), ZigBee, and NB-IoT (Narrowband IoT) stand out for their distinct advantages in serving various requirements. LoRaWAN is known for its long-range capabilities, making it suitable for widespread sensor deployment across vast areas, such as rural environments and disaster-prone regions [40]. In contrast, ZigBee offers low-power, cost-effective communication for short-range applications, which might be helpful for localized water quality monitoring tasks [33].

NB-IoT has emerged as a powerful alternative, particularly for environments requiring deep indoor coverage and connectivity for a high number of devices [41]. Its design, optimized for low-power, low-data-rate applications, allows sustainable energy consumption while ensuring robust connectivity, which is paramount for long-term deployments in water monitoring systems [42]. Additionally, 5G-enabled systems promise to enhance real-time communication capabilities further, offering faster data transfer rates that could revolutionize how water quality data is collected and analyzed, particularly during critical events such as natural disasters [41]. Hence, the choice of wireless communication protocol significantly impacts the effectiveness of IoT applications in the field of water monitoring.

13.3.3 Integration of IoT in Remote and Real-Time Monitoring

The integration of IoT technologies into remote and real-time monitoring of water resources has substantial implications in both rural and disaster-prone areas. In rural settings, the deployment of IoT-enabled sensor networks can facilitate the continuous monitoring of water quality and availability, providing essential data to local authorities for informed management [32]. These systems can collect and transmit data autonomously, reducing the need for manual checks, thereby enabling efficient resource management and immediate response to any detected anomalies, such as contaminants in drinking water supplies [33]. Reports indicate successful applications of such monitoring systems in regions struggling with water scarcity or contamination issues, demonstrating the potential for IoT to revolutionize water management [43].

In disaster-prone areas, IoT monitoring solutions provide early warning mechanisms that could mitigate the impacts of floods and droughts. With the ability to collect real-time data on changing environmental conditions, authorities can receive alerts and respond rapidly to threats, potentially saving lives and resources [33]. For instance, smart river monitoring systems leverage sensor networks to evaluate water levels and flow rates continuously, providing crucial information about flood risks and enabling timely evacuations [32]. Coupled with cloud computing technologies, these water monitoring applications can provide analytical insights that guide long-term planning and infrastructure development to enhance community resilience against natural hazards [42, 44, 45].

Smart river monitoring initiatives are crucial for analyzing ecosystem health and ensuring that water bodies are not being polluted by industrial runoff [35]. These smart systems often utilize NB-IoT and LoRaWAN technologies to ensure connectivity across challenging terrains, emphasizing accessibility and operational efficiency in data collection [40]. In the domain of irrigation, IoT has wound its way into modern agriculture, allowing farmers to monitor soil moisture levels and irrigate crops intelligently based on real-time data [38]. This integration enhances productivity while conserving water resources. The data accrued from these IoT systems could lead to refined agricultural practices, reducing waste and improving yield quality, which is essential for food security in a climate-challenged world [35].

Municipal water supply management is another area where IoT has made significant strides, helping cities modernize their water services and providing stakeholders with valuable insights into consumption patterns. By implementing real-time water quality monitoring systems, municipalities can detect and respond to system leaks or contamination threats instantaneously [43]. These innovations embody the critical intersection of technology and water management, leading to sustainable urban development while addressing pressing environmental challenges [33].

13.4 Remote Sensing and Satellite Technologies

Remote sensing technologies play a pivotal role in water quality assessment by utilizing spectral signatures to detect indicators such as chlorophyll concentrations, turbidity, and algal blooms. Chlorophyll *a*, in particular, serves as a crucial biomarker in evaluating aquatic health and productivity. The effectiveness of remote sensing in delineating the spectral properties associated with chlorophyll-a has been demonstrated through various methods, including the use of specific spectral bands that absorb light effectively at distinct wavelengths, typically in the red (around 680–690 nm) and near-infrared (around 730–740 nm) regions of the electromagnetic spectrum [46, 47]. This spectral analysis allows for the characterization of plant health and the quantification of chlorophyll content in both terrestrial and aquatic environments [48].

Advanced techniques in hyperspectral remote sensing have enhanced our ability to detect variations in chlorophyll content under different conditions, such as stress from environmental factors and nutrient variations. For instance, studies show that chlorophyll content can be predicted by analyzing the spectral reflectance patterns associated with changes in nitrogen levels in grasses [49]. Moreover, the relationship between chlorophyll concentration and environmental conditions has been corroborated by the ability of spectral indices like the Normalized Difference Vegetation Index (NDVI) to estimate chlorophyll levels from satellite data effectively, enhancing our understanding of phytoplankton dynamics in aquatic systems [50, 51].

Satellite missions, particularly the Landsat, Sentinel, and MODIS platforms, have made significant contributions to monitoring water quality on a global scale. Each of these platforms provides varying spatial, temporal, and spectral resolutions that are critical for different aspects of water quality management and evaluation. For example, the Sentinel-2 satellite's Multispectral Imager (MSI) has been shown to deliver detailed observations of chlorophyll *a* and turbidity in aquatic systems due to its high-resolution capabilities [50, 52]. This satellite not only offers insights into chlorophyll dynamics but also helps in tracking the spatio-temporal distribution of aquatic pollution sources. The synergy between satellite data and in-situ measurements, as noted in several studies, provides a robust framework for assessing and modeling water quality parameters [53–55].

At the forefront of technological advancements, drone and UAV-based monitoring has emerged as a flexible and cost-effective alternative to traditional water quality assessment methods. Drones equipped with high-resolution sensors can be deployed to gather data that complements satellite imagery, capturing fine-scale variations in water quality attributes such as turbidity and chlorophyll concentration [56]. The ability of these aerial platforms to quickly respond to environmental changes and provide timely data has proven beneficial for ongoing ecological assessments and the management of water resources [56]. Such drone technologies not only enhance data collection efficiency but also facilitate real-time monitoring and immediate decision-making in water management practices.

Integration with Geographic Information Systems (GIS) further amplifies the potential of remote sensing technologies. The combination of remote sensing data and GIS allows for sophisticated spatio-temporal mapping of pollution sources and trends across different water bodies. By spatially analyzing the distribution of various pollutants along with their temporal changes, researchers can better understand human impacts on aquatic ecosystems and devise effective management strategies [50, 52]. This integration presents a comprehensive approach to assessing water quality, enabling stakeholders to visualize the interplay between land-use patterns, anthropogenic activities, and water quality indicators.

Research utilizing remote sensing for water quality monitoring has shown that specific algorithms and machine learning techniques are becoming increasingly crucial for processing the vast amounts of data collected from various platforms. For instance, studies have indicated that machine learning models can improve predictions of chlorophyll levels based on remote sensing reflectance data, effectively reducing the complexity involved in determining water quality indices from high-resolution datasets [49, 53]. As the field continues to evolve, the collaboration between advanced remote sensing technologies and computational techniques will be essential for enhancing water quality assessment capabilities.

Furthermore, innovative methodologies for measuring water quality indicators from remote platforms have advanced significantly. Techniques combining chlorophyll *a* indices with field-collected water quality metrics, such as turbidity and dissolved oxygen levels, provide valuable insights into the health of aquatic ecosystems [50]. Such integrative approaches have enabled the development of predictive models that can forecast water quality conditions based on historical data and current environmental parameters, thereby assisting in preventive measures against water pollution and ecosystem degradation.

Emerging studies that explore the application of remote sensing in monitoring algal blooms highlight the critical importance of this technology in safeguarding water bodies. The detection of harmful algal blooms (HABs) through spectral analysis has been particularly effective, with several spectral metrics proving successful in identifying conditions conducive to bloom formation [47]. Notably, hyperspectral remote sensing has demonstrated significant capabilities in discriminating between different algal species based on their unique spectral signatures, allowing for enhanced management of water resources and recreation areas impacted by algal proliferation.

The realization that different phytoplankton species exhibit distinct spectral characteristics has led to refined methodologies in remote sensing, providing insights into the composition and abundance of aquatic communities [57]. This knowledge is crucial for addressing water quality issues associated with nutrient loading and understanding ecosystem responses to anthropogenic stressors. As remote sensing technologies continue to improve, their integration with environmental monitoring frameworks will likely become a cornerstone of sustainable water management strategies.

13.5 Artificial Intelligence and Machine Learning Applications

The advent of artificial intelligence (AI) and machine learning (ML) has significantly transformed water quality monitoring, providing valuable tools for data interpretation, model development, early warning systems, and practical applications across various aquatic environments. Employing these technologies allows for more efficient management and protection of water resources while enhancing the ability to respond to pollution and quality changes primarily driven by human activity and climate change.

13.5.1 Role of AI in Data Interpretation

AI plays a crucial role in interpreting complex datasets involved in water quality monitoring. One of the most impressive capabilities of AI is its ability to perform predictive analytics, which entails using historical data to forecast future water quality patterns. For instance, Guan et al. have demonstrated the effectiveness of Long Short-Term Memory (LSTM) networks in modeling water quality changes, leveraging the predictive capabilities of AI to anticipate potential pollution events before they occur, thus allowing for timely responses and better management of water resources [58]. Additionally, predictive analytics aids in identifying trends in water quality parameters over time, providing crucial insights for sustainable water resource management [4, 59].

Anomaly detection is another critical aspect where AI excels, as it allows for the identification of unusual patterns within water quality data that may indicate contamination or equipment failure. For example, automated systems can analyze streaming data from sensors to flag deviations from normal operational parameters, triggering alerts for immediate investigation [60]. This proactive approach enhances the reliability of monitoring systems and provides an invaluable early warning mechanism that directly protects public health and the environment.

Trend forecasting, powered by AI, helps stakeholders understand the long-term trajectories of water quality metrics, thereby influencing policy decisions and resource allocation. According to [61], the integration of machine learning models enables the identification of significant trends in water quality data that are essential for addressing environmental challenges proactively. The effective combination of AI's capabilities in predictive analytics, anomaly detection, and trend forecasting promises a paradigm shift in the management of water quality.

13.5.2 Machine Learning Models

Several machine learning models have emerged as effective tools for water quality prediction and assessment. Support Vector Machines (SVM), Random Forests, and Neural Networks stand out as particularly robust methodologies in this respect. SVMs, for instance, have demonstrated effectiveness in classifying water quality based on various physicochemical parameters, delivering high accuracy even in complex datasets [62, 63]. Their ability to model nonlinear relationships allows for a nuanced understanding of the interactions between different water quality variables.

Random Forest algorithms also show promise, especially in cases where robustness against overfitting is a concern. Kotwal et al. noted that Random Forest models could be employed in real-time assessments, significantly improving the prediction of water potability and enhancing the safety of drinking water [7]. Moreover, ensemble methods and decision trees are instrumental in optimizing predictive accuracy while maintaining interpretability, which is crucial for stakeholders who need actionable insights from complex data.

Neural networks, particularly deep learning architectures like Convolutional Neural Networks and LSTM networks, have exhibited superior performance in forecasting water quality parameters [64, 65]. These advanced models utilize layers of interconnected nodes to capture intricate relationships among variables, making them suitable for processing large, multi-dimensional datasets typical in water quality monitoring. As evidenced by various studies, including those of Fan and Zhao, deep learning techniques are capable of comprehensively modeling the temporal characteristics of water quality, leading to more accurate predictions and enhanced operational efficiencies [64, 66].

13.5.3 AI for Early Warning Systems

AI-powered early warning systems are crucial innovations in water quality monitoring, providing automated alerts for conditions indicative of contamination events or equipment failures. These systems typically integrate IoT sensors with AI analytics to monitor water quality parameters in real time, enabling rapid detection of adverse changes [59, 67]. As captured by the work of [7], the deployment of automated systems that leverage big data analytics enhances the reliability of water quality assessments and contributes to safe drinking water initiatives through timely notifications.

The ability to provide immediate feedback through automated alerts facilitates swift responses, essential in preventing or mitigating public health hazards [10]. The studies conducted by [67] harness IoT technologies to enable remote monitoring of water quality, significantly improving the efficiency and effectiveness of early warning systems. These systems not only reduce reliance on manual sampling but also streamline the data flow from sensors to decision-makers, allowing for data-driven interventions when needed.

Furthermore, the synergy between AI predictions and automated alert systems fosters an environment where stakeholders can engage in more proactive water management strategies. For example, integrating AI-driven predictive analytics with IoT frameworks can result in a holistic monitoring approach that anticipates water quality issues and prepares responses, thereby enhancing resilience against climate impacts and anthropogenic pressures [7, 60].

In river management scenarios, machine learning models have been utilized to predict water quality parameters based on historical data, yielding critical insights into the potential effects of environmental changes and human activities [68, 69]. The significant findings from studies indicate that methodologies such as the WASP model for hydrodynamic and water quality simulation can effectively incorporate data-driven approaches to enhance predictive capabilities within diverse aquatic settings [69]. Specific implementations, such as those reported by [70], showcase the development of water quality prediction models using variants of recurrent neural networks, which have proved effective in monitoring river water contaminants over time. Comparable outcomes have been observed in treatment facilities where IoT and AI technologies converge to optimize water treatment processes, leading to significant improvements in operational management and resource allocation [60, 67].

In bodies of water like lakes, the integration of AI technologies supports not only real-time monitoring but also the evaluation of pollution reduction efforts, as seen in projects focused on the Great Barrier Reef [71]. Similarly, predictive modeling illustrated in applications within aquaculture contexts highlights the relevance of machine learning in optimizing environmental conditions, such as dissolved oxygen levels, effectively demonstrating AI's versatility across various water bodies [72, 73, 74].

13.6 Nanotechnology and Microfluidic Devices

Emerging technologies in water quality monitoring have witnessed significant advancements through the integration of nanotechnology and microfluidic devices. Specifically, the utilization of nanomaterials such as nanoparticles, nanotubes, and quantum dots has substantially enhanced the sensitivity and specificity of detection systems. For instance, [75] employed a plasmonic liquid crystal biosensor that successfully identified *Escherichia coli* at low concentrations in potable water, demonstrating its suitability for early detection of biological contaminants in water systems, effectively enabling swift response measures in case of contamination events. Similarly, nanostructured materials such as titania nanotubes utilized in microfluidic reactors offer photocatalytic capabilities, facilitating the degradation of various organic and biological contaminants [76].

Microfluidic technologies have further revolutionized water quality monitoring by enabling portable and rapid diagnostics directly in the field. Devices incorporating microfluidic paper analytical methods (μPADs) have emerged as cost-effective and scalable solutions for environmental monitoring. Research by [77] showcases the applicability of μPADs for trace copper detection in drinking water, emphasizing their robust potential beyond traditional medical diagnostics. Microfluidic platforms combined with smartphone applications have enabled real-time tracking of water quality indices, highlighting the potential of mobile technology and lab-on-a-chip systems to advance public health and environmental safety [78].

In resource-constrained settings, the need for point-of-use water testing remains critical. Microfluidic devices specifically tailored for detecting waterborne pathogens and pollutants are vital for ensuring safe drinking water in underserved locations. [79] discuss the pressing need for periodic water quality assessments, driven by increasing contamination levels and the expanding demand for diverse water sources. This sentiment is echoed in the work of [80], who positioned microfluidic technologies as a solution for the in situ detection of chemical and biological contaminants, underscoring existing challenges and future opportunities for innovation in the field.

Advanced microfluidic systems have enabled the detection of various toxic substances, including heavy metal ions, which pose significant health risks. [81] stress the urgency for on-site, real-time monitoring solutions for heavy metals, citing multiple microfluidic-based devices developed for this purpose. Moreover, recent advancements have led to the integration of electrochemical and optical sensors within microfluidic platforms, thus enhancing their capability to perform simultaneous pollution analyses [82]. With advancements in material science, microfluidic devices have transitioned from traditional silicon-based methods to various fabrication techniques, including 3D printing and inkjet technology, improving detection capabilities [83].

Microfluidics also enables the development of portable detection devices that operate with minimized sample volumes and reduced analysis times, which are particularly advantageous in emergency scenarios and field applications. The study by [84] highlights a portable microfluidic detection system for residual chlorine, demonstrating the broader applicability of miniaturized analytical techniques beyond conventional laboratory settings. These technologies not only provide rapid results but also ensure lower overall costs by reducing reagent consumption and wastage, promoting environmental sustainability.

The specific applications of nanotechnology in enhancing the detection capabilities of microfluidic devices are noteworthy. For instance, quantum dots have shown promise for their luminescent properties that allow for sensitive detection of trace contaminants. Park et al. [85] describe a microfluidic platform that utilizes photoluminescent graphene oxide quantum dots for detecting trace lead in water samples, showcasing the effective combination of sample preconcentration and detection technologies. Additionally, advancements in resistive microfluidic sensors for salinity and other ion concentrations exemplify the potential for targeted sensing applications, especially in regions with salinity challenges in drinking water [86].

The integration of citizen science with microfluidic devices presents another innovative avenue to enhance water quality monitoring efforts. As demonstrated by [87] simple paper microfluidic devices enable non-experts to engage in environmental monitoring, thus promoting community involvement and awareness regarding water quality issues. Such participatory approaches can significantly broaden data collection and facilitate proactive measures against pollution.

Furthermore, the exploration of bacterial detection within microfluidic frameworks highlights the technology's versatility in addressing emerging health threats. The capacity for rapid analysis of microbial contaminants directly from water samples can aid in mitigating public health issues related to waterborne illnesses. Kim et al. [88] outlined a microfluidic device capable of label-free detection of *E. coli*, emphasizing the significance of such technologies in clinical and environmental settings (Table 13.2).

13.7 Blockchain and Data Security in Water Monitoring

The application of blockchain technology in water quality monitoring offers significant improvements in data security, integrity, and transparency. Security in the context of water quality data is paramount as it ensures that the information used to assess water safety and compliance is protected from unauthorized alterations. Blockchain technology is noteworthy for its immutable and decentralized nature, enabling secure data transmission and storage for monitoring systems. A decentralized approach facilitates the validation and verification of information without depending on a central authority, thus enhancing trust among stakeholders involved in water resource management [89].

13.7.1 Importance of Secure Data Transmission and Storage

Among the challenges faced by water monitoring systems, ensuring the integrity and traceability of data is of utmost importance. Data integrity refers to the accuracy and consistency of data over its lifecycle, while traceability allows stakeholders to track data history from its creation to its current state. These features can be indispensable for compliance with regulatory standards and managing emergencies such as contaminations or infrastructure failures. In the context of blockchain, each

Table 13.2 Examples of nanotechnology and microfluidic devices for water quality monitoring and assessment

Technology	Detection principle	Target analytes/ parameters	Advantages	Applications
Nanoparticle-based Sensors	Surface plasmon resonance, fluorescence, electrochemistry	Heavy metals (e.g., Pb^{2+}, Hg^{2+}), pesticides, and pathogens	High sensitivity, fast response, and low detection limit	Detection of toxic pollutants in drinking and surface water
Carbon Nanotube (CNT) Sensors	Electrochemical conductivity, impedance	Nitrates, bacteria, and organic compounds	High electrical conductivity, and small size	Point-of-use monitoring in rural and remote areas
Quantum Dot-based Sensors	Fluorescent signaling	Metal ions, pathogens, and endocrine disruptors	High brightness, and tunable emission spectra	Real-time environmental and groundwater monitoring
Nanoshells/ Nanorods	Optical resonance or colorimetric change	Pesticides, toxins, and microbes	Label-free detection, and multiplexing capability	On-site testing kits for agricultural runoff and contamination
Nanozyme Sensors	Enzyme-mimicking catalytic reactions	Phenols, pesticides, and urea	Stable under harsh conditions, and cost-effective	Industrial wastewater monitoring
Microfluidic Paper-based Analytical Devices (μPADs)	Capillary action, colorimetric/ electrochemical readout	pH, nitrates, phosphates, and pathogens	Low-cost, portable, and minimal reagent use	Field testing in resource-limited settings
Lab-on-a-Chip Devices	Integrated microchannels for multiplexed biochemical assays	Multiple contaminants (chemical + microbial)	Rapid analysis, automation, and miniaturization	Comprehensive water quality assessment in small sample volumes
Microfluidic Electrochemical Cells	Micro-scale electrochemical sensing	COD, BOD, heavy metals, and organic pollutants	High sensitivity, real-time monitoring, and small footprint	Industrial process water and effluent monitoring
Integrated Nano-Microfluidic Systems	Combination of nanomaterials with microfluidics	Multi-analyte detection (chemical, physical, and biological)	Enhanced sensitivity and selectivity, portable and automated	Smart water quality monitoring stations and IoT-enabled environmental sensors

transaction or data entry is stored in a shared ledger, ensuring that all parties can access a single source of truth that they can trust [90, 91]. This level of transparency is particularly beneficial in sectors like agriculture and urban water management, where managing and sharing information effectively can lead to better decision-making and resource allocation [92, 93].

13.7.2 Use of Blockchain in Water Management Systems

Blockchain's capabilities extend beyond security, providing tools like smart contracts that automate operational processes within water management systems. For instance, a smart contract could automatically enforce water usage limits based on data received from IoT devices monitoring water levels or quality, triggering alerts or actions in case of deviations [8, 94]. Additionally, by enabling decentralized data validation, blockchain reduces the risk of data manipulation and increases trust among participants in the water management ecosystem. Every stakeholder, from policymakers to consumers, can verify data independently, ensuring a more equitable distribution of water resources [91, 95].

Moreover, blockchain has the potential to improve the efficiency of irrigation systems in agriculture. By integrating with IoT sensors, blockchain can help manage water allocation dynamically, ensuring that water resources are used optimally without wasting water due to poor management practices [93, 96]. In situations where water rights trade is necessary, blockchain-based solutions can facilitate transparent and fair transactions, thereby enhancing sustainability in agricultural practices [96].

13.7.3 Potential and Limitations

While blockchain technology presents numerous advantages, it is not without its limitations. The scalability of blockchain networks remains a crucial concern, particularly when applied to extensive and complex water management systems. As the number of transactions increases, maintaining efficiency and response times can become challenging [97, 98]. Furthermore, integrating blockchain with existing systems can also pose interoperability issues, requiring significant investment in both technology and training for personnel who must navigate these new tools [99].

In addition, the successful implementation of blockchain in water management contexts necessitates a robust infrastructure and comprehensive data governance frameworks that are currently not universally available. Stakeholders must also address concerns regarding data privacy and the sharing of sensitive information, especially in regions where data access could impact water rights or economic interests [92, 100]. Despite the exciting potential of these technologies, they must be approached with careful consideration of these challenges to ensure effective deployment and successful outcomes.

13.8 Integration and Interoperability of Emerging Technologies

The continuous advancement in the integration and interoperability of emerging technologies has transformed the landscape of water quality monitoring systems. A systems-of-systems approach that amalgamates sensors, Artificial Intelligence (AI), the Internet of Things (IoT), and Geographic Information Systems (GIS) establishes a more robust framework for holistic water quality assessments. Each of these technologies contributes unique functionalities, enhancing the ability to conduct comprehensive monitoring, timely data analysis, and effective decision-making.

In recent years, the application of remote sensing and GIS has been recognized as a pivotal innovation within water quality monitoring mechanisms. Remote sensing techniques facilitate the retrieval of crucial environmental data at spatial and temporal resolutions that are often not achievable with traditional methods. This is particularly relevant in developing regions, where financial constraints make periodic, in-situ measurements prohibitively expensive [101, 102]. The incorporation of these technologies allows for the monitoring of critical indicators such as turbidity and chlorophyll levels, thereby enabling more efficient management of aquatic resources [103]. Moreover, [104], pointed out that the integration of real-time remote sensing data into GIS platforms enhances the detection and analysis of pollution sources, thereby supporting evidence-based decision-making for sustainable water resource management.

Data standards and interoperability are fundamental to the seamless operation of these integrated systems. Open data formats and Application Programming Interfaces (APIs) are essential for facilitating communication between the myriad devices and systems involved in water quality monitoring. Existing strategies for water quality management increasingly emphasize interoperability to avoid isolated data systems, which complicate analysis and slow down response times. Standards established in the EU Water Framework Directive advocate for harmonized coding systems to support interdisciplinary knowledge integration and improved decision-making processes surrounding water quality management [105].

Furthermore, the convergence of smart water grids with digital twin technologies epitomizes the future of water quality monitoring. A smart water grid leverages IoT technologies to facilitate real-time data collection and analysis, thereby optimizing resource management and maintaining water quality standards [106]. Digital twins, i.e., the virtual representations of physical water systems, enable real-time simulation scenarios, allowing stakeholders to assess the potential impact of various changes or interventions within the system [107]. This simulation capability, combined with GIS functionality, enhances the optimization of water distribution networks and contributes to improving water quality outcomes in urban environments.

Cloud-based platforms offer robust solutions for enhancing data accessibility and management in water quality systems. They allow for the aggregation of vast amounts of data collected from different sensors and remote sensing devices, enabling efficient data storage, analysis, and visualization [108]. The implementation of such platforms facilitates extensive data analysis and supports collaborative efforts in research and management, allowing for better-informed decisions and resource allocations in real-time [109]. For instance, the integration of web service frameworks within water quality monitoring systems has proven effective in facilitating the management and analysis of real-time data from multiple sources, demonstrating the efficiency of cloud solutions in modern water governance [110].

Emerging technologies foster innovative approaches to traditional water quality monitoring by integrating multiple systems for enhanced outcomes. The use of hydrological modeling combined with GIS allows for insightful analyses of landscape features and their impacts on water quality indicators. Research has shown that integrating geomorphic data with GIS significantly augments our understanding of hydrological processes influencing water resources [111]. Such integrated models facilitate improved assessments of pollution sources and encourage dynamic responses to emerging water quality crises.

In light of climate change and increasing demand for water resources, real-time monitoring and smart grid integration will be crucial for effective water quality management. Notably, the collaborative use of GIS with remote sensing provides valuable insights into watershed conditions, thereby supporting integrated watershed management initiatives [112]. The attention to both upstream and downstream interactions emphasize the importance of holistic approaches—balancing the various factors influencing water quality across geographic and temporal scales [113]. Such frameworks not only support compliance with environmental regulations but also assist in effectively addressing the complexities surrounding urban water quality problems.

13.9 Implementation Challenges and Ethical Considerations

Implementing advanced water quality monitoring systems, particularly in low-resource settings, is hindered by significant cost and accessibility issues. High initial costs for quality sensors and maintenance often render these technologies inaccessible to communities that would benefit most from them [114]. The affordability of water quality sensors is crucial, for example, traditional sensors can range from $3,000 to $5,000 each, making extensive deployments financially unfeasible for many regions [114, 115]. However, recent advancements in low-cost sensors integrated with the Internet of Things (IoT) present a promising avenue for improving accessibility, as they can provide real-time monitoring without extensive investment [115, 116].

In addition to financial constraints, the integration of new technologies must consider local infrastructure capabilities. Many regions suffer from inadequate power supply and network coverage required for constant data transmission [117, 118]. The reliance on wireless sensor networks (WSNs), while beneficial for real-time monitoring, still faces challenges with energy efficiency and connectivity, especially in remote or rural areas where infrastructure is often lacking [117, 119]. These issues necessitate the development of systems robust enough to operate effectively under constraints typical of low-resource settings, such as the use of low-power, long-range technologies like LoRa, which have shown potential for aquatic applications [120, 121].

13.9.1 Cost and Accessibility: Barriers to Adoption in Low-Resource Settings

The first step in addressing these challenges is a comprehensive understanding of the various barriers faced by low-resource settings in adopting water quality technologies. The high cost of sophisticated monitoring equipment, alongside inadequate technical support, creates barriers that are difficult to overcome without targeted investment and innovative solutions [116]. Existing studies indicate that integrating cost-effective sensor solutions, while ensuring data reliability and quality, is vital to fostering greater accessibility [115, 122]. Additionally, establishing community-based approaches to maintenance and operation may alleviate some of the burdens associated with these innovations [123].

13.9.2 Technical and Infrastructure Limitations: Power Supply, Network Coverage, Maintenance

Technical disparities further complicate the deployment of water quality monitoring systems. In many cases, the consistency of power supply and sufficient networking capabilities are requisites that are not met, particularly in developing regions [124]. The ability of systems to function under intermittent power conditions or to operate efficiently in areas with poor cellular coverage remains a critical challenge [117, 124]. Furthermore, ongoing maintenance and calibration are necessary to ensure the accuracy and reliability of sensor data. Yet, such needs often fall outside the scope of what local communities can manage independently [122]. Moreover, the design and implementation of monitoring systems must consider the lifetime and operational durability of devices, especially in harsh environmental conditions. As noted by [122], maintaining long-term stability and reliability of low-cost sensors in urban rivers is an ongoing challenge that must be adequately addressed for effective monitoring.

13.9.3 *Ethical Issues: Privacy, Data Ownership, Consent, and Surveillance Concerns*

The deployment of water quality monitoring systems raises ethical questions surrounding data privacy, ownership, and consent. As with any data collection endeavor, sensitive information may be inadvertently captured, including user locations or private data linked to water usage [117]. There is an urgent need for clear guidelines governing the ownership of data collected by these technologies, as well as the execution of informed consent protocols that respect individual privacy rights [125]. Additionally, the potential for surveillance arises when monitoring systems are deployed without adequate transparency and community engagement. Stakeholders must navigate the delicate balance between public health imperatives and individual privacy concerns to cultivate trust within the communities served by these technologies [117, 116]. It will be vital for future technologies to incorporate ethical designs that prioritize human rights alongside technological advancements.

13.10 Examples from Around the World

Emerging technologies play a transformative role in water quality monitoring systems globally, revolutionizing how urban environments manage and monitor their water resources. This paper outlines different approaches worldwide regarding innovations in urban water systems, community-based technologies, AI-driven pollution tracking, and remote monitoring initiatives, particularly emphasizing the importance of integrating advanced technologies into urban water governance.

13.10.1 *Urban Smart Water Systems (e.g., Singapore, and Netherlands)*

Urban environments such as Singapore and the Netherlands exemplify sophisticated smart water systems that leverage technologies to optimize water management. These cities utilize digital information and communication technologies (ICTs) that contribute to smart urban water networks, facilitating the real-time monitoring and management of water resources. As noted by Nardo et al. these systems enhance efficiency and sustainability in urban water management, with capabilities for real-time data collection through an array of smart sensors and automated monitoring systems [126]. In Singapore, for example, the integration of big data analytics in urban governance supports the formulation of policies that address sustainability challenges directly, showcasing a robust application of innovative ICT solutions [127]. The smart urbanism practices in Singapore further emphasize a proactive environment in using technology to manage water service delivery and anticipate future challenges associated with climate change and urbanization [128].

In the Netherlands, similar strategies are prevalent, reflecting a commitment to creating "smart water cities" that minimize resource waste and enhance service delivery efficiency through advanced technological applications. The notion of smart water cities encapsulates the implementation of technologies that enable the concerted management of water networks, significantly reducing leakage and optimizing the use of existing resources [129, 130]. It entails a comprehensive approach that includes harmonizing infrastructure with technological innovations to achieve resilience in urban water systems [131].

13.10.2 Community-Based Monitoring with Low-Cost Tech (e.g., India, and Kenya)

In regions like India and Kenya, community-based monitoring initiatives have emerged, utilizing low-cost technological solutions that empower local communities. These grassroots strategies leverage accessible technology to facilitate real-time water quality assessments, crucial in regions plagued by inadequate infrastructure and resource constraints. For instance, [132], highlight how integrating performance indicators in monitoring water systems can enable communities to engage actively in managing their water quality. This model not only provides immediate data but also fosters a sense of accountability and ownership among residents, allowing them to contribute to local governance effectively.

In India, the push for digitalization in the water sector has created pathways for communities to adopt mobile and internet-based platforms that can monitor water quality swiftly and efficiently [129]. These platforms utilize low-cost sensors and IoT devices, enabling citizens to convey data on water quality and availability [133]. Similarly, in Kenya, initiatives harness community participation by deploying simple yet effective monitoring devices that help detect pollutants and inform residents of potential hazards, making significant strides towards improving public health outcomes through informed citizen engagement [134].

13.10.3 AI-Driven Pollution Tracking (e.g., China, and USA)

The application of artificial intelligence (AI) in pollution tracking has emerged as one of the most promising innovations across countries like China and the USA. AI-driven technologies enhance the capability to detect, analyze, and report environmental pollutants swiftly and accurately, thus facilitating proactive management strategies. In China, research conducted by [135], underscores how AI is integrated into smart urban systems, leading to significant improvements in environmental monitoring and pollution reduction through data analytics and smart resource management. Additionally, the deployment of AI in smart cities not only bolsters pollution tracking efforts but also streamlines urban infrastructure management, ultimately contributing to enhanced quality of life for their inhabitants [136].

In the USA, substantial investment in AI technologies is transforming the environmental monitoring landscape. For instance, the implementation of AI-driven detection systems empowers local governments and organizations to predict pollution patterns, assess risks, and develop strategic responses [133]. This synergy of AI technologies with environmental management practices creates a robust framework for urban planning and decision-making, addressing pollution precisely and effectively while elevating public health standards [133].

13.10.4 Remote and Indigenous Water Quality Monitoring Programs

Remote monitoring programs utilizing cutting-edge technologies present substantial opportunities for enhancing water quality assessments in marginalized and indigenous communities. Incorporating technological advancements into traditional ecological knowledge facilitates more inclusive and comprehensive water management strategies. Smart water protocols, as developed in various regions, can be adapted to reflect indigenous knowledge systems, thereby ensuring that water management practices respect cultural nuances while being technologically relevant [137]. Moreover, remote monitoring technological solutions often involve satellite imagery, mobile applications, and cloud-based data management systems that allow for regularized water quality assessments without the requirement of extensive on-ground infrastructure [138]. Such systems immensely benefit remote communities, enabling them to engage in data collection and monitoring while relying on infrastructure they typically could not afford.

The integration of AI and IoT enhances the efficacy of these programs, making it easier for indigenous communities to collaborate with governmental and non-governmental organizations in environmental monitoring [133]. For instance, local stakeholders can leverage data collected from remote sensors to inform their management practices, streamline decision-making processes, and advocate for their water rights effectively [137].

13.11 Future Prospects and Research Needs

Emerging technologies for water qualitys monitoring are converging on innovative frontiers such as CRISPR-based diagnostics, which show substantial potential for rapid and precise detection of pathogens and contaminants. The CRISPR/Cas systems are acclaimed for their structurally programmable nature, enabling the development of specialized sensors capable of identifying specific nucleic acids associated with various microorganisms and pollutants in water ecosystems [139–141]. Recent advancements illustrate the integration of CRISPR technology with nanotechnology, allowing for the construction of highly sensitive biosensing platforms that can detect waterborne pathogens effectively [141, 142]. Moreover, the application of CRISPR/Cas-based diagnostic tools has extended beyond mere pathogen identification, facilitating extensive monitoring of water quality parameters through the detection of specific biomarkers [143, 144].

The interdisciplinary alliance between technology, environmental science, and public health forms a cornerstone for future developments in water quality monitoring. Integrating artificial intelligence (AI) with environmental science has proved transformative, as it allows for enhanced data analysis and predictive modeling, which are crucial for accurately monitoring water resources [145, 146]. Notably, AI systems can optimize the interpretation of diverse data streams from CRISPR diagnostics, enabling real-time assessments of water safety and contributing to effective management strategies. This nexus of disciplines not only propels innovation in water monitoring technologies but also fosters sustainable solutions to pressing environmental issues [145].

To fully realize the capabilities of these emerging technologies, supportive policy frameworks and governance structures are required. Enabling regulations are essential to facilitate the uptake of innovations such as CRISPR and AI in environmental monitoring. Effective governance can ensure that these technologies are employed responsibly and ethically, maximize public health benefits, and minimize ecological impacts [147, 148].

13.12 Conclusion

Water quality monitoring is undergoing a paradigm shift, driven by the convergence of emerging technologies such as smart sensors, IoT, remote sensing, artificial intelligence, and nanotechnology. These innovations offer unprecedented opportunities for real-time, accurate, and spatially extensive assessments of water quality, supporting early detection of contaminants, enhancing decision-making, and promoting proactive environmental management. Unlike traditional approaches, which are often limited by cost, labor, and temporal gaps, the new wave of technological tools allows for continuous, automated, and integrative monitoring that aligns with the complex demands of modern water systems.

This chapter has demonstrated how these technologies are not only reshaping the technical landscape but also creating new avenues for community engagement, policy innovation, and cross-sectoral collaboration. However, the adoption and scaling of these solutions are not without challenges. Issues related to infrastructure, cost, data interoperability, privacy, and capacity building, particularly in low-resource settings, remain significant barriers. Furthermore, ethical and governance considerations must accompany technological advancement to ensure equitable access and responsible use. Collaborative research, investment in digital infrastructure, and the development of supportive policy frameworks will be critical to realizing the full potential of these innovations. Ultimately, embracing emerging technologies in water quality monitoring can significantly contribute to achieving global water security and advancing Sustainable Development Goal 6: clean water and sanitation for all.

References

1. Priyanka K (2024) GIS constructed water monitoring network system via IoT. JES 20(2):2671–2678. https://doi.org/10.52783/jes.2038
2. Saab C, Farah E, Shahrour I, Chehade F (2017) Field study of real-time water quality control, vol 1, pp 237–248. https://doi.org/10.2495/ws170231
3. Shaik F, AL-Siyabi S, Lakkimsetty N, Syed M, Nayeemuddin M (2021) Assessment of Falaj water quality. Asian J Water Environ Pollut 18(1):25–30. https://doi.org/10.3233/ajw210004
4. Zhao J, Wei S, Wen X, Qiu X (2020) Analysis and prediction of big stream data in real-time water quality monitoring system. J Ambient Intell Smart Environ 12(5):393–406. https://doi.org/10.3233/ais-200571
5. Li S, Tao J, Yu L (2013) Research and application of real-time remote monitoring and early warning system of source water quality. Adv Mater Res 779–780:1408–1413. https://doi.org/10.4028/www.scientific.net/amr.779-780.1408
6. Lu T, Zhang G, Huang P, Hou D (2013) Design of an automatic three-dimensional sampling device for water quality emergency monitoring. Appl Mech Mater 316–317:674–677. https://doi.org/10.4028/www.scientific.net/amm.316-317.674
7. Mohialden Y, Hussien N, Salman S (2024) Automated water quality assessment using big data analytics. MesopN J Big Data 211–222. https://doi.org/10.58496/mjbd/2024/015
8. Chang Y, Xu J, Ghafoor K (2021) An IoT and blockchain approach for the smart water management system in agriculture. Scalable Comput Pract Exp 22(2). https://doi.org/10.12694/scpe.v22i2.1869
9. Kuncoro C, Asyikin M, Amaris A (2021) Development of an automation system for nutrient film technique hydroponic environment. https://doi.org/10.2991/aer.k.211106.069
10. Lakshmi V, Hemanth K, Settipalli A, Nanda A, Jasim L (2024) Harmonizing environmental protection, integration of monitoring and purification systems. E3s Web Conf 529:03006. https://doi.org/10.1051/e3sconf/202452903006
11. Behmel S, Damour M, Ludwig R, Rodríguez M (2019) Optimization of river and lake monitoring programs using a participative approach and an intelligent decision-support system. Appl Sci 9(19):4157. https://doi.org/10.3390/app9194157
12. Feng W, Sun X (2010) The remote monitoring system for key sewage displacements is based on GPRS and GIS. Key Eng Mater 439–440:172–177. https://doi.org/10.4028/www.scientific.net/kem.439-440.172
13. Saby L, Goodall J, Band L, Bowes B, Fults M (2021) Enhancing efficacy of water quality trading with automation: a case study in Virginia's nutrient trading program. Jawra J Am Water Resour Assoc 57(3):374–390. https://doi.org/10.1111/1752-1688.12903
14. Naresh V, Lee N (2021) A review on biosensors and recent development of nanostructured materials-enabled biosensors. Sensors 21(4):1109. https://doi.org/10.3390/s21041109
15. Cui Y, Lai B, Tang X (2019) Microbial fuel cell-based biosensors. Biosensors 9(3):92. https://doi.org/10.3390/bios9030092
16. Maas M, Perold W, Dicks L (2017) Review: biosensors for the detection of E*scherichia coli*. Water Sa 43(4):707. https://doi.org/10.4314/wsa.v43i4.17

17. Jin X, Angelidaki I, Zhang Y (2016) Microbial electrochemical monitoring of volatile fatty acids during anaerobic digestion. Environ Sci Technol 50(8):4422–4429. https://doi.org/10.1021/acs.est.5b05267
18. Plante M, Bérubé È, Bissonnette L, Bergeron M, Leclerc M (2013) Polythiophene biosensor for rapid detection of microbial particles in water. ACS Appl Mater Interfaces 5(11):4544–4548. https://doi.org/10.1021/am400162h
19. Yu D, Zhai J, Liu C, Zhang X, Bai L, Wang Y, Dong S (2017) Small microbial three-electrode cell-based biosensor for online detection of acute water toxicity. Acs SensS 2(11):1637–1643. https://doi.org/10.1021/acssensors.7b00484
20. Bose S, Maity S, Sarkar A (2021) Review of microbial biosensors for the detection of mercury in water. Environ Qual Manag 31(4):29–40. https://doi.org/10.1002/tqem.21742
21. Pham H (2018) Biosensors based on lithotrophic microbial fuel cells in relation to heterotrophic counterparts: research progress, challenges, and opportunities. Aims Microbiol 4(3):567–583. https://doi.org/10.3934/microbiol.2018.3.567
22. Ahuekwe E, Akinyele A, Benson A, Ikhiwili O, Oziegbe O (2024) A review of fabrication techniques and optimization strategies for microbial biosensors. IoP Conf Ser Earth Environ Sci 1342(1):012015. https://doi.org/10.1088/1755-1315/1342/1/012015
23. Coronado-Apodaca K, González-Meza G, Aguayo-Acosta A, Araújo R, González-González R, Oyervides-Muñoz M, Martinez-Ruiz M, Melchor-Martinez EM, Barcelo D, Parra-Saldivar R, Sosa-Hernández J (2023) Immobilized enzyme-based novel biosensing system for recognition of toxic elements in the aqueous environment. Top Catal 66(9–12):606–624. https://doi.org/10.1007/s11244-023-01786-8
24. Odobašić A, Šestan I, Begić S (2019) Biosensors for determination of heavy metals in waters. https://doi.org/10.5772/intechopen.84139
25. Fdez-Sanromán A, Bernárdez-Rodas N, Rosales E, Pazos M, González-Romero E, Sanromán MÁ (2025) Biosensor technologies for water quality: detection of emerging contaminants and pathogens. Biosensors 15(3):189. https://doi.org/10.3390/bios15030189
26. Chen J, Chen S, Fu R, Li D, Jiang H, Wang C, Peng Y, Jia K, Hicks B (2022) Remote sensing big data for water environment monitoring: current status, challenges, and future prospects. Earth's Futur 10(2). https://doi.org/10.1029/2021ef002289
27. Yang W, Wei X, Choi S (2016) A dual-channel, interference-free, bacteria-based biosensor for highly sensitive water quality monitoring. IEEE Sens J 16(24):8672–8677. https://doi.org/10.1109/jsen.2016.2570423
28. Grattieri M, Hasan K, Minteer S (2016) Bioelectrochemical systems as a multipurpose biosensing tool: present perspective and future outlook. ChemElectroChem 4(4):834–842. https://doi.org/10.1002/celc.201600507
29. Besmer M, Weissbrodt D, Kratochvil B, Sigrist J, Weyland M, Hammes F (2014) The feasibility of automated online flow cytometry for in situ monitoring of microbial dynamics in aquatic ecosystems. Front Microbiol 5. https://doi.org/10.3389/fmicb.2014.00265
30. Cho J, Gao Y, Ryu J, Choi S (2020) Portable, disposable, paper-based microbial fuel cell sensor utilizing freeze-dried bacteria for in situ water quality monitoring. ACS Omega 5(23):13940–13947. https://doi.org/10.1021/acsomega.0c01333
31. Pai J, Davey D, Hsu H (2011) Essential elements of biosensor development for water quality monitoring. https://doi.org/10.1109/issnip.2011.6146622
32. Jamil M, Lessy M, Samuda S (2018) The utilization of internet of things for multi-sensor data acquisition using Thingspeak. https://doi.org/10.4108/eai.24-10-2018.2280578
33. He Y, Yang Y, He T, Lai Y, He Y, Chen B (2024) Small and micro-water quality monitoring based on the integration of a full-space real 3d model and IoT. Sensors 24(3):1033. https://doi.org/10.3390/s24031033
34. Wiryasaputra R, Huang C-Y, Lin Y-J, Yang C-T (2024) An IoT real-time potable water quality monitoring and prediction model based on cloud computing architecture. Sensors 24(4):1180. https://doi.org/10.3390/s24041180
35. Routray S, Mohanty S (2022) Narrowband IoT: a greener solution for large-scale deployments. Eai Endorsed Trans Internet Things 7(27):1–7. https://doi.org/10.4108/eai.10-3-2022.173607

36. Kosoe EA, Osumanu IK, Ogwu MC (2025) Reimagining rural and small-town water supply in Ghana through inclusive governance and sustainable models. J Sustain 1(2):1–21. https://doi.org/10.55845/jos-2025-1234
37. Kosoe EA, Ogwu MC, Duniah J, Ndigun FB (2025) Sustaining indigenous water governance systems in Ayimaye and Lungni, Northern Ghana, through hybrid models. J Sustain 1(2). https://doi.org/10.55845/jos-2025-1235
38. Farooq M, Riaz S, Abid A, Abid K, Naeem M (2019) A survey on the role of IoT in agriculture for the implementation of smart farming. IEEE Access 7:156237–156271. https://doi.org/10.1109/access.2019.2949703
39. AlGhamdi R, Sharma SK (2022) IoT-based smart water management systems for residential buildings in Saudi Arabia. Processes 10(11):2462. https://doi.org/10.3390/pr10112462
40. Ballerini M, Polonelli T, Brunelli D, Magno M, Benini L (2020) Nb-IoT versus LoRaWAN: an experimental evaluation for industrial applications. IEEE Trans Industr Inf 16(12):7802–7811. https://doi.org/10.1109/tii.2020.2987423
41. Mwakwata C, Malik H, Alam M, Moullec Y, Pärand S, Mumtaz S (2019) Narrowband internet of things (NB-IoT): from physical (PHY) and media access control (MAC) layers perspectives. Sensors 19(11):2613. https://doi.org/10.3390/s19112613
42. Sultania A, Mahfoudhi F, Famaey J (2020) Real-time demand response using NB-IoT. IEEE Internet Things J 7(12):11863–11872. https://doi.org/10.1109/jiot.2020.3004390
43. Liu T, Qin F (2025) Study on industrial wastewater pollution monitoring technology based on NB-IoT wireless communication technology. Int J Grid High Perform Comput 17(1):1–20. https://doi.org/10.4018/ijghpc.374211
44. Ogwu MC, Kosoe EA (2024) Place of cultural diversity in sustainable water resource management in Ghana. In: Izah SC, Ogwu MC, Loukas A, Hamidifar H (eds) Water crises and sustainable management in the global south. Springer, Singapore. https://doi.org/10.1007/978-981-97-4966-9_14
45. Ogwu MC, Kosoe EA (2025) Integrating green infrastructure into sustainable agriculture to enhance soil health, biodiversity, and microclimate resilience. Sustainability 17(9):3838. https://doi.org/10.3390/su17093838
46. Atta B, Saleem M, Ali H, Arshad H, Ahmed M (2018) Chlorophyll as a biomarker for early disease diagnosis. Laser Phys 28(6):065607. https://doi.org/10.1088/1555-6611/aab94f
47. Slonecker E, Bufford B, Graham J, Carpenter K, Opstal D, Simon N, Hall N (2021) Hyperspectral reflectance characteristics of cyanobacteria. Adv Remote Sens 10(03):66–77. https://doi.org/10.4236/ars.2021.103004
48. Cai Y, Miao Y, Wu H, Wang D (2021) Hyperspectral estimation models of winter wheat chlorophyll content under elevated CO_2. Front Plant Sci 12. https://doi.org/10.3389/fpls.2021.642917
49. Sánchez M, Silva C, Demattê J, Mendonça F, Silva M, Romanelli T, Fiorio P (2024) Predictive modelling of chlorophyll in mombaça grass leaves by hyperspectral reflectance data and machine learning. Grass Forage Sci 79(4):543–556. https://doi.org/10.1111/gfs.12689
50. Alegria C, Albuquerque T (2023) Remote sensing for water quality monitoring—a study case for the Marateca reservoir, Portugal. https://doi.org/10.20944/preprints202304.0056.v1
51. Mohamed N, El-Shanhorey N, Elsayed S (2018) Remote identification and differentiation of Ficus species in Alexandria City using spectral reflectance measurements. Alex Sci Exch J 39:629–641. https://doi.org/10.21608/asejaiqjsae.2018.19584
52. Stengel V, Trevino J, King T, Ducar S, Hundt S, Hafen K, Churchill C (2023) Near real-time satellite detection and monitoring of aquatic algae and cyanobacteria: how a combination of chlorophyll-a indices and water-quality sampling was applied to north Texas reservoirs. J Appl Remote Sens 17(04). https://doi.org/10.1117/1.jrs.17.044514
53. Chu H, He Y, Chusnah W, Jaelani L, Chang C (2021) Multi-reservoir water quality mapping from remote sensing using spatial regression. Sustainability 13(11):6416. https://doi.org/10.3390/su13116416
54. Izah SC, Ogwu MC (2025) Modeling solutions for microbial water contamination in the global south for public health protection. Front Microbiol 16:1504829. https://doi.org/10.3389/fmicb.2025.1504829

55. Ogwu MC, Izah SC, Sawyer WE, Amabie T (2025) Environmental risk assessment of trace metal pollution: a statistical perspective. Environ Geochem Health 47:94. https://doi.org/10.1007/s10653-025-02405-z
56. Nmecha M, Okereke C, Egwuonwu C, Okorafor O (2024) Assessment of water quality of the Otamiri River along selected reaches using drone technology. Curr Res Env Sci Eco Lett 1(1):01–19. https://doi.org/10.33140/cresel.01.01.05
57. Martin F, Müllerová J, Borgniet L, Dommanget F, Breton V, Évette A (2018) Using single- and multi-date UAV and satellite imagery to accurately monitor invasive knotweed species. Remote Sens 10(10):1662. https://doi.org/10.3390/rs10101662
58. Guan G, Wang Y, Yang L, Yue J, Li Q, Lin J, Liu Q (2022) Water-quality assessment and pollution-risk early-warning system based on web crawler technology and LSTM. Int J Environ Res Public Health 19(18):11818. https://doi.org/10.3390/ijerph191811818
59. Forhad H, Uddin M, Chakrovorty R, Ruhul A, Faruk H, Kamruzzaman S, Sharmin N, Jamal AS, Haque MM, Morshed A (2024) Iot-based real-time water quality monitoring system in water treatment plants (WTPs). Heliyon 10(23):e40746. https://doi.org/10.1016/j.heliyon.2024.e40746
60. Kotwal V, Patil S, Patil J (2025) Iot-enabled water quality assessment and demand prediction for sustainable water management. https://doi.org/10.21203/rs.3.rs-6470418/v1
61. Pandey J, Verma S (2021) Water quality analysis and prediction techniques using artificial intelligence, pp 279–290. https://doi.org/10.1007/978-981-16-4177-0_29
62. Sun Z, Fan Y (2022) A combined water quality pollution prediction model based on the spark big data platform. J Water Supply Res Technology—aqua 71(9):963–974. https://doi.org/10.2166/aqua.2022.036
63. Zheng Y, Wang J, Ryzhkov S, Nechai O, Topalov A, Zivenko O, Babchuk S, Harasymiv T (2024) Computer system for forecasting water quality parameters based on machine learning. Proc Bulg Acad Sci 77(11). https://doi.org/10.7546/crabs.2024.11.06
64. Fan B, Zhao Z (2023) An efficient water quality prediction and evaluation method based on the DBN-LSTM model. https://doi.org/10.21203/rs.3.rs-2663527/v1
65. Li L, Gu Z, Lu D (2023) Research on surface water quality prediction based on a CNN-BiLSTM-attention combined deep learning approach. E3s Web Conf 393:01007. https://doi.org/10.1051/e3sconf/202339301007
66. Yang H, Liu S (2022) Water quality prediction in sea cucumber farming based on a GRU neural network optimized by an improved whale optimization algorithm. Peerj Comput Sci 8:e1000. https://doi.org/10.7717/peerj-cs.1000
67. Alshami A, Ali E, Elsayed M, Eltoukhy A, Zayed T (2024) Iot innovations in sustainable water and wastewater management and water quality monitoring: a comprehensive review of advancements, implications, and future directions. IEEE Access 12:58427–58453. https://doi.org/10.1109/access.2024.3392573
68. Dheda D, Cheng L (2020) A multivariate water quality parameter prediction model using recurrent neural network. https://doi.org/10.48550/arxiv.2003.11492
69. Obin N, Tao H, Ge F, Liu X (2021) Research on water quality simulation and environmental capacity in the Lushui River based on the WASP model. Water 13(20):2819. https://doi.org/10.3390/w13202819
70. Nair J (2023) Design and development of efficient water quality prediction models using variants of recurrent neural networks. Eur Chem Bull 12(si5). https://doi.org/10.48047/ecb/2023.12.si5.0143
71. Lloyd-Jones L, Kuhnert P, Lawrence E, Lewis S, Waterhouse J, Gruber R, Kroon F (2022) Sampling re-design increases power to detect change in the great barrier reef's inshore water quality. Plos One 17(7):e0271930. https://doi.org/10.1371/journal.pone.0271930
72. Kuang L, Shi P, Hua C, Chen B, Zhu H (2020) An enhanced extreme learning machine for dissolved oxygen prediction in wireless sensor networks. IEEE Access 8:198730–198739. https://doi.org/10.1109/access.2020.3033455
73. Iyiola AO, Izah SC, Morya S, Akinsorotan A, Ogwu MC (2023) Owalla reservoir in south-western nigeria: assessment of fish distribution, biological diversity, and water quality index. Indones J Agric Res 6(02):106–125. https://doi.org/10.32734/injar.v6i2.12207

74. Iyiola AO, Kolawole AS, Setufe SB, Bilikoni J, Ofori E, Ogwu MC (2024) Fish as a sustainable biomonitoring tool in aquatic environments. In: Izah SC, Ogwu MC, Hamidifar H (eds) Biomonitoring of pollutants in the global south. Springer, Singapore.https://doi.org/10.1007/978-981-97-1658-6_12
75. Sforza M, Petronella F, Biase D, Zaccagnini F, Lim S, Butt U, d'Alessandro A, Godman NP, Evans DR, McConney M, Jeong KU, Sio L (2024) Cascade structured plasmonic liquid crystal biosensor for the rapid detection of harmful bacteria dispersed in potable water. Adv Sens Res 3(8). https://doi.org/10.1002/adsr.202300201
76. Jayamohan H, Smith Y, Gale B, Mohanty S, Misra M (2016) Photocatalytic microfluidic reactors utilizing titania nanotubes on titanium mesh for degradation of organic and biological contaminants. J Environ Chem Eng 4(1):657–663. https://doi.org/10.1016/j.jece.2015.12.018
77. Quinn C, Cate D, Miller-Lionberg D, Reilly T, Volckens J, Henry C (2018) Solid-phase extraction coupled to a paper-based technique for trace copper detection in drinking water. Environ Sci Technol 52(6):3567–3573. https://doi.org/10.1021/acs.est.7b05436
78. Xiong X, Guo,C, Yan G, Han B, Wu Z, Chen Y, Xu S, Shao P, Song H, Xu X, Han J (2022) Simultaneous cross-type detection of water quality indexes via a smartphone-app integrated microfluidic paper-based platform. Acs Omega 7(48) 44338–44345. https://doi.org/10.1021/acsomega.2c05938
79. Jaywant S, Arif K (2019) A comprehensive review of microfluidic water quality monitoring sensors. Sensors 19(21):4781. https://doi.org/10.3390/s19214781
80. AlMashrea B, Almehdi A, Damiati S (2024) Simple microfluidic devices for in situ detection of water contamination: a state-of-art review. Front Bioeng Biotechnol 12. https://doi.org/10.3389/fbioe.2024.1355768
81. Filippidou M, Kanaris A, Aslanidis E, Rapesi A, Tsounidi D, Ntouskas S, Skodatis E, Tsekenis G, Tsoukalas D, Tserepi A, Chatzandroulis S (2023) Integrated plastic microfluidic device for heavy metal ion detection. Micromachines 14(8):1595. https://doi.org/10.3390/mi14081595
82. Alhalaili B, Popescu I, Rusănescu C, Vidu R (2022) Microfluidic devices and microfluidics-integrated electrochemical and optical (bio) sensors for pollution analysis: a review. Sustainability 14(19):12844. https://doi.org/10.3390/su141912844
83. Rai P, Islam M, Gupta A (2022) Microfluidic devices for the detection of contamination in water samples: a review. Sens Actuators A 347:113926. https://doi.org/10.1016/j.sna.2022.113926
84. Wang T, Niu J, Pang H, Meng X, Sun R, Xie J (2024) Development of a portable residual chlorine detection device with a combination of microfluidic chips and LS-BP algorithm to achieve accurate detection of residual chlorine in water. Micromachines 15(8):1045. https://doi.org/10.3390/mi15081045
85. Park M, Ha H, Kim Y, Jung J, Kim S, Kim D, Seo T (2015) Combination of a sample pretreatment microfluidic device with a photoluminescent graphene oxide quantum dot sensor for trace lead detection. Anal Chem 87(21):10969–10975. https://doi.org/10.1021/acs.analchem.5b02907
86. Heydari M, Tabatabaei N, Rezai P (2022) Low-cost resistive microfluidic salinity sensor for high-precision detection of drinking water salt levels. ACS Omega 7(18):15529–15539. https://doi.org/10.1021/acsomega.2c00268
87. Richardson S, Iles A, Rotchell J, Charlson T, Hanson A, Lorch M, Pamme N (2021) Citizen-led sampling to monitor phosphate levels in freshwater environments using a simple paper microfluidic device. Plos One 16(12):e0260102. https://doi.org/10.1371/journal.pone.0260102
88. Kim M, Jung T, Kim Y, Lee C, Woo K, Seol J, Yang S (2015) A microfluidic device for label-free detection of Escherichia coli in drinking water using positive dielectrophoretic focusing, capturing, and impedance measurement. BiosensS Bioelectron 74:1011–1015. https://doi.org/10.1016/j.bios.2015.07.059
89. Satilmsioglu T, Sermet Y, Demir İ (2023) Blockchain applications and opportunities for water resources and hydrology: a systematic review. https://doi.org/10.31223/x5594k

90. Chen F, Yang B, Peng S, Lin T (2020) Applying a deployment strategy and data analysis model for water quality continuous monitoring and management. Int J Distrib Sens Netw 16(6):155014772092982. https://doi.org/10.1177/1550147720929825
91. Patil S (2024) The use of blockchain for water rights in irrigation management. IWI 44(1):01–06. https://doi.org/10.52783/iwi.v44i1.82
92. Huseynov F, Mitchell J (2023) Blockchain for environmental peacebuilding: application in water management. Digital Policy Regulation and Governance 26(1):55–71. https://doi.org/10.1108/dprg-06-2023-0080
93. Zeng H, Dhiman G, Sharma A, Sharma A, Tselykh A (2021) An IoT and blockchain-based approach for the smart water management system in agriculture. Expert Syst 40(4). https://doi.org/10.1111/exsy.12892
94. Arridha R (2018) Classification extension based on IoT-big data analytic for smart environment monitoring and analytic in real-time system. https://doi.org/10.31227/osf.io/stmg6
95. Nayal K, Raut R, Narkhede B, Priyadarshinee P, Panchal G, Gedam V (2021) Antecedents for blockchain technology-enabled sustainable agriculture supply chain. Ann Oper Res 327(1):293–337. https://doi.org/10.1007/s10479-021-04423-3
96. Liu Y, Shang C (2022) Application of blockchain technology in agricultural water rights trade management. Sustainability 14(12):7017. https://doi.org/10.3390/su14127017
97. Satilmisoglu T, Sermet Y, Kurt M, Demir İ (2024) Blockchain opportunities for water resources management: a comprehensive review. Sustainability 16(6):2403. https://doi.org/10.3390/su16062403
98. Song L, Luo Y, Chang Z, Jin C, Merveille N (2022) Blockchain adoption in agricultural supply chain for better sustainability: a game theory perspective. Sustainability 14(3):1470. https://doi.org/10.3390/su14031470
99. Kara S, Karadirek I, Muhammetoğlu A, Muhammetoğlu H (2016) Real-time monitoring and control in water distribution systems for improving operational efficiency. Desalin Water Treat 57(25):11506–11519. https://doi.org/10.1080/19443994.2015.1069224
100. Hamdi M, Rehman A, Alghamdi A, Nizamani M, Missen M, Memon M (2021) Internet of Things (IoT) based water irrigation system. Int J Online Biomed Eng (IJOE) 17(05):69–80. https://doi.org/10.3991/ijoe.v17i05.22081
101. Sunaina A (2025) Bridging the gap: application of earth observational data set and remote sensing techniques for monitoring of water quality. Ecol Environ Conserv 31(2):506–516. https://doi.org/10.53550/eec.2025.v31i02.019
102. Zaidi A (2012) Water quality management using GIS and RS tools, pp 842–848. https://doi.org/10.1061/9780784412312.086
103. Mahadik S (2025) Review on developing integrated water quality management systems based on GIS and remote sensing mechanisms. Voice Creat Res 7(2):284–288. https://doi.org/10.53032/tvcr/2025.v7n2.35
104. Habeeb N, Weli S (2021) Combination of GIS with different technologies for water quality: an overview. Hightech Innov J 2(3):262–272. https://doi.org/10.28991/hij-2021-02-03-10
105. Christophoridis C, Bizani E, Fytianos K (2013) Environmental quality monitoring, using GIS as a tool of visualization, management, and decision-making, pp 1559–1586. https://doi.org/10.4018/978-1-4666-2038-4.ch094
106. Weston S, Scheli A, Behmel S, Rodríguez M (2022) Water quality in drinking water distribution systems: a whole-systems approach to decision making. https://doi.org/10.4995/wdsa-ccwi2022.2022.14007
107. Shomar B, Fakher S, Yahya A (2010) Assessment of groundwater quality in the Gaza Strip, Palestine using GIS mapping. J Water Resour Prot 02(02):93–104. https://doi.org/10.4236/jwarp.2010.22011
108. Aulakh D, Patil S, Kumar M, Bhardwaj U (2025) Integration of GIS and geomorphic data to assess the impact of landscape features on river water quality. Nat Eng Sci 10(1):290–300. https://doi.org/10.28978/nesciences.1646470
109. Mukhtorov U, Kakhorov B, Khafizova Z, Murodova D, Egamberdiev R (2024) Study of monitoring of water bodies using remote sensing data and GIS technologies (Talimarjan

water reservoir). Iop Conf Ser Earth Environ Sci 1420(1):012007. https://doi.org/10.1088/1755-1315/1420/1/012007
110. Basin Q, Samah M (2018) Groundwater quality assessment of the quaternary aquifer, Wadi Qena Basin, Eastern Desert, Egypt. Egypt J Geol 62(1):21–34. https://doi.org/10.21608/egjg.2018.216372
111. Wu C, Liang H, Yin Q, Zhang Y (2015) Research and application of online monitoring system in sewage treatment based on web service. https://doi.org/10.2991/asei-15.2015.245
112. Ali M, Rahman N, Khalid K (2014) Discharge assessment by using integrated hydrologic model for environmental technology development. Adv Mater Res 911:378–382. https://doi.org/10.4028/www.scientific.net/amr.911.378
113. Hussain A, Madan R, Kamboj V (2022) Applications of remote sensing and GIS in hydrological and hydrogeological studies, pp 237–251. https://doi.org/10.4018/978-1-7998-8331-9.ch012
114. Hu C, Ren G, Liu C, Li M, Jie W (2017) A spark-based genetic algorithm for sensor placement in large-scale drinking water distribution systems. Clust Comput 20(2):1089–1099. https://doi.org/10.1007/s10586-017-0838-z
115. Camargo E, Spanhol F, Slongo J, Silva M, Pazinato J, Lobo A, Coutinho FR, Pfrimer FW, Lindino CA, Oyamada MS, Martins L (2023) Low-cost water quality sensors for IoT: a systematic review. Sensors 23(9):4424. https://doi.org/10.3390/s23094424
116. Manjakkal L, Mitra S, Pétillot Y, Shutler J, Scott E, Willander M, Dahiya R (2021) Connected sensors, innovative sensor deployment, and intelligent data analysis for online water quality monitoring. IEEE Internet Things J 8(18):13805–13824. https://doi.org/10.1109/jiot.2021.3081772
117. López-Ramírez G, Aragón-Zavala A (2023) Wireless sensor networks for water quality monitoring: a comprehensive review. IEEE Access 11:95120–95142. https://doi.org/10.1109/access.2023.3308905
118. Dong J, Wang G, Yan H, Xu J, Zhang X (2015) A survey of smart water quality monitoring systems. Environ Sci Pollut Res 22(7):4893–4906. https://doi.org/10.1007/s11356-014-4026-x
119. Zeng D, Gu L, Lu L, Guo S, Yao H, Hu J (2016) On cost-efficient sensor placement for contaminant detection in water distribution systems. IEEE Trans Industr Inf 12(6):2177–2185. https://doi.org/10.1109/tii.2016.2569413
120. Li M, Lin C, Jia R, Jiang F (2019) A wireless ecological aquaculture water quality monitoring system based on LoRa technology. https://doi.org/10.2991/wcnme-19.2019.2
121. Taha S, Talip M, Mohamad M, Azizul Z, Izam T (2024) Evaluation of Lora network performance for water quality monitoring systems. https://doi.org/10.20944/preprints202405.1699.v1
122. Naloufi M, Abreu T, Souihi S, Thérial C, Rodrigues N, Goff A, Saad M, Vincon-Leite B, Dubois P, Delarbre M, Kennouche P, Lucas F (2024) Long-term stability of low-cost IoT system for monitoring water quality in urban rivers. Water 16(12):1708. https://doi.org/10.3390/w16121708
123. Adu-Manu K, Katsriku F, Abdulai J, Engmann F (2020) Smart river monitoring using wireless sensor networks. Wirel Commun Mob Comput 2020:1–19. https://doi.org/10.1155/2020/8897126
124. Zhao C, Gao Q, Song J, Yue-guo W, Sun F (2021) Research on design of the safety supervision system for desalinated seawater entering urban water supply network. Water 13(15):2017. https://doi.org/10.3390/w13152017
125. Pule M, Yahya A, Chuma J (2017) Wireless sensor networks: a survey on monitoring water quality. J Appl Res Technol 15(6):562–570. https://doi.org/10.1016/j.jart.2017.07.004
126. Nardo A, Boccelli D, Herrera M, Creaco E, Cominola A, Sitzenfrei R, Taormina R (2021) Smart urban water networks: solutions, trends and challenges. Water 13(4):501. https://doi.org/10.3390/w13040501
127. Zhao Y (2024) Evaluating the achievement of Singapore's environmental policy goals for enhanced sustainability and livability through big data analysis. Highlights Sci Eng Technol 83:568–578. https://doi.org/10.54097/60re5h86

128. Kong L, Woods O (2018) The ideological alignment of smart urbanism in Singapore: critical reflections on a political paradox. Urban Stud 55(4):679–701. https://doi.org/10.1177/0042098017746528
129. Banerjee C, Bhaduri A, Saraswat C (2022) Digitalization in urban water governance: a case study of Bengaluru and Singapore. Front Environ Sci 10. https://doi.org/10.3389/fenvs.2022.816824
130. Demirel D (2023) The impact of managing diversity on building the smart city: a comparison of smart city strategies: cases from Europe, America, and Asia. Sage Open 13(3). https://doi.org/10.1177/21582440231184971
131. Rojek I, Studziński J (2019) Detection and localization of water leaks in water nets supported by an ICT system with artificial intelligence methods as a way forward for smart cities. Sustainability 11(2):518. https://doi.org/10.3390/su11020518
132. Dasallas L, Lee J, Jang S, Jang S (2024) Development and application of technical key performance indicators (KPIs) for smart water cities (SWCs) global standards and certification schemes. Water 16(5):741. https://doi.org/10.3390/w16050741
133. Popescu S, Mansoor S, Wani O, Kumar S, Sharma V, Sharma A, Arya VM, Kirkham MB, Hou D, Bolan N, Chung Y (2024) Artificial intelligence and IoT-driven technologies for environmental pollution monitoring and management. Front Environ Sci 12. https://doi.org/10.3389/fenvs.2024.1336088
134. Chen Y, Han D (2018) Water quality monitoring in smart city: a pilot project. Autom Constr 89:307–316. https://doi.org/10.1016/j.autcon.2018.02.008
135. Shiping X, Wang L (2023) Do green information and communication technologies (ICT) and smart urbanization reduce environmental pollution in China? Sustainability 15(19):14492. https://doi.org/10.3390/su151914492
136. Vitry M, Schneider M, Wani O, Manny L, Leitão J, Eggimann S (2019) Smart urban water systems: what could possibly go wrong? Environ Res Lett 14(8):081001. https://doi.org/10.1088/1748-9326/ab3761
137. Adelani F, Okafor E, Jacks B, Ajala O (2024) Exploring theoretical constructs of urban resilience through smart water grids: case studies in African and U.S. cities. Eng Sci Technol J 5(3):984–994. https://doi.org/10.51594/estj.v5i3.952
138. Nwokediegwu Z, Ugwuanyi E (2024) Implementing AI-driven waste management systems in underserved communities in the USA. Eng Sci Technol J 5(3):794–802. https://doi.org/10.51594/estj.v5i3.903
139. Kostyusheva A, Brezgin S, Babin Y, Vasilyeva I, Glebe D, Kostyushev D, Chulanov V (2022) CRISPR-CAS systems for diagnosing infectious diseases. Methods 203:431–446. https://doi.org/10.1016/j.ymeth.2021.04.007
140. Liu P, Aodeng G, Ga L, Ai J (2023) Application of CRISPR/Cas technology in biological nano-analysis. Chemistryselect 8(45). https://doi.org/10.1002/slct.202303133
141. Mukherjee S, Sarkar S, Sarkar S, Ghosh S, Ghosh B (2024) CRISPR-CAS-based nano-sensors in water pathogen detection. Int J Res Adv Electron Eng 5(1):07–10. https://doi.org/10.22271/27084558.2024.v5.i1a.32
142. Yee B, Zakaria S, Chandrawati R, Ahmed M (2024) Detection of tetracycline with a CRISPR/Cas12a aptasensor using a highly efficient fluorescent polystyrene microsphere reporter system. ACS Synth Biol 13(7):2166–2176. https://doi.org/10.1021/acssynbio.4c00200
143. Tian Z, Yan H, Zeng Y (2024) Solid-phase extraction and enhanced amplification-free detection of pathogens integrated by multifunctional CRISPR-Cas12a. ACS Appl Mater Interfaces 16(12):14445–14456. https://doi.org/10.1021/acsami.3c17039
144. Zhu Y, Zhang M, Guo S, Hong X, Jie Z, Tao S (2023) CRISPR-based diagnostics of different biomolecules from nucleic acids, proteins, and small molecules to exosomes. Acta Biochim Biophys Sin 55(10):1539–1550. https://doi.org/10.3724/abbs.2023134
145. Akter M (2024) Interdisciplinary insights: integrating artificial intelligence with environmental science for sustainable solutions. JAIGS 1(1). https://doi.org/10.60087/jaigs.v1i1.28

146. Shuford J (2024) Interdisciplinary perspectives: fusing artificial intelligence with environmental science for sustainable solutions. JAIGS 1(1):106–123. https://doi.org/10.60087/jaigs.v1i1.87
147. Yurttaş A, Kartal E, Çağlar A (2021) Investigation of environmental-related course gains at the first level of primary education in terms of an interdisciplinary approach. Kastamonu Eğitim Dergisi 29(2):306–325. https://doi.org/10.24106/kefdergi.699761
148. Zhang T, Wang Z, Chen X, Zhang X, Zhu J, Li S, Wu C, Tian Z, Sui G (2022) Fully automated CRISPR-LAMP platform for SARS-CoV-2 Delta and Omicron variants. Anal Chem 94(44):15472–15480. https://doi.org/10.1021/acs.analchem.2c03607s
149. Chen H, Meng X, Liu D, Wang W, Xing X, Zhang Z, Dong C (2022) Closed-loop microbial fuel cell control system designed for online monitoring of TOC dynamic characteristics in public swimming pool. Int J Environ Res Public Health 19(20):13024. https://doi.org/10.3390/ijerph192013024
150. Gong S, Zhang S, Lü F, Pan W, Li N, Tang B (2021) CRISPR/Cas-based in vitro diagnostic platforms for cancer biomarker detection. Anal Chem 93(35):11899–11909. https://doi.org/10.1021/acs.analchem.1c02533
151. Konwarh R (2020) Can CRISPR/CAS technology be a felicitous stratagem against the COVID-19 fiasco? Prospects and hitches. Front Mol Biosci 7. https://doi.org/10.3389/fmolb.2020.557377
152. Lau A, Ren C, Lee L (2020) Critical review on where crispr meets molecular diagnostics. Prog Biomed Eng 3(1):012001. https://doi.org/10.1088/2516-1091/abbf5e
153. Ogwu MC (2025) Sustainable development in Africa: challenges, innovations, and policy pathways for the future. Afr Res Rep 1(2):55–69. https://doi.org/10.5897/ARR2025.0005

Chapter 14
Advancing Water Quality and Safety in the Global South Through International Partnerships and Sustainable Development Goals

Matthew Chidozie Ogwu, Rehema Ulimboka, and Sylvester Chibueze Izah

Abstract Ensuring equitable access to clean and safe water remains a critical global challenge, particularly in the Global South, where systemic inequalities, climate vulnerability, and infrastructural deficits converge. This chapter explores how international partnerships, under the framework of the Sustainable Development Goals (SDGs), notably SDG 6 (Clean Water and Sanitation) and SDG 17 (Partnerships for the Goals), are advancing water quality and safety in low- and middle-income countries. It synthesizes interdisciplinary evidence from case studies, institutional frameworks, and policy mechanisms to evaluate how global cooperation facilitates technology transfer, capacity building, financial innovation, and knowledge sharing. Drawing on examples from East and Southern Africa, Southeast Asia, and Latin America, the chapter examines how public–private partnerships, South-South collaborations, and multilateral initiatives address water governance through inclusive and locally responsive approaches. It critically interrogates the risks of dependency, unequal power dynamics, and short-termism that often undermine impact, while highlighting successful strategies rooted in community engagement, gender and youth inclusion, digital innovation, and decentralized governance. Furthermore, the chapter underscores the necessity of aligning water partnerships with climate resilience, biodiversity protection, and justice-oriented transitions. In advancing both

M. C. Ogwu (✉)
Goodnight Family Department of Sustainable Development, Appalachian State University, Boone, NC, USA
e-mail: ogwumc@appstate.edu

R. Ulimboka
Department of Biology, The University of Dodoma, Dodoma, Tanzania
e-mail: rehema.ulimboka@udom.ac.tz

S. C. Izah
Department of Community Medicine, Faculty of Clinical Sciences, Bayelsa Medical University, Yenagoa, Bayelsa State, Nigeria
e-mail: Sylvester.izah@bmu.edu.ng

Department of Microbiology, Faculty of Science, Bayelsa Medical University, Yenagoa, Bayelsa State, Nigeria

M. C. Ogwu and S. Chibueze Izah (eds.), *Water Quality and Safety in the Global South*, Springer Water, https://doi.org/10.1007/978-3-032-10602-5_14

conceptual and practical insights, this chapter contributes to reimagining water partnerships as transformative instruments—capable not only of achieving SDG targets, but also of fostering enduring systems of solidarity and sustainability. The findings provide actionable guidance for scholars, policymakers, development practitioners, and communities striving to build inclusive, accountable, and future-ready models of water governance in the Global South.

Keywords Water governance · Public–private partnerships · Climate resilience · Technology transfer · Inclusive development · Transboundary water management

14.1 Introduction

Water is a fundamental resource for human health, economic growth, and ecological sustainability. Yet, across the Global South, water quality is increasingly threatened by scarcity, pollution, climate change, and poverty. Approximately 2 billion people live in countries experiencing high water stress, a situation expected to intensify due to climate pressures and population growth [1]. In urban informal settlements such as those in Dar es Salaam, Tanzania, inadequate water infrastructure and pollution prevent access to safe drinking water, resulting in severe health and productivity losses [2]. The 2030 Agenda for Sustainable Development offers a global framework to address these interlinked challenges. SDG 6 focuses on ensuring the availability and sustainable management of water and sanitation, while SDG 17 emphasizes strengthening international partnerships as a means of achieving all development goals [3]. For regions in the Global South, where water challenges are acute, international cooperation is essential for advancing water safety and equitable access.

Persistent inequities in water access remain a significant concern. Urban areas often suffer from supply shortages, while rural communities depend on unsafe sources [4, 5, 6. In Uganda's Mbarara district, despite access to water, residents expressed ongoing concerns about its safety, highlighting the importance of community involvement in water governance [7]. Poor sanitation infrastructure contributes to waterborne disease transmission, with studies confirming the health impacts of mismanaged water resources [8]. These challenges necessitate partnerships that go beyond domestic capacities. Governments, NGOs, and private entities must combine efforts to strengthen water governance [9]. For example, public–private partnerships (PPPs) have demonstrated potential in improving water infrastructure and services by pooling expertise and resources [10]. Such collaborations support the development of systems that are resilient to demographic shifts and climate variability.

The SDG framework places water at the center of sustainable development. SDG 6 calls for universal and equitable access to safe, affordable drinking water and adequate sanitation by 2030 [11]. Achieving this depends heavily on fostering partnerships at all levels, particularly those that empower communities to participate in decision-making processes, which ensures responsiveness to local needs [12]. SDG

17 reinforces the need for inclusive and accountable partnerships. These collaborations link stakeholders such as governments, civil society, and the private sector and enable the sharing of technologies, knowledge, and financial support essential to water security [13], 95]. Because water challenges vary by region, localized approaches are needed to reflect diverse socio-environmental contexts.

Effective partnerships require robust frameworks. PPPs, in particular, have improved water infrastructure by allowing for shared investment and risk [14]. In South Africa, municipal governments have adopted PPPs to upgrade services, demonstrating how collaboration can mobilize technical and financial capacity [10]. Community involvement is also key. Local participation increases the likelihood of sustained use and maintenance of water systems [6]. Participatory approaches foster a sense of ownership and ensure that water solutions reflect local realities [3, 15, 16]. Additionally, interdisciplinary collaboration drawing on hydrology, engineering, and social science enables more holistic approaches. Capacity-building programs further strengthen the ability of local actors to manage resources effectively [17]. Funding limitations remain a barrier in many parts of the Global South [18]. Blended financing, which combines public and private capital, can expand the pool of available resources [19]. Results-based financing connects funding to outcomes, enhancing accountability and encouraging efficiency [20, 21]. Climate finance tools such as the Green Climate Fund also play a role in supporting water resilience projects, ensuring alignment with broader environmental goals [22].

In Uganda, the United Nations Development Program partnerships have strengthened community-managed water systems through training and resource mobilization [8]. Kenya's Water Fund brings together multiple actors to reduce water risks through integrated planning and community monitoring [18, 23]. In South Africa, a public–private initiative in the Eastern Cape improved service quality and access [10]. These partnerships must remain flexible in the face of climate and urban pressures [17]. Long-term success depends on inclusive engagement, capacity building [11], public awareness [24], data-driven tools [9], and policy harmonization across regions [20].

This chapter aims to examine how international partnerships and the Sustainable Development Goals (SDGs), particularly SDG 6 (Clean Water and Sanitation) and SDG 17 (Partnerships for the Goals), are shaping efforts to improve water quality and safety in the Global South. It explores the mechanisms through which global collaborations facilitate technology transfer, capacity building, financial support, and knowledge exchange to address persistent water challenges in low- and middle-income countries. The chapter contributes to the growing body of interdisciplinary literature on sustainable water governance by synthesizing diverse case studies, institutional frameworks, and policy instruments that exemplify both the successes and limitations of international cooperation. It also highlights the importance of inclusive, equitable, and locally responsive approaches to ensure that global partnerships lead to meaningful and sustainable improvements in water access and safety.

14.2 International Cooperation for Sustainable Water Governance

International cooperation plays a critical role in sustainable water governance, with various forms and modalities addressing the diverse challenges associated with water resource management. One notable typology includes bilateral, multilateral, public–private, South-South, and triangular partnerships, illustrating the multifaceted nature of collaborative efforts in water and sanitation projects (Fig. 14.1). Figure 14.1 illustrates the multi-level architecture of international cooperation aimed at advancing sustainable water governance in the Global South and highlights how global goals such as SDG 6 and SDG 17 translate into coordinated actions by multilateral institutions, bilateral partnerships, public–private collaborations, and community-led initiatives culminating in equitable, resilient, and inclusive water systems. Bilateral cooperation often involves partnerships between two countries, emphasizing dialogue, resource sharing, and mutual benefit in managing transboundary water resources [25, 26]. Multilateral efforts leverage the strengths of multiple nations and foster broader dialogues, agreements, and initiatives that transcend national boundaries, evident in frameworks such as the United Nations Water Law (2014) and the League of Arab States (2013). Public–private partnerships (PPPs) are increasingly recognized for their potential to harness private sector efficiency and innovation while ensuring public accountability in water delivery services [27]. South-South cooperation emphasizes knowledge exchange and capacity building between developing countries, facilitating tailored solutions for localized water management challenges [28, 29]. Triangular cooperation combines resources and expertise from both developed and developing nations, creating synergies that can significantly enhance project efficacy [27].

The principles of effective partnerships in water governance are paramount in ensuring that collaborative efforts yield equitable and sustainable outcomes. Equity is a foundational element, ensuring that all stakeholders, particularly marginalized communities, have an adequate voice in decision-making processes [30, 31]. Mutual accountability fosters trust and transparency among partners, enhancing the legitimacy of collaborative governance frameworks [30]. Transparency is integral to building public trust and ensuring that stakeholders have access to information about water management activities and decision-making processes [32]. Local relevance further emphasizes the importance of adapting strategies to fit the unique social, cultural, and ecological contexts of specific watersheds, ensuring that initiatives are not only practical but also culturally sensitive and locally accepted [26, 32]. Effective partnerships thus necessitate a comprehensive approach that incorporates these principles, fostering an environment conducive to sustainable practices.

Key international institutions play a pivotal role in facilitating global water governance by providing support, expertise, and resources for water-related programs. The United Nations Water (UN Water) coordinates the efforts of numerous UN agencies in addressing water issues globally, emphasizing the need for integrated approaches that align with the Sustainable Development Goals (SDGs) [33, 34]. The World

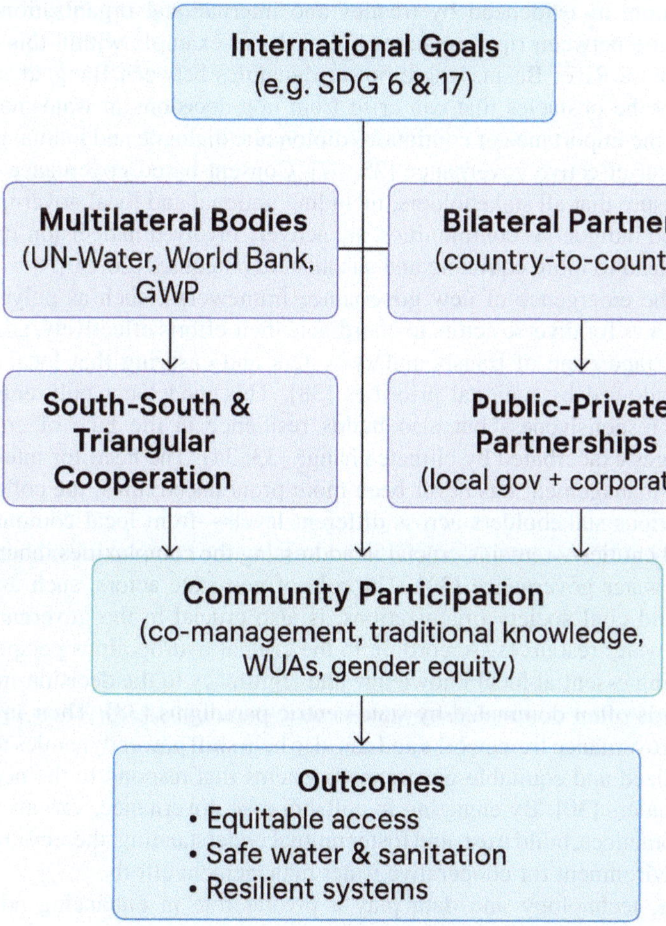

Fig. 14.1 International cooperation framework for sustainable water governance

Health Organization (WHO) focuses on ensuring safe drinking water and sanitation, recognizing their fundamental role in public health [30, 35]. The World Bank serves as a significant financial institution, providing critical funding and technical assistance for water infrastructure projects, with a strong emphasis on sustainability and inclusion [32]. Regional development banks, such as the Asian Development Bank (ADB) and the African Development Bank (AfDB), are also instrumental in financing water management initiatives, highlighting the necessity of localized approaches that resonate with specific regional challenges [25, 36].

In the context of international cooperation for sustainable water governance, transboundary water management represents a significant challenge, given the complexities of competing national interests and ecological necessities. Effective transboundary governance requires a legal framework that promotes cooperation and conflict resolution, as evidenced by treaties and international organizations that facilitate dialogue between riparian states [26, 33]. An example within this realm is the Brahmaputra River Basin, where power dynamics between Bangladesh and India exemplify the obstacles that can arise from non-decisions in water negotiations, stressing the importance of continuous diplomatic dialogue and mutual understanding to foster effective governance [25, 37]. Consent-based governance structures, which ensure that all stakeholders, including national and local governments, civil society, and indigenous communities, are actively involved in decision-making processes, can lead to more equitable and sustainable outcomes [26, 30].

Moreover, the emergence of new governance frameworks, such as polycentric governance, allows for diverse actors to coordinate their efforts effectively, adapting to the intricate landscape of transboundary waters and ensuring that local needs are not overshadowed by national priorities [38]. This model not only enhances flexibility and responsiveness but also builds resilience in the face of environmental challenges exacerbated by climate change [33, 34]. The need for integrated water resource management has never been more pronounced, thus, the collaboration among various stakeholders across different levels—from local communities to international entities—remains crucial in addressing the complexities inherent in transboundary water governance [32]. The role of non-state actors, such as local communities and civil society organizations, is also crucial in the governance of transboundary water resources. According to the critical hydropolitics perspective, these actors bring essential local knowledge and legitimacy to the decision-making process, which is often dominated by state-centric paradigms [38]. Their involvement enriches governance frameworks and can also help shift power dynamics toward more decentralized and equitable governance systems that respond to the needs of diverse stakeholders [30]. By engaging in collaborative governance, various actors can share best practices, build trust, and foster mutual understanding, thereby creating a conducive environment for cooperative water management efforts.

Furthermore, technology and data play a pivotal role in enhancing adaptive management practices within water governance. Information systems can facilitate better monitoring and assessment of water resources, contributing to improved decision-making and accountability [27, 35]. The integration of Geographic Information Systems (GIS) and remote sensing technologies allows for comprehensive spatial analyses of water resources, enabling stakeholders to visualize problems and devise practical solutions collaboratively [31]. Additionally, participatory approaches harnessing social media and digital platforms can engage broader audiences, promoting transparency and inclusiveness in governance [39].

However, the realization of such synergistic partnerships is contingent upon addressing existing barriers, including entrenched power hierarchies and institutional inertia that may impede innovative approaches to water governance [40]. The challenge remains to cultivate an atmosphere of trust and shared responsibility, where collaborative approaches to governance reflect the interests and needs of all stakeholders involved [32, 41]. This often involves a significant cultural shift within institutions that prioritize top-down approaches and may resist the incorporation of community input into governance processes.

The interplay of globalization and local governance dynamics also shapes water management practices, with local governments becoming increasingly involved in international environmental cooperation. They serve as crucial intermediaries between communities and higher-level authorities, facilitating dialogues that translate local realities into broader governance strategies [42]. Such connections enhance local capacities and ensure that policies resonate with the lived experiences of affected communities, thereby creating more impactful water governance frameworks [31, 35]. The efficacy of these partnerships hinges on core principles such as equity, mutual accountability, transparency, and local relevance. At the same time, the role of key international institutions solidifies the foundation needed to support these endeavors. As collective action is crucial in confronting emerging threats to water security, ongoing dialogue and cooperation among governmental, non-governmental, and community stakeholders will remain indispensable in shaping effective governance structures for future water management challenges.

14.3 SDG 17 and the Architecture of Water Partnerships

Achieving Sustainable Development Goal 17 (SDG 17), which emphasizes partnerships for the goals, particularly in the context of water management, necessitates a well-structured architecture of collaboration. The complexity of this goal encompasses core targets, including technology access, financial support, capacity-building, and policy coherence. These targets are pivotal for ensuring that water partnerships are effective, equitable, and sustainable.

14.3.1 Core Targets and Indicators Related to Water

To construct a robust framework for water partnerships, one must prioritize core targets that align with SDG 17. Access to technology is critical for enhancing water management practices. As outlined by Pradhan et al. [43], the interplay between SDG 6 (Clean Water and Sanitation) and technology access illustrates how advancements can support sustainable water use. By fostering technological innovations through partnerships that leverage academic and community resources, stakeholders can create solutions that are both practical and sustainable. The collaborative models suggested by [44] propose a structured approach where partnerships develop shared work plans that are vital for facilitating technology transfer in various contexts, although primarily referenced in educational settings.

Financial support constitutes another crucial component of effective water partnerships. According to [45], financial resources coupled with capacity-building initiatives enhance the performance of water service providers within the framework of

Water Operator Partnerships (WOPs). Effective financial frameworks in partnerships must ensure that resources are allocated efficiently, supporting both infrastructure development and technological implementation. As highlighted in Islam and Rahman's exploration of impact investment [46], aligning financing with SDGs can bridge resource gaps critical for sustainable water practices.

Building capacity through training and development ensures that all stakeholders are equipped with the necessary skills and knowledge to operate efficiently within the partnership. Mazer et al. [47] underscore the importance of continuous education and knowledge-sharing mechanisms in fostering successful water research partnerships. This emphasis on capacity-building aligns with ongoing evaluations of partnerships, which help identify areas needing improvement and the best practices to disseminate among members. Policy coherence is essential to harmonize efforts across different sectors involved in water management. The interaction of various SDGs necessitates comprehensive policies that address systemic challenges. As identified by [48], dual approaches integrating traditional knowledge with scientific insights can lead to more inclusive policy frameworks in water resource management. Establishing clear, coherent policies enhances the ability to implement collaborative initiatives effectively, thereby reinforcing the structure of water partnerships.

14.3.2 Monitoring and Evaluating Partnership Impact

For water partnerships to truly reflect their potential impacts, effective monitoring and evaluation (MandE) frameworks must be put in place. The complexity of interlinkages between SDGs presents challenges in tracking outcomes, particularly in low-resource contexts, as emphasized by [49]. A systematic approach to establishing MandE indicators, as discussed by [50], is vital for evaluating partnerships oriented toward water and sanitation goals. These indicators can help measure not just outputs but also long-term impacts on the communities served.

Moreover, the importance of adaptive evaluation processes that incorporate stakeholder feedback is critical, as highlighted in the qualitative analyses by [51]. Engaging stakeholders from the outset ensures that evaluation metrics are relevant and can track the diverse impacts of water partnerships on different communities. The ability to adapt based on feedback allows these partnerships to remain dynamic and responsive to the evolving needs of the populations they serve. Tracking outcomes in low-resource contexts poses additional challenges, primarily due to the lack of robust data. Plummer et al. [52] assert that foundational priority must be given to establishing reliable data collection systems that inform both policymakers and practitioners in the water sector. This reliance on data ensures that efforts are not only directed toward the most appropriate interventions but also aligned with broader goals of equity and accessibility. Furthermore, as illustrated by [53], collaborative networks among stakeholders can significantly amplify the effectiveness of monitoring initiatives. By involving multiple entities, including civil society, governments, and academic institutions, water partnerships can enhance their capability to track progress effectively while sharing critical insights and methodologies that improve overall impact.

14.3.3 Multi-Stakeholder Engagement for Water and Sanitation

Multi-stakeholder engagement is vital for the success of water and sanitation initiatives. The collaborative nature of partnerships necessitates the involvement of various actors: governments, academic institutions, NGOs, and private sector entities, each contributing unique resources and expertise. Wright-Contreras et al. [54] emphasize that water operator partnerships (WOPs) serve as a model for this multi-stakeholder engagement, enhancing utility performance in urban water supply systems. Moreover, the integration of diverse stakeholders fosters innovation and sustainability in water management solutions. Effective collaboration requires structured frameworks that facilitate communication and resource sharing among partners. Shoniwa et al. [55] point to the significance of establishing clear roles and responsibilities within partnerships, which can lead to more efficient and effective outcomes. By ensuring that all voices are heard and that contributions are recognized, partnerships can maintain active engagement and commitment from all stakeholders.

The commitment to inclusivity in multi-stakeholder approaches can illuminate pathways for addressing disparities in water access and quality. The insights gained from sophisticated engagement strategies, as proposed by [47], not only improve project design but also bolster long-term commitment among stakeholders. Additionally, the focus on institutional development within WOPs, as seen in [56], highlights how multi-stakeholder networks can adaptively respond to socio-political challenges that hinder water access.

Moreover, the role of academia in these partnerships cannot be understated. Research institutions can provide critical data, analytic capabilities, and evaluation frameworks that enhance the effectiveness of collaborative efforts—the twinning partnership model illustrated by [44] exemplify how academic institutions can coalesce with community organizations, creating peer-to-peer relationships that drive innovation and community resilience. The architecture of water partnerships must be fortified by dynamic engagement processes that promote resilience and adaptability. As highlighted by [52], aligning the interests of diverse stakeholders while fostering a collaborative spirit is essential for achieving the broader objectives of SDG 17 and delivering sustainable solutions to global water challenges.

14.4 Technology Transfer and Knowledge Sharing for Water Safety

Discussion surrounding technology transfer and knowledge sharing for water safety in the Global South often includes areas such as innovations in system monitoring, data transparency, and collaborative educational endeavors. Each of these arenas plays a critical role in fostering effective water management strategies that can bridge the North–South technology divide.

14.4.1 Addressing the North–South Technology Divide

Innovations in water purification, testing, treatment, and remote monitoring have become essential tools in addressing the disparities between the Global North and South. For instance, intelligent water monitoring systems based on the Internet of Things (IoT) have emerged as practical solutions that enhance both the management and sustainability of water resources. These systems enable remote monitoring, allowing real-time tracking of water conditions and resource utilization across various sectors, such as agriculture, industry, and domestic use [57]. The scalability of these technologies permits adaptation for diverse contexts, particularly in regions with limited resources or variable environmental conditions, underscoring their relevance in promoting sustainable water resource management in the Global South.

Additionally, understanding the complexities of water governance is essential for effective technology transfer. Frameworks that enhance data collection and monitoring can deter unsustainable practices in water use while increasing accountability among stakeholders [58]. Transparency in governance practices empowers local entities, such as Water User Associations (WUAs), to negotiate better resource-sharing agreements and advocate for sustainable practices [59]. Facilitating the adaptation of such governance frameworks can promote equitable access to water resources, thereby addressing underlying issues associated with the North–South technology divide.

14.4.2 Open Science and Open Data for Water Management

Open Science and open data platforms play a critical role in enhancing transparency and accountability in water management practices. Platforms offering access to data and fostering transparency are essential for collaborative governance and informed decision-making. The use of volunteer water quality data has expanded the scope of monitoring efforts, increasing participation and contributions from local communities [60]. This participatory approach not only enriches the quality of data but also enhances local engagement with water governance. Such engagement is particularly beneficial in the Global South, as local populations often possess valuable indigenous knowledge about water resources.

Moreover, the digital landscape allows stakeholders to share real-time data related to water resource management. For example, digital platforms can integrate satellite and in-situ data to support comprehensive monitoring of water quality in coastal areas [61]. By facilitating real-time data sharing, stakeholders can promptly respond to changes in water quality and quantity, leading to improved management practices that mitigate risks associated with water safety.

Successful models of transparency utilized in public utilities can provide crucial insights for water management in developing nations. Implementations emphasizing transparency can enhance utility responsiveness to community demands, thus fostering a cooperative environment between service providers and users [62, 63]. The effective use of transparency tools underscores the necessity for achieving public value through improved governance structures, which are indispensable for promoting sustainable water management practices.

14.4.3 Global Academic and Training Collaborations

Fostering global academic and training collaborations is vital to enhancing capacities in water management within the Global South. Initiatives that bring together universities and educational institutions can pave the way for joint research, knowledge exchange, and shared best practices among countries. Collaborative work among universities focusing on water governance has shown potential for localized innovation that is culturally sensitive and technically proficient [64, 65]. Such partnerships can create knowledge networks that serve as platforms for discussing pressing water management challenges and exploring adaptive governance strategies.

The potential of fellowships and training programs focusing on water management is significant. These initiatives can offer targeted skill development in areas such as remote sensing, ecological monitoring, and community engagement strategies. This comprehensive capacity building is crucial for equipping local stakeholders with the necessary skills and knowledge to implement effective water management practices [59, 66]. Additionally, understanding adaptive water governance frameworks can aid in managing climate change and environmental degradation complexities more effectively, ensuring resilience in the face of evolving ecological challenges [64].

Joint research ventures also play an instrumental role in addressing local and global water issues. Integrating traditional methodologies with modern technological advancements can yield hybrid solutions that respect local knowledge while promoting innovation [64, 65]. Collaborative research efforts in water resource management not only enhance the understanding of local practices but also yield findings that can inform policy at national and regional levels.

14.5 Financing Sustainable Water Solutions Through Partnerships

Achieving universal access to safe and sustainable water requires innovative financing models and multi-stakeholder collaboration. Strategic partnerships spanning public, private, community, academic, and international actors enable the pooling of financial, technical, and social resources to address water challenges at multiple scales. Table 14.1 outlines various partnership models, funding mechanisms, key actors, and the benefits they offer for delivering inclusive, resilient, and efficient water solutions. Effective financing mechanisms have become increasingly critical in the context of water and sanitation (WASH) programs intended to meet the Sustainable Development Goals (SDGs), specifically SDG 6, which aims to ensure the availability and sustainable management of water and sanitation for all by 2030. This task necessitates a comprehensive review of various financing methods, including official development assistance (ODA), innovative financing mechanisms such as blended finance and SDG-linked bonds, and the contributions made by private foundations and philanthropic organizations.

14.5.1 Official Development Assistance (ODA) and Water Programs

Official Development Assistance has traditionally played a vital role in financing water and sanitation initiatives in developing regions, particularly in the Global South. Donor countries often allocate substantial resources to support WASH programs, motivated by humanitarian concerns and developmental objectives. Recent trends indicate that while ODA for water and sanitation has shown fluctuations, there is a noticeable push towards integrating ODA with sustainable development frameworks. For example, [67] emphasize the importance of multi-stakeholder participation in developing strategies aimed at achieving SDG 6, pointing out that blended financing modalities can enhance the effectiveness of water programs. Similarly, [68] highlight that even established nations have yet to fully formalize comprehensive structures for financing public infrastructure, indicating a persistent gap in both the availability and application of ODA for water and sanitation initiatives.

Moreover, the necessity of adapting financing strategies to local contexts is imperative for maximizing impact. The evolution of community-managed water supply projects, as explored by Behailu et al., indicates that localized approaches can yield better sustainability outcomes compared to traditional top-down management systems [69]. The delivery of ODA funding, therefore, must also adapt to promote community involvement and self-management, which directly contributes to the sustainability of water systems in the long term.

Table 14.1 Financing sustainable water solutions through strategic partnerships

Partnership model	Funding mechanism	Key actors involved	Targeted solutions	Benefits
Public–private partnerships (PPPs)	Co-investment, user fees, service contracts	Government, private sector, and utility companies	Water supply infrastructure, wastewater treatment	Shared risk, improved efficiency, access to private capital
Public-public partnerships (PUPs)	Public grants, inter-agency transfers	Local governments, and public utilities	Capacity building, rural water access	Knowledge sharing, community focus, cost-effectiveness
Community-based partnerships	Community contributions, microfinancing	Local communities, NGOs, and microfinance institutions	Small-scale water projects, maintenance systems	Ownership, social cohesion, local empowerment
Development finance institutions (DFIs)	Concessional loans, blended finance	World Bank, AfDB, and IFIs	Large-scale infrastructure, sanitation systems	Long-term funding, reduced financial risk
Corporate SOCIAL RESPONSIBILIty (CSR)	CSR funds, philanthropic donations	Corporations, and local NGOs	Water access in underserved areas	Brand value, community impact, sustainability credentials
Non-governmental partnerships	Grants, donor funding	NGOs, foundations, and international donors	Hygiene education, clean water provision	Innovation, flexibility, targeted interventions
Research and academic collaborations	Research grants, joint ventures	Universities, research institutes, and public agencies	Water quality monitoring, smart technologies	Evidence-based solutions, innovation, capacity building
Green bonds and climate funds	Environmental bonds, adaptation funds	Governments, investors, and climate finance institutions	Nature-based solutions, resilient infrastructure	Climate-resilient investment, scalability, accountability

14.5.2 *Innovative Financing Mechanisms*

The search for innovative financing mechanisms has gained momentum as traditional funding sources prove inadequate to meet the escalating demands of water and sanitation service delivery. Among these creative strategies, blended finance stands out as a critical solution. This approach leverages different sources of capital—combining grants, concessional loans, and philanthropic investments with private sector financing to address the significant financing gap in the WASH sectors.

According to [70], blended finance facilitates a more attractive investment climate for both public and private stakeholders, catalyzing necessary funds for water projects, especially in low-income and marginalized communities. The implementation of blended finance has demonstrated tangible impacts, as illustrated in Ethiopia, where community initiatives have effectively harnessed mixed funding models to improve water access [69].

SDG-linked bonds represent another promising financial mechanism, enabling countries to attract investments tied directly to their achievement of specified SDG targets. Such bonds can mobilize significant capital influx into water infrastructure projects, as they tie the financial benefits of investment returns to the measurable success of water-related outcomes. The heightened interest in these financial instruments has encouraged development agencies and governments to engage more actively with the investment community, seeking innovative solutions that align financial return on investment with societal benefits [71].

Furthermore, climate resilience financing has also emerged as an essential consideration in water solutions. The intersection of climate finance and water management is critical as climate change increasingly threatens water resources. Investments in climate-resilient infrastructure can help mitigate the risks associated with extreme weather events, thus ensuring long-term sustainability in water access and sanitation systems. The assertions made by [72] regarding the link between climate change and water financing further establish the urgent need for comprehensive strategies that incorporate climate resilience into all water sector financing.

14.5.3 Role of Private Foundations and Philanthropy

The role of private foundations and philanthropic organizations in financing sustainable water solutions cannot be understated. Institutions such as the Bill and Melinda Gates Foundation and Aqua for All have made significant contributions to the water sector, reflecting a growing trend of private investments in public goods. These foundations often provide initial funding that can stimulate local economies and catalyze broader investment from both public and private sectors [73]. Private philanthropic contributions can also fund innovative pilot projects that pave the way for scalable solutions in water and sanitation, as noted by [71], who underscore the need for foundational relationships in mobilizing finance for WASH initiatives.

Furthermore, the integration of private sector dynamics into public–private partnerships (PPPs) has emerged as a viable route for financing water and sanitation infrastructure, particularly in informal settlements where government resources are insufficient. With the increasing complexity of urban water needs, focusing on innovative financing through PPPs allows for diversified funding sources, thus broadening access to sanitation and water services [74]. In contexts where informal settlements are prevalent, this collaborative approach can be efficient, combining public resource allocation with innovative private financing strategies to address unique local challenges [75].

14.6 International Partnerships in Practice: Case Studies from the Global South

In recent discussions related to climate-water action partnerships, the linkages established by the Paris Agreement are pivotal as they provide explicit frameworks for collaborative governance. The Conference of the Parties (COP) has facilitated numerous initiatives aimed at reinforcing water security, particularly through cooperative agreements that address the dual challenges of climate change and water access. Partnerships emerging from COP deliberations illustrate a growing recognition that climate impacts such as altered precipitation patterns and increased water scarcity necessitate robust collective action among countries [76, 77]. This mutually beneficial relationship fosters environmental sustainability and amplifies the potential for achieving the UN SDGs, particularly Goal 6, which focuses on clean water and sanitation.

Furthermore, case studies from organizations such as Water.org and the Global Water Partnership illuminate the practical applications of these frameworks in addressing water and sanitation challenges. Water.org, with its innovative financing mechanisms and community-led projects, showcases how financial inclusivity can lead to sustainable water management practices. Initiatives such as the Water Credit model have provided vulnerable populations in low-income regions with easier access to sanitary facilities and safe drinking water, effectively transforming local health outcomes [78, 79]. Similarly, the African Water Facility has been instrumental in promoting investments in water infrastructure across African nations, underscoring the commitments made through international climate agreements as vehicles for real change [79].

In terms of public health collaborations, the WASH programs have been pivotal, particularly during the COVID-19 pandemic. The WHO's water safety plans have been vital in directing attention to the essential role of safe water in public health, particularly in the wake of infectious disease outbreaks like COVID-19, where access to safe water significantly mitigates health threats [80]. The ongoing pandemic has exacerbated existing water access inequalities, with many communities in the Global South facing heightened risks due to inadequate sanitation infrastructure. Strategies that prioritize WASH, particularly in urban and peri-urban environments, have become increasingly essential in maintaining public health standards during crises [79, 81].

Moreover, the agriculture-water nexus illustrates the intricate relationship between irrigation practices and food security. The Consultative Group on International Agricultural Research (CGIAR) initiatives aimed at improving agricultural sustainability amidst changing climate conditions highlight the integration of water management in food production systems. The linkage between agriculture and water security is crucial, as efficient water usage strategies are vital for improving crop yields and reducing resource waste [82, 83]. These programs demonstrate how adaptive management practices can enhance resilience against climate variability while securing food systems in the Global South. The complexity surrounding water access during the pandemic reveals that systemic challenges require integrated solutions addressing both health and ecological sustainability. Leaders and stakeholders utilized innovative strategies to ensure a consistent water supply, as many regions witnessed severe disruptions in service due to lockdowns and resource constraints [84]. Wastewater management systems experienced heightened scrutiny as emerging pathogens necessitated a re-evaluation of sanitation practices. Targeted interventions such as the establishment of community water points and enhanced sanitation facilities have been crucial in preventing disease transmission [77, 80].

The role of public institutions and local governance in facilitating these collaborative efforts cannot be overstated. Effective partnerships between governmental bodies, NGOs, and local communities play a significant role in ensuring that water policies align with societal needs [85], 86]. Collaborative governance frameworks that prioritize stakeholder engagement and transparency can enhance the resilience of water systems, particularly as communities adapt to the multifaceted impacts of changing climatic conditions.

14.7 Capacity Building and Institutional Strengthening for Water Governance

Capacity building and institutional strengthening are critical components of effective water governance, particularly in the context of decentralized governance, which aims to engage local institutions more robustly in water management initiatives. A comprehensive understanding of capacity building highlights its evolution and application, especially in developing contexts such as North Kalimantan, Indonesia, where efforts to enhance governance capabilities focus on empowering local entities to manage water resources more efficiently and sustainably [87]. The emphasis on local ownership in capacity building further underscores the necessity of developing institutional competencies to mitigate reliance on external bodies, which has been problematic in regions like Nepal, where local authorities face challenges due to over-dependence on NGOs for technical assistance [88].

Decentralization is increasingly recognized as pivotal for effective governance, particularly in the management of natural resources like water. The shift towards localized governance presents unique challenges and opportunities for adaptation, creating a need for strategies that foster both governmental and community engagement [89]. Studies have shown that successful water governance relies on the active engagement of various local stakeholders, thereby enhancing their capacity to impact decision-making processes significantly [90]. A framework that fosters collaboration and stakeholder engagement can lead to substantial improvements in the management of water resources, as seen through community-managed initiatives that allow for participatory approaches to governance and resource allocation [91].

The integration of youth and gender perspectives into water governance frameworks is essential for promoting equitable access to clean water and fostering inclusive programming. Initiatives aimed at empowering youth can facilitate innovative solutions to address water challenges, as their unique perspectives often lead to dynamic problem-solving approaches [92]. Gender-inclusive partnerships also enhance the effectiveness of water governance systems by ensuring that women's voices and needs are represented, which contributes to more equitable access to resources. This is particularly significant in contexts where women are traditionally responsible for water procurement in households, making their involvement in governance processes critical for enhancing community resilience and sustainability [92].

Furthermore, the concept of building resilience in fragile and conflict-affected areas underscores the importance of water resource management as a tool for peacebuilding. In post-disaster scenarios, water interventions can play a vital role in enhancing stability and cooperation among communities, as access to safe water is fundamental for health and recovery [93]. Programmatic approaches that prioritize both environmental sustainability and conflict resolution can leverage water supply as a means to encourage collaboration among diverse groups, thus contributing to peacebuilding efforts [94]. Such integrated strategies benefit from frameworks that highlight the adaptive capacity of local institutions, enabling them to respond dynamically to the challenges posed by both natural disasters and socio-political conflicts [95].

More broadly, the ongoing challenges related to climate change further necessitate the strengthening of water governance capacities. Effective climate adaptation strategies must be embedded within local governance frameworks to manage the complexities associated with climate variability [89]. The recognition of climate change as a critical driver of water resource challenges highlights the essentiality of adaptive governance principles that integrate scientific knowledge with local experiences to formulate robust water management policies [94]. The emergent understanding is that governance models must evolve from traditional, top-down approaches to more collaborative, bottom-up strategies, allowing for greater adaptability in response to environmental stresses [95].

Additionally, the advancement of technological solutions is pivotal for enhancing the capacity of governance frameworks. Digital tools that enable better data management and foster information-sharing platforms can augment the institutional capacities needed for effective public service delivery [96]. The integration of technology in governance processes not only streamlines operations but also empowers local stakeholders by providing them with the knowledge and tools necessary to engage effectively with water governance systems [97]. To achieve effective water governance, particularly in decentralized contexts, a multifaceted approach that integrates capacity building, youth engagement, gender inclusion, and resilience strategies is imperative. By promoting a collaborative ethos within local governance structures, stakeholders can enhance their operational capabilities and drive significant progress towards sustainable water resource management. Through partnerships that bridge resource gaps and foster inclusive participation, communities can create a more resilient and equitable framework for water governance, ultimately leading to improved health and well-being for all [92].

14.8 Challenges and Critiques of International Water Partnerships

14.8.1 Unequal Power Dynamics and Dependency Risks

International water partnerships frequently exhibit unequal power dynamics, posing significant challenges, particularly regarding donor-driven agendas. The actors involved often reflect disparities in authority and influence, leading to dependencies that undermine local agency. Countries in the Global South, for instance, may find their strategic needs overshadowed by the priorities of wealthier donor nations. Mazer et al. [47] discussed how these partnerships, especially those involving academic institutions, can inadvertently impose external methodologies that do not align with the local context, thus curtailing genuine local participation and ownership in water governance initiatives. This dynamic can foster a sense of disempowerment among local stakeholders who may feel their insights and essential knowledge regarding their water resource management are overlooked or undervalued.

Furthermore, Iran's experience with regional water partnerships underscores the complexities surrounding unequal access to shared water resources. The lack of clear rules and stable agreements among transnational partners intensifies these dynamics, often leaving developing nations in precarious positions where they are heavily reliant on agreements shaped by more powerful entities [98]. Such dependency jeopardizes the sustainability of water resources and can lead to broader geopolitical tensions, particularly in arid regions. These unequal power dynamics are exacerbated by the nature of funding models in these partnerships. Short-term financial commitments often prioritize immediate results over long-term sustainability, which discourages meaningful local agency [99]. In the case of Da Nang, Vietnam, partnerships with Dutch water operators demonstrated that while external support can improve infrastructure, it can also create dependency if local operators do not develop the necessary competencies and autonomy to sustain improvements independently [54]. Therefore, seeking balance in partnerships is crucial for fostering equitable power dynamics that truly empower local stakeholders and promote sustainable water management.

14.8.2 Transparency, Accountability, and Impact Measurement

Gaps in transparency and accountability frequently compromise the effectiveness of international water partnerships. These deficiencies complicate the evaluation of long-term sustainability and overall impact. Alford and Koeman discuss the crucial role of interdisciplinary approaches and transparency in ensuring that partnerships effectively address global water challenges [100]. Their case study from Liberia illustrates how limited transparency in partnerships often leads to mistrust and hampers collaborative efforts aimed at enhancing community water access.

Additionally, effective impact measurement within these partnerships is challenging. Chen and Yen emphasize that tracking and highlighting achievements in international collaborations are hindered by difficulties in quantifying success and establishing relevant indicators [101]. This inadequacy can obscure the actual benefits derived from these partnerships, leading stakeholders to question the effectiveness of their engagements. Data collection and analysis frameworks remain insufficiently integrated, which complicates accountability [102]. The lack of universal standards for measuring success within water partnerships often results in discrepancies in reporting and understanding partner contributions, undermining the credibility of these initiatives.

Moreover, there is a pressing need for better-defined frameworks that not only evaluate immediate outcomes but also assess long-term sustainability and community impact post-engagement. Wright-Contreras et al. [54] highlight the importance of aligning the objectives of local partners with those of external stakeholders. Misalignment can lead to oversight of local needs, resulting in wasted resources and failed initiatives [54]. Evaluating partnerships requires continuous engagement with local communities to ensure that impact measures developed are relevant and practical, addressing genuine challenges faced by these communities.

14.8.3 Partnership Fragility and Funding Volatility

The fragility of international water partnerships is exacerbated by funding volatility, which often leads to short-term engagements and a lack of continuity in efforts to manage water resources sustainably. This pattern is particularly concerning in the context of the Lancang-Mekong Cooperation, where fluctuating financial commitments complicate long-term planning and cooperative management of transboundary water resources [103]. The constant negotiation for funding can divert focus away from achieving substantial results towards merely securing finances, compromising the integrity and longevity of water partnerships.

Short-term project cycles hinder the establishment of trust and shared commitment required for effective collaboration. As highlighted by [103], interventions tend to be episodic rather than integrative, creating a fragmented approach to water resource management. Furthermore, without sustained financial support, initiatives that demonstrate potential success may dissipate, undermining the positive contributions previously made by all stakeholders and generating skepticism towards future collaborative efforts.

Additionally, the evaluation of these partnerships often reveals gaps where funding issues intersect with governance challenges. The lack of ongoing assessment systems leads to an inability to capture lessons learned from prior initiatives, perpetuating a cycle of short-lived partnerships. Tyukhtenko et al. [104] argue that effective governance frameworks are essential to mitigating such risks, proposing that inclusivity and active stakeholder engagement can improve resilience against funding fluctuations.

14.9 Future Pathways for Strengthening International Water Partnerships

Addressing global water challenges requires coordinated international collaboration grounded in equity, innovation, and shared responsibility. Strengthening partnerships across borders, sectors, and disciplines can enhance resource governance, promote knowledge exchange, and mobilize sustainable financing. Table 14.2 outlines key strategic focus areas, actionable steps, and expected outcomes for building more resilient and effective international water partnerships.

Table 14.2 Sustainable pathways for strengthening international water partnerships

Strategic focus	Key actions	Expected outcomes
Policy harmonization	Align national and regional water policies	Reduced conflicts, improved transboundary cooperation
Technology sharing	Promote joint RandD, open-source water tech platforms	Accelerated innovation, cost-effective solutions
Capacity building	Train stakeholders across countries and sectors	Stronger local implementation, knowledge transfer
Data and information exchange	Establish global water data hubs and real-time sharing tools	Enhanced monitoring, informed decision-making
Inclusive governance	Engage civil society, women, and indigenous groups	Equitable access, social sustainability
Climate-resilient financing	Mobilize blended finance, climate funds, and water bonds	Scaled investment in resilient infrastructure
Conflict resolution mechanisms	Set up cross-border legal frameworks and mediation platforms	Peaceful water resource management
South–south and triangular cooperation	Foster partnerships among developing countries	Mutual learning, locally adaptable solutions

14.9.1 Decolonizing Water Development

Decolonizing water development entails challenging colonial legacies that have historically marginalized Indigenous perspectives on water governance. The shift towards empowering local leadership and co-creation of water solutions is essential for fostering genuine partnerships. Decolonization emphasizes the incorporation of traditional knowledge and the recognition of water as a living entity, a perspective often upheld by Indigenous communities [105]. Scholars argue for a restructuring of water governance systems to include these ontological views, which contrast with the commodity-focused lens of mainstream water management [106].

Research indicates that the engagement of local communities in decision-making processes can enhance water governance outcomes. For instance, studies have shown that institutional frameworks neglecting Indigenous ways of knowing can lead to unsustainable water management practices [107]. The reconstruction of trust through collaborative approaches has been identified as a fundamental step in decolonizing water governance. This involves active listening to and partnership with Indigenous communities [108]. These partnerships must move beyond tokenistic involvement to enact meaningful change rooted in equity and respect for local knowledge systems [106].

Moreover, incorporating southern perspectives in water management allows for adaptive, community-driven solutions that respect local customs and practices. As noted, the Calgary Regional Partnership illustrates a collaborative approach that integrates Indigenous knowledge with regional water management strategies [109]. Such partnerships can help bridge the gap between conventional resource management and the rich ecological wisdom of Indigenous cultures, ultimately leading to more sustainable and equitable water solutions.

14.9.2 Leveraging Digital Technologies and AI

The potential of digital technologies and artificial intelligence (AI) in enhancing water access and management practices is increasingly recognized. Technologies such as smart monitoring systems and predictive analytics can play a significant role in addressing the complex dynamics of water distribution and availability. These tools not only facilitate real-time data collection but also enable more informed decision-making, which is crucial in water-scarce regions [110]. The digital mapping of water resources has become an essential practice for understanding water flows and resource distribution, allowing stakeholders to manage scarcity more effectively [111].

Digital innovation aligns with the need for systemic resilience in the face of climate change impacts. For instance, integrating environmental data into decision support systems can improve planning and response strategies for water-related disasters. The capacity of these tools to process vast amounts of data can help identify patterns that inform better management practices, ultimately promoting sustainability [112]. However, it is imperative that these technologies are deployed equitably and do not exacerbate existing inequalities in water access and governance. The smart use of technology must also prioritize local community involvement, ensuring that the benefits of technical advancements are equitably distributed. As emphasized in various studies, the integration of technology into existing water governance frameworks should complement and support local knowledge rather than replace it [111, 113]. Building local capacity to utilize these technologies effectively is essential to empower communities and enhance their resilience against the challenges of climate variability.

14.9.3 Integrating Water Partnerships Within Planetary Boundaries and just Transitions

Aligning water partnerships with planetary boundaries introduces a critical dimension to sustainable water management. The recognition that water management cannot be disassociated from broader environmental goals, such as climate justice and biodiversity conservation, calls for an integrated approach [114]. The concept of just transitions embeds equity within the sustainability agenda, advocating for governance structures that support both environmental and social outcomes [115]. Research indicates that the interconnectedness of ecosystems necessitates collaboration across various sectors, including water, energy, and land use [116]. Holistic approaches enable stakeholders to address water management challenges within the context of sustainable development goals (SDGs) and the increasing pressures of climate change [110, 112]. Therefore, partnerships must consider ecological limits and adopt principles of collective stewardship to ensure that water resources are managed sustainably for future generations.

The focus on biodiversity as a nexus component further highlights the importance of integrating ecological considerations into water management practices. Collaborative efforts that prioritize habitat conservation and restoration can yield significant benefits for both water quality and biodiversity [117]. Acknowledging the intricate relationships between water resources and ecosystem health can guide policy-making and action toward more sustainable outcomes. Additionally, innovative financing mechanisms such as blended funds and payments for ecosystem services can further enhance water security while supporting conservation efforts [118]. The global community has recognized the need for a diverse range of financing sources to tackle pressing water challenges, emphasizing that financial models must be designed to promote inclusivity and sustainability [118].

14.10 Conclusion

Access to clean and safe water remains one of the most pressing development challenges of our time, especially in the Global South, where inequalities, infrastructure deficits, and environmental vulnerabilities intersect. International partnerships, guided by the Sustainable Development Goals (particularly SDGs 6 and 17), hold transformative potential for addressing these challenges. From technology transfer and financing mechanisms to collaborative research and policy development, global partnerships offer a robust framework for bridging resource gaps, accelerating innovation, and strengthening institutional capacity. Yet, while partnerships have catalyzed notable progress, they are not without their shortcomings. Power asymmetries, lack of contextual sensitivity, short-term funding cycles, and fragmented coordination can undermine impact and erode trust. To move from transactional arrangements to transformative cooperation, future partnerships must be grounded in equity, co-creation, and local ownership. Efforts must also align with broader climate resilience, justice, and biodiversity agendas to ensure water systems are not just improved but sustained over time.

The path to universal water safety and quality in the Global South demands more than good intentions; it requires strategic, accountable, and future-ready partnerships. As the global community enters the final stretch toward the 2030 Agenda, this chapter calls for a renewed commitment to integrative, inclusive, and resilient international cooperation that not only delivers measurable progress on water-related targets but also reshapes the very architecture of global solidarity. In doing so, we can turn shared challenges into shared solutions, advancing both water justice and sustainable development for all.

References

1. Grossman D, Humphreys B, Ruseski J (2017) Out of the outhouse: the impact of place-based policies on dwelling characteristics in Appalachia. SSRN Electron J. https://doi.org/10.2139/ssrn.2952400
2. Dakyaga F, Kyessi A, Msami J (2018) Households' assessment of the water quality and services of the multi-model urban water supply system in the informal settlements of Dar es Salaam, Tanzania. J Civ Eng Arch 12(5). https://doi.org/10.17265/1934-7359/2018.05.005
3. Crawford S, Cofalla C, Aumeier B, Brinkmann M, Classen E, Esser V, Ganal C, Kaip E, Haussling R, Lehmkuhl F, Letmathe P, Hollert H (2017) Project house water: a novel interdisciplinary framework to assess the environmental and socioeconomic consequences of flood-related impacts. Environ Sci Eur 29(1). https://doi.org/10.1186/s12302-017-0121-1
4. Kosoe EA, Osumanu IK, Ogwu MC (2025b) Reimagining rural and small-town water supply in Ghana through inclusive governance and sustainable models. J Sustain 1(2):1–21. https://doi.org/10.55845/jos-2025-1234
5. Kosoe EA, Ogwu MC, Duniah J, Ndigun FB (2025) Sustaining indigenous water governance systems in Ayimaye and Lungni, Northern Ghana, through hybrid models. J Sustain 1(2). https://doi.org/10.55845/jos-2025-1235
6. Osei L, Amoyaw J, Boateng G, Boamah S, Luginaah I (2015) The paradox of water accessibility: understanding the temporal and spatial dimensions of access to improved water sources

in Rwanda. J Water SanitIon Hyg Dev 5(4):553–564. https://doi.org/10.2166/washdev.2015.029

7. Abaasa C, Ayesiga S, Rukundo G, Lejju J, Byarugaba F, Tamwesigire I (2023) Community perceptions and practices on quality and safety of drinking water in Mbarara City, South Western Uganda. Plos Water 2(5):e0000075. https://doi.org/10.1371/journal.pwat.0000075
8. Rudolph M, Muchesa E (2023) Integrated and sustainable water and sanitation systems at two rural sites in South Africa. Water 15(22):3974. https://doi.org/10.3390/w15223974
9. John S, Olukotun S, Kupi T, Mathuthu M (2024) Health risk assessment of heavy metals and physicochemical parameters in natural mineral bottled drinking water using ICP-MS in South Africa. Appl Water Sci 14(9). https://doi.org/10.1007/s13201-024-02267-3
10. Ruiters C, Matji M (2016) Public–private partnership conceptual framework and models for the funding and financing of water services infrastructure in municipalities from selected provinces in South Africa. Water Sa 42(2):291. https://doi.org/10.4314/wsa.v42i2.13
11. Graham C, Akligoh H, Ori J, Adzaho G, Salekwa L, Campbell P, Saba CK, Landrain TE, Santolini M (2023) Education-based grant programmes for bottom-up distance learning and project catalysis: antimicrobial resistance in sub-Saharan Africa. Access Microbiol 5(3). https://doi.org/10.1099/acmi.0.000472.v3
12. Wantzen K, Alves C, Badiane S, Bala R, Blettler M, Callisto M, Cao Y, Kolb M, Kondolf GM, Leite MF, Macedo DR, Zingraff-Hamed A (2019) Urban stream and wetland restoration in the global south—a DPSIR analysis. Sustainability 11(18):4975. https://doi.org/10.3390/su11184975
13. Antohi V, Zlati M, Ionescu R, Neculiţă M, Rusu R, Constantin A (2020) Attracting European funds in the Romanian economy and leverage points for securing their sustainable management: a critical auditing analysis. Sustainability 12(13):5458. https://doi.org/10.3390/su12135458
14. Fujiwara N (2019) International city network and public-private cooperation for urban water-environment management: a study of Japanese public water services' overseas expansion, 2019/09. https://doi.org/10.25518/ciriec.wp201909
15. Ogwu MC, Kosoe EA (2024) Place of cultural diversity in sustainable water resource management in Ghana. In: Izah SC, Ogwu MC, Loukas A, Hamidifar H (eds) Water crises and sustainable management in the global south. Springer, Singapore. https://doi.org/10.1007/978-981-97-4966-9_14
16. Ogwu MC, Kosoe EA (2025) Integrating green infrastructure into sustainable agriculture to enhance soil health, biodiversity, and microclimate resilience. Sustainability 17(9):3838. https://doi.org/10.3390/su17093838
17. Mathuthu M, Olobatoke R (2016) Assessment of heavy metals and radionuclide concentrations in the Mafikeng wastewater treatment plant. J Water Resour Prot 08(01):93–99. https://doi.org/10.4236/jwarp.2016.81008
18. Oketch J, Mugalavai E, Nyandiko N (2021) Determinants of water-funded projects in mitigating water risks in Vihiga County, Kenya. Int J Pathog Res 15–26. https://doi.org/10.9734/ijpr/2021/v6i330162
19. Lolagari H, Daneshvar A, Zaj M, Roodposhti F (2022) Sustainable financing model considering project risk. Discret Dyn Nat Soc 2022(1). https://doi.org/10.1155/2022/2838913
20. Hagarsu A, Wanyonyi L, Kikwatha R (2020) Investigating major drivers of performance in community water projects: a case of water projects in Saku Sub-County, Marsabit County, Kenya. Eur J Bus Manag Res 5(5). https://doi.org/10.24018/ejbmr.2020.5.5.530
21. Neely K, Walters J (2016) Using causal loop diagramming to explore the drivers of the sustained functionality of rural water services in Timor-Leste. Sustainability 8(1):57. https://doi.org/10.3390/su8010057
22. Klagge B, Nweke-Eze C (2020) Financing large-scale renewable-energy projects in Kenya: investor types, international connections, and financialization. Geogr Ann Ser B Hum Geogr 102(1):61–83. https://doi.org/10.1080/04353684.2020.1729662
23. Izah SC, Ogwu MC (2025) Modeling solutions for microbial water contamination in the global south for public health protection. Front Microbiol 16:1504829. https://doi.org/10.3389/fmicb.2025.1504829

24. Roy A, Ghosh S, Dan K, Sai G, Maity S, Das M, Gupta D, Paul J, Pal K, Bandhyopadhay S, Haldar S (2022) Are we paying more to consume substandard drinking water? Crowd-sourced water education study in a South Asian megacity. Eur J Dev Stud 2(5):52–60. https://doi.org/10.24018/ejdevelop.2022.2.5.172
25. Barua A, Deka A, Gulati V, Vij S, Liao X, Qaddumi H (2019) Re-interpreting cooperation in transboundary waters: bringing experiences from the Brahmaputra basin. Water 11(12):2589. https://doi.org/10.3390/w11122589
26. Leb C (2014) One step at a time: international law and the duty to cooperate in the management of shared water resources. Water Int 40(1):21–32. https://doi.org/10.1080/02508060.2014.978972
27. Kloke-Lesch A (2020) The untapped functions of international cooperation in the age of sustainable development, pp 127–163. https://doi.org/10.1007/978-3-030-57938-8_7
28. Ogwu MC (2025) Sustainable development in Africa: challenges, innovations, and policy pathways for the future. Afr Res Rep 1(2):55–69. https://doi.org/10.5897/ARR2025.0005
29. Tjandradewi B, Marcotullio P, Kidokoro T (2006) Evaluating city-to-city cooperation: a case study of the Penang and Yokohama experience. Habitat Int 30(3):357–376. https://doi.org/10.1016/j.habitatint.2004.10.001
30. Mirumachi N, Wyk E (2010) Cooperation at different scales: challenges for local and international water resource governance in South Africa. Geogr J 176(1):25–38. https://doi.org/10.1111/j.1475-4959.2009.00344.x
31. Pei P, Zhang J (2013) Research for transboundary water pollution cooperation. Adv Mater Res 864–867:1525–1530. https://doi.org/10.4028/www.scientific.net/amr.864-867.1525
32. Vo T, Tran T, Duy L (2017) Water governance for sustainable development: international practices and implications for the Mekong Delta region. J Econ Dev 24(4):99–120. https://doi.org/10.24311/jed/2017.24.4.6
33. Leb C (2013) Cooperation in the law of transboundary water resources. https://doi.org/10.1017/cbo9781139565752
34. Orme M, Cuthbert Z, Sindico F, Gibson J, Bostic R (2015) Good transboundary water governance in the 2015 sustainable development goals: a legal perspective. Water Int 40(7):969–983. https://doi.org/10.1080/02508060.2015.1099083
35. Önder Z, Özyıldırım S (2011) Political connection, bank credits, and growth: evidence from Turkey. World Econ 34(6):1042–1065. https://doi.org/10.1111/j.1467-9701.2011.01342.x
36. Galindo A, Panizza U (2018) The cyclicality of international public sector borrowing in developing countries: does the lender matter? World Dev 112:119–135. https://doi.org/10.1016/j.worlddev.2018.08.007
37. Vij S, Warner J, Biesbroek R, Groot A (2019) Non-decisions are also decisions: power interplay between Bangladesh and India over the Brahmaputra River. Water Int 45(4):254–274. https://doi.org/10.1080/02508060.2018.1554767
38. Suhardiman D, Giordano M, Molle F (2012) Scalar disconnect: the logic of transboundary water governance in the Mekong. Soc Nat Resour 25(6):572–586. https://doi.org/10.1080/08941920.2011.604398
39. Lin J, Tian Y, Yao Q, Shi Y (2023) Structural characteristics of intergovernmental water pollution control cooperation networks using social network analysis and GIS in the Yangtze River Delta urban agglomeration, China. Sustainability 15(18):13655. https://doi.org/10.3390/su151813655
40. Fliervoet J, Geerling G, Mostert E, Smits A (2015) Analyzing collaborative governance through social network analysis: a case study of river management along the Waal River in the Netherlands. Environ Manag 57(2):355–367. https://doi.org/10.1007/s00267-015-0606-x
41. Lee J (2014) The governance of wetland ecosystems and the promotion of transboundary water cooperation—opportunities presented by the Ramsar convention. Water Int 40(1):33–47. https://doi.org/10.1080/02508060.2014.989681
42. Nakamura H, Elder M, Mori H (2011) The surprising role of local governments in international environmental cooperation. J Environ Dev 20(3):219–250. https://doi.org/10.1177/1070496511415644

43. Pradhan P, Costa L, Rybski D, Lucht W, Kropp J (2017) A systematic study of sustainable development goal (SDG) interactions. Earth's Futur 5(11):1169–1179. https://doi.org/10.1002/2017ef000632
44. Siraj S, Momand B, Brunton G, Dubrowski A (2024) Identification of a partnership model between a university and not-for-profit organization to address health professions education and health inequality gaps through simulation-based education: a scoping review. PLoS ONE 19(10):e0311349. https://doi.org/10.1371/journal.pone.0311349
45. Malemba E, Chiumya D, Anyanwu I, Camkin J (2024) Enabling policy environment for water operator partnerships in the Southern African development community. World Water Policy 10(2):511–523. https://doi.org/10.1002/wwp2.12173
46. Islam S, Rahman A (2023) Impact investment deal flow and sustainable development goals: mind the gap? Account Financ 63(4):3797–3813. https://doi.org/10.1111/acfi.13068
47. Mazer K, Erwin A, Popovici R, Delgado E, Bowling L, Ma Z, Prokopy J Contemp Water Res Educ LS, Zeballos-Velarde C (2020) Creating a collaboration framework to evaluate international university-led water research partnerships. J Contemp Water Res Educ 171(1):9–26. https://doi.org/10.1111/j.1936-704x.2020.3342.x
48. Baldwin C, Bradford L, Carr M, Doig L, Jardine T, Jones P, Bharadwaj L, Lindenschmidt K (2017) Ecological patterns of fish distribution in the Slave River Delta region, Northwest Territories, Canada, as relayed by traditional knowledge and western science. Int J Water Resour Dev 34(2):305–324. https://doi.org/10.1080/07900627.2017.1298516
49. Fu B, Wang S, Zhang J, Hou Z, Li J (2019) Unravelling the complexity in achieving the 17 sustainable development goals. Natl Sci Rev 6(3):386–388. https://doi.org/10.1093/nsr/nwz038
50. Ibrahim G, Cadotte D, Bernstein M (2015) A framework for the monitoring and evaluation of international surgical initiatives in low- and middle-income countries. PLoS ONE 10(3):e0120368. https://doi.org/10.1371/journal.pone.0120368
51. Rosewarne E, Chislett W, McKenzie B, Reimers J, Jolly K, Corben K, Trieu K, Webster J (2021) Stakeholder perspectives on the effectiveness of the Victorian salt reduction partnership: a qualitative study. BMC Nutr 7(1). https://doi.org/10.1186/s40795-021-00414-6
52. Plummer R, Witkowski S, Smits A, Dale G (2021) The issue of performance in higher education institutions—community partnerships: a Canadian perspective. J High Educ Policy Manag 43(5):537–556. https://doi.org/10.1080/1360080x.2020.1858386
53. Beck A (2018) Water operator partnerships: peer learning and the politics of solidarity in water and sanitation service provision. Wiley Interdiscip Rev Water 6(1). https://doi.org/10.1002/wat2.1324
54. Wright-Contreras L, Perkins J, Pascual M, Soppe G (2019) Water operators' partnerships and their supporting role in the improvement of urban water supply in Da Nang. Int J Water Resour Dev 36(1):1–26. https://doi.org/10.1080/07900627.2019.1625753
55. Shoniwa B, Terera K (2024) Enhancing the public-private-community partnerships in the banana value chain in Mutasa District, Zimbabwe. Res World Agric Econ 420–436. https://doi.org/10.36956/rwae.v5i4.1252
56. Milner K, Salazar R, Bhopal S, Brentani A, Britto P, Dua T, Gladstone M, Goh E, Hamadani J, Hughes R, Kirkwood B, Lawn J (2019) Contextual design choices and partnerships for scaling early child development programmes. Arch Dis Child 104(Suppl 1):S3–S12. https://doi.org/10.1136/archdischild-2018-315433
57. Zehui W, Chunzhi S, Yanbin M, Wang Y, Jingxue Y (2023) An intelligent water monitoring system based on the internet of things. Adv Comput Signals Syst 7(4). https://doi.org/10.23977/acss.2023.070403
58. Bea M, López-Gunn E, Caño L (2014) Forensic water governance? Analyzing land and water use changes in north-western doñana (Spain). Int J Water GovAnce 2(2):133–152. https://doi.org/10.7564/14-ijwg62
59. Richards N (2019) Water users associations in Tanzania: local governance for whom? Water 11(10):2178. https://doi.org/10.3390/w11102178

60. Loperfido J, Beyer P, Just C, Schnoor J (2010) Uses and biases of volunteer water quality data. Environ Sci Technol 44(19):7193–7199. https://doi.org/10.1021/es100164c
61. Arabi B, Salama M, Pitarch J, Verhoef W (2020) Integration of in-situ and multi-sensor satellite observations for long-term water quality monitoring in coastal areas. Remote Sens Environ 239:111632. https://doi.org/10.1016/j.rse.2020.111632
62. Douglas S, Meijer A (2016) Transparency and public value—analyzing the transparency practices and value creation of public utilities. Int J Public Adm 39(12):940–951. https://doi.org/10.1080/01900692.2015.1064133
63. Ogwu MC, Izah SC, Sawyer WE, Amabie T (2025) Environmental risk assessment of trace metal pollution: a statistical perspective. Environ Geochem Health 47:94. https://doi.org/10.1007/s10653-025-02405-z
64. Akamani K (2023) The roles of adaptive water governance in enhancing the transition towards ecosystem-based adaptation. Water 15(13):2341. https://doi.org/10.3390/w15132341
65. Hornidge A, Oberkircher L, Tischbein B, Schorcht G, Bhaduri A, Manschadi A (2011) Reconceptualizing Water Management in Khorezm, Uzbekistan. Nat Resour Forum 35(4):251–268. https://doi.org/10.1111/j.1477-8947.2011.01391.x
66. Sdiri A, Pinho J, Ratanatamskul C (2018) Water resource management for sustainable development. Arab J Geosci 11(6). https://doi.org/10.1007/s12517-018-3411-z
67. Quynh V, Anh H, Hoang T, Minh H (2020) Review of public financing for water, sanitation, and hygiene sectors in Vietnam. Environ Health Insights 14. https://doi.org/10.1177/1178630220938396
68. Mundonde J, Makoni P (2024) Framework model for financing sustainable water and sanitation infrastructure in Zimbabwe. Water 16(12):1691. https://doi.org/10.3390/w16121691
69. Behailu B, Suominen A, Katko T (2015) Evolution of community-managed water supply projects from 1994 to the 2010s in Ethiopia. Public Works Manag Policy 20(4):379–400. https://doi.org/10.1177/1087724x15593955
70. Godfrey S, Wambugu M, Parikh P, Tunhuma F (2022) Validation of the UNICEF fiscal diagnostic tool for SDGs 6.1 and 6.2 in East and Southern Africa using the analytical hierarchy process (AHP). J Water SanitIon Hyg Dev 12(10):721–736. https://doi.org/10.2166/washdev.2022.098
71. Pories L, Fonseca C, Delmon V (2019) Mobilising finance for wash: getting the foundations right. Water 11(11):2425. https://doi.org/10.3390/w11112425
72. Alaerts G (2019) Financing for water—water for financing: a global review of policy and practice. Sustainability 11(3):821. https://doi.org/10.3390/su11030821
73. Agarwal M, Saha R (2021) Water and sanitation: achievements in large Indian states. Indian J Hum Dev 15(1):82–99. https://doi.org/10.1177/09737030211001767
74. Atika S, Walubuka E, Nguta M (2024) Factors affecting financing of communal sanitation facilities by the public sector in informal settlements: a case of Nyalenda informal settlement, Kisumu City. Afr J Sci Technol Soc Sci 2(2):67–73. https://doi.org/10.58506/ajstss.v2i2.161
75. Simiyu S, Swilling M, Rheingans R, Cairncross S (2017) Estimating the cost and payment for sanitation in the informal settlements of Kisumu, Kenya: a cross-sectional study. Int J Environ Res Public Health 14(1):49. https://doi.org/10.3390/ijerph14010049
76. Angoua E, Dongo K, Templeton M, Zinsstag J, Bonfoh B (2018) Barriers to access improved water and sanitation in poor peri-urban settlements of Abidjan, côte d'Ivoire. PLoS ONE 13(8):e0202928. https://doi.org/10.1371/journal.pone.0202928
77. Warner M, Zhang X, Rivas M (2020) Which states and cities protect residents from water shutoffs in the COVID-19 pandemic? Utilities Policy 67:101118. https://doi.org/10.1016/j.jup.2020.101118
78. Kumpel E, Billava N, Nayak N, Ercümen A (2021) Water use behaviors and water access in intermittent and continuous water supply areas during the covid-19 pandemic. J Water Health 20(1):139–148. https://doi.org/10.2166/wh.2021.184
79. Smiley S, Agbemor B, Adams E, Tutu R (2020) COVID-19 and water access in sub-Saharan Africa: Ghana's free water directive may not benefit water-insecure households. Afr Geogr Rev 39(4):398–404. https://doi.org/10.1080/19376812.2020.1810083

80. Desye B (2021) COVID-19 pandemic and water, sanitation, and hygiene: impacts, challenges, and mitigation strategies. Environ Health Insights 15. https://doi.org/10.1177/11786302211029447
81. Bauza V, Sclar G, Bisoyi A, Majorin F, Ghugey A, Clasen T (2021) Water, sanitation, and hygiene practices and challenges during the COVID-19 pandemic: a cross-sectional study in rural Odisha, India. Am J Trop Med Hyg 104(6):2264–2274. https://doi.org/10.4269/ajtmh.21-0087
82. Kent K, Alston L, Murray S, Honeychurch B, Visentin D (2022) The impact of the COVID-19 pandemic on rural food security in high-income countries: a systematic literature review. Int J Environ Res Public Health 19(6):3235. https://doi.org/10.3390/ijerph19063235
83. Vanegas S (2025) Native American resilience to protect family nutrition during a pandemic: a qualitative analysis. https://doi.org/10.1101/2025.06.04.25329007
84. Pfaar O, Agache I, Bonini M, Brough H, Chivato T, Giacco S, Gawlik R, Gelincik A, Hoffman-Sommergruber K, Jutel M, Klimek L, Álvaro-Lozano M (2021) COVID-19 pandemic and allergen immunotherapy—an EAACI survey. Allergy 76(11):3504–3516. https://doi.org/10.1111/all.14793
85. Chin E, Leung K, Simas T, Kumaraswami T (2023) Changes in postpartum contraception utilization rates during the early stage of the COVID-19 pandemic. Women's Health 19. https://doi.org/10.1177/17455057231175311
86. John-Henderson N, Oosterhoff B, Johnson L, Lafromboise M, Malatare M, Salois E (2022) COVID-19 and food insecurity in the Blackfeet tribal community. Food Secur 14(5):1337–1346. https://doi.org/10.1007/s12571-022-01292-x
87. Daniel O (2023) Human resources professionalism in public administration: efforts to achieve good governance and organizational performance improvement in the era of bureaucratic reform. Kne Soc Sci. https://doi.org/10.18502/kss.v8i17.14155
88. Malla S, Dahal R, Hasegawa S (2020) Analyzing the disaster response competency of the local government officials and the elected representatives in Nepal. Geoenvironmental Disasters 7(1). https://doi.org/10.1186/s40677-020-00153-z
89. Ojwang L, Rosendo S, Celliers L, Obura D, Muiti A, Kamula J, Mwangi M (2017) Assessment of coastal governance for climate change adaptation in Kenya. Earth's Futur 5(11):1119–1132. https://doi.org/10.1002/2017ef000595
90. Megdal S, Eden S, Shamir E (2017) Water governance, stakeholder engagement, and sustainable water resources management. Water 9(3):190. https://doi.org/10.3390/w9030190
91. Das P (2016) Uncharted waters: navigating new configurations for urban service delivery in India. Environ Plan A: Econ Space 48(7):1354–1373. https://doi.org/10.1177/0308518x16640529
92. Grant M, Nguyen T, Vieira A, Niner S, Roche C (2023) Working together: a study of civil society partnerships between WASH (water, sanitation, and hygiene) and GESI (gender equality and social inclusion) organisations in Timor-Leste. Front Water 5. https://doi.org/10.3389/frwa.2023.1047955
93. Mgoba S, Kabote S (2020) Effectiveness of participatory monitoring and evaluation on achievement of community-based water projects in Tanzania. Appl Water Sci 10(8). https://doi.org/10.1007/s13201-020-01273-5
94. Cosens B, Chaffin B (2016) Adaptive governance of water resources shared with indigenous peoples: the role of law. Water 8(3):97. https://doi.org/10.3390/w8030097
95. Zary A, Bjørnlund H, Xu W (2015) A framework for assessing capacity in water governance. https://doi.org/10.2495/wrm150281
96. Dzulkifli NHN, Nurdin ZAP, Purna (2023) Unleashing the power of capacity building: transforming governance and policy implementation in the digital era. In: 1st doctoral international conference 2023. Knowledge E pp 274–288. https://doi.org/10.18502/kss.v8i17.14121
97. Thakur P, Aggarwal S, Nikam B, Garg V, Chouksey A, Dhote P (2018) Training, education, research, and capacity building needs and future requirements in applications of geospatial technology for water resources management. Int Arch Photogramm Remote Sens Spat Inf Sci 42(5):29–36. https://doi.org/10.5194/isprs-archives-xlii-5-29-2018

98. Amiraslani F, Dragovich D (2023) Iran's regional transnational water partnerships: unclear rules, unstable partnerships, and an unsettled future. Sustainability 15(15):11889. https://doi.org/10.3390/su151511889
99. Tang X, Sassa K, Brasseur G, Cullmann J, Fang Z (2020) Collaboration in mhews through an integrated way, pp 479–497. https://doi.org/10.1007/978-3-030-60196-6_39
100. Alford K, Koeman J (2019) Forging relationships between faith-based and secular organizations to address the global water crisis: a case study from Liberia. Christ J Glob Health 6(1):35–42. https://doi.org/10.15566/cjgh.v6i1.243
101. Chen C, Yen K (2023) Developing international collaboration indicators in fisheries remote sensing research to achieve SDG 14 and 17. Sustainability 15(18):14031. https://doi.org/10.3390/su151814031
102. Pathak S, Pradhan N, Guragai S, Baksi B, Azizi F, Shrestha A (2022) Complexities and opportunities of multi-stakeholder partnerships: a case study of water resource management in Afghanistan. Sustainability 14(23):15496. https://doi.org/10.3390/su142315496
103. Ding K (2023) Obstacles and countermeasures for China to promote the cooperative management of Lancang-Mekong water resources. Highlights Bus Econ Manag 6:287–293. https://doi.org/10.54097/hbem.v6i.6330
104. Tyukhtenko N, Serezhenkova O (2024) Partnership as an important factor in ensuring peace for sustainable development in a globalised world. Vìsnik Marìupol's'kogo Deržavnogo Unìversitetu
105. Wilson N, Inkster J (2018) Respecting water: indigenous water governance, ontologies, and the politics of kinship on the ground. Environ Plan E Nat Space 1(4):516–538. https://doi.org/10.1177/2514848618789378
106. Hurlbert M, Acharibasam J, Datta R, Strongarm S, Starblanket E (2024) Decolonizing indigenous drinking water challenges and implications: focusing on indigenous water governance and sovereignty. Water 16(5):748. https://doi.org/10.3390/w16050748
107. Kelly S, Valdés-Negroni J (2020) Tracing institutional surprises in the water–energy nexus: stalled projects of Chile's small hydropower boom. Environ Plan E Nat Space 4(3):1171–1195. https://doi.org/10.1177/2514848620945936
108. Vogel B, Yumagulova L, McBean G, Norris K (2022) Indigenous-led nature-based solutions for the climate crisis: insights from Canada. Sustainability 14(11):6725. https://doi.org/10.3390/su14116725
109. Islam M, Quinn M (2021) A composite graph theoretic approach to modeling landscape connectivity for wildlife movement in western Canada. J Environ Inform Lett. https://doi.org/10.3808/jeil.202100057
110. Truong V, Saunders S (2022) Guest editorial. J Soc Mark 12(1):1–4. https://doi.org/10.1108/jsocm-12-2021-246
111. Wibowo A, Suwarto S, Lestari E, Rahayu R (2024) Challenges and opportunities in building environmental awareness in the Keduang watershed region in Indonesia. Iop Conf Ser Earth Environ Sci 1302(1):012073. https://doi.org/10.1088/1755-1315/1302/1/012073
112. Twardek W, Nyboer E, Tickner D, O'Connor C, Lapointe N, Taylor M, Gregory-Eaves I, Smol JP, Reid AJ, Creed IF, Nguyen VM, Cooke S (2021) Mobilizing practitioners to support the emergency recovery plan for freshwater biodiversity. Conserv Sci Pract 3(8). https://doi.org/10.1111/csp2.467
113. Hansen H, Bergman E, Cowx I, Lind L, Pauna V, Willis K (2022) Resilient rivers and connected marine systems: a review of mutual sustainability opportunities. Glob Sustain 6. https://doi.org/10.1017/sus.2022.19
114. Estoque R (2024) Some key considerations for implementing the nexus approach in biodiversity conservation research and practice. Sci Prog 107(1). https://doi.org/10.1177/00368504231223023
115. Chiba S, Batten S, Martin C, Ivory S, Miloslavich P, Weatherdon L (2018) Zooplankton monitoring contributes to addressing global biodiversity conservation challenges. J Plankton Res 40(5):509–518. https://doi.org/10.1093/plankt/fby030

116. Lazăr L, Boicenco L, Pantea E, Timofte F, Vlas O, Bişinicu E (2024) Modeling dynamic processes in the black sea pelagic habitat—causal connections between abiotic and biotic factors in two climate change scenarios. Sustainability 16(5):1849. https://doi.org/10.3390/su16051849
117. Martel P, Sutherland C (2019) Governing river rehabilitation for climate adaptation and water security in Durban, South Africa, pp 355–387. https://doi.org/10.1007/978-3-030-04873-0_13
118. Brears R (2023) Best practices and conclusion, pp 187–202. https://doi.org/10.1093/oso/9780192847843.003.0012
119. Nurdin N, Purna Z (2023) Unleashing the power of capacity building: transforming governance and policy implementation in the digital era. Kne Soc Sci. https://doi.org/10.18502/kss.v8i17.14121
120. Ruiters C, Matji M (2016) Water institutions and governance models for the funding, financing, and management of water infrastructure in South Africa. Water Sa 41(5):660. https://doi.org/10.4314/wsa.v41i5.09
121. Serìâ Ekonomìka 14(27):38–48. https://doi.org/10.34079/2226-2822-2024-14-27-38-48

Correction to: Water Scarcity and Insecurity: Causes and Consequences

Sylvester Chibueze Izah and Matthew Chidozie Ogwu

Correction to:
Chapter 1 in: M. C. Ogwu and S. Chibueze Izah (eds.), *Water Quality and Safety in the Global South*, Springer Water, https://doi.org/10.1007/978-3-032-10602-5_1

In the original version of this chapter, the reference number [13] and its citation have been revised to include links. The chapter has been updated with the changes.

The updated version of this chapter can be found at
https://doi.org/10.1007/978-3-032-10602-5_1

M. C. Ogwu and S. Chibueze Izah (eds.), *Water Quality and Safety in the Global South*,
Springer Water, https://doi.org/10.1007/978-3-032-10602-5_15

MIX
Papier aus verantwortungsvollen Quellen
Paper from responsible sources
FSC® C105338

If you have any concerns about our products,
you can contact us on
ProductSafety@springernature.com

In case Publisher is established outside the EU,
the EU authorized representative is:
Springer Nature Customer Service Center GmbH
Europaplatz 3, 69115 Heidelberg, Germany

Printed by Libri Plureos GmbH
in Hamburg, Germany